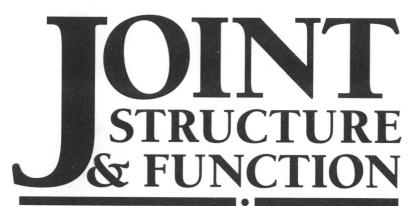

JOINT
STRUCTURE
& FUNCTION

A Comprehensive Analysis

Second Edition

JOINT
STRUCTURE
& FUNCTION
A Comprehensive Analysis
Second Edition

Cynthia C. Norkin, Ed.D., P.T.
Director and Associate Professor
School of Physical Therapy
Ohio University
Athens, Ohio

Pamela K. Levangie, M.S., P.T.
Assistant Professor
Department of Physical Therapy
Sargent College of Allied Health Professions
Boston University
Boston, Massachusetts

 F. A. DAVIS COMPANY • Philadelphia

F. A. Davis Company
1915 Arch Street
Philadelphia, PA 19103

Printed in the United States of America

Last digit indicates print number: 10 9 8 7 6

NOTE: As new scientific information becomes available through basic and clinical research, recommended treatments and drug therapies undergo changes. The author(s) and publisher have done everything possible to make this book accurate, up to date, and in accord with accepted standards at the time of publication. The authors, editors, and publisher are not responsible for errors or omissions or for consequences from application of the book, and make no warranty, expressed or implied, in regard to the contents of the book. Any practice described in this book should be applied by the reader in accordance with professional standards of care used in regard to the unique circumstances that may apply in each situation. The reader is advised always to check product information (package inserts) for changes and new information regarding dose and contraindications before administering any drug. Caution is especially urged when using new or infrequently ordered drugs.

Library of Congress Cataloging-in-Publication Data

Norkin, Cynthia C.
  Joint structure and function : a comprehensive analysis / Cynthia
C. Norkin, Pamela K. Levangie. — 2nd ed.
    p.   cm.
  Includes bibliographical references and index.
  ISBN 0-8036-6577-6 (alk. paper)
  1. Human mechanics.   2. Joints.   I. Levangie, Pamela K.
II. Title.
  [DNLM: 1. Joints—anatomy & histology.   2. Joints—physiology.
WE 300 N841j]
QP303.N59   1992
612.7′5—dc20
DNLM/DLC
for Library of Congress   91-42518
                                                     CIP

We would like to dedicate this book to the following family members: Carolyn G. Clair and Joseph Levangie, who provided support and sustenance during this endeavor, and our children and grandchildren, Alexandra Norkin Field, Ann-Michelle and Jeremy Levangie, and Taylor and Kimberly Field, who had to do without us during the long periods of time when we were revising the text.

Preface to the Second Edition

■ ■ ■

Our goals in writing the second edition have been threefold. One, we wished to ensure that this edition included the important aspects of the large volume of research published in recent years. New information and revised theories are more notable in the areas of tissue composition and response, muscle physiology, and in the specific reactions of regional structures to the application of normal and abnormal forces. Two, we tried to respond to the comments and suggestions of both students and instructors. Three, we attempted to maintain the fairly basic level of the text so that it remained an appropriate resource for those who wish to have a simple reference that gives a comprehensive overview of the principles needed to understand human function and dysfunction.

Our pursuit of the first two of our three goals, you will find, has caused the text to grow substantially. We have not only updated content but also have expanded explanations to improve clarity and supplemented the text with more than 250 new figures and summary tables. Responding to the clinical needs of our readers, we have added chapters on the temporomandibular joint and the chest wall. In meeting our third goal of maintaining the basic level of the text, we have had to make difficult decisions about limiting the inclusion of new information. Details that might be useful to experienced evaluators of human function have not been included unless, in our judgment, they enhanced a basic understanding of the content without overwhelming the reader. Occasionally we have chosen to introduce complex material at a superficial level with the intent of at least exposing the reader to advanced concepts. Readers who wish to pursue such topics in greater depth are encouraged to continue their reading using the reference lists at the ends of the chapters.

We would like to thank the readers who have been so responsive to our efforts to develop a readable and comprehensive text on human musculoskeletal function and would like to encourage you to continue your dialogue with us as we prepare for our third edition.

CYNTHIA C. NORKIN
PAMELA K. LEVANGIE

Preface to the First Edition

■ ■ ■

The prototype of this text was developed 8 years ago in response to a perceived need for a single source that would provide entry-level knowledge in biomechanics, muscle physiology, joint structure, and coordinated muscular function for physical therapy students. Through the years the content was modified and broadened in response to feedback from students and practitioners, as well as reviews of recent literature. What evolved was a *transdisciplinary* text that not only encompasses basic theory required to understand normal and pathologic function, but also provides the foundation for understanding current trends in musculoskeletal evaluation and treatment.

The text is organized around general principles of structure and function that are then applied to individual joint complexes using a cephalo-caudal, proximal-distal approach. The concepts developed in the earlier chapters are integrated with and applied to total body function by examining the complex tasks of posture and gait. Educational features of the text include learning objectives and review questions for each chapter, models that take the reader through the process of identifying the relationship between normal and abnormal function, and liberal use of diagrams.

The authors would like the process of feedback and modification to continue as the reading audience widens, and we hope that subsequent editions can respond to the changing needs of those involved in human evaluation and treatment.

CYNTHIA C. NORKIN
PAMELA K. LEVANGIE

Acknowledgments

■ ■ ■

We wish to express our gratitude to friends and colleagues who gave their time and support to the completion of this project.

We also wish to extend thanks to the entry-level physical therapy students at both Boston University, Sargent College of Allied Health Professions and Ohio University, School of Physical Therapy for their contributions to the development of this edition. In addition, we would like to acknowledge the contributions of the following individuals who reviewed the second edition: Richard Clemente, Ph.D., P.T., Slippery Rock State University, Slippery Rock, PA; Leonard Elbaum, M.M., P.T., Florida International University, Miami, FL; Alta Hansen, Ph.D., P.T., Pacific University, Stockton, CA; Debbie Nawoczenski, M.Ed., P.T., University of Iowa, Iowa City, IA; Lynn Palmer, Ph.D., P.T., Simmons College, Boston, MA; and James Zachazewski, M.S., P.T., A.C.S., A.T.C., Massachusetts General Hospital, Boston, MA.

We would particularly like to recognize the important role that Timothy Malone played not only in providing the artwork for this edition but also for his many other contributions. Finally, a special thanks to Jean-François Vilain and the editorial staff at F.A. Davis for their assistance in the timely production of this book.

Contributors

∎ ∎ ∎

Linda D. Crane, M.M.Sc., P.T., C.C.S.
Professor
Division of Physical Therapy
University of Miami School of Medicine
Coral Gables, Florida

Jan F. Perry, Ed.D., P.T.
Associate Professor and Chairman
Department of Physical Therapy
School of Allied Health
Medical College of Georgia
Augusta, Georgia

Contents

■ ■ ■

Chapter 1

■   ■   ■

Basic Concepts in Biomechanics

■

OBJECTIVES
Following the study of this chapter, the reader should be able to:

Define
1. The terminology used in biomechanics.

Describe
1. Four types of motion.
2. The plane in which a given joint motion occurs, and the axis around which the motion occurs.
3. The location of the center of gravity of a solid object; the location of the center of gravity of a segmented object; the location of the center of gravity of the human body.
4. The action line of a single muscle.
5. The name, point of application, direction, and magnitude of any interaction force, given its reaction force.
6. A linear force system, a concurrent force system, a parallel force system.
7. The relationship among torque, moment arm, and rotatory force component.
8. Two methods of determining torque for the same given set of forces.
9. How an anatomic pulley changes muscle action lines, moment arms, and torque of muscles.
10. In general terms, the point in the range of motion at which a muscle acting over that joint is biomechanically most efficient.
11. How external forces can be manipulated to maximize torque.
12. Friction, and its relationship to contacting surfaces and to the shear forces.

Determine
1. The identity (name) of diagramed forces on an object.
2. The new center of gravity of an object when segments are rearranged, given the original centers of gravity.
3. The resultant vector in a linear force system, a concurrent force system, and a parallel force system.
4. If a given object is in linear and rotational equilibrium.
5. The magnitude and direction of acceleration of an object not in equilibrium.
6. Which forces are joint distraction forces and which are joint compression forces. What are the equilibrium forces for each?
7. The magnitude and direction of friction in a given problem.
8. The class of lever in a given problem.

Compare
1. Mechanical advantage in a second- and third-class lever.
2. Work done by muscles in a second- and third-class lever.
3. Stability of an object in two given situations in which location of the center of gravity and the base of support of the object vary.

Draw
1. The action line of a muscle.
2. The rotatory force component, the translatory force component, and the moment arm for a given force of a lever.

The human body is a highly sophisticated machine composed of a large but finite number of components. These components can combine to produce an infinite variety of postures and movements. It is the intention of this text to investigate the nature of this machine, with the goal of understanding how joint structure and muscle function fulfill the needs of the human body for both mobility and stability. A knowledge of the physical principles that govern the body and of the forces that affect the body is prerequisite to examination of the structure and function of individual components. This knowledge is gained through the study of mechanics. The study of mechanics in the human body is referred to as **biomechanics** and consists of the area of kinematics and kinetics. Kinematics is the area of biomechanics that includes descriptions of motion without regard for the forces producing the motion. Kinetics is the area of biomechanics concerned with the forces producing motion or maintaining equilibrium.

KINEMATICS: DESCRIPTION OF MOTION

The human skeleton is, quite literally, a system of components or levers. A lever can have any shape, and each long bone can be visualized as a rigid bar that can transmit and modify force and motion. Kinematics involves terms that permit description of human movement. Kinematic variables for a given movement may include (1) the type of motion that is occurring, (2) the location of the movement, (3) the magnitude of the motion, and (4) the direction of the motion.

Types of Motion

There are four types of movement that can be attributed to any object or four pathways through which an object can travel. The human body is a large object made up of smaller levers. One can describe the path taken by the body as a whole, or describe the path taken by one or more of its component levers.

Rotatory (angular) motion is movement of an object or segment around a fixed axis in a curved path. Each point on the object or segment moves through the same angle, at the same time, at a constant distance from the axis of rotation. Since all human movement must occur at joints, the goal of most muscles would appear to be to rotate a bony lever around a relatively fixed axis (Fig. 1–1). In actuality, few if any joints in the human body move around truly fixed axes. Even so, for purposes of simplicity, joint motions are commonly described as if they were pure rotatory movements.

Translatory (linear) motion is the movement of an object or segment in a straight line. Each point on the object moves through the same distance, at the same time, in parallel paths. True translatory motion of a bony lever without concomitant joint rotation can occur to a very small extent when a bone is pulled directly away from its joint or pushed directly toward its joint. When one flat joint surface translates along a contiguous flat joint surface, this is also a translatory motion and is referred to as **gliding**. Although we think of muscles most commonly as structured to produce joint rotation, it shall be seen later that most forces exerted on the body (including muscles) have components that tend to produce both translatory and rotatory motions. Translatory forces across human joints, even when very small in magnitude, are important for understanding joint stress and joint stability.

A translatory motion of greater magnitude than that found within a joint can be illustrated by the forward movement of the hand and forearm to grasp an object (Fig. 1–2). In this example, the translation of the forearm/hand segment is actually produced by rotation of more proximal joints. That is, the translation of the forearm/hand is created by rotation at both the shoulder and the elbow joint.

Rotatory and translatory motions frequently combine to produce the third pathway of motion that can be taken by an object, **curvilinear motion.** The classic example is that of a thrown ball, where the ball both moves through space and

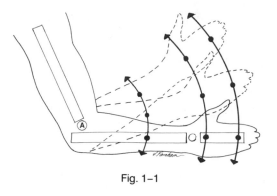

Fig. 1–1

Figure 1–1. Rotary motion. Each point in the forearm/hand segment moves through the same angle, in the same time, at a constant distance from the axis of rotation (A).

Figure 1–2. Translatory motion. Each point on the forearm/hand segment moves through the same distance, at the same time, in parallel paths.

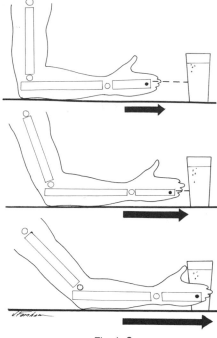

Fig. 1–2

rotates on its own axis concomitantly. In this context, curvilinear motion is rotation of a rigid object through space.

A subtle form of curvilinear motion is produced when rotation around a joint axis is accompanied by a slight translation (gliding) of bony articular surfaces. This combination of rotation and gliding results in a moving axis and, therefore, in curvilinear motion of the bony segment. Curvilinear movement is, therefore, the most common form of motion produced at a joint since all human joint axes shift slightly during movement. Given the relatively small axial shift, however, most motions appear to be pure rotatory motions and are commonly described as if this were the case.

The fourth path of motion that can be taken by an object has been described as **general plane motion**.[1] General plane motion might be considered a special case of curvilinear motion where the object is segmented and free to move rather than rigid or fixed. In general plane motion an object rotates about an axis while the axis is translated in space by motion of an adjacent segment. For example, as the person in Figure 1–3 brings the cup to his mouth, the humerus is translating forward while the elbow is rotating the forearm/hand segment. The forearm/hand segment describes a parabolic path of motion toward the mouth. Most movements of the upper and lower extremities in humans are accomplished by combined rotation and translation of adjacent joint segments resulting in curvilinear displacement of the distal segment. True translation of body segments such as exemplified by motion of the head as one walks are the exception rather than the rule.

Location of Motion

A kinematic description of motion must include the segments and joints being moved, as well as the place, or plane, of the movement. Borrowing from the universal three-dimensional coordinate system used in mathematics, motion at a joint

Figure 1-3. General plane motion. The forearm/hand segment moves in a parabolic path as it rotates around the elbow joint. The elbow joint is moved through space by shoulder joint rotation.

may be described as occurring in the transverse, frontal, or sagittal planes. Motion in any one of these planes means that a body segment is being rotated about its axis or translated in such a way that the segment is moving through a path that is parallel to one of the three cardinal planes. While human motion is by no means limited to these paths, the system of planes and axes provides a simple way of describing movement at a given joint. Since the plane of a movement could theoretically change if the position of the body changed (e.g., standing versus lying down), it is traditional to refer to motions as if they were occurring with the person standing in what is known as **anatomic position.** In anatomic position, a person stands, looking forward, with the palms of the hands facing forward.

The universal x-coordinate corresponds to the cardinal **transverse (horizontal) plane.** This plane divides the body into upper and lower halves (Fig. 1–4). Movements in the transverse plane occur parallel to the ground. For example, in rotation of the head, the nose moves parallel to the ground. Rotatory motions in the transverse plane occur around a vertical or longitudinal axis of motion. The term **longitudinal axis** is used when the axis of motion passes through the length of a long bone. The axis of a movement is always found perpendicular to its corresponding plane.

The y-coordinate corresponds to the **frontal (coronal) plane.** The frontal plane divides the body into front and back halves (Fig. 1–5). Movements in the frontal plane occur as side-to-side movements such as bringing the head to each of the shoulders. Rotatory motion in the frontal plane occurs around an **anterior-posterior (A-P) axis.**

The z-coordinate corresponds to the **sagittal plane** and divides the body into right and left halves (Fig. 1–6). Movements in this plane include forward and backward motions such as nodding of the head. Rotatory motion in the sagittal plane occurs around a coronal axis.

Direction of Motion

Narrowing movement down to a single plane does not indicate the direction of movement in that plane. We need further descriptors. For rotatory motions, the direction of movement of a lever around an axis can be described as occurring in a clockwise or counterclockwise direction. However, these terms are dependent on the perspective of the viewer (as viewed from the left side, bending the elbow is a clockwise movement; if the subject turns around and faces the opposite direction,

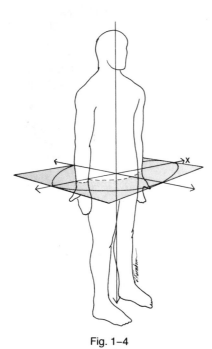

Fig. 1-4

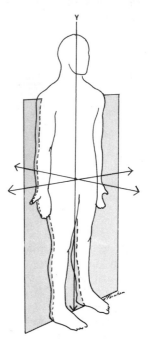

Fig. 1-5

Figure 1-4. Transverse plane.
Figure 1-5. Frontal plane.

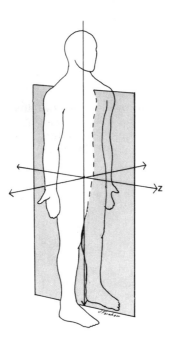

Figure 1-6. Sagittal plane.

the same movement is now seen by the viewer as a counterclockwise movement.) Positive and negative signs are arbitrarily assigned to clockwise and counterclockwise movements. Anatomic terms describing human movement are independent of viewer perspective and therefore more useful to us. **Flexion** refers to rotation of one or both bony levers around a joint axis so that ventral surfaces are being approximated. Rotation in the same plane in the opposite direction (approximation of dorsal surfaces) is termed **extension.** Flexion and extension generally occur in the sagittal plane around a coronal axis, although exceptions exist (carpometacarpal flexion and extension of the thumb).

Abduction is rotation of one or both segments of a joint around an axis so that the distal segment moves away from the midline of the body. **Adduction** occurs in the same plane, but in the opposite direction (movement of the distal lever of the joint occurs toward the midline of the body). When the segment that is moving is part of the midline of the body (e.g., the trunk and the head), the movement is commonly termed **lateral flexion.** Abduction/adduction and lateral flexion generally occur in the frontal plane around an A-P axis, although, again, some exceptions exist (carpometacarpal abduction and adduction of the thumb).

Motion of a body segment in the transverse plane around a vertical or longitudinal axis is generally termed medial or lateral rotation. **Medial** (or internal) **rotation** refers to rotation toward the body's midline, while **lateral** (or external) **rotation** refers to the opposite motion. When the segment is part of the midline, the movement in the transverse plane is simply called rotation to the right or rotation to the left. The exceptions to the general rules for naming motions must be learned on a joint-by-joint basis.

Descriptions of direction of translatory movements are conventionally given signs. Motions that are up or to the right are given positive values; motions down or to the left are given negative values. As will be described in further detail later, we can also refer to translatory movement of a segment toward its joint as **compression,** while translatory motion of a segment away from the joint can be termed **distraction.**

Quantity of Motion

The quantity or magnitude of a rotatory motion (range of motion) can be given either in degrees or in radians. If a segment describes a complete circle, it has moved through 360° or 6.28 radians. A radian is the ratio of an arc to the radius of its circle (Fig. 1–7). One radian is equal to 57.3°; 1° is equal to 0.01745 radians. The most widely used standardized method of clinical joint range measurement is goniometry, with units in degrees. Magnitude of motion may also be given as the number of degrees through which an object rotates per second (angular speed).

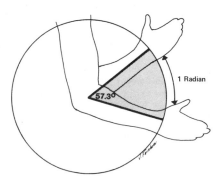

Figure 1–7. An angle of 57.3° describes an arc of one radian.

Translatory motions are quantified by the linear distance (displacement) through which the object or segment has moved. Units may vary but will be given in this text as pounds/inches/seconds, utilizing the English system of measurement. Displacement per unit time (speed) may also be considered as a description of magnitude of motion.

KINETICS: ANALYSIS OF FORCES

Definition of Forces

Kinematic descriptions of human movement permit us to visualize motion but do not give us any understanding of how or why the motion is occurring. This requires a study of forces. Whether a body or body segment is in motion or at rest is dependent on the forces exerted on that body. A **force,** simplistically speaking, is a push or a pull exerted by one material object or substance on another.[2] All forces can, in fact, be described as the push or pull of object A on object B. The concept of a force as a push or pull also can be used to describe the forces encountered in evaluating human motion.

External forces are pushes or pulls on the body that arise from sources outside the body. **Gravity** is a force that under normal conditions constantly affects all objects and, for that reason, should always be the first external force on the human body to be considered. Gravity is the pull of the earth on objects within its sphere of influence, or more specific to our purposes, it is the *pull of the earth on a body* (or its segments). Gravity is only one of an infinite number of external forces that can affect the human body. Other objects or substances that may exert a push or pull on the human body or its segments are (to name only a few) wind (push of air on the body), water (push of water on the body), other people (push of Mr. Jones' shoulder, pull of Mr. Jones' hand on Mr. Smith's hand), and other objects (the push of floor on the feet, the pull of a briefcase on the arm).

Internal forces are forces that act on the body but arise from sources within the human body. Examples are muscles (pull of the biceps brachii on the radius); ligaments (pull of a ligament on bone); and bones (the push of one bone on another bone). Internal forces are essential to human function since external forces cannot be relied on to create human movement. More importantly, internal forces serve to counteract those external forces that jeopardize the integrity of human joint structure. There are also some forces, such as friction and atmospheric pressure, which can act both external to and within the body.

Force Vectors

All forces, despite the source or the object acted on, are **vector** quantities and can be defined by:
- A point of application on the object being acted on
- An action line and direction indicating a pull toward the source or a push away from the source
- A magnitude, that is, the quantity of force being exerted

A vector is traditionally represented by an arrow, so a force is represented by an arrow that: (1) has a base on the object being acted on (point of application); (2) has a shaft and arrowhead in the direction of the force being exerted (action line, direction); and (3) has a length drawn to represent the amount of force being exerted (magnitude). Figure 1–8 shows a hand pushing on a book. The force, which can be called hand-on-book, is represented by vector **HB**. The point of application is on the book; the action line and direction indicate the direction of the push; and

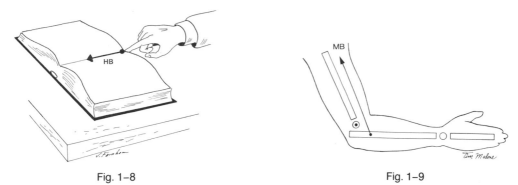

Fig. 1–8   Fig. 1–9

Figure 1–8. Vector HB represents the push of the hand on the book with a magnitude of 15 lb (0.25 in = 10 lb).

Figure 1–9. Vector MB represents the pull of a muscle on a bone, with a magnitude of 35 lb.

the length is drawn to represent the magnitude of the push. The length of a vector is usually drawn proportional to the magnitude using a given scale. For example, if the scale is specified as 0.25 = 10 lb of force, an arrow of 0.375 in would represent 15 lb of force. The length of a vector, however, need not be drawn to scale. The action line of any vector can be considered infinitely long. That is, any vector can be extended if this is useful in determining relationships of the vector to other vectors or objects. The length of a vector should not be arbitrarily drawn, however, if a scale has been specified.

An example of a vector that depicts the force of a muscle acting on a bony lever in the body is shown in Figure 1–9. The force can be named muscle-on-bone (**MB**). The point of application of the force is *on the bone*, which is the object being acted on; the action line and direction are in the direction of pull of the muscle; and the magnitude of the muscle force (using a scale of 0.5 in = 10 lb) is 35 lb.

Naming Forces

When the naming convention of "object-on-object" is used to identify forces, the first object named will always be the *source* of the force; the second object named will always be the *object being acted on*. This means that the point of application will always be found on the second object named (the object to which the force is applied will always be the "last name" of the force). The action line and direction of a force will be toward the source in the case of a pull, or away from the source in the case of a push. The source of the push or pull will always be the "first name" of the force.

Figure 1–10 shows a man holding a 30-lb box in both hands. The two vectors in the figure can be identified using the naming convention described. The point of application of vector BR is on the right hand and vector BL on the left hand. Preliminarily, BR can be named "object-on-right-hand" and BL can be named "object-on-left-hand."

The second step in the process of identification is to determine the source of the force. This is considerably easier when it is recognized that the source of most forces that may act on an object must *touch* or *contact* the object. The major exception to this rule when considering forces on the human body is the force of gravity. If permitted the exaggeration that gravity "contacts" all objects on earth, we can circumvent this exception and maintain the rule that any force acting on a segment of the human body must come from something touching the segment.

In Figure 1–10, the hands are clearly being contacted by the box. Each hand

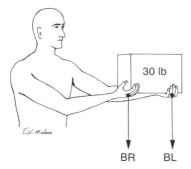

Figure 1–10. Vectors BR and BL represent the contact (or push) of the 30-lb box on each hand. BR = **box-on-right-hand**; BL = **box-on-left-hand.**

is also a segment of the body that is being contacted by the adjacent segment of the forearm. In addition to the contact of the box and the forearm, each hand is also being "contacted" by gravity. We now have three possible first names for the unidentified vectors. We can further narrow down the source by recognizing that each contact must be a push (away from the source) or pull (toward the source). A push or pull by the forearm segments would have to be in line with the forearms. Since BR and BL are not in line with the forearms, we can eliminate the forearms as the source of the vectors. The box would push on the hands and, therefore, is a possible source of vectors BR and BL. Gravity is also a possibility, since it is the pull of the earth on the hands. Without further information, it could not be determined whether the source of vectors BR and BL were gravity on the left and right hands or box on the left and right hands.

If the scale and length of vectors BR and BL were known, it would be possible to make the final determination about the identity of the vectors. If we specify the scale as 1 in = 20 lb, and measure the vectors as 0.75 in long, the magnitude of each force would then be 15 lb. Since the box would exert a force of 15 lb on each hand (30 lb with half distributed to each hand), the box is the likely source of vectors BR and BL. It is unlikely that the hands would each weigh 15 lb (gravity is the pull of the earth on an object and is better known as the weight of an object). Vectors BR and BL can now be named as box-on-right-hand and box-on-left-hand.

Force of Gravity

Gravity is the attraction of the mass of the earth for the mass of other objects and, on earth, has a magnitude of 32 ft/s². The force of gravity gives an object **weight,** which is actually the mass of the object times the acceleration of gravity.

$$\text{Weight} = \text{mass} \times 32 \text{ ft/s}^2$$

It should be noted that the proper unit for mass is the slug (lb $\times$ s²/ft). The slug is a scalar unit (without action line or direction), while the pound is a force unit (having vector characteristics). Weight, as a vector quantity produced by the force of gravity, is given in units of pounds.

Gravity is the most consistent force encountered by the human body and behaves in a predictable manner. As a vector quantity, it can be fully described by point of application, action line/direction, and magnitude. While gravity acts at all points on an object or segment of an object, its point of application is given as the **center of gravity (COG)** or **center of mass** of that object or segment. The COG is a hypothetical point at which all mass would appear to be concentrated and is the point at which the force of gravity would appear to act.

In a symmetrical object, the COG is located in the geometric center of the object (Fig. 1–11a). In an asymmetrical object, the COG will be located toward the

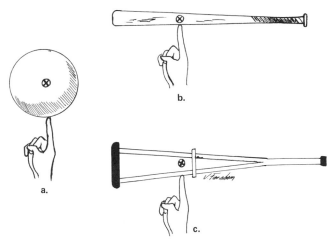

Figure 1–11. a. Center of gravity of a symmetric object. b. Center of gravity of an asymmetric object. c. The center of gravity may lie outside the object.

heavier end at a point at which the mass is evenly distributed around the point (Fig. 1–11b). The crutch in Figure 1–11c demonstrates that since the COG is only a hypothetical point, it need not lie within the object being acted on. Even when the COG lies outside the object, it is still the point at which gravity appears to act. The location of the COG of any object actually can be determined by a number of methods not within the scope of this text. However, the COG of an object can be approximated if one considers it as the balance point of the object (assuming you could balance the object on one finger).

The action line and direction of the force of gravity on an object are always vertically downward toward the center of the earth regardless of the orientation in space of the object. The gravity vector is commonly referred to as the **line of gravity** (**LOG**). The length of the LOG can be drawn to scale, or may be extended somewhat arbitrarily when other relationships are being explored. The LOG can best be visualized as a string with a weight on the end (such as a plumb line), with the string placed at the COG of an object. This gives an accurate representation of the point of application, direction, and action LOG, although not the magnitude.

Segmental Centers of Gravity

Each segment in the body is acted on by the force of gravity and has its own COG. One can group together two or more adjacent segments if they are going to move together as a single solid segment. In such a case, gravity acting on the combined segments can be represented by a single COG. Figure 1–12a shows the location of the gravity vectors at the centers of the arm (GA), the forearm (GF), and the hand (GH), considering the hand as a single solid segment. The COGs approximate those identified in studies done on cadavers and on in vitro body segments that have yielded standardized data on centers of mass and segmental body weights of individual and combined body segments.[3]

When two adjacent segments are combined and considered as one solid segment, the new larger segment will have a COG that is located between and in line with the original two COGs. When the segments are not equal in weight, the new COG will lie closer to the heavier segment. Figure 1–12b shows vector GA on the arm and new vector GFH on the forearm/hand segment. The forearm and hand have been combined into a single segment. The new COG for the combined fore-

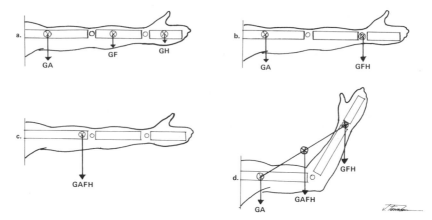

Figure 1–12. a. Gravity acting on the arm (GA), the forearm (GF), and the hand (GH). b. Gravity acting on the arm and the forearm/hand (GFH). c. Gravity acting on the arm/forearm/hand segment (GAFH). d. GAFH relocates when segments are rearranged.

arm/hand segment is located between the original two COGs with a magnitude equal to the sum of GF and GH. Figure 1–12c has combined all three segments into a single object, with the force of gravity (GAFH) acting at the new COG located between GA and GHF. The magnitude of GAFH is equal to the sum of the magnitudes of its components.

The COG for any solid object or fixed series of segments will remain unchanged despite the position of that object in space. However, when an object has two or more adjacent segments, the location of the COG of the combined unit will change when the segments are rearranged relative to each other. In Figure 1–12d the arm segment and the forearm/hand segment have been rearranged. The magnitude of the force of gravity will not change since the segments have not changed their mass, but the location of vector GAFH is now different from that in Figure 1–12c. The new location of the COG is still found on a line between the original two. Here, we have another example where the COG lies outside the solid arm/forearm/hand segment.

Center of Gravity of the Human Body

When all the segments of the body are combined and the body is taken as a single solid object in anatomic position, the COG of the body lies approximately anterior to the second sacral vertebra (Fig. 1–13). The precise location of the COG for a person is dependent on the proportions of that person, with the magnitude equal to the weight of the individual. People, of course, do not remain in anatomic position but constantly rearrange their segments. With each rearrangement, the location of the individual's COG will potentially change. The amount of change in the COG is dependent on how disproportionately the segments are rearranged.

If the body is considered to be composed of solid upper-body and lower-limb segments, the COGs for each of the two segments will be located approximately as given in Figure 1–14a. The combined COG for these two segments as aligned will still lie at S-2. When the trunk is inclined forward, however, the location of the new COG lies outside the body (Fig. 1–14b). Figure 1–15 shows a more disproportionate rearrangement of segments. The two lower-limb and single upper-body segments have produced a new COG located at point **ABC**.

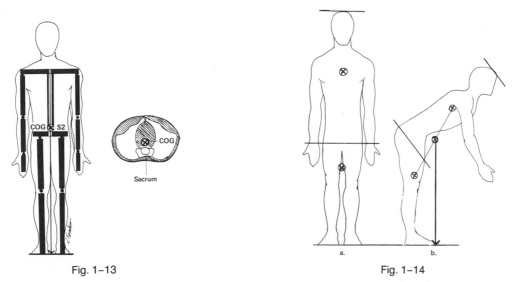

Fig. 1–13   Fig. 1–14

Figure 1–13. The center of gravity (COG) of the human body lies approximately at S-2, anterior to the sacrum (inset).

Figure 1–14. a. Location of the COGs of the upper trunk and lower limb segments: b. Rearrangement of segments produces a new combined COG.

Stability and the Center of Gravity

In Figure 1–15, the LOG (GABC) falls outside the base of support (the foot) of the football player. The LOG has been extended to indicate its relationship to the base of support (it no longer represents the magnitude of the football player's weight, although the point of application, action line, and direction remain accurate). In this instance, the player is unstable. For an object to be stable, the LOG

Figure 1–15. COG of the left leg (A) and the right leg (B) combine to form the COG for the lower limbs (AB). AB combines with the upper trunk center of gravity (C) to produce the COG for the entire body (ABC).

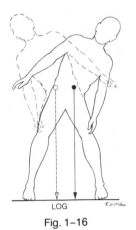

Fig. 1–16

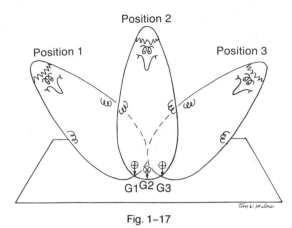

Position 1   Position 2   Position 3

G_1 G_2 G_3

Fig. 1–17

Figure 1–16. A wide base of support permits a wide excursion of the line of gravity (LOG) without the LOG falling outside the base of support.

Figure 1–17. Given the very low COG of the punching bag, the LOG remains within the base of support regardless of the tipping of the bag from one position to another.

must fall within the base of support. When the LOG falls outside the base of support, the object will tend to fall. Given that the LOG must fall within the base of support, two additional factors will affect the stability of an object:

- The larger the base of support of an object, the greater the stability of that object.
- The closer the COG is to the base of support, the more stable is the object.

In the case of the football player in Figure 1–15, the LOG falls anterior to his base of support; his COG has risen above S-2, and his base of support has been reduced from the area between and including his two feet to the area of one foot. It would be impossible for the player to maintain this pose.

When the base of support of an object is large, the LOG has more freedom to move without passing beyond the limits of the base. When a person stands with his or her legs spread apart, the base is larger and the trunk can move more without displacing the LOG from the base of support (Fig. 1–16). When the COG is low, movement of the object in space is less likely to cause the COG (and LOG) to fall outside the base. Figure 1–17 shows a punching bag as it moves from side to side. The base of the punching bag is filled with sand and the remainder is air. This creates a COG that nearly lies on the ground. *The position of the COG of the punching bag remains the same within the punching bag regardless of how tipped the bag might be. The LOG shifts just as it did in Figure 1–16.* However, the bag is extremely stable because it is nearly impossible to get the LOG to fall outside the base of support.

Relocation of the Center of Gravity

The location of the COG is dependent not only on the arrangement of segments in space but also on the distribution of mass of the object. While people certainly gain weight and may gain it disproportionately in the body, the most common redistribution of mass in the body that results in shifts of the body's COG has to do with the addition of external mass. That is, every time we add an object to the body by wearing it or carrying it, the new COG for the combined body and external mass will shift toward the additional weight; the shift will be proportional to the weight added.

The man in Figure 1–18 has had a cast applied to the right lower limb. This has resulted in the shift of the COG down and to the right. Since his COG is now lower,

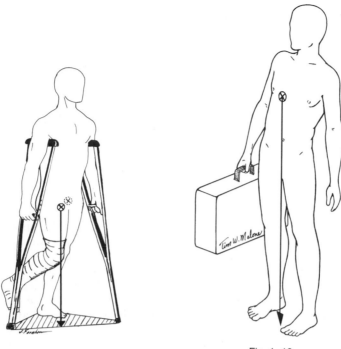

Fig. 1–18   Fig. 1–19

Figure 1–18. The addition of the weight of the cast has shifted the COG. Addition of crutches enlarges the base of support to improve stability.

Figure 1–19. The weight of the suitcase added to the shoulder girdle causes the COG to shift up and to the right. The man laterally leans to the left to bring the LOG back to the middle of his base of support.

he is theoretically more stable. However, his base of support has also been reduced to consist only of the left foot, since his right leg is now non-weight-bearing. Rather than require the patient to shift his LOG considerably to the left, crutches have been added. The crutches and the left foot combine to form a much larger base of support, adding to the patient's stability and avoiding a large compensatory weight shift.

In Figure 1–19, the man is holding a heavy suitcase in his right hand (suspending the suitcase from his shoulder girdle). This results in a shift of the COG up and to the right. Since the LOG would move toward the right foot (and potentially to the right of the right foot if the suitcase is of sufficient weight), the man laterally leans to the left to compensate. Rearrangement of segments through the lateral lean does little to relocate the COG; the main effect is to bring the LOG back to the middle of the base of support. The body segments are reoriented in space, not to relocate the COG, but to bring the LOG closer to the center of the base of support.

Reaction Forces

Newton's Law of Reaction

When studying the source and application of forces, one must take into account a critical property of forces; that is, that *forces always come in pairs.*[2] Newton's third law, the **law of reaction**, reflects this concept. The law is commonly stated as: For every action there is an equal and opposite reaction. This statement commonly

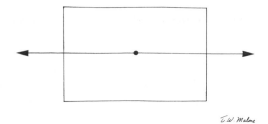

T.W. Malone

Figure 1–20. Newton's third law (''for every action there is an equal and opposite reaction'') is commonly *but incorrectly* represented by two vectors acting on the same object.

causes confusion, leading to the *incorrect* interpretation shown in Figure 1–20. Cromer restated Newton's third law more clearly: When one object applies a force to the second object, the second object simultaneously applies a force equal in magnitude and opposite in direction to the first object.[1] These two factors constitute an interaction pair, or **reaction forces.** The simplicity of Newton's third law can better be appreciated when we examine the concept that forces on an object always arise from things that contact that object. If a force *on object A* is due to the contact of object B, is it not also true that object A must also be contacting and therefore exerting a force *on object B?* Would it be possible for object B to touch A without A also touching B? When you clap hands, doesn't your right hand contact the left with the same force that the left contacts the right? Since all forces come from things that touch, and all touches must occur between two objects at a time, all forces come in pairs that are equal in magnitude, opposite in direction, and applied to adjacent contacting objects. As we will see, this concept includes the ''distant touch'' of gravity to an object.

In Figure 1–21 a book is resting on the table. Whenever two objects are in contact, each exerts a force on the other. The book must exert a force on the table, and the table must exert a force on the book. The magnitudes will be equal and the vectors opposite in direction. The forces exerted would be named book-on-table (BT) and table-on-book (TB). Vector BT is applied to the table; its source is a push from the book directed downward with a magnitude equal, in this instance, to the weight of the book. Vector TB is applied to the book; it results from a push of the table upward with a magnitude equal to that of BT.

It is extremely important to note that in any interaction pair the points of application are on *different objects,* not on the same object as shown incorrectly in Figure 1–20. In Figure 1–21, the book acted on the *table* while the table simultaneously acted on the *book.* The interaction pair or reaction forces thus always bear names that are the reverse of each other. In Figure 1–10, the vectors were named box-on-left-hand and box-on-right-hand. Since forces come in pairs and objects that are in contact exert forces on each other, we must also have two additional forces called

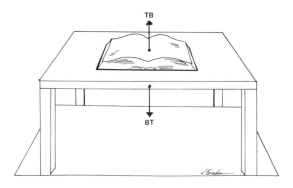

Figure 1–21. Reaction forces **book-on-table** (BT) and **table-on-book** (TB).

right-hand-on-box and left-hand-on-box. These vectors would each be applied to the *box* in an upward direction with a magnitude of 15 lb.

Since the force of gravity does not actually touch an object, one might think that gravity, or the force of earth on object, does not necessarily have a mate. Gravity, too, must have a reaction. Although the earth exerts an attraction for all objects with mass, likewise these objects exert an attraction for the earth equal in magnitude and opposite in direction. Since the attraction of a small object for the large earth seems inconsequential compared to the attraction of the earth for the small object, object-on-earth tends to be ignored (but does exist!). We can continue to consider that anything touching the body or its segments has a reaction or interaction pair.

Whenever one is trying to account for forces on an object or set of objects, it helps to remember that:

- Gravity exerts a force on all objects (is always touching an object).
- Forces on an object are exerted by things that touch that object.
- Whenever two objects contact, they exert a force on each other.
- Forces come in pairs.

Equilibrium

The primary concern, when looking at forces that act on an object or the body in particular, is the effect that these forces will have on the object or body. Whether an object is in translatory, rotatory, or curvilinear motion is dependent on the forces acting on that object. It is also possible to have forces applied to an object without causing movement of the object. Statics is the study of the conditions under which objects remain in **equilibrium,** or at rest, as a result of the forces acting on them.

Law of Inertia

Newton's first law, the **law of inertia,** deals with objects in equilibrium. The law states that an object will remain at rest or in uniform motion unless acted on by an unbalanced force. Uniform motion occurs when an object is moving with a constant velocity; when that constant velocity is zero, the object is at rest. **Inertia** is the property of an object that makes the object resist both the initiation of motion and a change in motion. **Velocity** is a vector quantity with both magnitude (speed) and direction. Constant velocity implies, therefore, both constant speed of an object and movement in a constant direction. Velocity can be linear (as for translatory motion) or can be angular (as for rotatory motion). When dealing with the human body and its segments, moving equilibrium (or uniform motion) occurs infrequently. Therefore, within the scope of this text, equilibrium can be simplified to mean an object at rest unless otherwise specified.

Newton's law of inertia (or law of equilibrium) can be restated: For an object to be in equilibrium, the sum of all the forces *applied to that object* must equal zero, $\Sigma F = 0$. An object cannot be in equilibrium if only one force is acting on that object, since there would be nothing to counteract that force. If a force exists, it must have magnitude; the magnitude of one force cannot be zero.

Determining Equilibrium of an Object

Using what has been reviewed so far, all the forces acting on a body at rest can be accounted for. Figure 1–22 shows a book resting on a table. Assuming the book

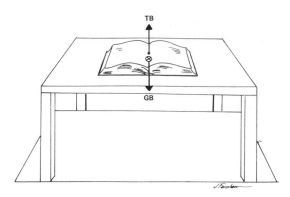

Figure 1–22. Equilibrium vectors **gravity-on-book** (GB) and **table-on-book** (TB).

is in equilibrium (i.e., it remains at rest on the table), the identity and the magnitude of all the forces acting on the book can be accounted for.

- Gravity acts on all objects. Therefore, gravity must be acting on the book at the book's COG, with a magnitude proportional to the mass of the book: vector **GB** = gravity-on-book.

Since an object with only one force cannot be in equilibrium, at least one other force must exist on the book.

- Any object in contact with another object exerts a force on the adjacent contacting object. The book is being contacted by the table; therefore, the table must be exerting a force on the book (**TB** = table-on-book).

As diagramed in Figure 1–22, nothing else is touching the book, so there are no other forces to account for. Since vectors GB and TB are applied to the same object and have action lines that lie in the same line, they are part of a linear force system. To prevent confusion in the figures, vectors with coinciding points of application and action lines will be diagramed as if they were next to each other. A **linear force system** exists whenever two or more forces act *on the same object* and in the same line. Vectors in the same linear force system will overlap if the vector lengths are extended. Vectors that overlap but are applied to *different* objects are not part of the same linear force system. Since linear forces cause translatory motion, the magnitudes of linear forces are given signs using the convention previously described for translatory forces: forces applied up or to the right are considered positive, while forces applied down or to the left are considered negative. Vectors in opposite directions should always have magnitudes of opposite sign. The net effect, or resultant, of all forces that are part of the same linear force system can be determined by finding the arithmetic sum of the magnitudes of each of the forces, taking into account its positive or negative value. For the sum of gravity-on-book and table-on-book to be zero, the magnitudes must be equal but opposite in sign and direction. If the book weighed 2.5 lb, GB would be −2.5 lb. Vector TB would then have to have a magnitude of +2.5 lb.

Shifting our attention from the book to the table in Figure 1–23, we can establish the equilibrium of the table in a similar manner. The table is being "touched" or contacted by gravity and by the book. Gravity will exert a downward force (GT) on the table that is the table's weight and is applied at the COG of the table. Presuming the table weighs 20 lb, GT has a magnitude of −20 lb. Since the table is touching the book and exerts a force on the book, the law of reaction states that the book must also exert a force (BT) on the table equal in magnitude and opposite in direction to vector TB. Consequently, BT must be −2.5 lb.

In Figure 1–23, presuming that the book is placed symmetrically in the middle of the table, there are now two forces acting on the table at the same point and in the same line (vectors GT and BT are diagramed separately to avoid overlap). Grav-

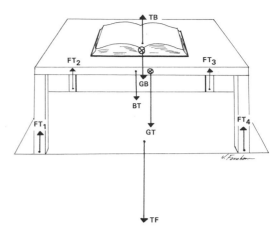

Figure 1-23. Equilibrium of the table is established with the forces of **gravity-on-table** (GT), **book-on-table** (BT), and **floor-on-table** ($FT_1 + FT_2 + FT_3 + FT_4$). **Table-on-floor** (TF) is the reaction to FT and does not have a direct effect on the equilibrium of the table.

ity-on-table and book-on-table are part of the same linear force system. The net effect on the table of GT and BT can be represented by a new vector applied to the COG of the table, with a magnitude of -22.5 lb. The table cannot be in equilibrium with such a net force acting on it. At least one other force must exist on the table.

The source of any additional force acting on the table must be something contacting the table. The only other contact on the table in Figure 1-23 is made by the floor. Since any contact between two objects creates a force, there must be a vector applied to the table called floor-on-table (FT). If the table is at rest, the sum of the forces acting on the table must be zero. Consequently, FT must be equal in magnitude and opposite in direction to the net effect of GT and BT. FT has a magnitude of $+22.5$ lb. The point of application of FT can be placed at the COG on the table, since that is the hypothetical point at which the mass of the table is concentrated. FT can also be applied at the actual points of contact; that is, at each leg of the table. Figure 1-23 shows FT distributed over the four legs to avoid redundant lines in the diagram, but it should be understood that FT can be considered to act at the COG of the table and is therefore part of the same linear force system as GT and BT. (FT $= FT_1 + FT_2 + FT_3 + FT_4$).

Figure 1-23 includes one other force vector. Table-on-floor (TF) is the reaction to floor-on-table (FT). TF must exist since the contact of the table and floor requires that they exert forces on each other. TF is equal in magnitude and opposite in direction to FT. It does *not*, however, have a direct influence on the equilibrium of the table, since only those forces that have a *point of application on an object* can contribute to the equilibrium of that object. Similarly, table-on-book and gravity-on-book, if diagramed in their precise locations, would overlap with gravity-on-table and book-on-table. Although coinciding, these four vectors are *not* part of the same linear force system, since they are applied to two different objects.

In Figure 1-24, the hand segment of the man holding a briefcase is in equilibrium. The hand segment is touched by gravity, by the briefcase, and by the distal forearm segment. Hand vectors GH (gravity-on-hand), BH (briefcase-on-hand), and FH (forearm-on-hand) each act on and have coinciding points of application and action lines on the hand. They are, therefore, part of the same linear force system. Gravity (GH) and the pull of the briefcase on the hand (BH) both act in a negative direction with magnitudes equal to the weight of the hand (2 lb) and the weight of the briefcase (8 lb), respectively. The magnitude of BH was determined to be 8 lb, since the pull of the briefcase on the hand (BH) must be equal in magnitude and opposite in direction to the pull of the hand on the briefcase (HB). If the briefcase is in equilibrium, HB must be equal in magnitude and opposite in

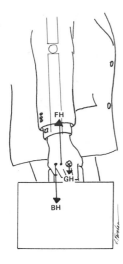

Figure 1–24. Linear force system formed by the forces of **gravity-on-hand** (GH), **briefcase-on-hand** (BH), and **forearm-on-hand** (FH).

direction to the only other contact on the briefcase, that of gravity (or the weight of the briefcase). Given the two downward forces (**BH** and **GH**), if the hand is in equilibrium, the pull of the forearm on the hand (**FH**) must be positive in direction and must be equivalent to the sum of **GH** and **BH** (or 10 lb).

Forces that result from the contact of objects always come in pairs and are always applied to different, and generally adjacent, objects. In some cases the reaction to a force can be ignored if it does not affect the problem at hand. In Figure 1–24, three forces were identified as being applied to the hand. Each of these *must* have a reaction force, although the reaction is not significant to the establishment of equilibrium of the hand. The reaction forces (interaction pairs) to each of these forces are:

Original Force Vector	Reaction Vector	Affected Object
Gravity (earth-on-hand)	Hand-on-earth (HE)	Earth
Briefcase-on-hand	Hand-on-briefcase (HB)	Briefcase
Forearm-on-hand	Hand-on-forearm (HF)	Forearm

Although the reaction forces **HE**, **HB**, and **HF** in Figure 1–24 can, for the most part, be ignored for our purposes, the reaction force or forces can sometimes be very important.

In Figure 1–25, a person is standing on a scale with the intent of recording his weight, **gravity-on-person** (GP). The scale cannot, however, measure or record a force applied to the person but only a force applied to the scale. What is actually being recorded on the scale is the *contact* of the **person-on-scale** (PS) and not the weight of the person. Under usual conditions of weighing oneself, this distinction is not a particularly important one. Vector **PS** will always be equal in magnitude and opposite in direction to its reaction, scale-on-person (**SP**). As long as the person is in equilibrium *and* nothing else is touching the person, scale-on-person and gravity-on-person will have equal magnitudes. Consequently, the magnitude of **PS** and **GP** are equivalent to each other.

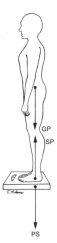

Figure 1–25. Although a scale is commonly thought to measure the weight of the person (**gravity-on-person** [GP]), it is actually recording the contact of the **person-on-scale** (PS). Vectors GP and PS are equal in magnitude as long as nothing else is touching the person.

The distinction between GP and the reaction force PS can be very important if something else is touching the person or the scale. If the person being weighed is holding something, the person's weight (GP) does not change, but the contact forces (PS and SP) will increase. Similarly, we are all familiar with the phenomenon where a gentle pressure down on the bathroom countertop as we weigh ourselves will result in a weight reduction. The pressure of the fingers down on the countertop creates an upward reaction of countertop-on-person (CP). Now scale-on-person equals (−**GP**) + (+**CP**), which is less than the magnitude of GP alone. The contact between the person and the countertop has created an additional reaction force acting on the person, resulting in what appears to be a weight reduction. It is not a decrease in GP, of course, but a decrease in PS. Since the scale is *not* actually measuring the person's weight but the reaction force of person-on-scale, PS will be equal to GP only if no other forces are applied to the person. In this example, the reaction force (PS) *cannot* be ignored because it is the variable of interest.

Objects in Motion

When a state of equilibrium exists, all forces applied to the object are balanced (the sum of all the forces applied to the object is zero). When unbalanced forces are applied to an object, **acceleration** of the object will necessarily result. Figure 1–26 shows three people pulling on a rope. Person A and person B are exerting a force of 100 lb each on the rope. Person C is exerting a force of 90 lb in an opposite or negative direction. Each force is acting on the same object (the rope) along the

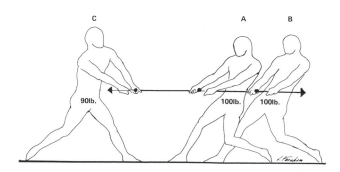

Figure 1–26. Unbalanced forces on the rope produce acceleration of the rope toward men (A) and (B).

same line and so are part of one linear force system. Since the people are the only objects touching the rope (the negligible effect of gravity is ignored—this is a very light rope), the resultant force R acting on the rope is: $(+100\text{ lb})+(+100\text{ lb})+(-90\text{ lb})$; that is, $R = +110\text{ lb}$. The net effect on the rope is a force of 110 lb acting to the right, causing the rope to move and accelerate in that direction. Once an object begins to accelerate, new forces are introduced and kinetic analysis becomes far more complex. We will maintain a fairly simple (and simplified) approach that will not include more advanced concepts in dynamics.

Law of Acceleration

The magnitude of acceleration of a moving object is defined by Newton's second law, the **law of acceleration.** Newton's second law states that the acceleration of an object is proportional to the unbalanced forces acting on it and inversely proportional to the mass of that object:

$$a = F/m$$

That is, a large push (F) applied to an object of constant mass (m) will produce more acceleration (a) than a small push. A push on an object of large mass will produce less acceleration than an equal push on an object of smaller mass. Acceleration may occur as a change in speed of an object and/or as a change in direction of movement. From the law of acceleration it can be seen that inertia, a body's or object's resistance to change in movement (or change in acceleration), is proportional to the mass of the body or object. The greater the mass of an object, the greater the magnitude of net force needed either to get the object moving or to change its motion. A very large woman in a wheelchair has more inertia than a small woman; an aide must exert a greater push on a wheelchair with a large woman in it than on the wheelchair with a small woman in it to obtain the same acceleration.

Examples of Dynamics

In Figure 1–27a, a 200-lb man is taking a step forward on the floor. At the particular moment in time captured in the figure, approximately half of the man's 200-lb body weight, gravity-on-man (G_1), is acting down the right leg. The right leg at that moment is in equilibrium only if there is another force equal in magnitude and opposite in direction applied to the leg by the contacting floor. Floor-on-leg (FL) is applied to the leg and is a result of the floor's ability to react to the force of leg-on-floor (LF). Since nothing else is touching the leg other than the floor and the

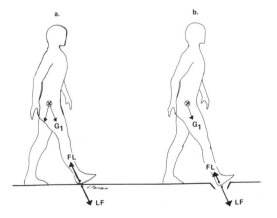

Figure 1–27. a. A man in equilibrium. Gravity (G_1) acting down the right leg is opposed by the equal and opposite force of **floor-on-leg** (FL). b. A man not in equilibrium. The push of FL is not equal to gravity; therefore the leg accelerates down through the floor.

body above, FL must be equal in magnitude to G_1. Since FL must equal LF, leg-on-floor must also be equal in magnitude to G_1, or 100 lb. Ordinarily, no matter how large G_1 and LF might be, the floor can exert an equal force back on the leg (we count on this whenever we cross the floor!). However, in Figure 1–27b, FL (and, therefore, LF) is smaller in magnitude (75 lb) than G_1. The floor has been weakened and is not capable of pushing back adequately. Since the sum of G_1 and FL is not zero, the person's leg cannot be in equilibrium. The leg will accelerate downward through the floor with an unbalanced force of 25 lb.

The magnitude of acceleration of the leg through the floor can be easily determined by using Newton's second law if the problem is treated as one in statics rather than dynamics. Once the floor actually begins to break, the force the floor exerts on the leg (FL) will diminish until it reaches zero, changing the unbalanced force acting on the leg continuously until that point. The problem can be simplified by considering only that point in time at which FL and LF equaled 75 lb.

The acceleration of the leg through the floor can be found using the equation $a = F/m$. One must determine both the magnitude of the unbalanced force (25 lb) and the mass of the accelerating segment (the leg). The direction (downward through the floor) will remain relatively unchanged and can be ignored. The leg's weight (or the force acting through the leg) was given as 100 lb. The force units (lb) will have to be converted to mass units (slugs); 100 lb of force is equal to 3.125 sl or 3.125 lb $\times$ s²/ft (1 lb = 1/32 sl).

Therefore,

$$a = \frac{25 \text{ lb}}{3.125 \text{ lb} \times \dfrac{s^2}{ft}}$$

$$a = 8 \text{ ft/s}^2$$

An acceleration of 8 ft/s² will occur only at the moment the floor breaks. As the floor gives less support, FL diminishes in magnitude, the unbalanced force increases, and the acceleration continues to increase until the floor no longer exerts any force on the leg at all. The leg will continue to accelerate until a new force is encountered that will restore equilibrium to the leg.

Although assuming a static situation by capturing a single moment in time simplifies an analysis, it ignores the additional forces found in a true dynamic situation and can, therefore, lead to incorrect conclusions. Figure 1–28 shows a body lying on a bed (apparently in equilibrium). All the forces acting on the body are shown. The body is touched by gravity (GM = 180 lb) and by the bed (BM). As long as nothing else touches the man but gravity and the bed, BM will also have a magnitude of 180 lb. The force of the bed, or more specifically the mattress, is distributed over the entire posterior aspect of the body; that is, bed-on-man can be represented by an infinite number of vectors (BM_1, BM_2, BM_3), each with an infinitely small magnitude and its own point of application but which together have a total magnitude of 180 lb. At any particular point in time, the man would certainly appear to be in equilibrium. However, it is possible over a period of time for disequilibrium to occur. The human body is not a solid object but is made up of sequential and con-

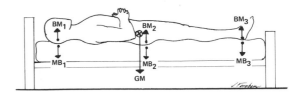

Figure 1–28. For the man on the bed to be in complete equilibrium, forces of **bed-on-man** (BM_1, BM_2, BM_3) must be equal and opposite to **gravity-on-man** (GM). Bed-on-man also has a reaction force to **man-on-bed** (MB_1, MB_2, MB_3).

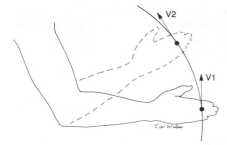

Figure 1-29. Although the magnitude of angular velocity (V_1, V_2) may be constant, the direction of angular velocity of a rotating limb segment changes as the limb segment moves around its axis. The angular velocity is always tangential to the arc of motion.

tiguous layers such as skin, adipose tissue, fascia, muscle, bone, and interposed blood vessels. Each layer could be considered as an object that must be in equilibrium, subject to the contact forces of the objects both superficial and deep to it. For example, bed-on-man (BM_2) is the force of the mattress on the skin over the ischial tuberosity. The skin is also receiving a force from the bone that lies deep to it. If the condition of the skin has been compromised by physiologic deficiencies (poor nutrition), the skin may not be able to "push back" on the mattress (MB_2) with sufficient magnitude to prevent what is, in effect, the acceleration of the mattress through the skin. The result is a bed sore, or decubitus ulcer. The static analysis masked what was actually a progressive dynamic situation.

Linear acceleration of an object as shown in these examples will produce new forces that can complicate an analysis (and make a static analysis inaccurate). Similarly, angular acceleration produced when a bony lever rotates about a joint axis also produces new forces that will complicate an analysis beyond what can be covered in this basic biomechanics unit. A segment can be rotating around an axis with a constant speed (degrees of movement per unit time). Although the magnitude of the angular velocity of the segment may be constant, the *direction* will not be. The direction of the angular velocity vectors (v_1 and v_2) is always tangential to the arc (or perpendicular to the moving segment) as shown in Figure 1–29. The constantly changing vector of angular velocity can make an analysis inaccurate if not taken into consideration. When torque (the magnitude of rotation of an object) is discussed later in this chapter, we will remind the reader again that a simple static analysis will underestimate the forces actually acting on the lever under consideration.

Joint Distraction in a Linear Force System

Knowledge of the principles of Newton's laws and of linear force systems can be used to understand how skeletal traction produces joint distraction. Figure 1–30 shows skeletal traction applied to the leg. How does hanging a 10-lb weight on this pulley system produce separation between the tibia and the femur at the knee joint? Follow each force carefully as it is described.

We will begin by assuming that the objects in Figure 1–30 are all in equilibrium. The one known force is gravity-on-weight (GW) which has a magnitude of -10 lb acting vertically downward. For the weight to be in equilibrium, a $+10$-lb

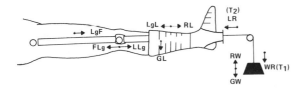

Figure 1-30. Traction applied to the foot sets up numerous forces that result in joint distraction at the knee.

force must also exist on the weight, coming from something touching the weight. Since the rope is the only other object contacting the weight, the $+10$-lb force must be rope-on-weight (RW). RW must have a reaction force weight-on-rope (WR), which is equal in magnitude and opposite in direction (-10 lb) to RW.

Assuming we have a frictionless pulley system, tension in the rope must be the same throughout the rope. That is, the magnitude of force applied to each end of the rope must be the same. Since WR represents the pull of the weight or tension (T_1) in the vertical rope segment, an equal tension vector (T_2) must exist at the other end of the rope. To create tension at both ends of the rope, T_2 must be applied in the opposite direction from T_1. Given that the pulley changes the direction of the force, T_2 in this instance is horizontal and to the left (with a magnitude of 10 lb). If the rope was not "bent" around the pulley, T_1 and T_2 would be directly opposite to each other.

Given that we know that T_2 exists, we must identify it by name. The force T_2 must originate from something that contacts the horizontal rope segment and is capable of exerting a force of 10 lb to the left. Both the pulley and the leg sling touch the horizontal rope segment. Both the sling and the pulley can only "pull" on the rope (a "push" on the rope would tend to make it slack—also, if a pulley could push, it wouldn't be called a pulley!). Since a pull is always in the direction of its source, only the sling could exert a pull to the left. Since the sling wraps around the leg, the leg and sling can be considered as one object and will be referred to as **leg.** Consequently, T_2 must be leg-on-rope (LR) and, as tension vector T_2, must have a magnitude of -10 lb. Leg-on-rope has a reaction, rope-on-leg (RL), applied to the leg with a magnitude of $+10$ lb.

The leg (the segment below the knee) is contacted by gravity (GL), the knee ligaments, and the rope. The force exerted by the rope has already been identified. Vector GL is vertically downward and not in line with RL. GL is not part of a linear force system with RL and can be ignored for the moment. We will only consider the horizontal forces acting on the leg or femur segments that we are trying to separate (distract). The force of knee-ligaments-on-leg (LgL) would potentially pull on the leg, creating a force to the left that would be part of the same linear force system as RL. (Although the joint capsule is also part of the ligamentous force acting on the leg, only the ligaments will be referenced for purposes of simplicity.) If the net effect of the two forces LgL and RL is no movement (equilibrium), the forces must be balanced and their sum must be zero. Therefore, LgL must have a magnitude of -10 lb.

Vector LgL must have a reaction force; the leg must exert an equal and opposite force of $+10$ lb on the ligaments, leg-on-ligaments (LLg). For the ligaments to be in equilibrium, an equal force of -10 lb must be exerted on the ligaments by the femur, femur-on-ligaments (vector FLg), since the femur is the only other object contacting the ligaments (again ignoring the vertical effect of gravity on the ligaments). Femur-on-ligaments will have a reaction force, ligaments-on-femur (LgF), of $+10$ lb. Assuming the femur is in equilibrium without examining all forces acting on it, all forces relevant to joint distraction have now been identified.

Equilibrium of each object identified in Figure 1–30 is dependent on the ability of the object to generate the required force. In the case of the suspended weight, the rope, and the leg, each is quite capable of exerting the required 10-lb force as long as no defects exist. The ligaments, however, cannot exert any significant force on the leg when they are slack. The ligaments will, in fact, be slack when the weight is first put on the traction system, since the femur and leg are close to each other (the goal of the traction is to separate them). Initially, the force of LgL (and LLg) might be as little as 1 lb, and a net force of 9 lb to the right will exist on the leg (the pull of the rope on the leg will not be completely offset by the pull of the ligaments on the leg).

The leg will accelerate to the right as the tension in the ligaments builds. The tighter the ligaments become, the less unbalanced force exists and the less acceleration there is. Once the ligaments are pulled taut by the movement of the leg, the ligaments are very capable of exerting the 10-lb force needed to establish equilibrium of the leg. Ligaments have a tremendous tensile strength and are capable of withstanding far greater forces than the force applied in this example for at least short periods of time. As shall be seen in subsequent chapters, prolonged loading may result in gradual elongation of the ligaments. The example of joint distraction created by acceleration of the leg to the right, directly away from the joint, is an example of true translatory motion without concomitant joint rotation.

Equilibrium of the leg has now been established in a position in which the ligaments are taut and the joint surfaces are separated by a distance permitted by the length of the ligaments. Note that in this example the muscles that cross the knee joint were assumed to be inactive, since no muscle forces were included. In the case of a bone fracture, the process of distraction of bone fragments required for proper alignment and healing is similar to that described in joint distraction. The fracture acts as a false joint and distraction is resisted by the muscles crossing the fracture site. Overactivity of the muscles initially accelerates the distal fragment toward the proximal one. Traction is applied to realign the bones. As the muscle spasm subsides, the muscle exerts less and less force, until the force of the traction rope on the distal fragment exceeds the force of the muscles on that fragment; the direction of movement will then cause the fragments to separate and realign. Equilibrium is reestablished when the structures crossing the fracture site become taut once again.

In the example of leg traction, we ignored the force of gravity on the leg and on the ligaments, stating that the vertical forces of gravity were not part of the same linear force system as the distraction forces and, therefore, could be ignored when calculating the magnitude of these forces. In fact, this can lead to a further refinement of Newton's first law or law of equilibrium:

- For an object to be in equilibrium, the sum of all the vertical forces on the object must be zero and, independently, the sum of all the horizontal forces must be zero.

$$\Sigma F_V = 0$$
$$\Sigma F_H = 0$$

We accounted for the sum of the horizontal forces, but ignored the sum of the vertical forces. Having identified only gravitational vertical forces (applied vertically downward), we did *not* take into consideration the vertical equilibrium of the leg in the traction example. If we conveniently ignored the negligible magnitude of gravity on rope, we would still have to consider the effect of gravity on leg. We would offset GL by resting the leg on a bed (creating a contact force, bed-on-leg) but would then potentially introduce a new force, the force of friction.

Force of Friction

The force of friction potentially exists whenever two objects contact. Friction will have magnitude whenever one object is made to slide or move over another contacting object. **Friction (Fx)** is a vector force.

- Friction may exist whenever two objects touch.
- Friction has magnitude whenever two contacting objects attempt to slide or move on each other.
- Friction forces also come in pairs, applied to each of the contacting objects, equal in magnitude and opposite in direction.

- The action line of friction forces always lies parallel to the contacting surfaces.
- The direction of the force of friction on an object is always opposite to the direction of potential or relative movement of the object.

The force of friction does not have magnitude unless there is attempted movement between the contacting objects. The *maximum* potential force of friction for an object that is *not* moving is the product of a constant known as the coefficient of static friction (μ_s) and the reaction or contact force (F_c) exerted by the contacting object. That is,

$$Fx \leq \mu_s F_c$$

The coefficient of static friction is a constant value for given materials. For example, μ_s for ice on ice is approximately 0.05; the value for wood on wood is approximately 0.25; as the contacting surfaces become softer or rougher, μ_s increases. F_c increases with the magnitude of contact of the adjacent objects. As we have seen, contact or reaction forces can relate at least indirectly to the weight of the objects. A heavier object generally has a greater contact with an object beneath it than a lighter object. The greater the reaction force on an object and the rougher the contacting surfaces, the greater the maximum potential force of friction. When using friction to warm your hands, the contact of the hands warms both of them (friction forces exist on both the right and the left hands). If you wish to increase the friction, you press your hands together harder as you rub. Increasing the pressure increases the contact between the hands and increases the maximum value of friction (the coefficient of friction remains unchanged since the surface remains skin on skin).

Friction is given as a *maximum potential* force, since it actually exists (or is "activated") only if an attempt is made to move one object on another. In Figure 1–31, a large box weighing 100 lb is resting on the floor. Assessing the equilibrium of the box, the maximum *potential* friction force on the box (FX) is a product of the coefficient of static friction of wood box on wood floor (0.25) and the reaction force of floor-on-box (FB) is 100 lb.

$$FX \leq (0.25)\,(100)$$
$$FX \leq 25\ lb$$

As long as no other forces are applied to the box, friction will not have magnitude. However, as an external force (the man pushing on the box) is applied to the

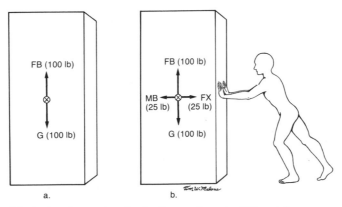

Figure 1–31. In (a), the box is acted on by the forces of gravity (G) and the contact of the floor (FB). The force of friction does not exist because it does not have magnitude. In (b), the force of the man pushing on the box (MB) causes an opposing friction force (FX).

left, friction will be activated as an opposing force. The force of friction can exert a *maximum* of only 25 lb of force to the right; an external force of more than 25 lb to the left will cause the box to accelerate to the left. Note that only the forces applied horizontally to the box affect its horizontal equilibrium.

Once an object is moving, friction is a constant value, equal to the product of the reaction force and the coefficient of *kinetic friction* (μ_k). μ_k is less than the maximum value of static friction for any contacting surfaces. The external force attempting to move the object must be larger than the maximum value of static friction before movement will begin (at least a small net unbalanced force must exist in the direction of movement). However, once movement is initiated, the value of friction drops from its maximum value to its smaller kinetic value, suddenly increasing the net unbalanced force if the external force has not changed since movement began. This results in the classic situation where the man pushes harder and harder to get the box moving along the floor and then suddenly finds himself and the box accelerating rapidly. Again calling attention to the more detailed restatement of Newton's law of equilibrium, note that the box being pushed along the floor is not in horizontal equilibrium, but is in *vertical* equilibrium (G and FB are balanced).

In the example of the leg in traction (Fig. 1–30), if the leg were to rest on the bed, we would add a vertical contact force (bed-on-leg), which would oppose gravity (GL), but would also add a potential friction force. The potential friction force would be parallel to the contacting surfaces (the leg and the bed) and so would be a horizontal force applied in the direction opposite to potential movement of the leg away from the femur. That is, friction-on-leg would be opposite to the pull of the traction (RL).

The maximum magnitude of the friction force on the leg (Fx_L) would be the product of bed-on-leg (the contact force) and an unknown coefficient of static friction. In this example, u_s would be fairly large, since the skin and the sheet do not slide easily on each other. We can assume the Fx_L is fairly large and could potentially completely offset rope-on-leg. The leg would not move since there would be no net unbalanced force acting on the leg. No distraction between the femur and the leg would occur and the ligaments would remain slack. To maintain the effectiveness of the traction the leg would either have to be supported off the surface of the bed or friction minimized in some other way.

CONCURRENT FORCE SYSTEMS

When identifying equilibrium forces and reaction pairs, the forces presented so far were part of one or more linear force systems applied to the object. More commonly, however, forces applied to an object are not in a line but have action lines that lie at angles to each other. Two or more forces acting at a common point of application but in divergent directions are part of a **concurrent force system**. In fact, to be part of a concurrent force system, the vectors do not have to have the same precise point of application but might, if extended, intersect at some point on the object under consideration. The net effect, or resultant, of all forces acting at a common point can be found by a process known as composition of forces.

Composition of Forces

In Figure 1–32 man A and man B are each pulling on the block at a right angle to each other with a force of 75 lb each. The action lines of man-A-on-block (AB) and man-B-on-block (BB) are in different directions but are commonly applied through the COG of the block (since vectors can be extended, the ropes do not need to be attached at the COG of the block). The net effect, or resultant action, of the

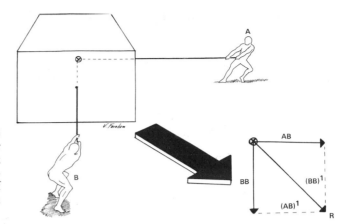

Figure 1–32. Man A and Man B pulling at angles to each other through the COG of the block represent a concurrent force system (Inset—Composition of forces from Man A [AB] and Man B [BB] give resultant R.)

two pulls will be in a line that lies between the men. Figure 1–32 (inset) shows a graphic solution of the problem by the polygon method. Vectors **AB** and **BB** are drawn to scale with a common point of application, maintaining the 90° angle between them. Line $(AB)_1$ is then drawn parallel to **AB** from the end of **BB**; line $(BB)_1$ is drawn parallel to **BB** from the end of **AB**, forming a polygon. The resultant force **R** is diagramed with its point of application at the intersection of **AB** and **BB**; its action line and magnitude are drawn so that the arrowhead lies at the intersection of lines $(AB)_1$ and $(BB)_1$. The resultant vector is always the diagonal of the polygon formed by the original two vectors. The net effect of forces **AB** and **BB** would be a force of 106.25 lb (scale: 0.5 in = 25 lb) in the direction of **R**. It is important to note that the magnitude of **R** is not equal to the sum of the magnitudes of **AB** and **BB**. The block can be pulled more efficiently if a single force of 106.25 lb is applied in the direction of **R**, rather than with two divergent forces of 75 lb each.

Muscle Action Lines
Total Muscle Force Vector

The force applied by a muscle to a bony segment is actually the resultant of the pull on a common point of attachment of all the fibers that compose the muscle. Since each muscle fiber can be represented by a vector (Fig. 1–33), the fibers taken together form a concurrent force system with a resultant that is the total muscle force vector (Fms). Fms has a point of application at the attachment of the muscle and an action line that is the direction of the resultant pull of all the muscle's fibers. An actual graphic solution could be attempted by taking the pull of two fibers ini-

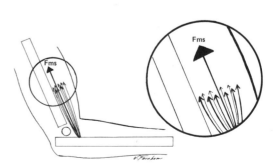

Figure 1–33. The total muscle force (Fms) is the resultant of all fiber pulls taken together.

tially, composing them into a single pull, and adding another fiber. This would have to be repeated until all the fibers have been taken into account. The resultant pull could also be approximated by putting the point of application on the segment being moved (at the muscle's attachment) and drawing a resultant action line symmetrically up the middle of the muscle's fibers. The direction of pull for all muscles is in the direction of the center of the muscle. The magnitude or length of Fms may be drawn arbitrarily unless a hypothetical magnitude is specified. The actual force of active pull of either individual muscle fibers or the total force muscle cannot be determined in the living person.

A contracting muscle exerts a force on both its proximal and distal segments. The point of application for the muscle vector should be placed either on the segment being moved or the segment under consideration. In actuality, however, every muscle creates a minimum of two force vectors, one on each of the bones to which the muscle is attached. Movement created by a muscle is dependent on the net forces acting on each of its levers, and *not* on the designation of the attachment as "origin" or "insertion." If consideration is being given to flexion of the forearm at the elbow joint, the forces (external and internal) acting on the *forearm* must be considered. Although muscles acting on the forearm will act on at least one other segment, the action on the other segment can be ignored until or unless that segment is analyzed.

- Muscles pull on all segments to which they are attached whenever a contraction occurs.
- Muscles create movement based on the net forces applied to the segment, *not* based on which is the distal end (insertion).

Divergent Muscle Pulls

The concept of a concurrent force system can be used to determine the resultant of two or more segments of one muscle, or two muscles when the muscles have a common attachment. Figure 1–34 shows the resultant force of the anterior portion of the deltoid muscle (AD) and the resultant force of the posterior portion of the deltoid muscle (PD) acting on the humerus. Using the polygon method of composition of forces, the resultant force vector for the constructed polygon (Fig. 1–34, inset) is vector R. Vector R, therefore, represents the vector sum of AD and PD. The deltoid muscle is also composed of a middle segment (MD) that is located between AD and PD. Since R and MD coincide and have a common point of application, they are part of a linear force system. The addition of MD to R will produce vector Fms, which is greater in magnitude but in the same direction as MD and R. Vector Fms represents the total pull of the three segments of the deltoid, producing abduction of the arm as the muscle exerts a force on the humerus.

Figure 1–35 shows another example of composing forces to find the total muscle pull. In this instance the resultant force of the clavicular portion of the pectoralis major muscle (CPM) and the resultant force of the sternal portion of the pectoralis major muscle (SPM) are shown with a common attachment on the humerus. When both portions of the pectoralis major contract simultaneously, the resultant force on the humerus Fms is produced. Fms is in a new direction and would produce adduction and medial rotation of the humerus.

Anatomic Pulleys

Frequently the fibers of a muscle or a muscle tendon wrap around a bone or are deflected by a bony prominence. When the direction of pull of a muscle is altered, the bone or bony prominence causing the deflection forms an anatomical pulley. Pulleys change the direction without changing the magnitude of the applied

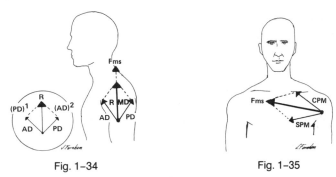

Fig. 1-34   Fig. 1-35

Figure 1-34. The pulls of the anterior deltoid (AD), the middle deltoid (MD), and the posterior deltoid (PD) form a concurrent force system with resultant pull Fms found by composition of forces. AD + PD = R; R + MD = Fms.

Figure 1-35. The clavicular portion of the pectoralis major (CPM) and the sternal portion of the pectoralis major (SPM) can be composed to find Fms.

force. When an anatomic pulley is crossed by a muscle, the muscle action line will not be parallel to the contracting muscle fibers.

In Figure 1-36, a schematic representation of the shoulder is shown on the left, treating the shoulder as a link between two straight levers. The deltoid muscle force vector (Fms) for abduction of the arm has been drawn as it would exist in this hypothetical situation. The right figure shows a more anatomical representation of the shoulder, including the rounded head of the humerus, and the overhanging acromion with the clavicle. These anatomical features change the direction of the fibers of the deltoid muscle. When Fms is drawn, the point of application will still be on the humerus. The action line is in the direction of the muscle fibers *at the point of attachment* and continues in a straight line regardless of any change in direction of muscle fibers. (Recall that the total pull of the deltoid in abduction of the humerus was determined to be in line with the middle deltoid.) The action line and direction of Fms are significantly different between the left portion of Figure 1-36 and the right portion, although the point of application and magnitude of the force are the same in each figure.

Since anatomical pulleys are commonly encountered by muscles in the body, the pull of a muscle (Fms) should be visualized for any given muscle or muscle segment in such a way that:

- The point of application is located on the segment being moved, at the point of attachment of the muscle on that segment.
- The action line is in the direction of pull that the fibers or tendons of the muscle create *at the point of application.*

Figure 1-36. Action line of the deltoid muscle on a schematic representation of the clavicle and humerus (left). Action line of the deltoid is deflected by the bony contours that form anatomic pulleys (right).

- Vectors are straight lines and do not change directions, regardless of any change in direction of muscle fiber or tendon; magnitude, unless given as a specific hypothetical value, is usually arbitrary.

PARALLEL FORCE SYSTEMS

In analyzing forces on an object and the effects produced on that object, forces thus far have had action lines that coincided or intersected. To evaluate the effects that forces have on the levers of the human body, it is necessary to analyze forces that act on the same objects (the lever) but have action lines that will never converge. **A parallel force system** exists whenever two or more parallel forces act on the same object but at some distance from each other.

The bones of the body are levers or rigid bars that rotate around an axis. Forces applied to a bar will produce either equilibrium (no motion) or movement such as rotation or translation. The effect of a given combination of forces acting at different points on a lever can be determined through an understanding of the principles of the levers.

First-Class Levers

A first-class lever is commonly exemplified by a seesaw (Fig. 1–37a). The lever, or seesaw, is being subjected to four forces (since four things touch the lever). These are the contact of man A, the contact of man B, the contact of the wedge, and the contact of gravity. Figure 1–37b shows a schematic representation of the lever and the forces acting on it. Wedge-on-seesaw and gravity are applied to the COG of the seesaw and are part of a linear force system. Since these two forces do not lie at a distance from each other, we will ignore them for the moment and focus on the effects of vectors A and B. **A first-class lever** system exists whenever two resultant forces are applied on either side of an axis at some distance from that axis, creating rotation in opposite directions.

In Figure 1–38, the first-class seesaw and its forces are again diagramed, but labeled differently than in Figure 1–37. Force B has been renamed the effort force (EF). The **effort force** is defined as the force that is causing or attempting to cause the motion. Force A has become the resistance force (R). The **resistance force** is the force that is opposing the movement. This assumes, of course, that man B is causing the rotation of the seesaw rather than man A. If the seesaw were balanced (in equilibrium), neither force would be causing motion since none is occurring. When

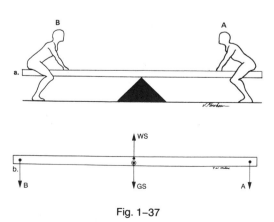

Fig. 1–37

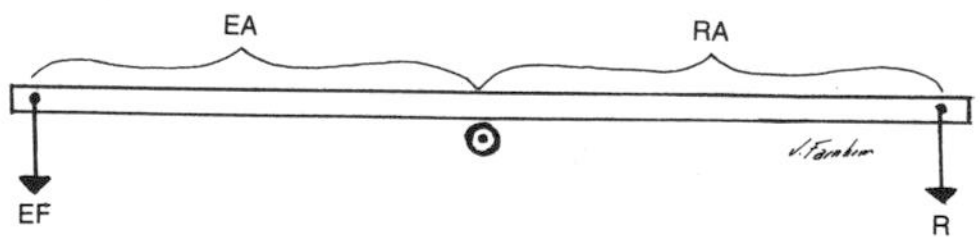

Fig. 1–38

Figure 1–37. a. A seesaw is commonly used to represent a lever to which forces are applied. b. The lever is contacted by man A, man B, gravity on the seesaw (GS), and the wedge (WS) that serves as its axis.

Figure 1–38. First-class lever system. EF is the effort force that lies at a distance EA (effort arm) from the axis. R is the resistance that lies at a distance RA (resistance arm) from the axis.

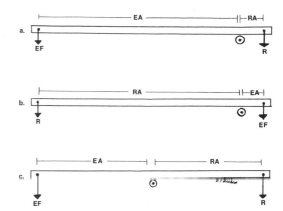

Figure 1-39 In a first-class lever system EA may be: (a) greater than RA, (b) smaller than RA, or (c) equal to RA.

there is equilibrium, it is arbitrary as to which force is identified as the effort and which as the resistance. Whenever equilibrium does not exist (whenever the lever is rotating), rotation of the lever will *always* occur in the direction exerted by the effort force.

The term **lever arm (LA)** is used to describe the distance from the axis to the point at which a force is applied to the lever. The **effort arm (EA)** is a term that refers specifically to the lever arm of the effort force. Similarly, the **resistance arm (RA)** is a specific term referring to the lever arm of the resistance force (Fig. 1–38). In a first-class lever, **EA** may be greater than, smaller than, or equal to **RA** (Fig. 1–39), since the axis may be located anywhere between the effort force and the resistance force without changing the classification of the lever.

In the human body, there are relatively few first-class levers. Figure 1–40 shows two forces acting on the forearm lever; one is the force of the triceps at the olecranon, creating a clockwise rotation; the other is a resultant external force pushing up on the forearm in a counterclockwise direction. This constitutes a first-class lever regardless of which force is the effort and which is the resistance since the axis lies between the two forces. If the forearm were to extend, the triceps brachii would be the effort force. If the external force overcame the triceps force and the forearm flexed, the external force would be the effort.

Second-Class Levers

A second-class lever exists whenever two resultant forces are applied so that the resistance lies between the effort force and the axis of rotation (Fig. 1–41). In a second-class lever EA is *always* greater than RA. Second-class levers in the human

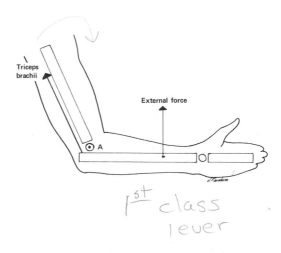

Figure 1-40. First-class lever in the human body. The force of the triceps muscle on the olecranon of the ulna and an external force on the ulna are separated by the axis (A).

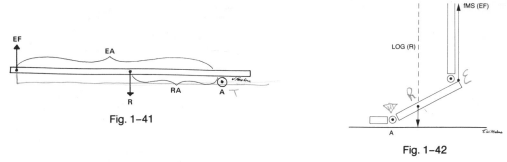

Fig. 1–41

Fig. 1–42

Figure 1–41. Second-class lever system. The effort arm (EA) is always larger than the resistance arm (RA).

Figure 1–42. As the triceps surae (Fms) actively contract, the weight of the body (LOG) is lifted around the metatarsophalangeal (MTP) axis of the toes. The force of Fms (EF) and gravity (R) act on a second-class lever with the axis at the MTP joints.

body commonly occur when gravity is the effort force and muscles are the resistance. There are also examples of second-class levers where the muscle is the effort force, but the distal segment to which the muscle is attached is weight-bearing. The result is movement of the proximal rather than distal lever. Figure 1–42 shows the action of the triceps surae (gastrocnemius, soleus, and plantaris) lifting the body around the axis of the toes (metatarsophalangeal joints). The superimposed body weight acting on the foot through the LOG is the resistance (R). Since the muscles are the effort and the body weight is the resistance, a second-class lever system is formed.

There are no apparent examples in the human body of second-class levers in which the muscle is the effort force, causing movement of its distal lever against the resistance of gravity. In such an instance the point of application of the muscle would have to lie farther from the axis than the COG of the limb being moved. There are no muscles with attachments that meet this requirement. Consequently, all second-class levers in the body occur either with a muscle eccentrically contracting against an external moving force, or with a muscle acting on its proximal segment while the distal is fixed.

Third-Class Levers

A third-class lever exists whenever the forces on a lever are applied so that the effort force lies closer to the axis of the lever than does the resistance (Fig. 1–43). In a third-class lever, EA will *always* be smaller than RA. In the human body, most muscles creating rotation of their distal segments are part of third-class-lever systems. The point of attachment of the muscle causing the motion is almost always closer to the joint axis than the external force, which is usually resisting the motion. Figure 1–44 shows an example of a third-class-lever system with the biceps brachii performing flexion of the forearm/hand segment against the resistance of gravity.

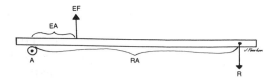

Figure 1–43. Third-class lever system. The effort arm (EA) is always smaller than the resistance arm (RA).

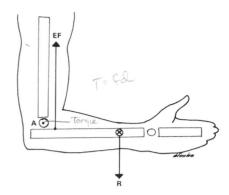

Figure 1–44. Third-class lever in the human body. The biceps brachii, as the effort force (EF), lies closer to the axis (A) than gravity (R).

Since the biceps force is labeled as **EF**, the rotation must be occurring in the direction of the biceps force. Gravity is opposing the motion and is labeled **R**.

Torque

Regardless of the class of the lever, the rotation of the segment is dependent both on the magnitude of force exerted by the effort and resistance forces and on the *distance from the axis* of these two forces. The ability of any force to cause rotation of the lever is known as **torque** or **moment of force**. Torque (T) is a product of the magnitude of the applied force (f) and the distance (d) that force lies from the axis of rotation. The distance (d) is the *shortest* distance between the action line of the applied force and the axis of the lever; it is the length of a line drawn perpendicular to the action line of the force and intersecting the axis. Therefore,

$$T = f^{\perp}d$$

In the schematic example of a second-class lever in Figure 1–41, $^{\perp}d$ for the effort force and the resistance force would correspond to the lever arms **EA** and **RA**, respectively. In this example **EA** and **RA** are perpendicular to their respective forces.

If hypothetical magnitudes and distances were given in Figure 1–44, the net torque acting on the lever could be determined. Let's assume that the biceps brachii (**EF**) is contracting with a force of 120 lb applied at a distance (**EA**) of 1 in from the axis. The weight of the forearm/hand segment (**R**) is 10 lb and the COG of the segment lies 10 in (**RA**) from the axis. Then

$$T_{EF} = f \times EA \qquad T_R = f \times RA$$
$$T_{EF} = (120\ lb)\ (1\ in) \qquad T_R = (10\ lb)\ (10\ in)$$
$$T_{EF} = 120\ in\text{-}lb \qquad T_R = 100\ in\text{-}lb$$

The biceps brachii exerts a torque of 120 in-lb on the forearm in a counterclockwise direction ($T_{EF} = -120$ in-lb). The torque exerted by gravity is in a clockwise or positive direction ($T_R = +100$ in-lb) and has been assigned a positive sign.

- The **net rotation** (or **resultant torque**) of a lever can be determined by finding the sum of all the torques acting on the lever (maintaining **appropriate positive and negative signs**).
- Forces that act through an axis (at no distance from it) cannot produce torque at that axis.

The second point above addresses why we were able to ignore the effect of the gravity and the contact of the wedge on the seesaw in Figure 1–37. Since both of these forces acted through the axis of rotation of the seesaw (the wedge), neither created a torque on the seesaw and was not relevant to the resultant torque applied to the seesaw.

When the sum of all the torques is zero, the torques are balanced and the lever will not rotate. The lever is in rotational equilibrium when

$$\Sigma_T = 0$$

In the example above where the torque of the biceps (**EF**) was -120 in-lb and the torque of gravity was $+100$ in-lb, the resultant torque is -20 in-lb or 20 in-lb of torque in a counterclockwise direction (flexion of the forearm/hand segment).

Composition of Forces in a Parallel Force System

The example shown in Figure 1–45 shows the forearm/hand segment being acted on by three forces: the muscle force of the biceps brachii on the forearm at its attachment (Fms = 120 lb), the force of gravity on the forearm/hand at the COG (G = 10 lb), and the contact of the weight ball on the forearm/hand at its point of contact (WH = 5 lb). The LAs of these forces are 1, 10, and 15 in, respectively. We can find the net torque acting on the lever by finding the sum of the torques created by each force. This does not help, however, if we are also interested in determining the class of the lever. The class of the lever is best determined by composing forces until there is a single effort force and a single resistance.

In a parallel force system, all forces causing rotation in one direction can be represented by a single vector acting in the same direction with a magnitude equal to the sum of the composing forces. Vectors G and WH can be combined and represented by GWH. The point of application of the resultant vector GWH will lie on a line between the original two vectors. This is how the new COG was located when the COGs of two composing segments were known. Now the precise point of application of the new resultant force can be determined:

- When two or more forces applied to a lever are composed into a single resultant force, the magnitude of the resultant force will be equal to the sum of the magnitudes of the original forces and will be applied in the same direction.
- When two or more forces applied to a lever are composed into a single resultant force, the torque produced by the resultant will be the same as the net torque produced by the original forces.
- The point of application of the resultant force can be found once its magnitude and the torque it produces are known:

$$^\perp d = T/f$$

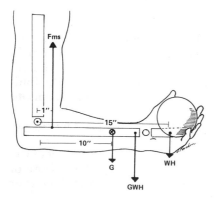

Figure 1–45. The class of the lever can be identified by determining the net torque acting on the lever. The net rotation will be in the direction of force EF.

GWH will be equal in magnitude to the magnitudes of the composing forces G and WH:

$$GWH = (10 \text{ lb}) + (5 \text{ lb})$$
$$GWH = 15 \text{ lb}$$

The torque of GWH will be the same as the net torques produced by vectors G and WH:

$$T_G = (+10 \text{ lb}) (10 \text{ in}) \qquad T_{WH} = (+5 \text{ lb}) (15 \text{ in})$$
$$T_G = +100 \text{ in-lb} \qquad T_{WH} = +75 \text{ in-lb}$$

Therefore, T_{GWH} must be:

$$T_{GWH} = (+100 \text{ in-lb}) + (+75 \text{ in-lb})$$
$$T_{GWH} = +175 \text{ in-lb}$$

Since $T = f \times {}^{\perp}d$ and both the torque and magnitude of GWH are known, the distance (${}^{\perp}d$) from the axis can be determined.

$${}^{\perp}d = T/f$$
$${}^{\perp}d = \frac{+175 \text{ in-lb}}{15 \text{ lb}}$$
$${}^{\perp}d = 11.67 \text{ in}$$

The resultant force GWH in Figure 1–45 has a magnitude of 15 lb, lies 11.67 in from the joint axis, and applies a torque of 175 in-lb in a clockwise direction.

Although the number of forces acting on the forearm/hand segment has been reduced to two (Fms and GWH), the class of the lever cannot be determined until the effort force and the resistance force can be labeled. To label the effort force, the direction of movement must be ascertained, since the effort force is the force producing the motion in the direction of the net torque. (In a first-class lever, the axis lies between the effort and the resistance, so the labeling is irrelevant to identifying the class of the lever.) To determine the net torque acting on the forearm/hand segment in Figure 1–45, the torque exerted by the muscle (Fms) must be found. The torque produced by Fms would be:

$$T_{Fms} = (-120 \text{ lb}) (1 \text{ in})$$
$$T_{Fms} = -120 \text{ in-lb}$$

The resultant torque on the lever will be the sum of the torques produced by GWH and Fms:

$$T = T_{Fms} + T_{GWH}$$
$$T = (-120 \text{ in-lb}) + (+175 \text{ in-lb})$$
$$T = +55 \text{ in-lb}$$

The forearm/hand segment is rotating in a clockwise direction (extension) with a magnitude of 55 in-lb of torque. Since rotation is occurring in the direction of GWH, GWH must be the effort force and Fms the resistance. The resistance lies between the axis and the effort, so a second-class lever exists.

The example just analyzed points out that simply describing that the lever is extending (kinematics) can lead to erroneous conclusions about what muscles might be active. In this instance, the elbow is extending in spite of the fact that there are no active *elbow extensors*. In fact, the only active muscle is an elbow flexor! An understanding of the muscles involved in any movement can only occur through a kinetic analysis of the motion, and not simply by observing the movement.

In the examples given for both Figures 1–44 and 1–45, the biceps brachii, an elbow flexor, is exerting a 120-lb force. In one instance, however, the elbow is flexing (Figure 1–44) and in one instance the elbow is extending (Figure 1–45). The force of the muscle, while consistently applied toward flexion, is losing to the combined forces of gravity and the weight in the latter example. The elbow flexor in effect is acting as a brake or *control* to the external forces. Since the muscle is pulling the lever in one direction, yet the lever is moving in the opposite direction, the muscle must be getting longer (its insertion is moving away from its origin). When an *active* muscle is lengthening, it is contracting **eccentrically.** When an *active* muscle is shortening (as it will when the muscle is the effort force), it is contracting **concentrically.** When the biceps brachii was the moving force *producing* elbow flexion, a third-class lever existed. When the muscle became the *resisting* force, the lever became second class. Since most muscles in the body act on third-class levers when contracting concentrically (distal lever free), the most common second-class lever in the human body is when the same muscles work eccentrically as brakes (the controlling resistance) to external forces.

- When an active muscle shortens (does a concentric contraction) and moves its lever in the direction of the muscular force, the muscle will be the effort.
- When an active muscle lengthens, it is contracting eccentrically and must be the resistance (control) to another force which is producing the motion. The pull of the muscle would be opposite to the direction of movement.
- When a lever is in rotational equilibrium, any muscle acting on the lever is neither shortening nor lengthening. Such a muscle would be performing an **isometric** contraction (iso = same; metric = length). In such a case whether the muscle is labeled as the effort force or as the resistance is arbitrary. (When a lever is in equilibrium, it is easiest to refer to the internal force as the effort force and the external force as the resistance.)

Mechanical Advantage

Mechanical advantage (M Ad) is a measure of the efficiency of the lever (the relative effectiveness of the effort force as compared to the resistance) and is related to the classification of a lever and can be used to develop an understanding of the relevance of the concept of classes of levers. Mechanical advantage is the ratio of the *effort arm* to the *resistance arm,* or

$$M\ Ad = \frac{EA}{RA}$$

When the effort arm (EA) is larger than the resistance arm (RA), the mechanical advantage will be greater than one. When the mechanical advantage is greater than one, the magnitude of the effort force can be smaller than the magnitude of the resistance. That is, a small effort force can create more torque and overcome a larger resistance. This can be appreciated when we realize that:

$$(EF)\ (EA) > (R)\ (RA)$$

The product of the effort force and its effort arm will always be greater than the product of the resistance and its resistance arm (the torque of the effort force by definition will always be greater than the torque of the resistance force). When EA is greater than RA, as it is when the mechanical advantage is greater than one, the EF can be smaller than R and still create more torque. The greater the ratio of EA to RA, the smaller EF can be as compared to R. The "advantage" to a lever with a mechanical advantage greater than one is simply that the effort can overcome the resistance without expending as much force as the resistance.

In the example shown in Figure 1–45, EF was vector GWH applied 11.67 in from the axis. **R** was the force of the biceps brachii applied 1 in from the axis. Therefore,

$$M \, Ad = \frac{11.64 \ \text{in}}{1 \ \text{in}} = 11.64$$

The forearm/hand segment, a second-class lever, was efficient since it took only 15 lb of force GWH to overcome 120 lb of force exerted by the biceps brachii. In every second-class lever, EA will always be larger than RA since the resistance is closer to the axis than the effort force. A second-class lever is always efficient (has a mechanical advantage greater than one), although the magnitude of the ratio of EA to RA will vary with the lever and the forces applied to it.

In third-class levers, EA is always smaller than RA since the effort force lies closer to the axis than the resistance. The mechanical advantage of a third-class lever, therefore, will always be less than one. A third-class lever is inefficient in that the magnitude of the effort required to move the lever will have to be *greater* than the magnitude of the resistance.

- In all second-class levers, the mechanical advantage of the lever will always be greater than one. The magnitude of the effort force can be less than the magnitude of the resistance.
- In all third-class levers, the mechanical advantage of the lever will always be less than one. The magnitude of the effort force will have to be *greater* than the magnitude of the resistance for the effort to produce greater torque.
- First-class levers follow no rules relative to mechanical advantage. EA can be greater than, less than, or equal to RA.

Trade-offs of Mechanical Advantage

It has already been observed that the majority of the muscles in the human body act as part of third-class lever systems when contracting concentrically (distal lever free). It would appear, then, that the human body is structured inefficiently. In fact, the muscles of the body are structured to take on the burden of "mechanical disadvantage" to achieve the goal of moving the lever through space.

Figure 1–46a shows the forearm/hand segment being flexed (rotated counter-clockwise) through space by a muscle (Fms) against the resistance of gravity (G).

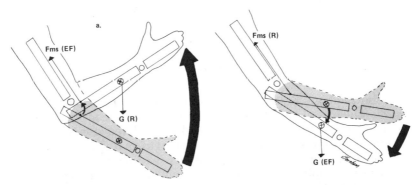

Figure 1–46. a. In a mechanically inefficient third-class lever, movement of the point of application of EF (Fms) through a small arc produces a large arc of movement of the lever distally. b. In a mechanically efficient second-class lever, movement of the point of application of EF (G) through a small arc produces little increase in the arc distally.

This is a third-class lever, since the force closer to the axis is creating the movement. It has already been shown that the magnitude of Fms must be much larger than the magnitude of G for flexion to occur. The system is, indeed, inefficient. However, as Fms pulls its point of application (on the proximal forearm/hand lever) through a very small arc, the distal portion of the lever is moving through a much greater arc. Although the muscle force needed to create the movement was considerable, the result was a large range of motion and speed for the distal portion of the segment. Since the goal in much of human function is movement through space and *not* mechanical efficiency, the use of third-class-lever systems achieves the desired goal. In fact, the shorter the lever arm of the effort force (the closer the effort force is applied to the axis), the greater is the movement of the distal end of the lever.

In second-class levers in the human body, the effort force is usually the external force. Given the mechanical advantage of the second-class lever, the effort can be (but need not necessarily be) smaller than the resistance. As shown in Figure 1–46b, the trade-off is that little is gained in speed or distance of the distal end of the segment. A small arc of movement at the point of application of the effort force (G) results in only a small movement of the more distal segment(s). In any second-class lever (and in a first-class lever where EA is greater than RA), the lever is efficient in terms of effort force output, but gains relatively little in movement through space.

- When the muscle is the effort force and the EA is smaller than the RA, the magnitude of the effort force must be large, but the expenditure of muscle force is offset by gains in speed and distance of the distal portions of the segment. This is true for all third-class levers and for first-class levers in which RA > EA.
- When the external force is the effort force and the EA is larger than the RA, the magnitude of the effort can be small as compared to the resistance, but relatively little is gained in speed and distance. This is true for all second-class levers and for first-class levers in which EA > RA.

When one is looking at the advantages and disadvantages of each lever class, it is important to keep in mind the role of the muscle. In a second-class lever the external force is the efficient force on the lever. The muscle is still expending a large amount of resisting force to control motion of the lever. In a third-class lever, the muscle overcomes the resistance, but only at great expense. Consequently, the muscle must be able to create large forces regardless of the class of the lever. As shall be shown in Chapter 3, the muscle is structured to be able to generate the large forces required both to produce large movements of the distal segments of the body in mechanically inefficient lever systems and to resist the external forces in mechanically efficient lever systems.

Work

A more thorough understanding of the functional significance of classification of levers and of the role of muscles in human lever systems can be achieved by looking at the mechanical concept of work and the related concept of energy. In mechanical terms, **work** is done by a force applied to an object whenever the force is applied in the direction of motion of the object. The magnitude of work is directly proportional to the applied force and to the magnitude of movement produced. Most simply, this relationship can be expressed by

$$W = f \times d$$

where the direction of the applied force is parallel to the movement produced.

Where this is not the case, only the portion of the force that *is* parallel to the direction of movement will contribute to work. That is,

$$W = F \, d \cos \theta$$

Work is given in units of foot-pounds (ft-lb). The sign given to work will be positive if the force is exerted in the same direction as the motion and negative if the force is exerted in a direction opposite to the motion. When the movement being produced is rotatory, as it is when examining levers moving around joint axes, then:

$$W = (\text{Torque}) \, \theta$$

The net work done on an object is always the sum of the work produced by each of the forces applied to the object.

In Figure 1–46a and b, the muscle (Fms) moves its point of application through a smaller distance than does the force of gravity, because the muscle is located closer to the axis than G. In Figure 1–46b, the muscle is doing negative work, since it exerts its force in a direction opposite to the motion (it is performing an eccentric contraction). This observation highlights the fact that ''work'' is a mechanical, rather than biomechanical, concept. It is clear that a muscle doing an eccentric contraction is still doing work as we humans use the word; that is, an eccentric contraction results in use of energy to activate the muscle. If the resistance in the second-class lever were a spring rather than a muscle, virtually no energy would be expended by the spring; in fact, it would store energy since it was having work done on it. While the active contractile components of a muscle are not analogous to a mechanical spring, some parts of the muscle (the passive connective tissue components) do behave as the spring does; these portions of the muscle store energy as they are elongated. The effect of negative work done by a muscle (although not due entirely to the springlike qualities of the muscle) is that it takes *less* energy to exert 120 lb of muscle force through a given distance in an eccentric contraction than it does to exert 120 lb of force through the same distance in a concentric contraction. Therefore:

- When a muscle exerts a moving force in a third-class lever, it expends energy and force to gain speed and distance.
- When a muscle exerts a resisting force in a second-class lever, little speed or distance is gained, but the muscle does negative work and uses less force and energy than in a third-class lever situation.

MOMENT ARM OF FORCE

In the examples in which torque ($T = f \times {}^{\perp}d$) has been computed thus far, ${}^{\perp}d$ has been equivalent to the lever arm of the force. Whenever the action line of the force is applied at 90° to the segment, ${}^{\perp}d$ will lie along the lever and coincide with the lever arm for that force. Most often in the human body, however, forces are applied at some angle other than 90°. This is certainly true of the forces produced by muscles. The action lines of muscles rarely approach an angle of 90°, since this would mean that the muscle would lie perpendicular to the bone. Humans would have an unusual shape if this were the rule, rather than the exception. Most muscles have actions lines that are closer to being parallel to the bones to which they attach. When the force is not applied at 90° to the lever, the lever arm is no longer the shortest distance between the action line of the force and the joint axis. The **moment arm (MA)** is always the shortest distance between the action line and the joint axis and is found by measuring the length of a line drawn perpendicular to the force vector, intersecting the joint axis. Consequently,

$$MA = {}^\perp d$$

and

$$T = f \times MA$$

When a force is applied at 90° to a segment, the MA and the lever arm are equivalent. At all other angles, as shall be seen, the MA will be smaller than the LA.

Moment Arm of a Muscle Force

In Figure 1–47 the force of the biceps brachii (Fms) is shown acting at a 65° angle to the forearm lever. (The angle of application of a force is the angle between the action line and the lever on the side of the joint axis.) The torque exerted on the lever by the muscle will be a product of the magnitude of Fms in pounds and the size of the MA in inches. Unlike the LA, which is a fixed distance for a given force, the MA of a force will change depending on the angle of application of the force vector.

In Figure 1–48 the muscle vector of the biceps brachii is shown schematically as it applies a force to the forearm with the elbow at 35°, 70°, 90°, and 145° of elbow flexion. The angle of pull of the action line of Fms on the forearm lever can be seen to change as the elbow flexion angle changes. As the *angle of application of the force* changes, so does the *length of the MA*. The MA would appear to be smallest in Figure 1–48a and largest in Figure 1–48c. If the force of the biceps contraction were a constant 50 lb throughout the elbow range of motion, the torque (T = F × MA) would have to change in direct proportion to the change in the MA of the force. The torque would be least at 35° of elbow flexion and greatest at 90° of elbow flexion. The ability of the muscle to rotate the joint would vary during the movement, even though the muscle is contracting with a constant force.

The MA is the shortest distance between a vector and a joint axis. This distance can be infinitely small (it is zero whenever the action line passes *through* the joint

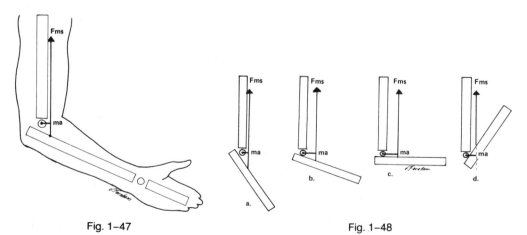

Fig. 1–47   Fig. 1–48

Figure 1–47. The moment arm of Fms is the length of a line drawn perpendicular to Fms and intersecting the joint axis.

Figure 1–48. Schematic representation of the biceps brachii (Fms) exerting a force on the forearm at: (a) 35°; (b) 70°; (c) 90°; and (d) 145° of elbow flexion.

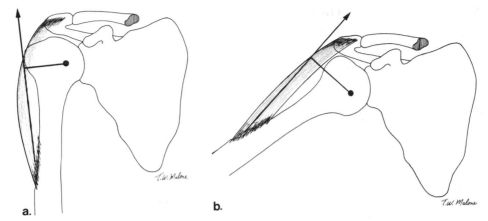

Figure 1–49. The action line of the deltoid muscle (Fms) at 0° of shoulder abduction (a) and at 60° of shoulder abduction (b). The moment arm (MA) of the deltoid is larger at 60° than at any point in the glenohumeral range.

axis), but it can never be larger than the distance between the point of application of the force and the joint axis (that is, the LA). Since the LA and the MA coincide whenever the force is applied at 90° to its lever, as in Figure 1–48c, the MA of any force will be largest when the force is applied at 90° to the lever. When the MA is greatest the torque produced by a muscle contraction of a given magnitude will be at its maximum.

When a force is applied to its lever at an angle *greater than* 90° as in Figure 1–48d, the MA is found by extending the vector (this is the only way a line can be drawn perpendicular to the vector and intersecting the joint axis). The extension of the vector is still part of the vector since vectors can be extended to assess their effect. The extended vector will get closer to the axis as the vector increases its angle of application. As the vector increases its angle of application past 90°, the MA will continue to get smaller.

- Given a constant force of contraction, the torque generated by a muscle is greatest at the point in the joint range of motion at which the muscle's action line lies farthest from the joint axis.
- The MA of any force is greatest when the action line is applied at 90° to its lever, or when the action line is as close to 90° as possible.

When the goal of a muscle is either to rotate a segment or to resist its rotation, the muscle is biomechanically most effective at the point in the range at which the muscle is capable of generating the greatest torque; that is, the point at which the MA of the muscle is greatest.

The critical factor in optimizing torque is in obtaining an angle of application of a muscle force that lies as close to 90° as possible. The *angle of application of the force* ordinarily is *not* directly related to the *joint angle*. The example of the biceps brachii in Figure 1–48 is unusual in that the angle of elbow flexion is similar to the angle of application of the force. This is true because the action line of the biceps lies consistently parallel to the humerus. For most muscles, the angle of application of the muscle to the bone is quite different than the angle between the two bones forming the joint. Figure 1–49 shows the MA of the deltoid at 0° of shoulder abduction and at 60° of shoulder abduction. The angle of application of the muscle never approaches 90° to the lever. However, at 60° of shoulder abduction, the action line of the deltoid muscle lies as far from the joint axis as it will get; the MA is greatest at this point in the range of shoulder abduction. Although the joint lies at 60° of abduction, the angle of application of the muscle is not even close to 60°. The MA of any force is greatest when the angle of application of the muscle force is greatest (or closest to 90°), which is unrelated to the value of the joint position.

Moment Arm of Gravity

Any force applied to a lever may change its angle of application as the lever moves through space. The change in angle of application will result in an increase or decrease in the MA of the force. Figure 1–50 shows the forearm at the same ranges of elbow flexion shown in Figure 1–48, with the force of gravity applied to the lever at its COG. As the angle of application of the LOG changes, so does the length of the MA. As is true for all forces, the MA of the force of gravity is greatest when the force is applied at 90° to the lever. Unlike muscle forces, however, the LOG is always vertical and is completely independent of joint position. For example, Figure 1–51 shows two different positions of the forearm and humerus levers in space. In both instances, the elbow is flexed to 135°. Figure 1–51a has a very large MA because the LOG is applied at 90° to the lever (MA is at its maximum value). Figure 1–51b has a much smaller MA because the LOG is approaching an angle of application of 0°.

- Since gravity always acts vertically downward, the force of gravity will be applied perpendicular to the lever whenever the lever is parallel to the ground.
- When a body lever is parallel to the ground, gravity acting on that segment exerts its maximum torque.

Figure 1–52 shows three graded exercises for trunk flexion. The vertebral space L-5 to S-1 is given as the hypothetical axis about which the segmented head/arms/trunk segment (HAT) rotates. In Figure 1–52a, the arms are raised above the

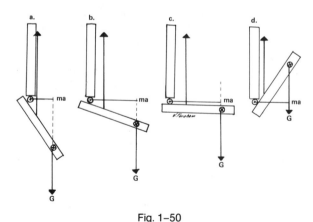

Fig. 1–50

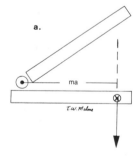

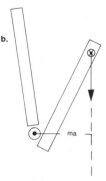

Fig. 1–51

Figure 1-50. Gravity (G) acting on the forearm at: (a) 35°; (b) 70°; (c) 90°; and (d) 145° of elbow flexion. The MA of gravity changes with the position of the forearm.
Figure 1-51. The LOG remains vertical regardless of the position of the forearm in space. As the angle of application of LOG changes with respect to the forearm, the MA also changes.

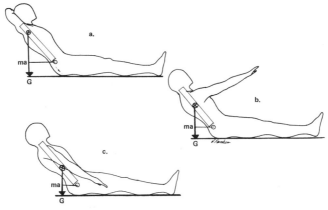

Figure 1–52. Changes in arm position in a sit-up cause the COG of the upper body segment to move, the MA to change, and the torque of gravity (G) to decrease from a to c.

head (the segments are rearranged) causing the COG of HAT to move toward the head (cephalad). The LOG lies at a great distance (MA) from the L-5/S-1 axis. The torque generated by gravity is counterclockwise and has a magnitude equivalent to the product of the weight of **HAT** (two thirds of body weight) and the **MA**. For the posture to be maintained, the abdominal muscles must generate an equal torque in the opposite direction (rotational equilibrium).

In Figure 1–52b and c, the COGs move caudally as the arms are lowered. The relocation of the COG of HAT (through rearrangement of the segments) brings the LOG closer to the L-5/S-1 axis and reduces the length of the MA. Since the weight of the upper body does not change when the arms are lowered, the magnitude of torque applied by gravity to the upper body diminishes in proportion to the reduction in the MA. The decreased gravitational torque requires less opposing torque by the abdominal muscles to maintain equilibrium.

The angle of pull of the abdominal muscles changes relatively little as the trunk lever moves through space. When increased torque production is needed by the abdominals, the increase will occur predominantly through an increase in the force of contraction of the muscles (Fms). When the abdominal muscles are weak, the muscles may not be capable of exerting the force needed to balance the magnitude of gravitational torque in Figure 1–52a. It may be possible, however, for the muscles to counteract the lesser gravitational torques produced in the alternative positions.

Anatomic Pulleys

It has already been noted that anatomical pulleys change the direction but not the magnitude of a muscle force. The change in direction of the muscle force results, however, in improved ability of the muscle to generate torque. The change in direction or deflection of the action line of a muscle is *away* from the axis of the joint being crossed. By deflecting the action line away from the joint axis, the **MA** of the muscle force is increased. Figure 1–53 shows the quadriceps femoris muscle acting on the tibia. Figure 1–53a shows a schematic representation of the action line of the quadriceps without the patella. The action line lies parallel to the femur and close to the knee joint axis; the MA is small. Figure 1–53b shows the deflection of the action line away from the joint axis when the patella is interposed; the MA is significantly larger. If the quadriceps femoris muscle contracted with equal mag-

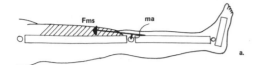

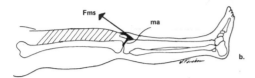

Figure 1–53. a. Schematic representation of the action line of the quadriceps femoris without the patella. b. The action line of the quadriceps femoris deflected by the patella, increasing the MA.

nitude both with and without the patella, the torque applied to the tibia by the muscle would be much greater with the patella since the force is applied at a greater distance from the joint axis.

- Anatomic pulleys change the direction of pull of a muscle.
- Anatomic pulleys deflect the line of pull of a muscle away from the joint axis, thus increasing the MA of the muscle and, consequently, the muscle's ability to produce torque.

FORCE COMPONENTS

We have shown that a force applied to a lever at some angle other than 90° will result in less torque than the same force applied at 90°. If the same magnitude of force can produce less torque in one instance than in another, some of the applied force must be "wasted" (not producing rotation) when applied at angles other than 90°. Torque is produced only by that portion of the force that is directed toward rotation.

While torque is the product of total force and its (perpendicular) distance from the joint axis, it is equivalently a product of the *portion of the force that is directed toward rotation* and its (perpendicular) distance from the joint axis. The portion of the force applied perpendicular to the lever is known as the **rotatory component of the force (f_r).**

Torque can now be expressed in three equivalent ways. Generically,

$$\text{Torque} = F \times {}^{\perp}d$$

When using the total force,

$$\text{Torque} = F \times MA$$

When using just the proportion of the total force expended toward rotating the lever:

$$\text{Torque} = f_r \times LA$$

The rotatory component is the proportion of the force applied perpendicular to the lever. Consequently, the LA (which lies along or parallel to the lever) is, by definition, perpendicular to f_r and is a measure of the shortest distance between f_r and the joint axis.

Figure 1–54a shows the biceps brachii force applied at an angle of approximately 80° to the forearm lever. The torque produced by Fms can be determined by calculating Fms × MA, or by finding the value of f_r and the LA. The magnitude

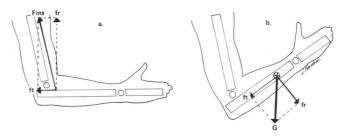

Figure 1–54. a. Resolution of a muscle action line (Fms) into rotatory (f_r) and translatory (f_t) components. b. Resolution of gravity (G) into rotatory and translatory components.

of f_r can be found graphically or mathematically by **resolution** of the vector force Fms into two components. Just as two concurrent forces can be composed into a single resultant vector, a single vector can be resolved into two concurrent components. In this case, the components will be specifically constructed so that one component lies perpendicular to the lever (f_r). Since both composition and resolution of forces involve construction of a parallelogram, the second component will automatically be parallel to the lever. The second component will be the **translatory component** (f_t). In fact, resolving Fms into perpendicular and parallel components will give both the proportion of Fms applied toward rotation *and* the portion of Fms affecting the translation of the segment.

To resolve the resultant vector Fms in Figure 1–54a into its concurrent components, a rectangle is constructed:

1. Starting at the point of application of the resultant force Fms, component f_r is drawn perpendicular to the *long axis of the moving lever.*
2. Starting again at the point of application of vector Fms, component f_t is drawn parallel to the *long axis of the moving lever.*
3. From the head of the resultant vector, a line is drawn that is parallel to f_r.
4. From the head of the resultant vector, a line is drawn parallel to f_t. In this way a rectangle is constructed for which the total force Fms is the diagonal.

Component vectors f_r and f_t are the sides of the constructed rectangle. If the scale of a diagram is known, graphical resolution permits measurement of the components and calculation of magnitudes. For example, if the scale in Figure 1–54a were $\frac{1}{16}$ in = 10 lb, then vector Fms would have a magnitude of 320 lb. The translatory component of Fms would have a magnitude of 50 lb and the rotatory component would have a magnitude of 190 lb. Note that the magnitude of the resultant (Fms) is not equal to the sum of its components f_r and f_t. Determination of the magnitude of the component vectors could also be obtained trigonometrically. If both the magnitude of the total force and the angle of application of the total force are known:

$$f_r = F \sin \theta$$
$$f_t = F \cos \theta$$

Regardless of how the magnitudes of f_r and f_t are determined, these components will always have a fixed proportional relationship to the resultant force at a fixed angle of application. Figure 1–54a indicates that f_r is almost four times larger than f_t. This relationship will remain constant regardless of the magnitude of Fms.

The translatory component of any force is that portion of the force applied toward linear movement of the lever. The translatory component of a force is not "wasted" but is simply applied in a direction that contributes to something other than rotation. In the human body, the translatory component of a force can be thought of as either being toward the joint being analyzed or away from the joint being analyzed. A force moving the lever toward the joint would result in moving the lever toward the adjacent joint segment. Since this would bring the two joint segments together, a translatory force applied in the direction of the joint is referred to as a **compression component.** A compression component generally contributes to the stability of a joint by maintaining contact between joint surfaces. A translatory force in the opposite direction would tend to separate the adjacent joint segments and is known as a **distraction component.** Distraction forces across the knee joint caused by the application of an external load were shown in the example of traction applied to the leg.

Figure 1–54b shows the graphical resolution of gravity acting on the forearm into its rotatory and translatory components. In this instance, the resultant force G is applying equal portions of its force toward rotation and translation (compression).

It has been shown that when a constant force is applied to a lever throughout the range of motion of the lever, the angle of application of force and the torque may change. Since the rotatory component of a force is the portion of the total force that produces torque, a change in torque applied by a constant force must mean that there has been a change in the magnitude of the rotatory component. Similarly, a change in the rotatory component must indicate a change in the proportion of the total force applied toward translation, since the magnitudes of the rotatory component and the translatory component are indirectly proportional to each other. That is, when there is an increase in the portion of the total force applied perpendicular to the lever, concomitantly there will be a decrease in the proportion of the total force applied parallel to the lever (and vice versa).

- When a force of constant magnitude is applied to a lever at different points in the range of motion of the lever, a change in torque produced by the force must indicate that the change in angle of application of the force has resulted in a change in the size of the rotatory and translatory components.

The changing torque produced by a constant force as a lever moves around its joint axis can be seen either as a function of the change in MA (distance of the force from the joint axis) or the change in the rotatory component (portion of the force applied perpendicular to the lever). Consequently, it should be seen that these two quantities are directly proportional to each other. Figure 1–48 showed the change in MA of the biceps force with the elbow in four different positions. Figure 1–55 shows the same biceps force at the same elbow positions (35°, 70°, 90°, and 145° of elbow flexion), this time indicating the changing magnitude of the rotatory component. At 90° of elbow flexion, Fms is already applied at 90° to the lever. That is, Fms and f_r coincide and all the force is applied to rotation without any translation. When any force is applied at 90° to a lever, f_r will be maximal since it will be equivalent to the total force. Similarly, the MA has already been shown to be greatest

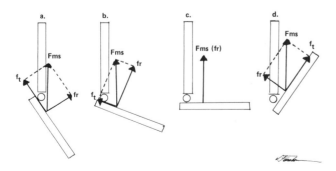

Figure 1–55. Resolution of the force of the biceps brachii (Fms) into rotatory (f_r) and translatory (f_t) components at (a) 35°; (b) 70°; (c) 90°; and (d) 145° of elbow flexion.

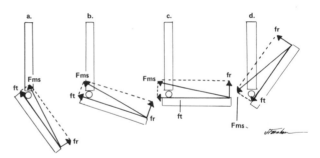

Figure 1–56. Resolution of the force of the brachioradialis (Fms) into rotatory (f_r) and translatory (f_t) components at (a) 35°; (b) 70°; (c) 90°; and (d) 145° of elbow flexion.

when a force is applied at 90°. When a force is applied at 90°, the two torque formulas (F × MA and f_r × LA) are virtually identical; F and f_r are equivalent and MA and LA are equivalent.

Figure 1–55 shows not only the changing rotatory component but also the change in the translatory component in each position. The translatory component is larger than the rotatory component both at 35° of elbow flexion and at 145° of elbow flexion. As the action line of Fms moves closer to the lever (closer to lying parallel to the lever), f_t increases in size. Being inversely proportional in size to the rotatory component, the translatory component is smaller than the rotatory at 70° of elbow flexion and completely absent at 90°. The translatory component, however, changes in more than magnitude. At 35° of elbow flexion the translatory force is toward the joint (compression), while at 145° of elbow flexion the translatory component is away from the joint (distraction).

The change of the translatory component from compression to distraction as seen in Figure 1–55 is unusual for a muscle force. In fact, most muscles lie close and nearly parallel to the lever, with action lines almost always directed toward the joint axis regardless of the position of the lever in space. The effect of this arrangement is that muscle forces generally have relatively small rotatory components, with large translatory components that are nearly always compressive. Most of the force generated by a muscle contributes to joint compression, rather than joint rotation! This enhances joint stability, but means that a muscle must generate a large total force to generate the rotatory force necessary to move the lever through space.

Figure 1–56 shows the resultant force of the brachioradialis muscle as it is applied to the forearm with the forearm in increasing flexion. This muscle is more typical of others in the human body than is the biceps brachii. The angle of application of the brachioradialis changes relatively little, with the rotatory component never exceeding the translatory component (although f_r is still maximal at 90° of elbow flexion as is true of f_r for the biceps). The translatory force remains compression regardless of the position of the limb.

- The action line of most muscles is more parallel to the lever than perpendicular to the lever.
- The rotatory component of a muscle force is rarely larger than the translatory component.
- The translatory component of most muscle forces contributes to joint compression.

The constraints found to exist on muscle forces in the body do not apply to external forces. Figure 1–57 shows the force of gravity applied to the leg with the leg at different points in space. In Figure 1–57a, the leg is parallel to the ground and force G is applied at 90° to the lever. The entire magnitude of G is applied toward rotation of the knee joint in a clockwise direction (G = f_r); that is, toward flexion.

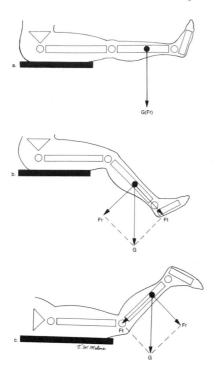

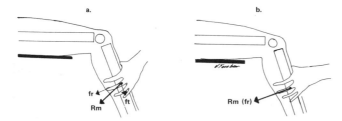

Figure 1-57. Resolution of the force of gravity (G) into rotatory (f$_r$) and translatory (f$_t$) components with the leg at different locations in space. a. Gravity is all rotatory. b. Gravity has rotatory and distraction components. c. Gravity has rotatory and compression components.

In Figure 1–57b and c, force G is applied to the leg such that its rotatory and translatory components are approximately equal in magnitude. The torque applied by f$_r$ to the knee joint is clockwise in both instances. However, the translatory component of vector G is distracting the knee joint in one instance while it is compressing the knee joint in the other. The change in direction of the translatory component is related to the change in position of the limb and, consequently, the change in angle of application of gravity. Vector G in Figure 1–57b is applied at 135° to the lever while it is applied at 45° to the lever in Figure 1–57c.

In Figure 1–58 a manual external force (R$_m$) is applied to the leg in an attempt to flex the leg against the resistance of the quadriceps femoris muscle. By creating a clockwise torque with the hand, the quadriceps femoris muscle is required to apply a countertorque to maintain the position of the leg. Figure 1–58a resolves the manual force (R$_m$) into rotatory and translatory components. While the majority of the force is applied to rotation (f$_r$ is substantially larger than f$_t$), some of the manual force is also wasted as a distraction force. This distraction force serves no useful purpose and actually represents wasted effort. The angle of application of the force applied by the hand, however, can be manipulated so that it lies perpendicular to the lever (Fig. 1–58b). For the same effort, more torque is being generated by the

Figure 1-58. a. A manual force (R$_m$) applied at an angle to the leg wastes force in the direction of distraction. b. A manual force applied perpendicular to the leg results in a force that is all rotation.

person applying the resistance, thus requiring that the quadriceps generate greater torque.

Further manipulation of the manual resistance being applied in Figure 1–58 can further increase the biomechanical effectiveness of the applied force. A force will produce the greatest torque when the LA or MA is as large as possible. If the hand were placed at the ankle, and the angle of application of the force were maintained at 90°, the lever arm and rotatory component would be maximal. In this way the greatest possible torque can be generated by R_m.

- The torque of an external force can be increased by increasing the magnitude of the applied force.
- The torque of an external force can be increased by applying the force perpendicular to the lever.
- The torque of an external force can be increased by increasing the distance from the point of application to the joint axis.

EQUILIBRIUM OF LEVERS

Rotational and Linear Equilibrium

Complete equilibrium of a lever will exist when all the perpendicular forces are balanced and all the parallel forces are balanced. However, forces or force components applied perpendicular to a lever at a distance from the axis produce rotation. The sum of the perpendicular forces on a lever, therefore, must be the sum of the torques applied to that lever. When the sum of the torques on a lever is zero, the lever is in **rotational equilibrium.** When the lever is *not* in rotational equilibrium, it will undergo **angular acceleration.** The magnitude of angular acceleration is proportional to the unbalanced torque applied to the lever and inversely proportional to the mass of the lever being moved. Angular acceleration, as noted earlier, introduces to the lever a new set of forces that are beyond the scope of analysis in this text; it is sufficient to acknowledge that the magnitude of unbalanced torque applied to a lever is an *underestimation* of the forces applied to that lever while it is undergoing angular acceleration.

Forces or force components applied parallel to a lever (presuming they are in line with the joint axis) do not affect the rotation of a lever but will cause the lever to translate. For the lever to be in **linear equilibrium,** the sum of all the parallel forces acting on the lever must be zero. The resultant of a group of linear forces is found by determining the arithmetic sum of the forces.

Figure 1–59a shows the force of gravity (G) on the dependent leg. The total force of G is parallel to the lever, causing a pure distraction force. Since there are

Figure 1–59. a. The dependent leg is in rotatory and linear equilibrium. Gravity is offset by the pull of the knee ligaments on the leg (LL). b. The quadriceps femoris is causing the leg to extend. Linear equilibrium is produced by the translatory force of the quadriceps (f_{tq}), the translatory force of gravity (f_{tg}), and the push of the femur on the tibia (FT).

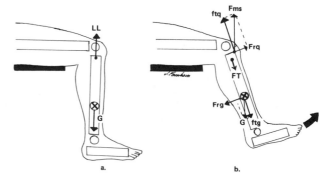

no perpendicular forces applied to the leg, the leg must be in rotational equilibrium. However, if the leg is not translating downward and disrupting the joint, an upward (compressive) force of equal magnitude must be applied to the lever by some other force or set of forces. Muscles predominantly produce compressive forces and a muscle might provide the balance. Presuming the person is completely relaxed, however, the force must come from something else that is touching the leg. The force can be provided by the upward pull of the knee joint ligaments that touch the leg and are capable of "pulling" on the leg with the needed amount of force. Since the ligaments surround the knee joint axis, the resultant ligamentous force can be considered to act through the knee joint, creating a pure translatory force through the joint.

Figure 1–59b shows the leg being extended by the quadriceps against the resistance of gravity (vectors are not drawn to scale). The leg cannot be in rotational equilibrium, since there must be a resultant counterclockwise torque (the quadriceps force serving as the effort force and gravity as the resistance). The force of the quadriceps (Fms) and G have been resolved into components. The pertinent values are

$$f_{rq} = 100 \text{ lb} \qquad f_{rg} = 10 \text{ lb}$$
$$f_{tq} = 50 \text{ lb } f_{tg} \qquad f_{tg} = 5 \text{ lb}$$
$$LA = 1 \text{ in} \qquad LA = 6 \text{ in}$$

The leg must be placed in linear equilibrium or the joint would be disrupted as the tibia accelerated upward past the femur. The sum of the parallel components acting on the lever must be zero. The translatory component of gravity (f_{tg}) is a distraction force. The translatory component of the quadriceps force (f_{tq}) is compression and has a magnitude greater than the magnitude of f_{tg}. The net compression force of 45 lb will cause the tibia to accelerate toward the femur until a balancing force is encountered. The balancing force will be the contact of the femur on the tibia (FT). The contact between the tibia and the femur creates a force of equal magnitude and opposite direction on each of the bony segments. FT will be equal in magnitude and opposite in direction to the net compression force *if* there are no other forces to account for. The force applied by one segment on its adjacent segment during joint compression is known as a **joint reaction force.** As with all reaction forces, these come in pairs. Femur-on-tibia is accompanied by tibia-on-femur, a force that must be taken into account when assessing the linear equilibrium of the femur.

Contact, Shear, and Friction Forces

It has already been shown that a force can be resolved into components that are perpendicular to and parallel to a lever. We have also seen that the components that lie in each of these two planes have different effects on the axis around which the lever rotates (one having a rotatory effect, the other a translatory effect). In fact, any force can be viewed not only as acting across a joint but also as acting between two contacting objects. When the force on an object (regardless of whether or not the object is a lever) is being created by the contact of another object, the component of the force that lies perpendicular to the contacting surfaces can be referred to as the **contact force** or **normal force.** The component that lies parallel to contacting surfaces is referred to as a **shear force** if it is attempting to cause movement between the two surfaces. Depending on the nature of the contacting surfaces, a net shear force may result in an opposing friction force between the two surfaces.

Figure 1–60 shows a block resting on a ramp. Gravity-on-block (GB) acts vertically downward and is opposed by the reaction force, ramp-on-block (RB). Each

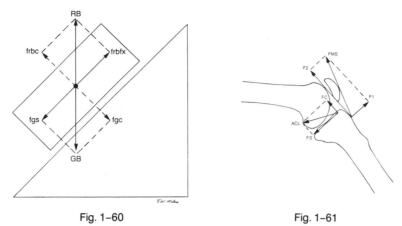

Fig. 1–60   **Fig. 1–61**

Figure 1-60. As the block sits on the ramp, gravity on the block (GB) and the force of the ramp on the block (RB) each have contact components (fgs and frbc, respectively) that are perpendicular to the contacting surfaces. Of the components that are parallel to the contacting surfaces, gravity's is a shear force (fgs), whereas RB's is a friction force (frbfx).

Figure 1-61. The rotatory force (F_1) of the quadriceps (Fms) not only rotates the tibia at the knee joint, but also creates a shear force between the tibia and the femur. The shear of the quadriceps may be offset in part by the opposing shear force (FS) of the tensed anterior cruciate ligament (ACL) of the knee. Both (F_2) of the quadriceps and FC of the ACL increase the contact between the tibia and the femur.

of these two forces lie at an angle to the contacting surfaces of the block and the ramp. There are two components to each force. The contact components (or normal forces) are those components that are perpendicular to the touching surfaces (fgc and frbc). The two components that lie parallel to the contacting surfaces (fgs and frbfx) are shear and friction forces, respectively. The shear force (arising from gravity) is attempting to slide the block down the ramp. The friction force (arising from contact RB) is resisting this motion. The equilibrium of the block is dependent on the relative magnitude of the shear and friction forces. Modifying an earlier statement of Newton's first law, we can say that the equilibrium of the block is dependent on the sum of the perpendicular forces and also on the sum of the parallel forces. These are each translatory forces for which the resultant can be found by determining the arithmetic sum.

Figure 1–61 shows the force of the quadriceps on the tibia. The force Fms is divided into its rotatory (f_1) and translatory (f_2) components, which cause joint rotation and joint compression, respectively. The action of the quadriceps on the tibia can also be evaluated in terms of its contact, or normal, component and its shear component relative to the effect of these forces on the contacting surfaces of the tibial plateau and the femoral condyles. We already know that f_2 is a translatory compression force at the knee joint. This force, while applied parallel to the lever, is simultaneously perpendicular to the surface of the tibia. That is, relative to the tibial plateau, f_2 is a contact or normal component. Component f_2 while creating contact between tibia and femur, is also applied at a distance from the axis of the joint. Since f_2 is perpendicular to the tibial plateau and at a distance from the joint axis, f_2 will cause a small amount of counterclockwise rotation (torque) of the tibial plateau.

Component f_1 in Figure 1–61 will not only rotate the tibia around the knee joint axis since it is perpendicular to the lever, but it will also attempt to slide (shear) the tibia anteriorly on the femur since this force simultaneously is parallel to the surface of the tibial plateau. The magnitude of the translatory shear force is the simple magnitude of the component; it is not affected by its distance from the joint axis. The shear across a joint surface created by a component perpendicular to a lever should never be equal in magnitude to the torque created by that same force. The shear force f_1 may be offset by an opposing shear from the perpendicular component of gravity (this will also create a flexion torque around the knee joint axis); by some small friction occurring between joint surfaces (usually minimal in a normal joint); or by the shear force created in tensed joint ligaments such as the anterior cruciate ligament (the ACL in Figure 1–61). Most joint ligaments are arranged not only to tense on pure joint distraction but also to resist net shear forces across the joint surfaces.

- A force or force component that is applied perpendicular to a bony lever *and* is parallel to contacting joint surfaces will create both a torque around the joint axis and a shear between joint surfaces.
- A force or force component that is applied parallel to a bony lever but does not intersect the joint axis will create not only compression or distraction but also a small amount of rotation.

Equilibrium of a lever must ultimately be assessed in terms of

1. The rotational forces applied around the joint axis (net torque).
2. The translatory forces applied across the joint and perpendicular to joint surfaces (resultant of the joint reaction forces, compressive forces, distractive forces).
3. The translatory forces applied across the joint and parallel to the joint surfaces (resultant of the shear forces, friction forces).

The physical principles that govern equilibrium or movement of the levers of the human body have been examined at a basic level. This is only a first step, however, in understanding human function. The next step includes a study of joint structures to determine the variable nature of the contacting surfaces and the range of movements permitted at the joints and the ligamentous structure supporting those joints. Also necessary is a study of the structure of muscles, focusing on the properties of a muscle that affect the force (Fms) that either a single muscle or many muscles acting in concert may apply to a lever.

REFERENCES

1. Cromer, AH: Physics for the Life Sciences, ed 2. McGraw-Hill, New York, 1977.
2. Webster's New Collegiate Dictionary. G. and C. Merriam Co. Springfield, Mass., 1961.
3. LeVeau, B: Williams and Lissner's Biomechanics of Human Motion, ed 2. WB Saunders, Philadelphia, 1977.

ADDITIONAL READINGS

Barnham, JN: Mechanical Kinesiology. CV Mosby, St Louis, 1978.
Gowitzke, BA and Miller, M: Scientific Bases of Human Movement, ed 2. Williams & Wilkins, Baltimore, 1980.
Schenck, JM and Cordova, FD: Introductory Biomechanics, ed 2. FA Davis, Philadelphia, 1980.

STUDY QUESTIONS
1. In what plane does flexion of the shoulder occur? Around what axis? What type of motion is this?
2. In what plane and around what axis does rotation of the head occur? What type of motion of the head is produced if you rotate your head while you walk?

3. Is naming the plane of motion considered part of kinetics or kinematics? Why?

4. To what is the force pencil-on-hand applied? What is its source?

5. What do the forces pencil-on-desk, book-on-desk, glass-on-desk, and blotter-on-desk all have in common?

6. What characteristic(s) does/do all gravity vectors have in common?

7. What generalizations can be made about the LOG (gravity vector) of all stable objects?

8. What happens to the COG of a solid object when the object is moved around in space?

9. What happens to the COG of the body when the body segments are rearranged? What happens to the COG if the right upper extremity is amputated?

10. Explain how to generally find the resultant (combined) COG for two adjacent objects.

11. A student is carrying all of her books for the fall semester courses in her right arm. What does the additional weight do to her COG? How will her body most likely respond to this change?

12. Why did your Superman punching bag always pop up again?

13. Describe the typical gait of a child just learning to walk. Why does the child walk this way?

14. Give the name, point of application, magnitude and direction of the reaction to body-on-bed when the body is a man weighing 200 lb.

15. What are the reaction forces to each of those forces named in question 5? To what are each of the reactions applied?

16. You see a woman in a waiting room sitting in a chair with a child on her lap. The woman's feet are not touching the floor. Disregarding her clothing, name all the forces responsible for maintaining her equilibrium.

17. Using the example in question 16, what would happen if the magnitude of the force floor-on-chair was equivalent to the sum of the woman's weight and the weight of the child?

18. Are the two forces of an action-reaction pair part of the same linear force system? Defend your answer.

19. What conditions must exist for friction to have magnitude on an object?

20. When is the magnitude of the force of friction always greatest?

21. You have a patient in leg traction similar to that described in the text (Buck's extension traction). There is a 10-lb weight suspended on the rope. The leg weighs 20 lb. Assume the leg is *not* contacting the bed. Given these forces, is the knee joint being distracted?

22. Repeat problem 21. Now, assuming the leg is resting on the bed, and that the coefficient of friction for skin on bed is 0.25, is the knee joint undergoing distraction?

23. A patient is lying in bed with traction applied to her leg. The patient is acted on by the forces of gravity, bed-on-patient, and traction-on-patient. The patient will not be in equilibrium. What additional force(s) is/are necessary to keep the patient in equilibrium?

24. What kind of force system do muscles form? Explain.

25. At which end of a muscle do you place the point of application of Fms?

26. How do you determine the net effect of two muscle pulls applied to the same spot? What is this process called?

27. How do anatomic pulleys affect the magnitude and direction of a muscle force (Fms)?

28. What are the three classes of levers? Give an example of each in the human body.

29. What is torque? Describe how it is determined using an example in the human body of two parallel forces acting on the same lever.

30. How does one determine which is the effort force and which is the resistance force?

31. What factors cause torque to change?

32. A 2-year-old has difficulty pushing open the door into MacDonald's. What advice will you give him as to how to perform the task independently?

33. What is always true of the mechanical advantage of a third-class lever? Of a second? Of a first?

34. What is the "advantage" to a lever with a mechanical advantage greater than one?

35. What kind of work (positive or negative) does a muscle do whenever it is acting in a second-class lever? Why?

36. Using the values below, identify the class of the lever, its mechanical advantage, what kind of contraction the muscle is doing, and the point of application of the resultant force of gravity-on-forearm and ball-on-forearm (the hand will be considered part of the forearm).
    Fms = muscle force, G = gravity-on-forearm, and B = ball-on-forearm (assume that all forces are applied perpendicular to the forearm lever).

    Fms = 50 lb (counter clockwise)   G = 7 lb (clockwise)   B = 5 lb (clockwise)
      LA = 1 in   LA = 10 in   LA = 12 in

37. How do you determine torque if the forces applied to the lever are not perpendicular to the lever or parallel to each other?

38. Describe how the angle of application of a force and the MA of that force are related. When is the MA potentially greatest?

39. Describe how you would position a limb in space so that gravity exerts the least torque on the limb. How would you position the limb to have gravity exert the greatest torque?

40. How do anatomic pulleys affect torque generated by the muscle that passes over the pulley?

41. If not all a muscle's force is contributing to rotation, what happens to the "wasted" force?

42. Describe how a rotatory force or force component applied across a joint can create a shear force at that joint. What are some possible sources of forces to offset that shear force?

Chapter 2

■ ■ ■

Joint Structure and Function

───────────────────────────── ■ ─────────────────────────────

───

OBJECTIVES
Following the study of this chapter the reader should be able to:

Recall
1. The elementary principles of joint design.
2. The five features common to all diarthrodial joints.
3. The two main types of joints.
4. Types of materials used in human joint construction.
5. Properties and functions of materials used in human joints.
6. Definitions of arthrokinematics and osteokinematics.

Identify
1. The axis of motion for any given motion at a specific joint (knee, hip, metacarpophalangeal).

2. The plane of motion for any given motion at a specific joint (shoulder, interphalangeal, wrist).
3. The degrees of freedom at a given joint.
4. The distinguishing features of a diarthrodial joint.
5. The structures that contribute to joint stability.

Compare
1. A synarthrosis with a diarthrosis on the basis of method of construction, materials, and function.
2. A closed kinematic chain with open kinematic chain.
3. The close-packed position of a joint with the loose-packed position.
4. Stress and strain to load and deformation.
5. Motion of a convex surface moving on a concave surface versus a concave surface moving on a convex surface.
6. Theories of joint lubrication.
7. Creep and hysteresis.

Diagram
1. A typical load-deformation curve for a ligament and identification of the various regions on the curve.
2. A typical stress/strain curve for a ligament or tendon.

JOINT DESIGN

The joints that are found in the human skeleton are similar to joints used in the construction of buildings, furniture, and machines. Both human and nonhuman joints adhere to the same basic design principles, and both types of joints illustrate the strong interrelationship that exists between structure and function. Function may be said to determine structure and structure in turn determines function. For example, if one were to design a joint for a piece of furniture or a machine, one would have to know the function of the joint in question to produce an appropriate design. Therefore, in this instance, function determines structure. However, once the joint was designed and constructed, the structure of the joint would determine its function. The joints of the human body already have been designed and constructed, and therefore, human joint structure determines function. However, the reader should keep in mind that human tissue can adapt to the stresses placed on it and thus function may determine structure to some extent. Although human joints are the main focus of this chapter, the task of designing a table joint has been selected as a way of introducing the reader to the basic principles of joint design. A table leg joint is easy to conceptualize and illustrates the basic principles of joint design.

Basic Principles

A joint (articulation) is used to connect one component of a structure with one or more other components. The design of a joint and the materials used in its construction depend partly on the function of the joint and partly on the nature of the components. If the function of a joint is to provide stability or static support, the design of the joint will be different from the design used when the desired function is mobility. Therefore, function, at least in part, will determine structure. If one wished to design a joint such as that found between the legs of a table and the tabletop, one would have to take into consideration the function of the joint and the

Figure 2-1. Folding table joint. a. The table leg is free to move, and the joint provides mobility when the brace is unlocked. b. The table leg is prevented from moving, and the joint provides stability when the brace is locked.

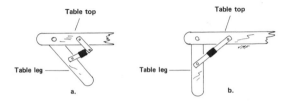

nature of the components. The function of the table joint is stability, and therefore the design must be such that the components are united to form a stable union. If one wished the legs of the table to fold, the joint would have to be designed to provide mobility in one situation and stability in another situation. The design of the folding table joint and the materials used would have to be different from and more complex than that of a purely stable joint. One possible method of designing a folding table joint would be by using a metal brace fitted with a locking device. When the brace is unlocked the leg would be free to move: when the brace is locked, the joint would be stable (Fig. 2-1). The use of such a design would require the application of an external force to produce the locking and unlocking. However, the table joint would not be able to provide the stability during motion that human joints often must provide. The table joints have been used as a way of introducing the reader to the elementary principles of joint design that follow, but one should remember that human joints are more complex.

- The design of a joint is determined by its function and the nature of its components.
- Once a joint is constructed, the structure of the joint will determine its function.
- Joints that serve a single function are less complex than joints that serve multiple functions.

Human Joint Design

An appreciation of the complexities that are involved in human joint design may be gained by considering the nature of the bony components and the functions that the joints must serve. There are approximately 200 bones in the human skeleton that must be connected by joints. These bones vary in size from the pea-sized distal phalanx of the little toe to the over-a-foot-long femur of the thigh. The shape of the bones varies from round to flat, and the contours of the ends of the bones vary from convex to concave. The task of designing a series of joints to connect these varied bony components to form a stable structure would be difficult. The task of designing joints that are capable of working together to provide both mobility and stability for the total structure represents an engineering problem of considerable magnitude.

Joint designs in the human body vary from simple to complex. Although generally the designs used are complex, the same principles used in the table joints are used in human joint design. The more simple human joints usually have stability as a primary function, while the more complex joints usually have mobility as a primary function. However, most joints in the human body have to serve a dual mobility/stability function and must also provide dynamic stability. The human stability joints are similar in design to the table joints in that the ends of the bones may be contoured either to fit into each other or to lie flat against each other. Bracing of human joints is accomplished through the use of joint capsules, ligaments, and tendons. Joints designed primarily for human mobility are called synovial joints. These joints are constructed so that the ends of the bony components are enclosed in a

synovial sheath (joint capsule). The capsules, ligaments, and tendons located around mobility (synovial) joints not only help to provide stability for the joint but also permit motion. Wedges of cartilage, called **menisci, disks, plates,** and **labrums** are used in synovial joints to increase stability, to provide shock absorption, and to facilitate motion. In addition, a lubricating fluid, called **synovial fluid,** is secreted at all mobility (synovial) joints to help reduce friction between the articulating surfaces.

JOINT CATEGORIES

In the traditional method of joint classification, the joints (**arthroses** or **articulations**) of the human body are divided into two broad categories based on the type of materials and the methods used to unite the bony components. Subdivisions of joint categories are based on materials used, the shape and contours of the articulating surfaces, and the type of motion allowed. The two broad categories of arthroses are **synarthroses** (nonsynovial joints) and **diarthroses** (synovial joints).[1]

Synarthroses

The material used to connect the bony components in synarthrodial joints is interosseus connective tissue (fibrous and cartilaginous). Synarthroses are grouped into two divisions according to the type of connective tissue used in the union of bone to bone: fibrous joints and cartilaginous joints. The connective tissue directly unites one bone to another (bone–solid connective tissue–bone).

Fibrous Joints

In fibrous joints, the fibrous tissue directly unites bone to bone. Three different types of fibrous joints are found in the human body: suture, gomphosis, and syndesmosis joints. A joint in which two bony components are united by a thin layer of dense fibrous tissue is a **suture joint.** The ends of the bony components are shaped so that the edges interlock or overlap one another. This type of joint is found only in the skull and early in life allows a small amount of movement. Fusion of the two opposing bones in suture joints occurs later in life and leads to the formation of a bony union called a **synostosis.**

Example: Coronal suture (Fig. 2–2). The serrated edges of the parietal and frontal bones of the skull are connected by a thin fibrous membrane to form the coronal suture. At birth the fibrous membrane allows some

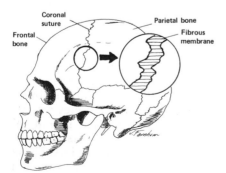

Figure 2–2. The coronal suture. The frontal and parietal bones of the skull are joined directly by fibrous tissue to form a synarthrodial suture joint.

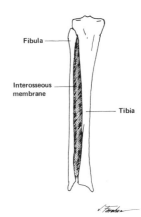

Figure 2-3. The shafts of the fibula and tibia are joined directly by a membrane to form a synarthrodial syndesmosis.

motion for ease of passage through the birth canal. Also during infancy slight motion is possible for growth of the brain and skull. In adulthood, the bones grow together to form a synostosis and little or no motion is possible.

A joint in which the surfaces of bony components are adapted to each other like a peg in a hole is a **gomphosis joint**. In this type of joint the component parts are connected by fibrous tissue. The only gomphosis joint that exists in the human body is the joint that is found between a tooth and either the mandible or maxilla.

Example: The conical process of a tooth is inserted in the bony socket of the mandible or maxilla. In the adult, the loss of teeth is, for the most part, due to disease processes affecting the connective tissue that cements or holds the teeth in approximation to the bone. Under normal conditions in the adult these joints do not permit motion between the components.

A type of fibrous joint in which two bony components are joined directly by a ligament, cord, or aponeurotic membrane is a **syndesmosis joint**.

Example: The shaft of the tibia is joined directly to the shaft of the fibula by a membrane (Fig. 2–3). A slight amount of motion at this joint accompanies movement at the knee and ankle joints.

Cartilaginous Joints

The materials used to connect the bony components in cartilaginous joints are either fibrocartilage or hyaline growth cartilage. These materials are used to directly unite one bony surface to another (bone-cartilage-bone). The two types of cartilaginous joints are symphyses and synchondroses.

In a **symphysis joint** the two bony components are directly joined by fibrocartilage in the form of disks or plates.

Example: The symphysis pubis (Fig. 2–4a). The two pubic bones of the pelvis are joined by fibrocartilage. This joint must serve as a weight-bearing joint and is responsible for withstanding and transmitting forces; therefore, under normal conditions very little or no motion is permissible or desirable.

During pregnancy when the connective tissues are softened, some slight separation of the joint surfaces occurs to ease the passage of the baby through the birth canal. However, the symphysis pubis is considered

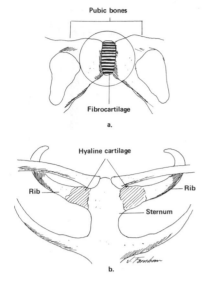

Figure 2–4. Cartilaginous joints. a. The two pubic bones of the pelvis are joined directly by fibrocartilage to form a symphysis joint called the *symphysis pubis*. b. The first rib and the sternum are connected directly by hyaline cartilage to form a synchondrosis joint called the *first chondrosternal joint.*

to be primarily a stability joint with the thick fibrocartilage forming a stable union between the two bony components.

Synchondrosis is a type of joint in which the material used for connecting the two components is hyaline growth cartilage. The cartilage forms a bond between two ossifying centers of bone. The function of this type of joint is to permit bone growth while also providing stability and allowing a small amount of mobility. Some of these joints are found in the skull and in other areas of the body at sites of bone growth. When bone growth is complete these joints ossify and convert to bony unions (synostoses).

Example: The first sternocostal joint (Fig. 2–4b). The adjacent surfaces of the first rib and sternum are connected directly by articular cartilage.

Diarthroses

The method of joint construction in diarthrodial or synovial joints differs from that used in synarthrodial joints. In synovial joints the ends of the bony components are free to move in relation to one another because no cartilaginous tissue directly connects adjacent bony surfaces. The bony components are *indirectly connected to one another by means of a joint capsule that encloses the joint.* All synovial joints are constructed in a similar fashion and all have the following features: (1) a joint capsule that is formed of fibrous tissue; (2) a joint cavity that is enclosed by the joint capsule; (3) a synovial membrane that lines the inner surface of the capsule; (4) synovial fluid that forms a film over the joint surfaces; and (5) hyaline cartilage that covers the joint surfaces (Fig. 2–5).

In addition, synovial joints often have accessory structures such as fibrocartilaginous disks, plates or menisci, and labrums, along with fat pads, ligaments, and tendons that are located within the joint capsule or immediately adjacent to the joint. Menisci, disks, and the synovial fluid help to prevent excessive compression of opposing joint surfaces. Ligaments and tendons associated with these joints play an important role in keeping joint surfaces together and may assist in guiding motion. Separation of joint surfaces is limited by passive tension in ligaments, the

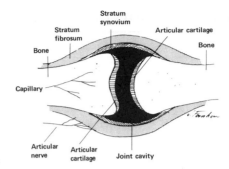

Figure 2-5. A typical diarthrodial joint.

joint capsule, and tendons. Active tension in muscles also limits the separation of joint surfaces.

The Joint Capsule

The joint capsule is composed of two layers, an outer layer called the **stratum fibrosum** and an inner layer called the **stratum synovium**. The outer layer, which is composed of dense fibrous tissue, completely encircles the ends of the bony components. The outer layer is attached to the periosteum of the component bones by Sharpey's fibers and is reinforced by ligamentous and musculotendinous structures that cross the joint. The outer layer is poorly vascularized but richly innervated by joint receptors. The receptors that are located in and around the joint capsule are able to detect the rate and direction of motion, compression and tension, and vibration and pain.[2] An overview of the joint receptors is presented in Table 2–1.[3]

Joint receptors transmit information about the status of the joint to the central nervous system (CNS). The CNS interprets the information sent by the joint receptors and responds by coordinating muscle activity around the joint to meet joint mobility and stability requirements. Information from joint receptors is necessary for the CNS to provide protection for joint structures, to produce controlled movement at the joint, and to provide a sense of joint position in static or dynamic postures. For example, a stretch to the capsule may indicate to the CNS that the joint is approaching the limits of the normal range of motion. The CNS can respond to this information by inhibiting the muscle(s) producing the motion and activating muscle(s) that will stop or oppose the motion.

The inner layer (stratum synovium) of the joint capsule is highly vascularized but poorly innervated.[4] It is insensitive to pain but undergoes vasodilation and vasoconstriction in response to heat or cold, respectively.[4] One of the most important features of the stratum synovium is that specialized cells called **synoviocytes** can synthesize the hyaluronate (hyaluronic acid) component of the synovial fluid.[4] The stratum synovium also produces matrix collagen and serves as an entry point for nutrients and as an exit point for waste materials. The synovial membrane and fluid of bursae located near joints may be continuous with the joint capsule. These bursae are called **communicating bursae**.[1]

The Synovial Fluid

The composition of synovial fluid is similar to blood plasma except that synovial fluid contains hyaluronate (hyaluronic acid) and a glycoprotein called lubricin.[5] The hyaluronate component of synovial fluid is responsible for the viscosity of the fluid and is essential for lubrication of the synovium. Hyaluronate reduces the fric-

Table 2–1. JOINT RECEPTORS*

Receptor	Location	Sensitivity	Distribution
Group I, Golgi ligament endings	Ligaments	Stretch of ligaments	Found in most joints except in vertebral column
Group I–II, Ruffini endings	Outer layer of joint capsule	Stretch of joint capsule; change in joint fluid pressure and changes in joint position	Found in highest concentrations in proximal joints
Group II, Pacinian corpuscles	Outer layer of joint capsule	High frequency vibration; acceleration; and high velocity changes in joint position	Found in highest concentrations in distal joints
Group II–III, Golgi-Mazzoni corpuscles	Inner layer of joint capsule	Compression of joint capsule	Found in knee joint and most likely in other joints
Group IV–V, free nerve endings	Throughout capsule and in ligaments	Mechanical stress or biomechanical stimuli	Found in many joints and ligaments

*Adapted from Rowinski.[3]

tion between the synovial folds of the capsule and the joint surfaces.[5] **Lubricin** is the component of synovial fluid that is responsible for cartilage-on-cartilage lubrication. Changes in the concentration of hyaluronate or lubricin in the synovial fluid will affect the overall lubrication and the amount of friction that is present. Many experiments have confirmed that articular coefficients of friction in synovial joints are lower than those that can be produced using manufactured lubricants.[5]

Normal synovial fluid appears as a clear, pale yellow viscous fluid that is present in small amounts at all synovial joints.[6] There is a direct exchange between the vasculature of the stratum synovium and the intracapsular space where nutrients can be supplied and waste products can be taken away from the joint by diffusion.[5] Usually, less than 0.5 mL of synovial fluid can be removed from large joints such as the knee.[1] However, when a joint is injured or diseased, the volume of the fluid may increase.[5]

The synovial fluid exhibits properties common to all viscous substances in that it has the ability to resist loads that produce shear.[7] The viscosity of the fluid *varies inversely* with the joint velocity or rate of shear. Thus the synovial fluid is thixotropic. When the bony components of a joint are moving rapidly, the viscosity of the fluid decreases and provides less resistance to motion.[7] When the bony components of a joint are moving slowly, the viscosity increases and provides more resistance to motion. Viscosity also is sensitive to changes in temperature. High temperatures decrease the viscosity while low temperatures increase the viscosity.[1] The thin film of synovial fluid that covers the surfaces of the inner layer of the joint capsule and the articular cartilage helps to keep the joint surfaces lubricated and thus reduces friction between the bony components. The fluid also provides nourishment for the cartilage.

Lubrication of Synovial Joints

A number of models have been proposed to explain how diarthrodial joints are lubricated under varying loading conditions. The general consensus is that no single model is adequate to explain human joint lubrication and that human joints are lubricated by two or more of the following types of lubrication used in engineering. The two basic types of lubrication are boundary lubrication and fluid lubrication.

Boundary lubrication occurs when each bearing surface is coated or impregnated with a thin layer of molecules that keeps the opposing surfaces from touching each other (Fig. 2–6a). The molecules slide on the opposing surface more readily than they are sheared off of the underlying surface. In human diarthrodial joints these molecules are composed of a special glycoprotein called lubricin, which is found in the synovial fluid. The lubricin molecules adhere to the articular surfaces.[3,8]

Fluid lubrication models include: hydrostatic (weeping) lubrication; hydrodynamic, squeeze-film lubrication; and elastohydrodynamic and boosted lubrication. Generally, fluid lubrication models include the existence of a film of fluid that is interposed between the joint surfaces. **Hydrostatic, or weeping, lubrication** is a form of fluid lubrication in which the load-bearing surfaces are held apart by a film of lubricant that is maintained under pressure (Fig. 2–6b). In engineering, the pressure is usually supplied by an external pump. In the human body the pump action can be supplied by contractions of muscles around the joint. Compression of articular cartilage causes the cartilage to deform and to ''weep'' fluid, which forms a fluid film over the articular surfaces. This is possible because an impervious layer of calcified cartilage keeps the fluid from being forced into the subchondral bone.[5] When the load is removed, the fluid flows back into the articular cartilage. This type of lubrication is most effective under conditions of high loading.

Hydrodynamic lubrication is a form of fluid lubrication in which a wedge of fluid is created when nonparallel opposing surfaces slide on one another. The resulting lifting pressure generated in the wedge of fluid and by the fluid's viscosity keeps the joint surfaces apart. In **squeeze-film lubrication**, pressure is created in the fluid film by the movement of articular surfaces that are perpendicular to one another.[8] As the opposing articular surfaces move closer together, they squeeze the fluid film out of the area of impending contact. The resulting pressure that is created by the fluid's viscosity keeps the surfaces separated. This type of lubrication is suitable for high loads maintained for a short duration.

In the **elastohydrodynamic model** the protective fluid film is maintained at an

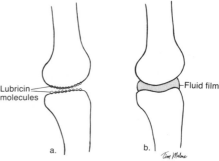

Figure 2-6. Joint lubrication models. a. Schematic drawing of the lubricin molecules coating the joint surfaces in boundary lubrication. b. Schematic drawing of the fluid film that keeps joint surfaces apart in hydrostatic lubrication.

Lubricin molecules

Fluid film

a.   b.

appropriate thickness by the elastic deformation of the articular surfaces. In **boosted lubrication** the model suggests that pools of concentrated hyaluronate molecules are filtered out of the synovial fluid and are trapped in the natural undulations and areas of elastic deformation on the articular surface just as the opposing surfaces meet.[8]

The joint lubrication models presented provide a number of possible options for explaining how diarthrodial joints are lubricated. The variety of conditions under which human joints function indicate that more than one of the lubrication models are operating. Until a unified model of joint lubrication is proposed, proved, and accepted, the exact mechanisms involved in human joint lubrication will be subject to speculation.[1]

Subclassifications of Synovial Joints

Traditionally, synovial joints have been divided into three main categories on the basis of the number of axes about which "gross visible" motion occurs. A further subdivision of the joints is made on the basis of the shape and configuration of the ends of the bony components. The three main traditional categories are uniaxial, biaxial, and triaxial. A **uniaxial joint** is constructed so that visible motion of the bony components is allowed in only one of the planes of the body around a single axis. The axis of motion usually is located near or in the center of the joint or in one of the bony components. Since uniaxial joints only permit visible motion in one plane, or around one axis, they are described as having 1° of freedom of motion.

The two types of **uniaxial diarthrodial joints** found in the human body are hinge joints and pivot (trochoid) joints. A **hinge joint** is a type of joint that resembles a door hinge. It permits motion around a single axis only.

> *Example:* Interphalangeal joints of the fingers (Fig. 2–7a). These joints are formed between the distal end of one phalanx and proximal end of another phalanx. The joint surfaces are contoured so that motion can occur in the sagittal plane only (flexion and extension) around a coronal axis (Fig. 2–7b).

A **pivot (trochoid) joint** is a type of joint constructed so that one component is shaped like a ring and the other component is shaped so that it can rotate within the ring.

> *Example:* The median atlantoaxial joint (Fig. 2–8). The ring portion of the joint is formed by the atlas and the transverse ligament. The odontoid process (dens) of the axis which is enclosed in the ring rotates within the osteoligamentous ring. Motion occurs in the transverse plane around a longitudinal axis.

Biaxial diarthrodial joints are joints in which the bony components are free to move in two planes around two axes. These joints have 2° of freedom. There are two types of biaxial joints in the body, condyloid and saddle. The joint surfaces in a condyloid joint are shaped so that the concave surface of one bony component is allowed to slide over the convex surface of another component in two directions.

> *Example:* Metacarpophalangeal joint (Fig. 2–9a). The metacarpophalangeal joint is formed by the convex distal end of a metacarpal bone and the concave proximal end of the proximal phalanx.

Flexion and extension at this joint occur in the sagittal plane around a coronal axis (Fig. 2–9b). Abduction, or movement away from the middle finger, and adduction, or movement toward the middle finger, occur through a side-to-side gliding of the

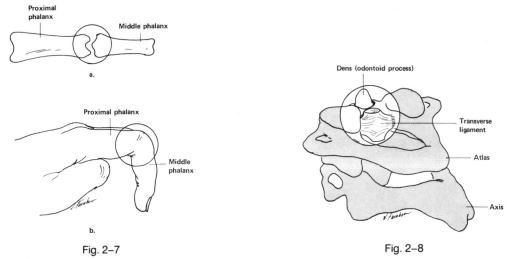

Fig. 2-7   Fig. 2-8

Figure 2-7. A uniaxial hinge joint. a. The interphalangeal joints of the fingers are examples of simple hinge joints. The joint capsule and accessory joint structures have been removed to show the bony components in the superior view of the joint. b. Motion occurs in one plane around one axis.

Figure 2-8. A pivot joint. The joint between the atlas, transverse ligament, and the dens of the axis is a uniaxial diarthrodial pivot joint called the *median atlantoaxial joint*. Rotation occurs in the transverse plane around a vertical axis.

phalanx on the metacarpal. Adduction and abduction occur in the frontal plane around an anterior-posterior axis (Fig. 2–9c).

A **saddle joint** is a joint in which each joint surface is both convex in one plane and concave in the other and these surfaces are fitted together like a rider on a saddle.

> *Example:* Carpometacarpal joint of the thumb. The carpometacarpal joint of the thumb is formed by the distal end of the carpal bone and the prox-

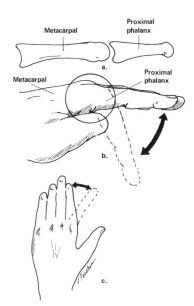

Figure 2-9. A condyloid joint. a. The metacarpophalangeal joints of the fingers are biaxial condyloid joints. The joint capsule and accessory structures have been removed to show the bony components. Motion at these joints occurs in two planes around two axes. b. Flexion and extension occur in the sagittal plane around a coronal axis. c. Abduction and adduction occur in the frontal plane around an A-P axis.

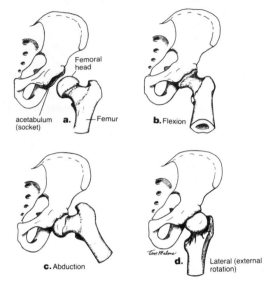

Figure 2–10. A ball-and-socket joint. a. The joint between the femoral head and the acetabulum is a triaxial diarthrodial joint called the *hip joint.* Motion may occur in three planes around three axes. b. Flexion/extension occur in the sagittal plane around a coronal axis. c. Abduction/adduction occur in the frontal plane around an A-P axis. d. Rotation occurs in the transverse plane around a longitudinal axis.

imal end of the metacarpal. The motions available are flexion and extension and abduction/adduction.

Triaxial or multiaxial diarthrodial joints are joints in which the bony components are free to move in three planes around three axes. These joints have 3° of freedom. Motion at these joints also may occur in oblique planes. The two types of joints in this category are **plane joints** and **ball-and-socket joints.**

Plane joints permit gliding between two or more bones.

Example: Carpal joints. These joints are found between the adjacent surfaces of the carpal bones. The adjacent surfaces may glide on one another or rotate with respect to one another in any plane.

Ball-and-socket joints are formed by a ball-like convex surface being fitted into a concave socket. The motions permitted are flexion/extension, abduction/adduction, and rotation.

Example: Hip joint. The hip joint is formed by the head of the femur and a socket called the acetabulum (Fig. 2–10a). The motions of the flexion/extension occur in the sagittal plane around a coronal axis (Fig. 2–10b). Abduction/adduction occurs in the frontal plane around an anterior-posterior axis (Fig. 2–10c) while rotation of the femur occurs in the transverse plane around a longitudinal axis (Fig. 2–10d).

JOINT FUNCTION

The structure of the joints of the human body reflects the functions that the joints are designed to serve. The synarthrodial joints are relatively simple in design and function primarily as stability joints, although motion does occur. The diarthrodial joints are complex and are designed primarily for mobility, although many of these joints must also provide stability. Effective functioning of the total structure is dependent on the integrated action of many joints, some providing stability and some providing mobility. Generally, stability must be achieved before mobility.

Kinematic Chains

Some of the joints of the human body are linked together into a series of joints in such a way that motion at one of the joints in the series is accompanied by motion at an adjacent joint. For instance, when a person in the erect standing position bends both knees, simultaneous motion must occur at the ankle and hip joints (Fig. 2–11a). However, when the leg is lifted from the ground, the knee is free to bend without causing motion at either the hip or ankle (Fig. 2–11b). The type of motion that occurs in the joints of the lower limb when a person is standing may be explained by using the concept of a **kinematic chain.**[9] Kinematic chains in the engineering sense are composed of a series of rigid links that are interconnected by a series of pin-centered joints. In engineering, the system of joints and links is constructed so that motion of one link at one joint will produce motion at all of the other joints in the system in a predictable manner. The kinematic chains in engineering form a closed system or **closed kinematic chain.**[9] In the human system of joints and links, the joints of the lower limbs and the pelvis function as a closed kinematic chain when a person is in the erect weight-bearing position because the ends of the limbs are fixed on the ground and the upper ends of the limbs are virtually fixed to the pelvis. However, the ends of human limbs frequently are not fixed but are free to move without necessarily causing motion at another joint. The motion of waving the hand may occur at the wrist without causing motion of the elbow or shoulder. When the ends of the limbs or parts of the body are free to move without causing motion at another joint, the system is referred to as an **open kinematic chain.** In an open kinematic chain, motion does not occur in a predictable fashion because joints may function either independently or in unison.

> *Example:* One may wave the whole upper limb by moving the arm at the shoulder (Fig. 2–12a) or one may move only at the wrist. In the first instance, all of the degrees of freedom of all of the joints from the shoulder to the wrist are available to the distal segment (hand). If the person is waving from the wrist, only the degrees of freedom at the wrist would be available to the hand, and motion of the hand in space would be more limited than in the first situation (Fig. 2–12b).

The concept of kinematic chains, which is useful for analyzing human motion and the effects of injury and disease on the joints of the body, will be referred to

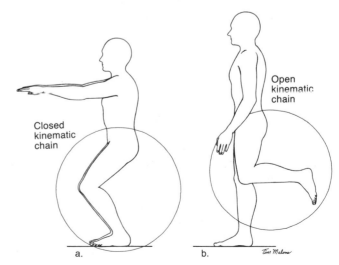

Figure 2–11. Closed and open kinematic chains. a. In a closed kinematic chain, knee flexion is accompanied by hip flexion and ankle dorsiflexion. b. Knee motion in an open kinematic chain may occur with or without motion at the hip and ankle. In the diagram, knee flexion is shown without simultaneous motion at the hip and the ankle.

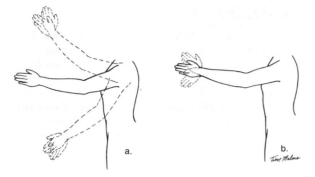

Figure 2–12. Open kinematic chain motion. a. When the entire upper limb is moving at the shoulder, 7° of freedom of motion (sum of the degrees of freedom at the shoulder, elbow, and wrist) are available to the hand. b. The hand only has the 2° of freedom at the wrist available, and the motion of the hand is limited in space.

throughout this text. Although the joints in the human body do not always behave in a predictable fashion in either a closed or open chain, the joints are interdependent. A change in the function or structure of one joint in the system will usually cause a change in the function of a joint either immediately adjacent to the affected joint or at a distal joint. For example, if the range of motion at the knee were limited, the hip and ankle joints would have to compensate so that the foot could clear the floor when walking to avoid stumbling.

Joint Motion

Arthrokinematics

Motion at a joint occurs as the result of movement of one joint surface in relation to another. The term **arthrokinematics** is used to refer to movements of joint surfaces. Usually one of the joint surfaces is more stable than the other and serves as a base for the motion, while the other surface moves on this relatively fixed base. The terms roll, slide, and spin are used to describe the type of motion that the moving part performs.[10] A **roll** refers to the rolling of one joint surface on another, as in a tire rolling on the road. In the knee, the femoral condyles roll on the fixed tibial surface. **Sliding,** which is a pure translatory motion, refers to the gliding of one component over another, as when a braked wheel skids. In the hand, the proximal phalanx slides over the fixed end of the metacarpal. The term **spin** refers to a rotation of the movable component, as when a top spins. Spin is a pure rotatory motion. At the elbow, the head of the radius spins on the capitulum of the humerus during supination and pronation of the forearm. Combinations of rolling, sliding, and spinning may also occur between segments.

The type of motion that occurs at a particular joint depends on the shape of the articulating surfaces. Most joints fit into either an ovoid or a sellar category. In an **ovoid** joint, *one* surface is convex and the other surface is concave (Fig. 2–13a). In a **sellar** joint, *each* joint surface is *both* convex and concave (Fig. 2–13b). In an ovoid joint, when a convex surface moves on a stable concave surface, the sliding of the convex articulating surface occurs in the *opposite direction* to the motion of the bony lever (Fig. 2–14a). When a concave surface is moving on a stable convex surface, sliding occurs in the *same direction* as motion of the bony lever (Fig. 2–14b).

The sliding that occurs between articular surfaces is an essential component of joint motion and must occur for normal functioning of the joint. The distal end of a bone cannot be expected to move if the articular end of the bone is not free to move (slide) in the appropriate direction.

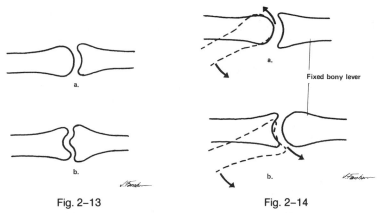

Fig. 2–13   Fig. 2–14

Figure 2–13. Ovoid and sellar joints. a. In an ovoid joint one articulating surface is convex and the other articulating surface is concave. b. In a sellar joint each articulating surface is concave and convex.
Figure 2–14. Motion at ovoid joints. a. When a convex surface is moving on a fixed concave surface, the convex articulating surface moves in a direction opposite to the direction traveled by the shaft of the bony lever. b. When a concave surface is moving on a fixed convex surface, the concave articulating surface moves in the same direction as the remaining portion of the bony lever (proximal phalanx moving on fixed metacarpal).

Example: Abduction of the distal end of the humerus must be accompanied by downward sliding (inferior movement) of the proximal end of the convex head of the humerus on the concave surface of the glenoid fossa in order for the distal end to elevate without damage to the joint (Fig. 2–15a). Superior gliding of the humeral head must occur for the distal end of the humerus to be brought downward (Fig. 2–15b).

In order to determine if the articular surfaces are free to slide in the direction of the desired movement of the distal end of the bone, the joint must have a certain amount of "joint play." This movement of one articular surface on another is not usually under voluntary control and must be tested for by the application of an external force. Superior gliding of the humeral head is an example of a joint play movement. In an optimal situation, a joint has a sufficient amount of play to allow normal motion at the joint. If the supporting joint structures are lax, the joint may have too much play and become unstable. If the joint structures are tight, the joint will have too little movement between the articular surfaces and the amount of motion will be restricted.

The combination of sliding and spinning or rolling produces curvilinear motion and a moving axis of motion. The axes of motion in Figure 2–14a and b change

Figure 2–15. Sliding of joint surfaces. a. Abduction of the humerus must be accompanied by inferior sliding of the head of the humerus in the glenoid fossa. b. Adduction of the humerus is accompanied by superior sliding of the head of the humerus.

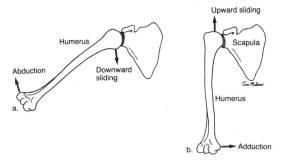

throughout the motion. Since an axis during rolling or sliding motions forms a series of successive points along a certain plane, the axis of rotation at any particular point in the motion is called the **instantaneous axis of rotation (IAR)**. In many joints the opposing surfaces are of unequal size. In some joints, such as the shoulder, the articulating surface of the moving bone is larger than the surface of the stabilized component. In other joints such as the metacarpophalangeal and interphalangeal joints of the fingers, the articulating surface of the moving bone is smaller than the surface of the stabilized component.

> *Example:* When the articulating surface of a moving component is larger than the stabilized component (the head of the humerus is larger than the articulating surface on the scapula), a pure motion such as rolling will result in the larger moving component rolling off the smaller articulating surface before the motion is completed (Fig. 2–16a).

Therefore, combination motions, wherein a moving component alternates rolling in one direction with sliding in the opposite direction, help to increase the range of motion available to the joints and keep opposing joint surfaces in contact with each other. Another method of increasing the range of available motion is by permitting both components to move at the same time. The humerus and the scapula move together during flexion/extension and during abduction/adduction at the glenohumeral joint. The rolling and sliding movements of the articular surfaces are not usually visible and thus have not been described in the traditional classification system of joint movement. However, these motions are considered in the 6°-of-freedom model described by White and Panjabi.[11] These authors have suggested that motion at the joints in the vertebral column occurs in six planes, around three axes. The implication is that motion at the joints of the body might be more thoroughly described by using a 6°-of-freedom model.

All synovial joints have a **locked** or **close-packed** position in which the joint surfaces are maximally congruent and the ligaments and capsule are maximally taut. The close-packed position is usually at the extreme end of a range of motion. In the locked position the joint surfaces are compressed and the joint possesses its greatest stability and is resistant to tensile forces that tend to cause distraction (separation) of the joint surfaces. The position of extension is the close-packed position for the humeroulnar, knee, and interphalangeal joints.[1,12] In the **unlocked** or **loose-packed** position of a joint, the articular surfaces are relatively free to move in relation to one another. The ligaments and capsule are relatively slack and the bony components may be moved actively or passively through the anatomic range of motion. The loose-packed position of a joint is any position other than the close-packed position, although in some positions the joint structures are more lax than in other positions. In the loose-packed position, the joint has a certain amount of "give" or "play." An externally applied force, such as that applied by a therapist or physician, can produce movement of one articular surface on another and enable the examiner to assess the amount of joint play that is present. Movement in and out of the close-packed position is likely to have a beneficial effect on joint nutrition

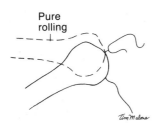

Pure
rolling

Figure 2–16. Rolling and sliding of joint surfaces. The larger head of the humerous rolls out of the glenoid fossa when pure rolling occurs. The head of the humerus remains in contact with the glenoid fossa when a combination of rolling and sliding occurs.

because of the squeezing out of the fluid during each compression and imbibing of fluid when the compression is removed.[12] In an injured joint that has swelling, the close-packed position is a position of discomfort. In the loose-packed position, the joint cavity has a greater volume and, therefore, is a position of comfort.

Osteokinematics

Osteokinematics refers to the movement of the bones rather than the movement of the articular surfaces. The normal range of motion (ROM) of a joint is sometimes called the anatomic or physiologic ROM, because the normal range refers to the amount of motion available to a joint within the anatomic limits of the joint structure. The extent of the anatomic range is determined by a number of factors, including the shape of the joint surfaces, the joint capsule, ligaments, muscle bulk, and surrounding musculotendinous and bony structures. In some joints there are no bony limitations to motion and the ROM is limited only by soft tissue structures. For example, the knee joint has no bony limitations to motion. Other joints have definite bony restrictions to motion in addition to soft tissue limitations. The humeroulnar joint at the elbow is limited in extension (close-packed position) by bony contact of the ulna on the olecranon fossa of the humerus.

A ROM is considered to be pathologic when motion at a joint either exceeds or fails to reach the normal anatomic limits of motion. When a ROM exceeds the normal limits the joint is **hypermobile.** When the ROM is less than what would normally be permitted by the structure, the joint is **hypomobile.**[13] Hypermobility may be caused by a failure to limit motion by either the bony or soft tissues and results in instability. Hypomobility may be caused by bony or cartilaginous blocks to motion or by the inability of the capsule or ligaments to elongate sufficiently to allow a normal ROM. A **contracture,** which is a term used to describe the shortening of soft tissue structures around a joint, is one cause of hypomobility. Either hypermobility or hypomobility of a joint may have undesirable effects, not only at the affected joint but also on adjacent joint structures.

MATERIALS USED IN HUMAN JOINTS

The fact that the materials used in human joints are composed of living tissue makes human joints unique and difficult to replicate. Living tissue is capable of changing its structure in response to changing environmental or functional demands. It requires nourishment to survive and is subject to disease processes, injury, and the effects of aging. Therefore, to understand the structure and function of the human joints it is necessary to have some knowledge of the nature of the materials that are used in joint construction and the forces that are acting at the joints.

Structure of Connective Tissue

The living material used in the construction of human joints is connective tissue in the form of ligaments, tendons, bursae, cartilage, disks, plates, menisci, labra, fat pads, and sesamoid bones (Fig. 2–17). The bony components are also composed of connective tissue. The gross anatomic structure and microarchitecture of connective tissues are extremely varied, and the biomechanical behaviors of specific ligaments, capsules, and tendons are still being investigated.[14–19] Generally, the structure of the connective tissue is characterized by the presence of a large extracellular matrix and a wide dispersion of cells. The extracellular matrix has both a nonfibrous component, referred to as the **ground substance,** and fibrous compo-

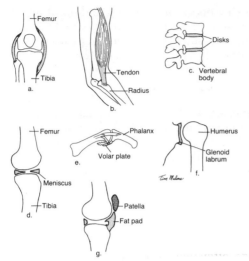

Figure 2-17. The shaded areas identify connective tissue structures. a. Collateral ligaments at the knee support the medial and lateral aspects of the joint. b. Tendon connects muscle to the bone. c. Intervertebral disks. d. Menisci located in the knee joint. e. Fibrocartilagenous plates at the metacarpophalangeal joints of the fingers. f. The glenoid labrum extends the area of the glenoid fossa. g. The patella (largest sesamoid bone in the body) and a fat pad.

nents. Table 2–2 provides a summary of the composition and function of connective tissue.

The **ground substance** is composed of hydrated networks of proteins: primarily glycoproteins and proteoglycans. The relative proportions of these proteins differ somewhat among the various connective tissue structures; approximately 5 percent in bone, and between 3 and 10 percent in articular cartilage.[20] These proteins are responsible for attracting and binding water and thus the proportion of these proteins in the extracellular matrix of a particular structure (bone, cartilage, tendon, or ligament) affects its hydration. In addition to their water-binding function, the proteins form a supporting substance for the fibrous and cellular components. The

Table 2–2. COMPOSITION AND FUNCTION OF CONNECTIVE TISSUE

Composition	Function
Cellular component	
Resident cells	
Fibroblasts	Synthesis and maintenance
Osteoblasts (bone)	of extracellular matrix
Chondroblasts (cartilage)	
Circulating cells	
Lymphocytes	Defense against infection,
Macrophages	secretion of antibodies,
	and cleanup of dead
	tissue
Extracellular matrix	
Nonfibrous component	Provides support, attracts
Glycoproteins	and binds water
Proteoglycans	
Fibrous component	
Collagen	Supporting framework,
Elastin	provides elasticity

Table 2–3. COLLAGEN TYPES AND TISSUE DISTRIBUTION

Collagen Type	Tissue Distribution
I	Skin, bone, tendon, synovium, labra (widespread)
I trimer	Tumors, skin, liver
II	Cartilage, vitreous, nucleus pulposus
III	Fetal skin, blood vessels, intestines
IV	Lamina densa of basement membrane
V	Often found in low concentration with type I
VI	Aortic intima, placenta, skin, kidney, muscle
VII	Amnion, anchoring fibrils
VIII	Endothelial cells
IX	Hyaline cartilage, IV disk, vitreous humor
X	Growth plate cartilage
XI	Hyaline cartilage, often found with type II

Adapted from Jimenez.[24]

nonfibrous component also plays an important role in protection of the connective tissue structure by contributing to the overall strength of the structure.

The **fibrous component** of the extracellular matrix contains two major classes of fibers, collagen and elastin.[1] The primary fibrous component of the intercellular substance in dense fibrous tissue is **collagen** (white fibrous tissue). Collagen is the most abundant protein in the human body[5] and accounts for 30 percent of all protein in mammals.[21] Collagen has a tensile strength that approaches that of steel[22] and is responsible for the functional integrity of connective tissue structures.[21,23] Eleven classes of collagen have been identified (Table 2–3), but the functions of all of these types have not been determined.[1,24,25]

Type I collagen is the most common and is found in almost all connective tissue including tendon, synovium, bones, labra, and skin.[24] Type II collagen is found in hyaline cartilage and in the nucleus pulposus in the center of the intervertebral disks.[1,24,25] Types I and II collagen are found in the annulus fibrosus of the intervertebral disks. Type III collagen is found in the skin and in the stratum synovium of joint capsules.[24] Collagen fibers may be arranged in many different ways and may also vary in size and shape. Collagen fibers are nonelastic, but the arrangement of the fibers in some structures allow a certain amount of elastic-type deformation.[26] In the relaxed state, collagen fibers assume a wavy configuration called **crimp.**

Elastin (yellow fibrous tissue), as the name implies, has elastic properties that allow elastin fibers to deform under an applied force and to return to their original state following removal of that force. The relative proportion of elastin to collagen fibers in different connective tissue structures varies considerably but generally, elastin fibers make up a much smaller portion of the fibrous component in the extracellular matrix than collagen fibers.

Ligaments and Tendons

Ligaments and tendons, like other connective tissue structures, are composed of a small amount of cells and a large extracellular matrix. The cellular component, which in ligaments and tendons consists of **fibroblasts,** accounts for about 20 percent of the total tissue volume. The remaining 80 percent comprises the extracellular component. The fibrous component of the extracellular matrix in both ligaments and tendons contains a larger collagen content than elastin content. However, the relative proportion of collagen to elastin fibers varies considerably among different ligaments. Some ligaments, such as the ligamentum flavum, which will be discussed in Chapter 4, have more elastin fibers than collagen fibers. In tendons, the collagen fibers predominate. The arrangement of the collagen fibers and the collagen-elastin fiber ratio in various ligaments and tendons determines the relative abilities of these structures to provide stability and mobility for a particular joint. Generally, the collagen fibers in tendons have a parallel arrangement to handle high unidirectional tensile (distractive) forces, while the collagen fibers in ligaments have a somewhat more varied arrangement depending on the function of the ligament.

In addition to their usual connective tissue components, tendons and ligaments are surrounded by loose areolar connective tissue that forms either complete or partial sheaths around these structures. The sheath formed around tendons is called **paratenon.** The paratenon protects the tendon and enhances movement of the tendon on contiguous structures. In tendons that are subjected to high friction forces, an additional synovial layer, called **epitenon,** is located directly beneath the paratenon. The epitenon produces synovial fluid, which helps to reduce friction. Tendons where epitenon can be found are at the wrist and the hand. The connective tissue sheath covering ligaments has not been given a name.[20]

Bursae

Bursae, which are similar in structure and function to tendon sheaths, are flat sacs of synovial membrane in which the inner sides of the sacs are separated by a fluid film. Bursae are located where moving structures are in tight approximation, that is, between tendon and bone, bone and skin, muscle and bone, or ligament and bone. Bursae located between the skin and bone, such as those found between the patella and the skin and the olecranon process of the ulna and the skin, are called **subcutaneous bursae**[1] (Fig. 2–18a). **Subtendinous bursae** lie between a tendon and bone. **Submuscular bursae** lie between muscle and bone[1] (Fig. 2–18b).

Cartilage

Cartilage has the same general structure as other connective tissues in that it is characterized by a large extracellular matrix and a relatively small cellular component. However, the composition of both the cellular component and extracellular matrix differ somewhat from that found in tendons and ligaments. The cellular component in cartilage contains **chondrocytes** and the fibrous component in artic-

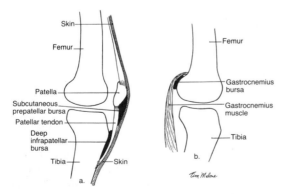

Figure 2–18. Bursae. a. The subcutaneous prepatellar bursa reduces friction between the bone and the overlying skin. The subtendinous deep infrapatellar bursa reduces friction between the patellar tendon and the tibia. b. A submuscular bursa reduces friction between the gastrocnemius muscle and the femur.

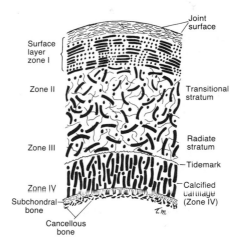

Figure 2–19. Structure of hyaline cartilage.

ular cartilage contains type **II** collagen and small quantities of types **IX** and **XI** collagens.[24,25]

Cartilage is usually divided into the following types: fibrocartilage (white and yellow) and hyaline cartilage (articular and growth). Cartilage also may be calcified. **White fibrocartilage** forms the bonding cement in joints that permit little motion. This type of cartilage also forms the intervertebral disks. In contrast to the other types of cartilage, white fibrocartilage contains type **I** collagen in the fibrous component of the extracellular matrix. **Yellow fibrocartilage** is found in the ears and epiglottis and differs from white fibrocartilage in that it has a higher ratio of elastin to collagen fibers than the white variety, which consists primarily of collagen fibers.[1]

Hyaline cartilage forms a relatively thin covering on the ends of many of the bones in the adult skeleton. It forms a smooth, resilient, low-friction surface for the articulation of one bone with another. These cartilaginous surfaces are found in freely movable joints and are capable of bearing and distributing weight over a person's lifetime. In the hyaline type of cartilage, the extracellular matrix is composed of a fibrous component that includes elastin and types **II, IX,** and **XI** collagens.[24,25] Although water is the most abundant component of hyaline cartilage, the ground substance forms a stiff gel.[20]

Three distinct layers or zones of hyaline cartilage are found on the ends of bony components of mobility joints[27] (Fig. 2–19). In the outermost layer (zone I) the collagen fibers are arranged parallel to the surface, while in the second and third zones they are randomly arranged and form an open latticework. In the third layer (radiate stratum) some fibers lie perpendicular to the surface and extend across the interface between uncalcified and calcified cartilage to find a secure hold in the calcified cartilage.[1,24] The calcified layer of cartilage, sometimes referred to as the fourth zone, lies adjacent to subchondral bone and anchors the cartilage securely to the bone.[28] The interface between the calcified and uncalcified cartilage is called the **tidemark**.[1,24] The tidemark area of the cartilage is important because of its relation to growth, healing,[29] and aging.[30] The smooth outermost layer of the cartilage helps to reduce friction between the opposing joint surfaces and to distribute forces. The second layer with its loose-coiled fiber network permits deformation and helps to absorb some of the force imposed on the joint surfaces.

During joint motion or when the cartilage is compressed, some of the fluid content of the cartilage is exuded through pores in the outermost layer. The fluid flows back into the cartilage after motion or compression ceases. The rate of fluid flow is affected by the magnitude and duration of the applied force. When the applied force is increased and sustained over a long period of time, the permeability of the cartilage is decreased and fluid flow, either in or out of the cartilage, is decreased accordingly.[31] Since hyaline cartilage is devoid of blood vessels and nerves in the adult, its nourishment is derived solely from the back and forth flow of fluid. The free flow of fluid is essential for the survival of cartilage and as an aid to reducing friction. The fact that hyaline cartilage often undergoes degenerative changes following prolonged immobilization may be related to an interference with the nutrition of the cartilage. The effects of immobilization, in which compression of joint surfaces is absent or diminished, are similar to the effects of prolonged high compressive forces in that fluid flow in and out of the cartilage is diminished, and hence cartilage nutrition is adversely affected.[32] The thin layer of hyaline cartilage rests on a plate of subchondral bone, which overlies the cancellous bone. The cancellous bone deforms under compression and the amount of bone deformation determines the amount of deformation that the overlying cartilage will undergo.

Bone

Bone is the hardest of all connective tissue found in the body. Like other forms of connective tissue it consists of a cellular component, a ground substance, and a fibrous component.[3] However, bone differs from other connective tissue structures in the composition of all three components and therefore is considered to be a specialized form of connective tissue. The cellular component consists of fibroblasts, fibrocytes, osteoblasts, osteocytes, osteoclasts, and osteoprogenitor cells. The fibroblasts and fibrocytes are essential for the production of collagen. The osteoblasts lay down bone while the osteoclasts are responsible for bone resorption. The ground substance in bone contains minerals in addition to glycosaminoglycans (GAGs), which are primarily in the form of proteoglycans (PGs), hyaluronic acid, and water. The mineral content, which consists mainly of calcium and phosphate crystals that are embedded within and between collagen fibrils, is referred to as the inorganic component of bone.[3,20] The inorganic component of bone helps to give bone its solid consistency.[1] The fibrous component of the extracellular matrix contains reticular fibers, in addition to type I collagen, and elastin fibers. The ground substance forms a supporting framework for the fibrous component.

The collagenous extracellular matrix in bones is highly calcified and takes different forms. In the innermost layer, called **cancellous (spongy) bone,** the calcified tissue forms thin plates called **trabeculae.** The trabeculae are laid down in response to stresses placed on the bone. The trabeculae undergo self-regulated modeling that not only maintains the shaft and other portions of the bone but also maintains a joint shape that is capable of distributing the load optimally. The loading history of the trabeculae, including loading from multiple directions, has been suggested as influencing the distribution of bone density and trabecular orientation.[33] Increases in bone density in some areas and decreases in density in other areas occur in response to the loads placed on bone. The cancellous bone is covered by a thin layer of dense compact bone called **cortical bone,** which is laid down in concentric layers.

The cortical bone, which appears to be solid, is covered by a tough fibrous membrane called the **periosteum.** The inner surface of the periosteum is composed of osteoblasts, which are essential for the growth and repair of bone. The periosteum is well vascularized and contains many capillaries that provide nourishment for the bones. In all synovial joints the periosteum at the ends of bones is replaced by hyaline cartilage.

Bone has the capacity for remodeling, which occurs normally throughout life, as it responds to external forces (or loads), such as the pull of tendons and the weight of the body during functional activities. Internal influences such as aging or various metabolic and disease processes also affect bone remodeling. Functional external forces (or loads) cause osteoblast activity to increase and, as a result, bone mass increases. Without these forces, osteoclast activity predominates and bone mass decreases. If the osteoclasts break down or absorb the bone at a faster rate than the osteoblasts can remodel or rebuild the bone, a condition called **osteoporosis** results. In osteoporosis, the bones have a decreased density (mass per unit volume) as compared to normal bone and thus are weaker (more susceptible to fracture) than bones with normal density.

Replacement of the calcified layer of articular cartilage by bone occurs by **endochondral ossification.** The calcification front advances to the noncalcified area of cartilage at a slow rate, which is in equilibrium with the rate of absorption of calcified cartilage by endochondral ossification.[34] Bone is considered to be a composite material because the properties of bone are the result of the combined properties of the different components that make up bone and these properties differ significantly from the properties of any one of its components.[3] Bone differs from cartilage in that bone receives its nourishment from blood supplies located within the bone, whereas cartilage receives its nourishment from sources outside the cartilage.

General Properties of Connective Tissue

All of the structures that have been described in the preceding section can be described as **heterogeneous** in that they are composed of a variety of solid and semisolid components including water, collagen, and other composite materials. Each of these materials has its own properties and thus the properties of the structure as a whole are a combination of the properties of the different tissues and the varying proportions of each tissue in the structure. The composition of the different structures reflect very specific functions. For example, in tendons the collagen fibers are oriented specifically to resist and transmit to bone the distractive forces produced by the attached muscles.

The heterogeneous nature of connective tissue structures causes these structures to exhibit properties (strength and elasticity) that vary according to their orientation in space when a constant force is applied. The label **anisotropic** is applied to structures demonstrating this behavior.

> *Example:* In the case of a long bone, which is a composite material, the properties exhibited by the bone will be different when a constant force is applied along the length of a section of bone than when forces are applied across the middle of the shaft of a bone.

Anisotropic materials differ from **isotropic** materials such as metal in that isotropic materials exhibit the same properties regardless of where the force is applied on the structure.[20,35]

Viscoelasticity

Although connective tissue appears in many forms throughout the body, all connective tissue exhibits the common property of **viscoelasticity.** The behavior of viscoelastic materials is a combination of the properties of elasticity and viscosity. **Elasticity** refers to a material's ability to return to its original state following deformation (change in dimensions, i.e., length or shape) after removal of the deforming load. When a material is stretched, it has work done on it and its energy increases. An elastic material stores this energy and keeps the energy available so that the stretched elastic material can recoil immediately to its original dimensions following removal of the distractive force. Elasticity implies that length changes or deformations are directly proportional to the applied forces or loads. **Viscosity** refers to a material's ability to dampen shearing forces. When forces are applied to viscous materials they exhibit time- and rate-dependent properties.

TIME-DEPENDENT PROPERTIES

Viscoelastic materials are capable of undergoing deformation under either a tensile (distractive) or compressive force and of returning to their original state following the removal of the force. However, under normal conditions viscoelastic materials do not return to their original state immediately. Viscoelastic materials, unlike pure elastic materials, have time-dependent mechanical properties. In other words, viscoelastic materials are sensitive to the duration of the force application. When a viscoelastic material is subjected to either a constant compressive or tensile load the material deforms and continues to deform over a finite length of time (hours, days, and months) even if the load remains constant. Deformation of the tissue continues until a state of equilibrium is reached. This phenomenon is called **creep** and is attributed to different mechanisms in different materials (Fig. 2–20). In bone, at the microscopic level, creep in compression has been attributed to the slip of lamellae within the osteons (haversian system) and the flow of the interstitial fluid.[23] In articular cartilage subjected to a compressive force, creep is attributed to the gradual loss of fluid from the tissue. Generally, the longer the duration of the applied force, the greater the deformation.

Increases in the magnitude of the applied load tend to increase the rate of creep. In some tissues, an acceleration of the rate of creep occurs after a prolonged time. Changes in temperature also have been found to affect the rate of creep. High temperatures increase the rate of creep and low temperatures decrease the rate. Theoretically, if one wishes to stretch out (elongate) a connective tissue structure one should heat it, and use a large load over a long period of time to produce creep.

RATE-DEPENDENT PROPERTIES

Viscoelastic materials respond differently to different rates of loading. When viscoelastic materials are loaded rapidly, they exhibit greater resistance to deformation than occurs if they are loaded more slowly. Generally, the higher the rate and the longer the duration of the applied force, the greater the deformation.

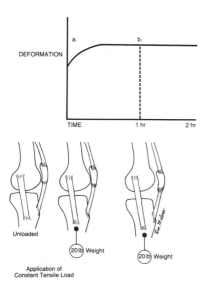

Figure 2-20. Creep response of a ligament or tendon when a constant load is applied. The tendon or ligament deforms rapidly at first as seen in the portion of the curve labeled (a) and then continues to deform (elongate) over time.

Figure 2-21. Hysteresis. a. A typical curve obtained from the measurement of the elasticity of cartilage. The arrow labeled (L) represents the loading cycle, and the curve represents the energy expended during the loading cycle. The arrow labeled (R) represents the unloading cycle, and the curve represents the energy regained. The shaded area represents the loss of energy or hysteresis (difference between energy expended and energy regained). The distance between (L) and (R), which is labeled (D), represents the deformation (change in dimensions). b. In this figure, which represents a more elastic material, the energy expended (L) and energy regained (R) are the same and, therefore, the material returns to its original dimensions following removal of the load.

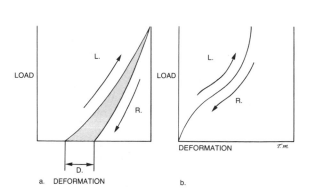

HYSTERESIS

Viscoelastic materials do not store all of the energy that is transferred to them when they are deformed by an applied force, and thus the transferred energy is not available for recovery. When a force is applied and then removed, some of the energy created during the stretching or compression of the material may be dissipated (lost) in the form of heat and therefore the material may not return to its original dimensions. The loss of energy (difference between energy expended and energy regained) is called **hysteresis**. Hysteresis is exhibited by viscoelastic materials when they are subjected to the application and removal of forces and is illustrated in Figure 2–21.

Mechanical Behavior

The materials used in the construction of human joints are subjected to continually changing forces during activities of daily living, and the ability of these materials to withstand these forces and thus provide support and protection for the joints of the body is of extreme importance. To understand how different materials and structures are able to provide support (the mechanical behavior of these structures), the reader must be familiar with the concepts and terminology used to describe their behavior, for example stress, strain, failure, and stiffness, among others. The types of tests that are used to determine the mechanical behavior of human building materials are the same as the type of tests used for nonhuman building materials.

STRESS AND STRAIN

The term **load** commonly is used to refer either to an external force or forces applied to a structure. Many examples of externally applied loads are given in chapter 1, including the forces exerted on the table by the book lying on the table, the forces exerted on the man's hand and arm by the suitcase he is carrying, and the forces exerted by a weight on the leg. The external forces exerted by the book, suitcase, and weight can all be designated as applied loads. When such loads (forces) are applied to a structure or material, forces are created within the structure or material that are called **mechanical stresses.** Stress can be expressed mathematically in the following formula, where S = stress; F = applied force; and A = area:

$$S = \frac{F}{A}$$

or stress equals the magnitude of the force applied to an object per unit area.[35]

In a solid or semisolid material, deformation (change in shape, length, or width) of the structure or material may accompany the stress and is referred to as **strain.** The type of stress and strain that develops in human structures is dependent on the nature of the material, type of load that is applied, the point of application of the load, direction, magnitude of the load, and the rate and duration of loading. When a structure can no longer support a load, the structure is said to have failed. **Ultimate stress** is the stress at the point of failure of the material while **ultimate strain** is the strain at the point of failure.

If two externally applied forces (applied loads) are equal and act along the same line and in opposite directions, they constitute a distractive or **tensile load** and will create **tensile stress** and **tensile strain** in the structure or material. Maximal (normal or principal) tensile stress occurs on a plane perpendicular to the applied load.

Tensile stress = tensile force/cross-sectional area
(perpendicular to the direction of the applied force)

When a tensile load is applied, the stress can be thought of as the intensity of the force and the strain as the amount of elongation and narrowing (deformation) that the structure sustains. The deformation is defined by comparing the original dimensions (L_0) of the object with change in dimensions (ΔL) brought about by the application of the force.

Elongation of the structure produced by a tensile stress is accompanied by a proportional amount of narrowing of the material (lateral strain).

If two externally applied forces are equal and act in a line toward each other on opposite sides of a structure, they constitute **compressive loading** and **compressive stress** and, as a result, **compressive strain** will develop in the structure. Maximal compressive stress occurs on a plane perpendicular to the applied load.

Compressive stress = compressive force/cross-sectional area
(perpendicular to the direction of the force)

When a compressive load is applied, the stress can be considered to be a measurement of the intensity of the force that develops on the plane surface of the structure and the strain as the amount of deformation (shortening and widening) that occurs in the structure (Fig. 2–22b).

$$\text{Compressive strain} = \frac{\text{decrease in length } (\Delta L)}{\text{original length } (L_0)}$$

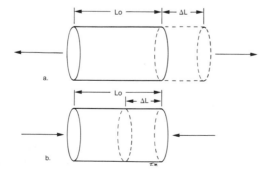

Figure 2–22. Strain. a. Tensile strain in the rod is evidenced by an increase in length. b. Compressive strain is evidenced by a decrease in length.

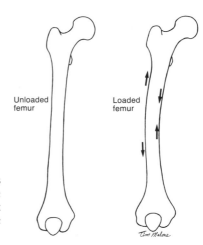

Figure 2–23. Stress and strain in a long bone. The arrows that point away from each other on the convex side of the bone indicate tensile stress and strain. The arrows that point toward each other on the concave side of the bone indicate compressive stress and strain in the structure.

Shortening of the structure is accompanied by a proportional amount of widening (lateral strain). When a structure such as a long bone is subjected to bending moments, tensile stress/strain develops on the convex side and compressive stress/strain develops on the concave side of the long axis of the bone (Fig. 2–23).

If two externally applied forces are equal, parallel, and applied in the opposite direction but are not in line with each other they constitute **shear loading,** which causes **shear stress and strain** in the structure (Fig. 2–24).

Shear stress = shear force/area parallel to the direction of the applied force[35,*]

LOAD/DEFORMATION AND STRESS/STRAIN CURVES

Load/deformation curves and stress/strain curves are employed to determine the strength of building materials, including human building materials such as bones, ligaments, tendons, joint capsules, and other structures that constitute and support human joints. The load/deformation curve in which the applied load (external force) is plotted against the deformation provides information regarding the strength properties of a particular material or structure (Fig. 2–25). The stress/strain curve, in which load is expressed in load per unit area and strain is expressed in deformation per unit length or percentage of deformation, is employed to compare the strength properties of one material with another material. Although each type of material has a unique curve, some characteristics of the curves are similar among materials that have similar properties.

The load/deformation curve depicted in Figure 2–25 provides information about the elasticity, plasticity, ultimate strength, and stiffness of the material as well as the amount of energy that the material can store before failure. The first region of the curve between point A and point B is the **elastic region.** In this region, deformation of the material will not be permanent and the structure will return to its original dimensions after removal of the load. Point B is the **yield point,** which indi-

---

*The reader may notice the similarity between the formula for stress and the formula for pressure. Pressure is defined as force per unit area or $P = F/A$ where the force is perpendicular to the area. Usually the term "pressure" is reserved for liquids and gases, but it may be used for other materials. Pressure applied to a confined liquid at rest follows Pascal's principle in that the pressure is transmitted equally to every point within the liquid.[36] The application of a uniform compression force to a confined liquid at rest produces uniform pressure in all directions and volumetric strain = change in volume/original volume. The joint disks found between the vertebral bodies in the vertebral column have a fluid-filled central portion that behaves according to Pascal's principle.

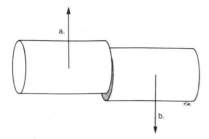

Figure 2–24. Shear stress and strain.

cates that at this point the material will no longer react elastically and some deformation will be evident after release of a load. Therefore, point **B** signifies the end of the elastic region or elastic limit. The next region on the curve from B to C is the **plastic region.** In this region deformation of the material will be permanent when the load is removed. If the load is removed at B_1, the amount of permanent deformation is represented by the distance from A to B_1. If loading continues in the plastic range, the material will continue to deform until it reaches the **ultimate failure point** C. At this point the material is no longer able to support a load and is described as having failed. In the case of a bone, failure signifies fracture while in a ligament failure signifies rupture. In hyaline cartilage, failure is evidenced by splitting and fissuring of the surface.

Materials are designated as *brittle, ductile,* or as a combination of the two, depending on the amount of deformation that they can undergo before reaching the ultimate failure point. Brittle materials such as glass do not have a plastic region and undergo very little deformation prior to failure. Ductile materials such as soft metal undergo considerable deformation prior to failure. Materials also are designated as having **resilience and toughness.** The resilience of a material is its ability when loaded to absorb and store energy within the elastic range and to release that energy and then return to its original dimensions immediately following removal of the load. Toughness is the ability of a material to absorb energy within the plastic range. Toughness reflects the material's resistance to failure or ability to absorb large amounts of energy prior to failure.

Young's modulus or **modulus of elasticity** of a material under compressive or tensile loading is represented by the slope of the curve from point **A** to point **B** in Figure 2–25. A value for stiffness can be found by dividing the load by the deformation at any point in the elastic range. The modulus of elasticity defines the mechanical behavior of the material and is a measure of the material's stiffness (resistance offered by the material to external loads).

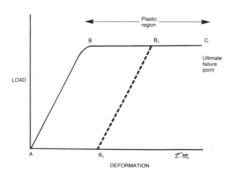

Figure 2–25. Load/deformation curve. The A-B area of the curve is the elastic region. Point B is the yield point. The area of the curve from B-C is the plastic region in which permanent deformation occurs. C is the ultimate failure point. The distance between A and B represents the amount of permanent deformation that would occur if the load was removed at B_1.

$$\text{Modulus} = \frac{\text{stress(load)}}{\text{strain(deformation)}}$$

$$\text{Young's modulus} = \frac{F/A}{\Delta L/L_0}$$

When the first portion of the curve is a straight line, the deformation (strain) is directly proportional to the material's ability to resist the load (Fig. 2–25). If the slope of the curve is steep and the modulus of elasticity is high, the material will exhibit a high degree of stiffness. If the slope of the curve is gradual and the modulus of elasticity is low, the material will exhibit a low degree of stiffness. Cortical bone has a high modulus of elasticity, while subcutaneous fat has a low modulus of elasticity.

Each type of material has its own unique curve, but a typical curve for tendons and extremity ligaments using a constant rate of loading is presented in Figure 2–26. The first region of the curve from O to A is called the **toe region,** and for tendons and ligaments is described as being the region wherein the wavy pattern (crimp) that exists in collagen fibers at rest is straightened out. In this region, a minimal amount of force produces a large amount of deformation (elongation). The toe region has been equated to the area in which an evaluator clinically tests the integrity of a ligament by the application of a tensile force.[20] The toe region also represents the slack in a tendon that must be taken up by the muscle before the muscle can apply a force to the bone through the tendon. The second linear A–B region of the curve is the elastic region in which elongation (strain) has a more or less linear relationship with the stress. The stiffness or resistance to deformation increases in this region, so more force is required to produce elongation. However, within this region, the ligament or tendon returns to its prestressed dimensions following the removal of a load. This region illustrates the type of stress and strain that occurs in normal physiologic motion. In the third region, B–C (the plastic range) progressive failure of collagen fibers occurs, and at D, the ultimate failure point, complete failure occurs and the structure is no longer able to support a load.

When connective tissue is subjected to sudden, prolonged, or excessive forces, the elastic limits of the tissue may be exceeded and the tissue may enter the **plastic range.** In the plastic range the tissue is permanently deformed or is no longer able to return to its original state following the removal of a deforming force. This situation is similar to what occurs when ligaments are overstretched and become lax. The ligaments are no longer capable of returning to their original length after being elongated and remain in a partial state of elongation. Ligamentous laxity places a joint at risk for injury because an important source of joint support and protection has been compromised. When the plastic range is exceeded, **failure** of the tissue occurs. In the case of a ligament or tendon, the failure may occur in the middle of

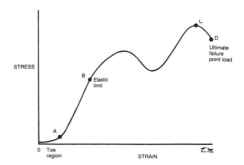

Figure 2-26. Load/deformation curve for a tendon or ligament. O-A area of the curve is the toe region. The A-B area of the curve is the elastic region. The area of the curve between B and C is the plastic range. The ultimate failure point is indicated by the letter D.

the structures through tearing and disruption of the connective tissue fibers; this is called a **rupture**. If the failure occurs through a tearing off of the bony attachment of the ligament or tendon it is called an **avulsion**. When failure occurs in bony tissue, it is called a **fracture**. Failure in cartilage was described previously.

Each type of connective tissue is able to undergo a different percentage of deformation before failure. This percentage varies not only among the types of connective tissue but also within the various types. Generally, ligaments are able to deform more than cartilage and cartilage is able to deform more than bone.

Bone. Stress/strain curves for bone demonstrate that cortical bone is stiffer than cancellous bone, meaning that cortical bone can withstand greater stress but less strain than cancellous bone. When cortical bone is loaded in compression, the strength of longitudinal sections of the bone have the greatest strength. In the femur, longitudinal sections have twice the modulus of elasticity of transverse sections.[3] The compressive stress and strain that cortical bone can withstand before failure is greater than the tensile stress and strain. In other words, bone can withstand greater stress in compression than it can in tension.

The application of high loads maintained for a short period of time or low loads held for a long period of time will produce high stress and strain. Like cortical bone, the compressive strength of trabecular bone is greater than the tensile strength, while the modulus of elasticity is higher with tensile loads than with compressive loads. The physiologic response of trabecular bone to an increase in loading is hypertrophy. Trabeculae become smaller and as a consequence less able to provide support when loading is decreased or absent. The rate, frequency, duration, and type of loading affects bone in that repeated loadings, either high repetition coupled with low load or low repetition with high load, can cause permanent strain and lead to bone failure. Bone loses stiffness and strength with repetitive loading as a result of creep strain. Creep strain occurs when a tissue is loaded repetitively during the time the material is undergoing creep.

Tendons and Ligaments. The physiologic response of tendons and ligaments to intermittent tension (application and release of a tensile force) is an increase in thickness and strength. Tendons are stronger when subjected to tensile stress than they are when subjected to compressive and shear forces. Ligaments are more variable than tendons in that they are designed to withstand compressive and shear forces as well as tensile forces.

GENERAL EFFECTS OF DISEASE, INJURY, AND IMMOBILIZATION

The design of the joints in the human body is such that each structure that is a part of a joint has one or more specific functions that are essential for the overall performance of the joint. Therefore, any process that disrupts any one of the parts of a joint will disrupt the total function of the joint. The complex joints are more likely to be affected by injury, disease, or aging than the simple joints. The complex joints have more parts and are subject to more wear and tear than stability joints. Also, the function of the complex joints is dependent on a number of interrelated factors. For example, the capsule must produce synovial fluid. The fluid must be of the appropriate composition and of sufficient quantity so that it can lubricate and provide nourishment for the joint. The hyaline cartilage must be smooth so that the joint surfaces can move easily and must be permeable so that it can receive some of its nourishment from the joint fluid. The cartilage also must undergo periodic compressive loading and unloading to facilitate movement of the fluid. The ligaments and capsules must be able to provide sufficient support for stability and yet be flexible enough to permit normal mobility.

Disease

The general effects of disease, immobilization, and aging may be postulated by using the normal function of a joint structure as a basis for analysis. For example, if the synovial membrane of a joint is affected by a collagen disease such as rheumatoid arthritis, one may assume that since the normal function of the synovial membrane is to produce synovial fluid, the production and perhaps the composition of the synovial fluid will be altered in this disease. One could also postulate that since fluid is altered, the lubrication of the joint also would be altered. The disease process and the changes in joint structure that occur in rheumatoid arthritis actually involve more than synovial fluid alteration, but the disease does change the composition and the quantity of the synovial fluid. In another type of arthritis, osteoarthritis, which is thought to be a genetic disorder, the cartilage rather than the synovium and the soft tissue is the focus of the disease process. Based on normal cartilage function, one can assume that the cartilage in osteoarthritic joints will not be able to withstand normal stress. Actually, erosion and splitting of the cartilage occur under stress. As a result, friction is increased between the joint surfaces, thus further increasing the erosion process.

Injury

If an injury has occurred, such as the tearing of a ligament, one may assume that there will be a lack of support for the joint. In the example of the table with an unstable joint between the leg and the table top, damage and disruption of function may occur as a result of instability. If a heavy load is placed on the damaged table joint, the joint surfaces will separate under the compressive load and the leg may be angled. The once-stable joint now allows mobility and the leg may wobble back and forth. This motion may cause the screws to loosen or the nails to bend and ultimately to be torn out of one of the wooden components.

Complete failure of the joint may result in splintering of the wooden components, especially when the already weakened joint is subjected to excessive, sudden, or prolonged loads. The effects of a lack of support in a human joint are similar to that of the table joint. Separation of the bony surfaces occurs and may result in wobbling or a deviation from the normal alignment of one of the bony components. These changes in alignment create an abnormal joint distraction on the side where a ligament is torn. As a result, the other ligaments, the tendons, and the joint capsule may become excessively stretched and consequently be unable to provide protection. The supported side of the joint may also be affected and subjected to abnormal compression during weight bearing or motion.

Immobilization

Any process or event that disturbs the normal function of a specific joint structure usually will set up a chain of events that eventually affects every part of a joint and its surrounding structures. Immobilization is particularly detrimental to joint structure and function. Immobilization may be externally imposed by a cast or self-imposed as a reaction to pain and inflammation. An injured joint or joint subjected to inflammation and swelling will assume a loose-packed position in which the pressure within the joint space is minimized. This position may be referred to as the position of comfort because pain is decreased in this position. Each joint has a posi-

Joint Structure and Function

Table 2–4. CHANGES IN JOINT STRUCTURE AND FUNCTION FOLLOWING PROLONGED IMMOBILIZATION

Structure	Changes Due to Prolonged Immbolization
Synovium	Proliferation of fibrofatty connective tissue into joint space and the formation of adhesions.
Cartilage	Adherence of fibrofatty connective tissue to surface of cartilage. Atrophy of cartilage and decrease in water and proteoglycan content.
Bone	Regional osteoporosis.
Ligament	Disorganization of parallel fiber arrangement and decrease in water and proteoglycan content.
Ligament insertion	Destruction of ligament fibers due to osteoclastic activity.

Adapted from Akeson et al.[38]

tion of minimum pressure. For the knee and hip joints, the position of comfort is between 30 and 45° of flexion, and for the ankle joint the position is at 15° of plantar flexion.[37] If the joint is immobilized for a few weeks in the position of comfort, contractures will develop in the surrounding soft tissues and as a consequence a normal range of joint motion will be impossible.

The effects of immobilization are not confined to the surrounding soft tissues but also may affect the articular surfaces of the joint and the underlying bone. Biochemical and morphologic changes that have been attributed to the effects of immobilization include proliferation of fibrofatty connective tissue within the joint space, adhesions between the folds of the synovium, atrophy of cartilage, regional osteoporosis, weakening of ligaments at their insertion sites due to osteoclastic resorption of bone and Sharpey's fibers, and a decrease in the proteoglycan and water content of articular cartilage.[38,39] Water and proteoglycans are also lost from the tendons, the ligaments, and the joint capsule.[40] As a result of these changes, the ROM available to the joint is decreased, the time between loading and failure is decreased, and the energy-absorbing capacity of the bone-ligament complex is decreased. Swelling and/or immobilization of a joint also inhibits and weakens the muscles surrounding the joint.[41–43] Therefore, the joint is unable to function normally and is at high risk of additional injury. A summary of the effects of prolonged immobilization is presented in Table 2–4.

Recognition of the adverse effects of immobilization has led to the development of the following strategies to help minimize the consequences of immobilization: (1) use of continuous passive motion (CPM) devices following joint surgery, (2) reduction in the duration of casting periods following fractures and sprains, and (3) development of dynamic splinting devices. The CPM is a mechanical device that is capable of moving joints passively and repeatedly through a specified portion of the physiologic range of motion. In these devices the speed of the movement as well as the range of motion can be controlled. The CPM devices are able to produce joint motion without the potentially deleterious compressive/tensile stresses and strains produced by active muscle contractions.

Overuse

While immobilization is detrimental, constant or repetitive loading of articular structures also has adverse effects. Constant loading, such as occurs in prolonged standing, sitting, or squatting, subjects the joints and their supporting structures to the effects of load deformation and creep. Ligaments subjected to constant tensile loads will creep and may undergo excessive lengthening. Cartilage subjected to constant compressive loading may creep and may undergo excessive deformation. Joints and their supporting structures subjected to repetitive loading may be injured and fail because they do not have time to recover their original dimensions before they are subjected to another loading cycle. Thus these structures are subjected to repeated loading while they are still deforming. The popular term for this type of injury is **overuse injury** or **overuse syndrome** and has been identified in athletes, dancers, musicians, and factory and office workers.

SUMMARY

The health and strength of joint structures and hence their function are dependent on a certain amount of stress and strain. Cartilage and bone nutrition and growth are dependent on joint movement and muscle contraction. Cartilage nutrition is dependent on joint movement through a full range of motion to ensure that all of the articular cartilage receives the nutrients necessary for survival. Ligaments and tendons are dependent on a normal amount of stress and strain to maintain and increase strength. Bone density and strength increase following the stress and strain created by muscle and joint activity. In contrast, bone density and strength decrease when stress and strain are absent.

In subsequent chapters the specific structure and function of each of the major joints in the body will be explored. Knowledge of the specific location, function, and structure will permit the reader to analyze the effects of injury, disease, or aging on these joint structures.

REFERENCES

1. Williams, P, et al: Gray's Anatomy, ed 37. Churchill Livingstone, Edinburgh, 1989.
2. Guyton, AC: Basic Human Physiology: Normal Function and Mechanisms of Disease. WB Saunders, Philadelphia, 1977.
3. Gould, JA and Davies, GJ: Orthopaedic and Sports Physical Therapy. ed 2. CV Mosby, St Louis, 1990.
4. Hettinga, DL: 1 Normal joint structures and their reaction to injury. J Orthop Sports Phys Ther 1:1, 1979.
5. Simkin, PA: In Schumacher, HR, Klippel, JH, and Robinson, DR (eds): Primer on the Rheumatic Diseases, ed 9. Arthritis Foundation, Atlanta, 1988.
6. Wolf, AW, et al: Current concepts in synovial fluid analysis. Clin Orthop 134:261, 1978.
7. Hettinga, DL: II Normal joint structures and their reaction to injury. J Orthop Sports Phys Ther 1:2, 1979.
8. Radin, EL and Paul, IL: A consolidated concept of joint lubrication. J Bone Joint Surg 54A:607.
9. Lehmkuhl, D and Smith, L: Brunnstrom's Clinical Kinesiology, ed 4. FA Davis, Philadelphia, 1984.
10. Gowitzke, BA and Milner, M: Understanding the Scientific Basis of Human Movement, ed 3. Williams & Wilkins, Baltimore, 1988.
11. White, AA and Panjabi MM: Clinical Biomechanics of the Spine, ed 2. JB Lippincott, Philadelphia, 1990.
12. Hertling, D and Kessler, RM: Management of Common Musculoskeletal Disorders, ed 2. JB Lippincott, Philadelphia, 1990.
13. Cookson, JC and Kent, BE: Orthopedic manual therapy—An overview. Part 1: The extremities. Phys Ther 59:2, 1979.
14. Noyes, FR, DeLucas, JL, and Torvik, PJ: Biomechanics of anterior cruciate ligament failure: An analysis of strain-rate sensitivity and mechanisms of failure in primates. J Bone Joint Surg 56A:2, 1974.
15. Decraemer, WF, et al: A non-linear viscoelastic constitutive equation for soft biological tissues based upon a structural model. J Biomech 13:559, 1980.

16. Goldstein, SA, et al: Analysis of cumulative strain in tendons and tendon sheaths. J Biomech 20:1, 1987.
17. To, SYC, Kwan, MK, and Woo, SL-Y: Simultaneous measurements of strains on two surfaces of tendons and ligaments. J Biomech 21:511, 1988.
18. Woo, SL-Y, et al: Mechanical properties of tendons and ligaments. Biorheology 19:397–408, 1982.
19. Rong, G and Wang, Y: The role of the cruciates ligaments in maintaining knee joint stability. Clin Orthop Rel Res 215:65, 1987.
20. Nordin, M and Frankel, VH: Basic Biomechanics of the Musculoskeletal System. ed 2. Lea and Febiger, Philadelphia, 1989.
21. Nimni, ME: The molecular organization of collagen and its role in determining the biophysical properties of connective tissues. Biorheology 17:51, 1980.
22. Widmann, FK: Pathobiology: How Disease Happens. Little, Brown, Boston, 1978.
23. Cailliet, R: Soft Tissue Pain and Disability. ed 2. FA Davis, Philadelphia, 1988.
24. Jimenez, SA: In Schumacher, HR, Klippel, JH, and Robinson, DR (eds): Primer on the Rheumatic Diseases, ed 9. Arthritis Foundation, Atlanta, 1988.
25. Eyre, DR: In Ghosh, P (ed): The Biology of the Intervertebral Disc, Vol 1. CRC Press, Boca Raton, FL 1988.
26. Hollinshead, WH: Functional Anatomy of the Limbs and Back, ed 4. WB Saunders, Philadelphia, 1976.
27. Hettinga, DL III: Joint structures and their reaction to injury. J Orthop Sports Phys Ther 1:3, 1980.
28. Ghadially, FH: Structure and function of articular cartilage. Clin Rheum Dis 7:3, 1980.
29. Mitchell, N and Shepard, N: Healing of articular cartilage in intra-articular fractures in rabbits. J Bone Joint Surg 62A:4, 1980.
30. Lane, LB and Bullough PG: Age-related changes in the thickness of the calcified zone and the number of tidemarks in adult human articular cartilage. J Bone Joint Surg 62B:3, 1980.
31. Mansour, JM and Mow, VC: The permeability of articular cartilage under compressive strain and at high pressures. J Bone Joint Surg 58A:4, 1976.
32. McDonough, AL: Effects of immobilization and exercise on articular cartilage–A review of literature. J Orthop Sports Phys Ther 2:5, 1981.
33. Carter, DR, Orr, TE, and Fyhrie, DP: Relationship between loading history and femoral cancellous bone architecture. J Biomech 22:231, 1989.
34. Bullough, PG and Jagannath, A: The morphology of the calcification front in articular cartilage. J Bone Joint Surg 65:72, 1983.
35. O'Dwyer, J: College Physics, ed 2. Wadsworth, California, 1984.
36. Buecke, F: Principles of Physics, ed 3. McGraw-Hill, New York, 1977.
37. Perry, J: Contractures: A historical perspective. Clin Orthop Rel Res 219:8, 1987.
38. Akeson, WH, et al: Effects of immobilization on joints. Clin Orthop Rel Res 219:28–37, 1987.
39. Enneking, WF and Horowitz, M: The intra-articular effects of immobilization on the human knee. J Bone Joint Surg 54A:973, 1972.
40. de Andrade, JR, Grant, C, and Dixon, St J: Joint distension and reflex muscle inhibition in the knee. J Bone Joint Surg 47A:313, 1965.
41. Young, A, Stokes, M, and Iles, JF: Effects of joint pathology on muscle. Clin Orthop Rel Res 219:21, 1987.
42. Spencer, JD, Hayes, KC, and Alexander, IJ: Knee joint effusion and quadriceps reflex inhibition in man. Arch Phys Med Rehabil 65:171, 1984.
43. Stokes, M and Young, A: The contribution of reflex inhibition to arthrogenous muscle weakness. Clin Sci 67:7, 1984.

STUDY QUESTIONS

1. How does the structure and function of synarthroses differ from diarthroses?

2. What is the composition and function of synovial fluid?

3. Name the plane and corresponding axis of motion for each of the following: flexion/extension, abduction/adduction, and rotation.

4. What is the difference between a closed-kinematic chain and an open-kinematic chain? Give at least one example of each type of chain.

5. What type of motion is available at a pivot joint? Give at least two examples of pivot joints.

6. Explain the movement of a bony lever during motion at an ovoid joint.

7. What is stress and strain? Give at least one example using a load/deformation curve.

8. What is the difference between arthrokinematics and osteokinematics? Give at least two examples of each.
9. What happens to a material when hysteresis occurs?
10. What are the properties of a material that is viscoelastic?
11. How does immobilization affect connective tissue?
12. How does an overuse injury occur?
13. Describe how joints are lubricated.
14. What is creep? How does it affect joint structure?

Chapter 3

■ ■ ■

Muscle Structure and Function

OBJECTIVES
Following the study of this chapter, the reader should be able to:

Describe
1. The structure and function of the contractile unit (sarcomere).
2. The structure and function of the functional unit (motor unit).
3. The connective tissue in a muscle.
4. The various muscle fiber types.

Define
1. Muscle tension including active and passive tension.
2. Active and passive insufficiency.
3. Concentric, eccentric, and isometric muscle action.

4. Reverse action.
5. Agonists, antagonists, and synergists.

Recall
1. Factors affecting muscle tension.
2. Characteristics of different muscle fiber types.
3. Characteristics of motor units.
4. Factors affecting angular velocity and torque.
5. The effects of immobilization.

Differentiate
1. A spurt from a shunt muscle.
2. A phasic from a tonic muscle.
3. Among antagonists, agonists, and synergists.
4. Active from passive insufficiency.
5. Active from passive tension.
6. Concentric from eccentric.
7. Normal muscle action from reverse action.

Compare
1. Tension development in eccentric versus concentric contractions.
2. Isokinetic exercise with concentric exercise. Isoinertial exercise with concentric exercise.
3. The effects of immobilization in a lengthened versus a shortened position.

INTRODUCTION

Mobility and Stability Functions of Muscles

The skeletal muscles, like the joints, are designed to contribute to the body's needs for mobility and stability. Muscle forces, like all forces that are applied to the levers of the body, have both rotatory (mobility) and translatory (usually stability) components. Muscles serve a mobility function by producing or controlling the movement of a bony lever around a joint axis; they serve a stability function by resisting movement of joint surfaces and through approximation of joint surfaces. The body is incapable of either supporting itself against gravity or of producing motion without muscle function.

A human skeleton without muscles will collapse when placed in the erect standing position. Persons with spinal cord injuries that result in a loss of muscle function in their lower extremities are unable to stand or walk without external support. The joint structures in the injured person are intact, but these structures are incapable of counteracting gravitational torques in a weight-bearing position. Therefore, when the injured person wishes to stand or walk, an external support such as a brace must be used to stabilize the joints of the lower extremities. The brace replaces some of the function of lost muscles by employing a locking device that is similar to the device used in the folding table joint. When the brace is locked, it provides joint stability. Under normal conditions, the skeletal muscles of the body provide forces that help human joints move freely into and out of close-packed and loose-packed positions. When a joint has attained a close-packed position, the stability role of the muscles is decreased. The noncontractile structures (ligaments and joint capsules) supporting the joint are taut and capable of contributing to joint stability. However, when a joint is in the loose-packed position, the muscles must assume a somewhat larger stability function because the passive supporting structures are lax and have reduced capacity for providing joint stability. The amount of force that a muscle can contribute either to joint stability or mobility is a function of its structure, contractile ability, and biomechanical characteristics. These factors must be explored to understand the interrelationships that determine human function.

ELEMENTS OF MUSCLE STRUCTURE

The two types of materials that are found in skeletal muscle are muscle tissue (contractile) and connective tissue (noncontractile). The properties of these tissues and the way in which they are interrelated give muscles their unique characteristics. Muscle tissue, like other biologic tissue, is viscoelastic.[1] Muscle tissue also possesses the properties of contractility and irritability. **Contractility** refers to the muscle's ability to develop tension. **Irritability** refers to a muscle's ability to respond to chemical, electrical, or mechanical stimuli.[1]

Composition of a Muscle Fiber

A skeletal muscle is composed of many thousands of muscle fibers. The arrangement, number, size, and type of these fibers may vary from muscle to muscle,[2] but each fiber is a single muscle cell that is enclosed in a cell membrane called the **sarcolemma** (Fig. 3–1a). Muscle cells (muscle fibers) are grouped together in bundles called **fasciculi** and a single muscle contains many fasciculi (Fig. 3–1b). Like other cells in the body, the muscle fiber is composed of cytoplasm, which in a muscle is called **sarcoplasm**. The sarcoplasm consists of structures called **myofibrils** (Fig. 3–1c), which are the contractile structures of a muscle fiber and nonmyofibrillar structures such as ribosomes, glycogen, and mitochondria, which are required for cell metabolism.

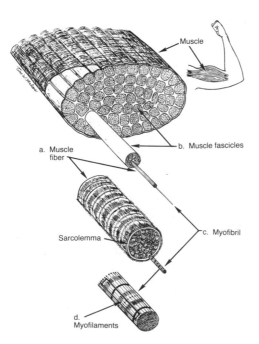

Figure 3–1. Composition of a muscle fiber. a. The muscle fiber is enclosed in a cell membrane called the **sarcolemma**. b. Groups of muscle fibers form bundles called **fasciculi**. c. The muscle fiber contains myofibrillar structures called **myofibrils**. d. The myofibril is composed of thick myosin and thin actin myofilaments.

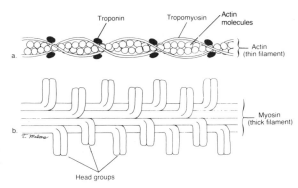

Figure 3–2. Myofilaments. a. The actin molecules are shown as circles. The troponin molecules are globular and are shown located in notches between the two strands of actin molecules. The tropomyosin molecules are thin and are shown lying along grooves in the actin strands. b. A myosin myofilament showing head groups or globular enlargements.

The myofibril is composed of tiny filaments called **myofilaments** (Fig. 3–1d). Some of the myofilaments are composed of the protein **actin**, while others are composed of the protein **myosin**. The binding together of these two myofilaments causes a muscle contraction. The actin myofilaments are thin and are formed by two chainlike strings of actin molecules wound around each other. Molecules of the globular protein troponin are found in notches between the two actin strings and the protein tropomyosin is attached to each troponin molecule (Fig. 3–2a). The troponin and tropomyosin molecules control the binding of actin and myosin myofilaments.

The myosin myofilaments are thick and composed of large myosin molecules that are arranged to form long molecular filaments (Fig. 3–2b). The myofilaments formed by the myosin molecules are not of equal diameter throughout their length but are wider in the middle portion. Each of the myosin filaments has globular enlargements called **head groups**.[3] The head groups, which are able to swivel and are the binding sites for attachment to the actin, play a critical role in muscle contraction and relaxation. When the entire myofibril is viewed through a microscope, the alternation of thick (myosin) and thin (actin) myofilaments forms a distinctive striped pattern as seen in Figure 3–1d. Therefore, skeletal muscle is often called **striated muscle**. A schematic representation of the ordering of the myofilaments in a myofibril is presented in Figure 3–3.

The Contractile Unit

In Figure 3–3, the portion of the myofibril that is located between two Z lines is called the sarcomere. In resting muscle the sarcomere is about 2.5 μm long. The Z lines, which are located at regular intervals throughout the myofibril, not only serve as boundaries for the sarcomere but also link the actin filaments together.

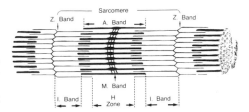

Figure 3–3. Ordering of myofibrils in a muscle at rest. The sarcomere is the portion of the myofibril that is located between the Z bands (or lines). The A band portion of the sarcomere contains an overlap of the myosin and actin filaments. The portion of the A band that contains only myosin filaments without overlap is called the H zone. The M band located in the central portion of the H zone contains transversely oriented myosin filaments that connect one myosin filament with another. The I band portion contains only actin fibers.

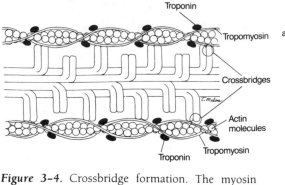

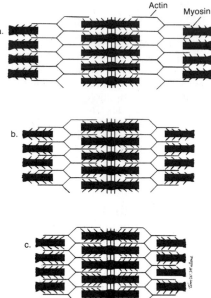

Figure 3-4. Crossbridge formation. The myosin head groups attach to binding sites on the actin that have been exposed by the movement of the tropomyosin away from these binding sites. The thin actin myofilament is pulled along past the myosin filament by movement of the head groups. Although only two crossbridges are shown in the illustration, crossbridges are formed at many sites during a muscle contraction.

Figure 3-5. Sarcomere shortening and crossbridge formation. a. Minimal overlap between the actin and myosin filaments allows only a few crossbridges to be formed. b. More crossbridge formation occurs as more overlap occurs between the actin and myosin. c. When maximal overlap occurs between the actin and myosin, the largest number of crossbridges may be formed.

Areas of the sarcomere called **bands** or **zones** help to identify the arrangement of the actin and myosin filaments. The portion of the sarcomere that extends over both the length of the myosin filaments and a small portion of the actin filaments is called the **anisotropic** or **A band.** The central portion of the myosin filament (A band area) in which there is no overlap with the actin filaments is called the **H zone.** The central portion of the H zone, which consists of the wide middle portion of the myosin, is called the **M band.** The M band is formed by myosin myofilaments that connect the central wider region of one myosin myofilament with the filament above or below. Areas that include only actin filaments are called **isotropic** or **I bands.**[3]

Voluntary activation of a muscle is initiated by the arrival of a nerve impulse at the motor end plate, which evokes an electric impulse or **action potential** that travels along the muscle fiber. The action potential initiates the release of calcium ions, and the calcium ions cause troponin to reposition the tropomyosin molecules so that receptor sites on the actin are free and the head groups of the myosin can bind with actin. This bonding of filaments is called a **crossbridge** and is considered to be the basic unit of active muscle tension[1,3] (Fig. 3–4).

After some crossbridges are formed, the myosin head groups swivel in an arc and pull the actin filament along the myosin. As the actin filaments are pulled along past the myosin by the swiveling head groups, old crossbridges are broken and new ones are formed (Figs. 3–5a and b). The formation, breaking, and re-formation of crossbridges may continue until there is maximum overlap of the actin and myosin filaments. At this point all of the possible sites for crossbridge formation will have

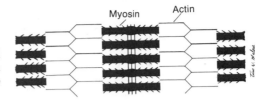

Figure 3-6. Sarcomere illustrated at a lengthened position in which little or no overlap between actin and myosin is occurring. When no overlap occurs, no crossbridges can be formed.

been filled. The sarcomere is shortened when there is complete overlap of the filaments (Fig. 3–5c).

When the actin and myosin filaments of the sarcomere are at a point where there is no overlap of filaments (Fig. 3–6), there can be no crossbridge formation (and thus no active tension developed) until sliding occurs to a position of overlap. Likewise when the muscle fiber is maximally shortened and there is maximum overlap of actin and myosin filaments, no additional crossbridges can be formed and no increase in active tension is possible. Tension will be generated in a muscle whenever crossbridges are formed. The sliding of the actin filaments toward and past the myosin filaments accompanied by the formation and re-formation of crossbridges in each sarcomere will result in shortening of the muscle fiber and the generation of tension. The entire muscle will shorten (contract) if a sufficient number of muscle fibers actively shorten and if either one or both ends of the muscle are free to move. The active shortening of a muscle is called a **concentric contraction** or **shortening contraction.** In contrast to a shortening contraction wherein the actin filaments are being pulled toward the myosin filaments, the muscle may undergo a lengthening or **eccentric contraction** in which case the actin filaments are pulled away from the myosin filaments and crossbridges are broken and re-formed as the muscle lengthens. Tension is generated by the muscle as crossbridges are re-formed. Eccentric contractions occur whenever a muscle actively resists motion created by an external force (such as gravity or, more rarely, by another muscle.)

The following is a summary of the important facts about muscle contraction at the sarcomere level:

- Tension is generated whenever crossbridges are formed.
- No crossbridges can be formed and no generation of active tension can occur unless there is some overlap of the actin and myosin myofilaments.
- The maximum number of crossbridges can be formed when there is maximum overlap of myofilaments.
- In a shortening contraction the actin myofilaments are pulled toward the myosin myofilaments and crossbridges are formed, broken, and re-formed.
- In a lengthening contraction the actin myofilaments are pulled away from the myosin myofilaments and crossbridges are broken, re-formed, and broken.

The Functional Unit

The stimulus that the muscle fiber receives that begins the contractile process is transmitted through a nerve called the **alpha motor neuron** (Fig. 3–7). The cell body of the neuron is located in the anterior horn of the gray matter of the spinal cord. A long fiber called the **axon** extends from the cell body to the muscle, where it divides into either a few or as many as thousands of smaller branches. Each of the smaller branches terminates in a motor end plate that lies in close approximation to the sarcolemma of a single muscle fiber. All of the muscle fibers on which a branch of the axon terminates are part of one motor unit, along with the cell body and the axon.

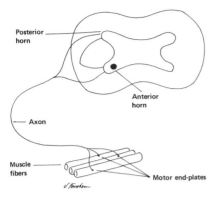

Posterior
horn

Anterior
horn

Axon

Muscle
fibers

Motor end-plates

Figure 3-7. An alpha motor neuron. The cell body is shown as a black dot in the anterior horn.

The nerve impulse transmitted from the cell body along the axon to the motor end plate causes depolarization of each muscle fiber's sarcolemma and generates an action potential that spreads both along the external surface of the sarcolemma and into the interior of the fiber by way of narrow tubular invaginations called **transverse tubules (T tubules)**. Two transverse tubules supply each sarcomere at the level of the junctions of the A and I bands. The **sarcoplasmic reticulum**, which has a large calcium-storage capacity, is composed of anastomosing membranous channels that fill the space between the myofibrils and form large sacs (terminal cisternae) in areas where the membranous channels are close to the T tubules. The combination of two terminal cisternae with a T tubule in the middle is called a **muscle triad.**[1,4] When the action potential sweeps down the T tubules, free calcium ions from the terminal cisternae are released into the myofibrils. However, the exact mechanics by which the action potential in the T tubules causes the release of calcium from the sarcoplasmic recticulum is unknown. The release of the calcium ions initiates actin-myosin crossbridge activity and causes muscle tension. When the sarcolemma becomes electrically stable after depolarization, calcium ions rebind to the sarcoplasmic reticulum and the muscle fiber relaxes.[1,3] (Motor units go through a latency or refractory period just after firing and require time to recover before the cycle of depolarization and tension generation can be repeated. Therefore, the frequency of firing of motor units is limited by the need for recovery time prior to reactivation.)

The **motor unit** consists of the alpha motor neuron and all of the muscle fibers it innervates. The contraction of the entire muscle is the result of many motor units firing *asynchronously* and *repeatedly*. The magnitude of the contraction of the entire muscle may be altered by altering the number of motor units that are activated or the frequency at which they are activated. The number of motor units in a muscle as well as the structure of these units varies from muscle to muscle.

Motor units may vary according to the size of the neuron cell body, diameter of the axon, number of muscle fibers, and type of muscle fibers. Each of these variations in structure affects the function of the motor unit. Some motor units have small cell bodies and others have large cell bodies. Units that have small cell bodies have small-diameter axons. Nerve impulses take longer to travel through small-diameter axons than they do through large-diameter axons. Therefore, in the small-diameter units, a stimulus will take longer to reach the muscle fibers than it will in a unit with a large-diameter axon.

The size of the motor unit is determined by the number of muscle fibers that it contains (Fig. 3-8). The number of fibers may vary from two to three to a few thousand. Muscles that either control fine movements or that are used to make

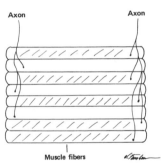

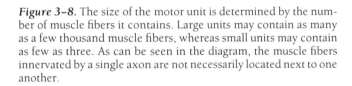

Figure 3–8. The size of the motor unit is determined by the number of muscle fibers it contains. Large units may contain as many as a few thousand muscle fibers, whereas small units may contain as few as three. As can be seen in the diagram, the muscle fibers innervated by a single axon are not necessarily located next to one another.

small adjustments have small-size motor units, and such motor units generally have small cell bodies and small-diameter axons. Muscles that are used to produce large increments of force and large movements usually have large-size motor units, large cell bodies, and large-diameter axons. The motor units of the small muscles that control eye motions may contain as few as six muscle fibers, while the gastrocnemius muscles have motor units that contain about 2000 muscle fibers.[3] Muscles with a predominantly large number of fibers per motor unit usually have a relatively smaller total number of motor units than muscles that have few fibers per motor unit. The platysma muscle in the neck has relatively small motor units consisting of approximately 25 muscle fibers, but the muscle has a total of 1000 of these motor units. The gastrocnemius, on the other hand, has relatively large motor units consisting of about 2000 muscle fibers per unit, but the muscle has a relatively small number (600) of such units. In most instances a muscle has at least some mix of small and large motor units.

Usually, when a muscle action is desired, the motor units with the small cell bodies and few motor fibers are automatically selected (recruited) first by the nervous system.[5] Small motor units generate less tension than large motor units and require less energy expenditure, and therefore this recruitment strategy is thought to be energy-conserving. If a few small motor units are capable of accomplishing the task, the recruitment of large motor units is unnecessary. If the task demands are such that the small motor units are unable to complete the task, larger motor units can be recruited. However, the recruitment strategy may be based not only on energy conservation but also on: previous experience; the nature of the task (how rapidly the muscle must respond); and a mechanism that takes into account the actions of all muscles around a joint, including such considerations as the muscle's mechanical advantage at a particular point in the range of motion (ROM). The recruitment strategy also may involve the selection of motor units from not just one but a variety of muscles surrounding a joint to accomplish a particular task.[6]

The following regarding motor units affect the function of the muscle:

- The number of muscle fibers (affects the magnitude of the response to a stimulus)
- The diameter of the axon (determines the conduction velocity of the impulse)
- The number of motor units that are firing at any one time (affects the total response of the muscle)
- The frequency of motor unit firing (affects the total response of the muscle)

In addition, the type of muscle fibers contained within a motor unit will affect the response of a muscle. All of the muscle fibers contained in a single motor unit are of one type, but the type of muscle fibers within a muscle may vary from one motor unit to another motor unit.

Muscle Fibers
Types

Three principal types of muscle fibers are found in varying proportions in skeletal muscles. These fiber types may be distinguished from one another histochemically, metabolically, morphologically, and mechanically. Since the reader may encounter different systems of nomenclature in different texts,[7] Table 3–1 presents not only the characteristics of these fibers but also a few of the different names that are used to designate the fibers. In this text the three muscle fiber types will be referred to as: (1) **fast-twitch glycolytic (FG)**, (2) **fast-twitch oxidative glycolytic (FOG)**, and (3) **slow-twitch oxidative (SO)**.[8] SO fibers may also be referred to as **type I fibers**, whereas FG fibers may be referred to as **type II fibers**. FOG fibers are referred to as **type IIA fibers** in some references.[1]

Each skeletal muscle in the body is composed of motor units of each of the three types of fibers, but wide variations exist among individuals in the number of motor units allocated to each fiber type in similar muscles. Therefore, in a hypothetical situation, person A's biceps might contain the following motor units: 75 percent FG fibers, 15 percent FOG fibers, and 10 percent SO fibers, whereas person B's biceps might have only 50 percent FG fibers, 35 percent FOG fibers, and 15 percent SO fibers. The variations in fiber types among individuals are believed to be genetically determined.The vastus lateralis, rectus femoris, deltoid, and gastrocnemius muscles have been found to be similar among individuals in that they have been found to contain about 50 percent FG and 50 percent SO fibers.[2]

The soleus muscle, on the other hand, has been found to contain twice as many SO fibers as FG fibers.[9] Muscles that have a relatively high proportion of SO fibers in relation to FG fibers, such as the soleus muscle, are able to carry on sustained activity because the SO fibers do not fatigue rapidly. These muscles are often called **stability, postural,** or **tonic** muscles because they help to maintain stability of the body. The relatively small SO motor units of the soleus muscle, which have small cell bodies, small-diameter axons, and a small number of muscle fibers per motor unit, are almost continually active during erect standing so as to make the small adjustments in muscle tension that are required to maintain body balance and counteract the effects of gravity. Muscles that have a high proportion of the FG fibers, such as the biceps brachii, are sometimes designated as **mobility, nonpostural,** or

Table 3–1. CHARACTERISTICS OF SKELETAL MUSCLE FIBERS

	Fast-twitch Glycolytic (FG)°	Fast-twitch Oxidative Glycolytic (FOG)†	Slow-twitch Oxidative (SO)‡
Diameter	Large	Intermediate	Small
Muscle color	White	Red	Red
Capillarity	Sparse	Dense	Dense
Myoglobin content	Low	Intermediate	High
Speed of contraction	Fast	Fast	Slow
Rate of fatigue	Fast	Intermediate	Slow
Motor unit size	Large	Intermediate to large	Small
Axon conduction velocity	Fast	Fast	Slow

°FG fibers are also referred to as fast-twitch white fibers or fast-twitch fibers.
†FOG fibers are also referred to as intermediate fibers or fast-twitch red fibers.
‡SO fibers are also referred to as slow-twitch or slow-twitch red fibers.

phasic muscles. These muscles are involved in producing a large ROM of the bony components.[10] The FG fibers respond more rapidly to a stimulus but also fatigue more rapidly than SO fibers. Following intermittent bouts of high-intensity exercise, muscles with a high proportion of FG fibers, which involve a large initial response, show greater fatigue and recover more slowly than muscles with a high proportion of SO fibers.[11]

Size, Arrangement, and Number

Muscle fiber length, fiber arrangement, and number of muscle fibers per muscle vary throughout the body. These structural variations affect not only the overall shape and size of the muscles but also the function of the various muscles.

Each muscle fiber is capable of shortening to approximately one-half of its total length. Consequently, a long muscle fiber is capable of shortening over a greater distance than a short muscle fiber. For example: a muscle fiber that is 6 in long is able to shorten 3 in, whereas a fiber that is 4 in long is able to shorten only 2 in. The significance of the preceding example is apparent if one considers that a hypothetical muscle with long fibers is able to move the bony lever to which it is attached through a greater distance than a muscle with short fibers. However, the relationship between the length of a muscle fiber and the distance it is able to move a bony lever is not always a direct relationship. The arrangement of the muscle fibers affects the length-shortening relationship and therefore must be taken into consideration.

Arrangement of fasciculi, like the length of the muscle fibers, varies among muscles. The fasciculi may be parallel to the long axis of the muscle (Fig. 3–9a), may spiral around the long axis (Fig. 3–9b), or may be at an angle to the long axis (Fig. 3–9c). Muscles that have a parallel fiber arrangement (parallel to the long axis and to each other) are designated as **strap** or **fusiform** muscles. In strap muscles, such as the sternocleidomastoid, the fasciculi are long and extend throughout the length of the muscle. However, in the rectus abdominis, which also is considered to be a strap muscle, the fasciculi are divided into short segments by fibrous intersections. In fusiform muscles, most but not all of the muscle fibers extend throughout the length of the muscle. Generally, muscles with a parallel fiber arrangement will produce a greater ROM of a bony lever at a joint than muscles with the same cross-sectional area but with a different fiber arrangement.

Muscles that have a fiber arrangement oblique to the muscle's long axis are called **unipennate, bipennate,** or **multipennate** muscles because the fiber arrange-

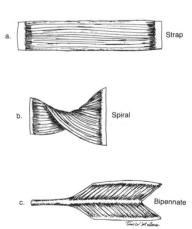

Figure 3–9. Arrangement of fasciculi in a muscle. a. Parallel arrangement. b. Spiral arrangement. c. Bipennate arrangement.

ment resembles that found in a feather, and the word "pennate" is derived from the Latin word for feather. The fibers that make up the fasciculi in pennate muscles are usually shorter and more numerous than the fibers in many of the strap muscles. In unipennate muscles such as the flexor pollicis longus, the obliquely set fasciculi fan out on only one side of a central muscle tendon. In a bipennate muscle, such as the gastrocnemius, the fasciculi are obliquely set on both sides of a central tendon. In a multipennate muscle, such as the deltoid, the oblique fasciculi converge on several tendons. The oblique set of the muscle fibers in a pennate muscle disrupts the direct relationship between the length of the muscle fiber and the distance that the total muscle can move a bony part. Only a portion of the force of the pennate muscles goes toward producing motion of the bony lever. However, because pennate muscles usually have a large number of muscle fibers, they are able to transmit a large amount of force to the tendon to which they attach.

Muscular Connective Tissue

Muscles and muscle fibers, like other soft tissues in the body, are surrounded and supported by connective tissue. The sarcolemma of individual muscle fibers is surrounded by connective tissue called the **endomysium,** and groups of muscle fibers (fasciculi) are covered by connective tissue called the **perimysium.** The endomysium and perimysium are continuous with the outer connective tissue sheath called the **epimysium,** which envelops the entire muscle (Fig. 3–10). Continuations of the outer sheath form the tendons that attach each end of the muscle to the bony components. Tendons are attached to bones by **Sharpey's fibers,** which become continuous with the periosteum.

Other connective tissue associated with muscles is in the form of fasciae, aponeuroses, and sheaths. Fasciae can be divided into the following two zones: superficial and deep. The zone of superficial fasciae, which is composed of loose tissue, is located directly under the dermis. This zone contributes to the mobility of the skin, acts as an insulator, and contains skin muscles such as the platysma in the neck. The zone of **deep fasciae** is composed of compacted and regularly arranged collaginous fibers. The deep fasciae attach to muscles and bones and may form tracts or bands and retinacula. For example, the deep femoral fasciae in the lower extremity forms a tract known as the **iliotibial tract or band.** This tract transmits the pull of two of the lower-extremity muscles to the bones of the leg (Fig. 3–11). Retinacula are formed by localized transverse thickenings of the fasciae, which form a loop that is attached at both ends to bone (Fig. 3–12a). The tunnels or osseofibrous channels formed by retinacula retain or prevent tendons from bowing out of position during muscle action (Fig. 3–12b). Sometimes deep fasciae is indistinguishable from aponeuroses, which are sheets of dense white compacted collagen fibers that attach directly or indirectly to muscles, fasciae, bones, cartilage, and other muscles. Aponeuroses distribute forces generated by the muscle to the structures to which they are attached.[1]

All of the connective tissue in a muscle is interconnected and constitutes the passive elastic component of a muscle. The connective tissues that surround the muscle fibers are in parallel with the muscle fibers. These tissues, plus the sarcolemma, intracellular elastic filaments made of the protein titin, and other structures (i.e., nerves and blood vessels), form the **parallel elastic component** of a muscle. When a muscle lengthens or shortens, these tissues also lengthen or shorten, that is, act in parallel with the muscle fibers. For example, the collagen fibers in the perimysium of fusiform muscles are slack when the sarcomeres are at rest but straighten out and become taut as sarcomere lengths increase. As the perimysium is lengthened, it also becomes stiffer (resistance to further elongation increases). The perimysium's increased resistance to elongation may prevent overstretching of the

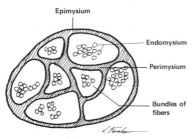

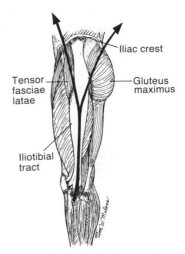

Figure 3-10. Muscular connective tissue. A schematic cross-sectional view of the connective tissue in a muscle shows how the perimysium is continuous with the outer layer of epimysium.

Figure 3-11. Iliotibial tract. A lateral view of the left lower limb showing the deep fascial iliotibial tract extending from the tubercle of the iliac crest to the lateral aspect of the knee. The right arrow represents the pull of the gluteus maximus. The left arrow represents the pull of the tensor fascia latae.

muscle fiber bundles.[12] When sarcomeres shorten from their resting position, the slack collagen fibers within the parallel elastic component buckle (crimp) even further. Whatever tension that might have existed in the collagen at rest is diminished by the shortening of the sarcomere. Given the many parallel elastic components of a muscle, the increase or decrease in passive tension can substantially affect the total tension output of a muscle.

The tendon of the muscle is considered to be in **series** with the contractile elements. This means that the tendon will be under tension when the muscle actively

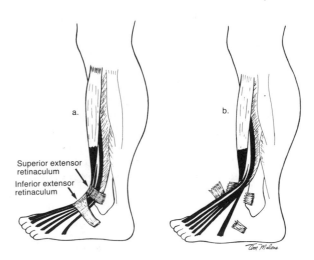

Figure 3-12. Retinacula. a. The superior and inferior retinacula are shown in their normal position, in which they form a tunnel for the tendons from the extensor muscles of the lower leg. b. When the retinacula are torn or removed, the tendons move anteriorly.

Superior extensor retinaculum
Inferior extensor retinaculum

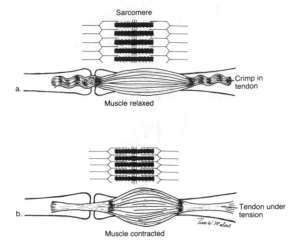

Figure 3–13. Series elastic component. a. The muscle is shown in a relaxed state with the tendon slack (crimping or buckling of collagen fibers has occurred). The sarcomere shown above the muscle shows minimal overlap of thick and thin filaments and little crossbridge formation. b. The muscle in an actively shortened position shows that the tendons are under tension and no crimp can be observed. The sarcomere shown above the muscle shows extensive overlap of filaments and crossbridge formation.

shortens (contracts). When the contractile elements in a muscle actively shorten, they exert a pull on the tendon. The pull must be of sufficient magnitude to take up the slack in the tendon so that the muscle pull can be transmitted through the tendon to exert a pull on the bony lever (Fig. 3–13a and b). The tendon also must be under tension when a muscle is exerting a pull on the bony lever as the muscle acts to control or brake the motion of the lever in an eccentric contraction. A tendon is under reduced tension when a muscle is completely relaxed and in a relatively shortened position. The tendon is one element of the series elastic component of a muscle (Fig. 3–13a and b).

MUSCLE FUNCTION

Muscle Tension

The most important characteristic of a muscle is its ability to develop tension and to exert a force on the bony lever. Tension can be either active or passive and the total tension that a muscle can develop includes both active and passive components. Total tension which was identified in Chapter 1 as Fms, is a vector quantity that has magnitude; two points of application (at the proximal and distal muscle attachments); an action line; and direction of pull.

Active Tension

Active tension refers to tension developed by the contractile elements of the muscle. Active tension in a muscle is initiated by crossbridge formation and movement of the actin and myosin. The amount of active tension that a muscle can generate depends on the frequency, number, and size of motor units that are firing, and at the sarcomere level, on the number of crossbridges that are formed.

- Tension may be increased by increasing the frequency of firing of a motor unit or by increasing the number of motor units that are firing.
- Tension may be increased by recruiting motor units with a larger number of fibers.
- The greater the number of crossbridges that are formed, the greater the tension.
- Muscles that have large physiologic cross sections are capable of producing more tension than muscles that have small cross sections.

Passive Tension

Passive tension refers to tension developed in the passive noncontractile components of the muscle. Passive tension in the connective tissue elements can be created by active and passive shortening or lengthening of a muscle. The connective tissue structures associated with the muscles either may add to the active tension produced by the muscle or may become slack and not contribute to the total tension. The total tension that develops during an active contraction of a muscle is a combination of the contractile (active) tension added to the noncontractile (passive) tension.

To produce motion of the bony component, the active tension must be sufficient to take up the slack in the tendon (Fig. 3–13). The total tension also must be sufficient to exert a force that is capable of overcoming any external resistance (i.e., gravity and/or a load) and inertia of the bony lever (to produce motion of a bony component). Other factors that determine the amount of tension exerted by a muscle are the speed and type of contraction. Total muscle tension and biomechanical variables such as the angle of pull, the moment arm (MA), and the length of the lever arm determine the muscles' ability to produce torque.

Length-Tension

There is a direct relationship between tension development in a muscle and the length of a muscle. There is an optimal length at which a muscle is capable of developing maximal tension.[12] The optimal length is close to what is known as the resting length of a muscle. Experimentally, the actual resting length is the length a muscle assumes when it is detached from the bone. Optimal length is slightly longer than resting length, about 1.2 times resting length. Muscles can develop maximal tension at optimal length because the actin and myosin filaments are positioned so that the maximum number of crossbridges can be formed. If the muscle is lengthened or shortened beyond optimal length, the amount of tension that the muscle is able to generate when stimulated diminishes (Fig. 3–14). When a muscle is lengthened beyond optimal length, there is less overlap between the actin and the myosin filaments and consequently fewer possibilities for crossbridge formation exist. However, the passive elastic tension in the parallel component may be increased when the muscle is elongated.

A similar loss of active tension or diminished capacity for developing tension occurs when a muscle is shortened from its optimal length. At the sarcomere level

Figure 3–14. Length-tension relationships at the sarcomere level. A loss of tension occurs whenever muscle fibers are excessively lengthened or shortened. When a muscle fiber is excessively lengthened there is no overlap of actin and myosin. When a muscle fiber is excessively shortened, overlap of actin and myosin takes place and tension decreases.

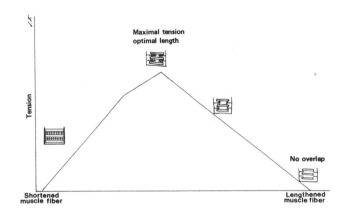

the distance between the Z bands is decreased and there is overlap of the filaments. The maximum number of crossbridges has been formed. Therefore no additional opportunities exist for crossbridge formation with further shortening. Consequently, no additional tension can be generated. The reasons that tension development actually declines are not totally clear. The optimal range in which a muscle fiber can develop maximum tension is very small, being in the vicinity of 0.2 μm just at or around optimal length. While muscle length is not the only factor affecting tension, the body unconsciously and/or consciously learns to place muscles at their optimal length for maximum tension development. Muscles are able to generate moderate tension in the lengthened range, maximum tension in the middle of the contractile range, and minimal tension in the shortened range during a concentric or active shortening contraction.

Active Insufficiency

Active insufficiency is the diminished ability of a muscle to produce or maintain active tension. At the sarcomere level, this state may occur either when a muscle is elongated to a point at which there is no overlap between the myofilaments or when the muscle is excessively shortened. Usually, this state occurs when a muscle has shortened to a point at which no further sliding of the filaments can take place. In many instances, muscles are arranged around a joint so that the muscle can be neither excessively elongated nor excessively shortened relative to its resting length. This arrangement is most effective for muscles that cross only one joint (one-joint muscles). Muscles that cross more than one joint (two- or multijointed muscles) may reach maximum elongation or shortening prior to the attainment of full ROM at all of the joints crossed by the muscle. Active insufficiency is most commonly encountered when the full ROM is attempted simultaneously at all joints crossed by a two- or multijoint muscle. Therefore, during active shortening a two-joint muscle will become actively insufficient at a point prior to the end of a joint range, when full ROM at all joints occurs simultaneously.[5] Active insufficiency also may occur in one-joint muscles, but is not as common.

> *Example:* (Fig. 3–15a). The finger flexors cross the wrist, carpometacarpal, metacarpophalangeal, and interphalangeal joints. When the finger flexors shorten they will cause simultaneous flexion at all joints crossed. If all of the joints are allowed to flex simultaneously, the finger flexors will be considerably shorter than optimal length and as a result will be actively insufficient. Normally, when the finger flexors contract the wrist is maintained in slight extension by the wrist extensor muscles (Fig. 3–15b). The wrist extensors prevent the finger flexors from flexing the wrist and therefore an optimal length of the flexors is maintained.

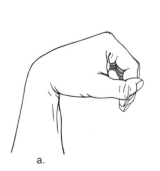

Figure 3–15. Active insufficiency. a. The individual is attempting to make a tight fist but cannot because the finger flexors are shortened over both the flexed wrist and fingers and have become actively insufficient. b. The length-tension relationship of the finger flexors has been improved by stabilization of the wrist in slight extension. The individual, therefore, is able to form a tight fist.

a.

b.

Figure 3-16. A concentric muscle contraction. When a muscle develops tension it exerts a pull on both its bony attachments.

Types of Muscle Action

Traditionally the word "contraction" has been used to describe different muscle actions, that is, isotonic contraction (constant tension); isometric contraction (constant length); concentric contraction (shortening contraction); and eccentric contraction (lengthening contraction). Recently, a suggestion has been made to replace the word contraction with the term "action."[13] The proposed change in terminology was made on the basis that the term "contraction" (which means "drawing together") does not adequately or accurately reflect what happens to the entire muscle during activity and is contradictory when used to describe eccentric muscle activity, that is, lengthening contraction. The term "contraction," however, is used so extensively in the literature that although the authors of this text agree that "muscle contraction" is probably not the best term to use, we will continue to do so. The use of the term "isotonic contraction" is not used by some authors, including the authors of this text, because it refers to equal or constant tension, which is unphysiologic. The tension generated in a muscle cannot be controlled or kept constant. Therefore the types of muscle actions that will be considered in the following section are concentric, isometric, and eccentric. Two other types of muscle action, isokinetic and isoinertial, which are sometimes referred to as types of muscle contraction, will be considered in a later section of this chapter.[14]

When bones into which muscles insert are free to move and the force generated by the muscle can produce torque sufficient to overcome the load, then the muscle will undergo a shortening or concentric contraction and both bones will be pulled toward each other and toward the center of the muscle (Fig. 3–16). Usually during functional activities, one attachment of a muscle is stabilized and the other attachment is free to move. If the proximal attachment is fixed and the distal attachment is free (as in an open kinematic chain), the muscle action will pull the distal bony component (Fig. 3–17a) toward the proximal bony component. If the distal attachment is fixed and the proximal bony component is free (in a closed kinematic chain), reverse action will occur (Fig. 3–17b) and the proximal bony component will be pulled toward the distal bony component. During concentric contractions

Figure 3-17. Normal and reverse action in a concentric contraction. a. Motion of the free distal bony lever occurs if the proximal bony lever is fixed and the torque of the muscle force (Fms) is sufficient to overcome the torque of the gravitational resistance. b. In reverse action, motion of the proximal bony lever occurs when the distal bony lever is fixed.

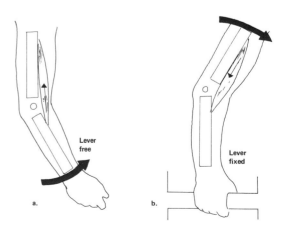

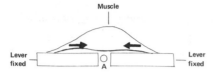

Figure 3–18. An isometric contraction. Both the distal and proximal bony levers are fixed and no visible motion occurs when the muscle develops active tension.

work is performed by the muscle as the muscle moves the bony lever through a distance in the direction of the muscle pull. The formula for work is

$$W = F \times d.$$

When both the distal and proximal attachments of a muscle are fixed and the torque produced by a muscle is *equal to* the torque of the resistance, the visible muscle length will remain unchanged, although shortening occurs at the myofibril level. When there is no visible change in the length of a muscle that is developing tension, the muscle action is called an **isometric contraction** (Fig. 3–18). No mechanical work is performed in an isometric contraction because work is equivalent to the product of $F \times d$. There is no distance involved in an isometric contraction because both bony components are fixed and do not move during the contraction. However, energy is being expended to produce crossbridge cycling.

When the force that a muscle generates is insufficient to offset an opposing force on a lever, the muscle will undergo a **lengthening or eccentric contraction**. In this type of contraction the muscle acts as a brake and controls the movement of a bony component. Crossbridges break and re-form as the muscle lengthens. When the hand holding a glass is moved from the face to the table, the flexor muscles of the elbow undergo a lengthening or eccentric contraction as they control the gravity-produced descent of the hand. In this activity the two ends of the muscle move apart or away from each other (Fig. 3–19). The muscle acts as the resistance to the gravitational effort and a second-class lever is formed. Eccentric muscle action like concentric muscle action can occur as motion of either the distal or proximal bony lever or normal and reverse action, respectively. The mechanical work that is done by a muscle during an eccentric contraction is called negative work because work is done *on* the muscle rather than *by* the muscle. The energy cost of an eccentric contraction is considerably less than that of a concentric contraction when equal loads are used.

The amount of tension that can be developed in a muscle varies according to the type of contraction. A greater amount of tension can be developed in an isometric contraction than in a concentric contraction, because in an isometric contraction none of the available energy is going into shortening of the muscle.[15] In

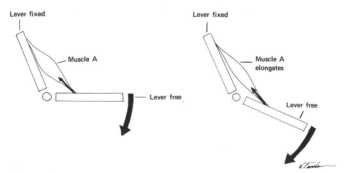

Figure 3–19. An eccentric contraction. The muscle elongates while continuing to produce active tension.

general, the tension developed in an eccentric contraction is greater than what can be developed in an isometric contraction. However, this relationship may not hold true for all muscles at all points in a joint's ROM.[16] The reason for greater tension development in a muscle during an eccentric contraction than in a concentric contraction is still under investigation.[17]

Angular Velocity and Torque

Another factor that affects the development of tension within a muscle is the speed of internal shortening of the myofilaments. The speed of internal shortening is the rate at which the myofilaments are able to slide past one another and form and re-form crossbridges. The maximum internal shortening speed occurs when there is no resistance to the sliding, such as occurs when a muscle fiber is separated from its bony attachment. However, in this situation there is no tension developed because there is no resistance. Conversely, tension may be developed when the resistance prevents visible shortening such as occurs in an isometric contraction. The greater the speed of shortening, the less tension. In a concentric muscle contraction, as the shortening speed increases, the tension decreases. In an isometric contraction the speed of shortening is effectively zero, and tension increases. In an eccentric contraction, as the speed of lengthening increases, the tension increases.[18]

Velocity is a vector quantity and has direction in addition to magnitude. When an object moves along a linear path in a certain direction, the velocity of the object may be calculated by determining the distance that the object travels in a certain period of time or, in other words, its rate of change of position. When an object is undergoing rotational motion, such as occurs when a bony component moves around a joint axis, the rate of change of position of the bony component is referred to as its **rotational (angular) velocity.**

The angular velocity that a bony lever attains during a dynamic muscle contraction (eccentric or concentric) is a function of the magnitude of the net torque applied to the lever. When the muscle force (Fms) and gravity and/or an external load (Fext) are applied in opposite directions (as they are commonly), net torque can be determined by the following equation:

$$\text{Net torque} = (\text{Fms} \times \text{MA}) - (\text{Fext} \times \text{MA})$$

The equation, however, can be seen as an oversimplification when one examines the components and subcomponents of the formula. These are

  I. Muscle torque
    A. Total muscle tension (Fms)
       1. Active tension
          a. Number of motor units recruited
          b. Fiber type of motor units recruited
          c. Type of contraction being performed
          d. Frequency of firing of motor units
       2. Passive elastic tension
          a. Parallel elastic tension
          b. Series elastic tension
       3. MA of the muscle
 II. Torque of the external force
    A. Magnitude of gravitational and/or external load (R)
    B. MA of gravitational and/or external load

A change in any of the components above can and will change the results of the formula and, consequently, net torque and angular velocity. During normal activities many and sometimes all of these components are changing simultaneously.

The impact of change in one component or subcomponent can be determined by holding the other components constant. In the following section only one factor will be changed at a time using as a reference the net torque formula.

The active tension component of total muscle tension is largely under the volitional control of the individual within physiologic limitations (i.e., conduction velocity of the axons, and rate of sacromere shortening). By increasing the number of motor units recruited or the frequency of firing of motor units, the force generated by the muscle can be increased. An increase in the muscle force will increase the net torque and the angular velocity if a concentric contraction is occuring. If a slower motion is desired, the person can reduce the number of motor units recruited and the frequency of firing, which decreases the muscle force and the net torque. When no motion is desired, such as in an isometric contraction, the force generated by the muscle can be reduced until the muscular force balances that of gravity and the applied load. If the person wishes to lower an object with control (an eccentric contraction), the muscle force can be minimized to a point where the torque of the external load exceeds the torque produced by the muscle. In this instance the net torque will have a negative value indicating that the angular velocity is in the direction opposite to the muscle pull. The larger the negative net torque, the faster the limb will move opposite to the muscle pull.

When the magnitude of the applied load on a bony lever is increased during a concentric contraction, the torque of the external force is increased and net torque and angular velocity will decrease if other contributors remain unchanged. A decrease in the magnitude of the applied load will decrease the torque of the external force and cause an increase in angular velocity. If the magnitude of the external force increases to the point where its torque is equivalent to the torque produced by the muscle, the net torque will be zero and an isometric contraction will occur. If the torque of the external force exceeds the torque produced by the muscle force, the muscle will undergo a lengthening or eccentric contraction.

In both the instance of the muscle force and the force of the external load, the MAs are generally changing as the lever moves and are position-dependent. The MA of the muscle and the MA of the external load may change in the same direction (both increase or decrease) or they may change in opposite directions. For this reason, the net torque of the moving lever will vary even when all other components are kept constant. If a constant velocity is desired, the changing MAs require a constant adjustment of active muscle tension. Active tension must be modified because the passive muscle tension is not under voluntary control.While the changing position of the joint will affect the MAs of both muscular and external forces, the muscular force also will be affected by the changes in the length of the muscle that accompany joint movement. Changing the length of a muscle will change both the active tension contribution of the muscle and the passive tension contribution.

The actual angular velocity achieved by a lever is affected by constantly changing variables. Consequently, changes in the velocity of the lever as it moves through the ROM are common.

Muscle Action Under Controlled Conditions

Isokinetics

Advances in technology have led to the development of testing and exercising equipment that provide for manipulation and control of some of the variables that affect muscle function. In **isokinetic exercise and testing**[19] or **isokinetic muscle contraction**,[15] the angular velocity of the bony component is preset and kept constant by a mechanical device throughout a ROM. The concept of an "isokinetic contraction" may not be so much a type of muscle action as it is a made-up variable. To maintain a constant velocity, the resistance produced by the isokinetic device is *directly proportional* to the torque produced by a muscle at all points in the ROM. Therefore, as the torque produced by a muscle increases, the magnitude of the resistance increases proportionately. Control of the resistance may be accomplished mechanically by using isokinetic devices such as a Biodex, Cybex, KINCOM, Orthotron, or others.

Experienced evaluators of human function may attempt to control manually the angular velocity of a bony component and apply resistance that is proportional to the torque produced by a muscle throughout the ROM. In manual-muscle testing the evaluator may apply manual resistance throughout the ROM to a concentric muscle contraction produced by the subject being tested. The evaluator's resistance must be proportional to the torque produced by the muscle being tested at each point in the ROM. If the evaluator successfully balances the torque output of the subject, a constant angular velocity is achieved. However, manually controlled angular velocity or manually applied resistance cannot be given with the same measure of precision or consistency that can be given by a mechanical device. Furthermore, manual resistance cannot be quantified as accurately as mechanical resistance.

The advantage of isokinetic exercise over free weight lifting through a ROM is that isokinetic exercise accommodates for the changing torques created by a muscle throughout the ROM. As long as the preset speed is achieved, the isokinetic device provides resistance that is proportional to the torque produced by a muscle at all points in the ROM. For example, the least amount of resistance is provided at a point in the ROM where the muscle has the least mechanical advantage, that is, at the extremes of the ROM. The resistance provided is greatest at the point in the ROM where the muscle has the largest mechanical advantage.[19]

The maximum isokinetic torques for concentric contractions obtained at high-angular velocities are less than the maximum isokinetic torques obtained at low-angular velocities. The slower the preset angular velocity of the device, the greater the torque output of the muscle. In fact, isometric torque values are higher than isokinetic concentric torque values at any other velocity for a particular point in the joint ROM. Therefore, the closer the angular velocity of a concentric isokinetic contraction approaches zero, the greater the isokinetic torque.[20,21]

Isokinetic equipment is used extensively for determining the amount of torque that a muscle can develop at different velocities, for strength training, and for comparing the relative strength of one muscle group with another. Some isokinetic devices permit quantification for testing eccentric muscle torque. In comparisons of isometric and isokinetic testing of the strength of back and arm muscles during lifting, significant differences have been found between peak isometric and peak isokinetic lifting strength capabilities of muscles throughout the lifting range. Isometric strength was found to be significantly higher than isokinetic concentric strength, and concentric isokinetic strength was found to decline as lifting speed increased.[22] Normal on-the-job lifting is not performed isokinetically and therefore isokinetic testing or exercising for the performance of lifting and other tasks may not be appropriate. The isometric strength, on the other hand, is an indicator of maximum capability at a particular posture during the lifting task.

Isoinertial

Another made-up variable, which has been called **isoinertial contraction**,[14] and a mechanical device (the B200 Isotation) for isoinertial testing and exercising have been developed to quantify dynamic muscle work. Isoinertial muscle action is defined as a type of muscle action in which muscles act against a constant load or resistance. If the torque produced by the muscle is equal to or less than the resistance, the muscle length does not change and the muscle contracts isometrically. If the torque produced by the muscle is greater than the resistance, the muscle short-ens and the muscle contracts concentrically. Isoinertrial muscle action is equivalent to normal muscle activity in which both isometric and concentric muscle contrac-tions are used in response to a constant load. This type of muscle action parallels normal functional activities.

Example: When a person lifts a constant external load the inertia of the load must be overcome at both extremes of the ROM. At the initial moment of lifting, the muscles contract isometrically as they attempt to develop the torque necessary to match inertial resistance. Once the resist-ance is matched, the muscles contract concentrically to produce sufficient torque to overcome the resistance and lift the load.[14]

This type of testing/exercising has been proposed as being superior to isokinetic exercise/testing because the isoinertial form of muscle action simulates real-life activities better than isokinetic exercise. The primary advantage of both isokinetic and the isoinertial devices is that they are capable of quantifying muscle activity.

Summary of Factors Affecting Active Muscle Tension

Active muscle tension is one component of total muscle tension, and total mus-cle tension is one component of a muscle's ability to move a bony lever. Active ten-sion is, however, the most variable of the components of total muscle tension.

How rapidly a muscle can develop maximal tension is affected by

- Recruitment order of the motor units: Units with slow conduction velocities are generally recruited first.
- Type of muscle fibers in the motor units: Units with FG muscle fibers can develop maximum tension more rapidly than units with SO muscle fibers; rate of crossbridge formation, breaking, and re-formation may vary.
- Size of motor units recruited: Units with a large number of muscle fibers can produce maximum tension more rapidly than units with a small number of fibers.

The magnitude of the active tension is affected by

- Size of motor units: Larger units produce greater tension.
- Number and size of the muscle fibers in a cross section of the muscle: The larger the cross section, the greater the amount of tension that a muscle may produce.
- The number of motor units firing: The greater the number of motor units firing in a muscle, the greater the tension.
- Frequency of firing of motor units: The higher the frequency of firing of motor units, the greater the tension.
- The length of a muscle: The closer the length to optimal length, the greater the amount of tension that can be generated.
- Number of crossbridges: The greater the number of crossbridges that are formed, the greater the amount of tension is generated.
- Fiber arrangement: A pennate fiber arrangement gives a greater number of muscle fibers, and therefore a greater amount of tension may be generated in a pennate muscle than in a parallel muscle.
- Type of muscle contraction: An isometric contraction can develop greater tension than a concentric contraction; eccentric contraction can develop greater tension than an isometric contraction.
- Speed: As the speed of shortening increases, tension decreases in a concen-tric contraction. As speed of active lengthening increases, tension increases in an eccentric contraction.

Classification of Muscles

Individual muscles may be named according to shape (rhomboids, deltoid) or location (biceps femoris or tibialis anterior); or by a combination of location and function (extensor digitorum longus or flexor pollicis brevis). Groups of muscles are categorized on the basis of either the actions they perform or the particular role they serve during specific actions. When muscles are categorized on the basis of action, muscles that cause flexion at a joint are categorized as **flexors**. Muscles that cause either extension or rotation are referred to as **extensors** or **rotators**. When muscles are categorized according to role, individual muscles or groups of muscles are described in terms that demonstrate the specific role that the muscle plays during action. When using this type of role designation, it does not matter what action is being performed (flexion, extension) but only what role the muscle plays.

> *Example:* The term **prime mover (agonist)** is used to designate a muscle that is responsible for producing a desired motion at a joint. If flexion is the desired action, the flexor muscles are the prime movers and the muscles (extensors) that are directly opposite to the desired motion are called the **antagonists**. The desired motion may not be opposed by the antagonists, but these muscles have the potential to oppose the action.

Ordinarily when an agonist is called to perform a desired motion, the antagonist is inhibited (reciprocal inhibition.) If, however, the agonist and the potential antagonist contract simultaneously, then **co-contraction** occurs (Fig. 3–20). Co-contraction of muscles around a joint can help to provide stability for the joint and represents a form of synergy that may be necessary in certain situations. Co-contraction of muscles with opposing functions can be undesirable when a desired motion is prevented by involuntary co-contraction such as occurs in disorders affecting the control of muscle function.

Muscles that help the agonist to perform a desired action are called **synergists.**

> *Example:* If flexion of the wrist is the desired action, the flexor carpi radialis and the flexor carpi ulnaris would be referred to as the "agonists" or "prime movers" because these muscles produce flexion. The wrist extensors would be the potential antagonists. The synergists that might directly help the wrist flexors would be the finger flexors.

Synergists may assist the agonist directly by helping to perform the desired action, such as in the wrist flexion example, or the synergists may assist the agonist indirectly either by stabilizing a part or by preventing an undesired action.

> *Example:* If the desired action is finger flexion, such as in clenching of the fist, the finger flexors, which cross both the wrist and the fingers, cannot function effectively (a tight fist cannot be achieved) if they flex the wrist and fingers simultaneously. Therefore, the wrist extensors are used synergistically to stabilize the wrist and to prevent the undesired motion of wrist flexion. By preventing wrist flexion, the synergists are able to maintain the finger flexors at an optimal length.

Sometimes the synergistic action of two muscles is necessary to produce a pure motion such as radial deviation (abduction) of the wrist. The radial flexor of the wrist acting alone produces wrist flexion and radial deviation. The radial extensor acting alone produces wrist extension and radial deviation. When the wrist extensor and the wrist flexor work together as prime movers to produce radial deviation of the wrist, the flexor action of the flexor is prevented or neutralized by the extensor action of the extensors, and pure motion of radial deviation results (Fig. 3–21).

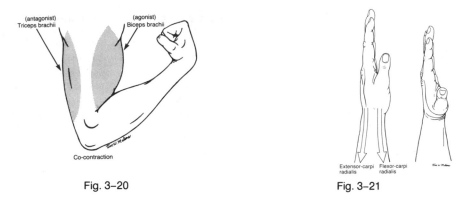

Fig. 3–20   Fig. 3–21

Figure 3–20. Co-contraction of agonist and antagonist.

Figure 3–21. Synergistic muscle activity. When the flexor carpi radialis and the extensor carpi radialis work synergistically, they produce radial deviation of the wrist.

In this example, the muscles that are the potential antagonists of the desired motion are the wrist extensor and flexor on the ulnar side of the wrist.

Placing muscles into functional categories such as flexor or agonists and antagonists helps to simplify the task of describing the many different muscles and of explaining their actions. However, muscles can change roles. A potential antagonist in one instance may be a synergist in another situation. Other systems that are used to group muscles are based on the characteristics of various muscles. In these systems an attempt has been made to divide muscles into basic types, for example, spurt and shunt, tonic and phasic, and fast twitch and slow twitch.

Spurt and Shunt Muscles

Spurt muscles have a proximal point of attachment that is far from the joint axis and a distal attachment that is close to the joint axis. These muscles have a relatively large rotatory component, although the rotatory component is not necessarily greater than the translatory component at all points in the range (Fig. 3–22a). **Shunt** muscles have a proximal attachment close to a joint axis and a distal attachment far away from the joint axis. These muscles help to maintain joint stability by

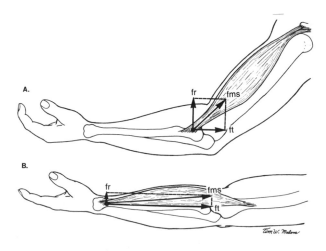

Figure 3–22. Spurt and shunt muscles. a. A spurt muscle. b. A shunt muscle.

providing a large amount of compression, since they have large translatory components (Fig. 3–22b). When muscle action is reversed (proximal lever moves and distal lever is fixed), the spurt muscle may act like a shunt muscle and vice versa.[1,5]

Most muscles have both spurt and shunt components, but some muscles have primarily spurt components and thus are designated as spurt muscles, whereas others have primarily shunt components and therefore are designated as shunt muscles. During very rapid swings of a bone, the tendency of centrifugal forces to separate joint surfaces may require the use of shunt synergists to assist the shunt component of a spurt muscle during the action.

Tonic and Phasic Muscles

Some of the characteristics used to differentiate between tonic and phasic muscles are fiber type and arrangement, and location and function of the muscle. Table 3–2 includes a few of the more commonly used distinguishing features.

Some of the characteristics of tonic and phasic muscles indicate that there is overlap between the spurt and shunt classification scheme and the tonic and phasic scheme. The **tonic** muscles appear to have some of the same characteristics as the shunt muscles, while **phasic** muscles appear similar to spurt muscles. Characteristics of the fast-twitch and slow-twitch muscles will not be presented here, because they are similar to the characteristics presented earlier under fiber types. The reader should refer to Table 3–1 to find out the similarities between the tonic and phasic classification and the fiber-type characteristics.

In addition to the features already presented, one group of researchers claims that tonic muscles react differently to training, overstress, and immobilization than phasic muscles. Tonic muscles react rapidly to training, while phasic muscles react slowly. Tonic muscles react to overstress by shortening (length less than normal) and tightness (muscle does not allow full ROM at joint), whereas phasic muscles react by a decrease in strength (weakness). Tonic muscles have a slow reaction to immobilization, whereas phasic muscles demonstrate a very fast or immediate reaction to immobilization.[24]

The basic structure, classification systems, and general properties of muscles have been introduced in the preceding sections. Although all skeletal muscles adhere to a general basic structural design, a considerable amount of variability exists among muscles in regard to the number, size, arrangement, and type of muscle fibers. Therefore, attempts to classify muscles into only two groups may be inappropriate. Based on the evidence that subpopulations of motor units from muscles rather than groups of muscles appear to work together for a particular motor task, a more appropriate way of describing muscle action might be in terms of motor

Table 3–2. DIFFERENTIAL CHARACTERISTICS OF TONIC AND PHASIC MUSCLES

	Tonic	Phasic
Fiber type	High proportion of SO fibers	High proportion of FG fibers
Fiber arrangement	Penniform	Parallel
Location	Deep and cross one joint	Superficial and cross more than one joint
Function	Stability	Mobility
Action	Extension, abduction, and external rotation	Flexion, adduction, and internal rotation

units.[6] However, more research needs to be performed in this area before a motor unit classfication system can be widely used and accepted.

Factors Affecting Muscle Function

In addition to the large number of factors that affect muscle function presented previously, there are a few other factors that need to be considered:

- Type of joint crossed by a muscle
- Location of the muscle relative to the joint
- Type and angle of attachment of the muscle relative to the joint
- Number of joints crossed by the muscle

Types of Joints and Location of Muscle Attachments

The type of joint affects the function of a muscle in that the structure of the joint determines the type of motion that will occur (flexion and extension) and the ROM. The muscle's location or line of action relative to the joint determines which motion the muscle will perform. Generally, muscles that cross the anterior aspect of the joints of the upper extremities, trunk, and hip are flexors, while the muscles located on the posterior aspect of these joints are extensors. Muscles located laterally and medially serve as abductors and adductors, respectively, and as rotators. Muscles that have their distal attachments close to a joint axis (spurt muscles) usually are able to produce a wide ROM of the bony lever to which they are attached. Muscles that have their distal attachments at a distance from the joint axis (shunt muscles), such as the brachioradialis, are designed to provide stability for the joint, since a large component of their force is directed toward the joint and this compresses the joint surfaces. A muscle's stability role may change throughout a motion with more compression being provided at one point than at another point within the range.

Usually one group of muscles acting at a joint is able to produce more torque than another group of muscles acting at the same joint. Disturbances of the normal ratio of agonist-antagonist pairs may create a muscle imbalance at the joint and may place the joint at risk for injury. Agonist-antagonist strength ratios for normal joints are often used as a basis for establishing treatment goals following an injury to a joint. For example, if the shoulder joint were to be injured, the goal of treatment might be to strengthen the flexors and extensors at the injured joint so that they had the same strength ratio as at the uninjured joint.

Number of Joints

Muscles that cross more than one joint are economic, since they are able to produce motion at more than one joint. These muscles, however, are susceptible to development of **active insufficiency.** Other muscles (synergists) help to prevent active insufficiency by acting to ensure that the two-joint muscle is maintained at the optimal length. Usually the multijoint muscle is able to function most effectively when it is shortened over one joint and lengthened over the other joint or joints. For example, the two-joint rectus femoris crosses the hip and the knee and is a hip flexor and knee extensor. The muscle is able to shorten and work more effectively at the knee if the muscle is lengthened over the hip (hip extended). If two joint muscles are required to actively shorten over one joint while being lengthened over the other joint (i.e., the rectus femoris in combined hip and knee extension), motor unit activity in the rectus femoris is decreased in comparison to motor unit activity when hip extension is not combined with knee extension. Hip extension apparently exerts an inhibitory control on the rectus femoris during combined hip and knee extension.[25]

During most activities the muscles work together to produce stability and mobility. However, it has been suggested that the body works on an efficiency principle in that it uses the least number of muscles able to perform a task. The same principle is operative in the selection of motor units in that the smaller units are usually selected first depending on the demands of the task. If a single joint motion is desired, a one-joint muscle is recruited first because recruitment of a multijoint muscle may require the use of either additional muscles or motor units from additional muscles to prevent motion from occurring at the other joint(s) crossed by the multijoint muscles.

Passive Insufficiency

Passive insufficiency occurs when an inactive, potentially antagonistic muscle is of insufficient length to permit completion of the full ROM available at the joints crossed by the passive muscle. Under normal conditions, one-joint muscles rarely, if ever, are passively insufficient. Two-joint or multijoint muscles, however, frequently are of insufficient length or extensibility to permit a full ROM to be produced simultaneously at all joints crossed by these muscles. When such a situation occurs, passive tension develops in the muscle. Sometimes, the passive tension developed in a passively insufficient muscle may be significant enough to cause the tendon to exert a pull on the bone. The passive tension developed in these muscles either may check further motion of the bony lever or, if the tension generated is sufficient, may actually pull the bony lever in the direction of the passive pull. When movement of the lever occurs as a result of the passive pull it is called the **tendon action** of a muscle or **tenodesis**. If the bone is not free to move, damage to the muscle being stretched may occur. Irreversible damage occurs at lengths beyond 1.5 times a muscle's resting length. Usually pain will signal a danger point in stretching and active contraction of the muscle will be initiated to protect the muscle.

> *Example:* If one's elbow is placed on the table with the forearm in a vertical position and the hand is allowed to drop forward into wrist flexion, one will notice that the fingers tend to extend (Fig. 3–23a).

Extension of the fingers is a result of the passive insufficiency of the finger extensors that are being stretched over the wrist. The increase in passive tension that results

Figure 3–23. Passive insufficiency. a. The finger extensors become passively insufficient as they are lengthened over the wrist and fingers during wrist flexion. The passive tension that is developed causes extension of the fingers (tenodesis). b. The finger flexors become passively insufficient as they are lengthened over the wrist and fingers during wrist extension. The passive tension developed in the finger flexors causes the fingers to flex.

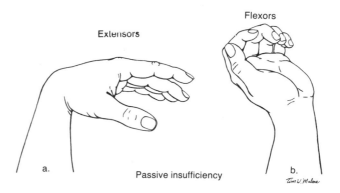

Extensors

Flexors

a.

Passive insufficiency

b.

from the stretching of the muscle over the wrist causes tenodesis. If the person moves his or her wrist backward into wrist extension, the fingers will tend to flex (Fig. 3–23b). Flexion of the fingers is a result of passive insufficiency of the finger flexors as they are stretched over the extended wrist. Although both active and passive insufficiency are related to the length of a muscle, each state involves very different elements in the muscle. Passive insufficiency involves the elastic or passive elements, while active insufficiency involves the active or contractile elements.

In spite of the different mechanisms involved in active and passive insufficiency these two phenomena are a common source of confusion because they often occur together. When a multijoint muscle on one side of a joint becomes actively insufficient due to excessive shortening, a multijoint muscle on the opposite side of the joint often becomes passively insufficient due to excessive lengthening.

> *Example:* When the wrist and fingers are simultaneously flexed, the formation of a tight fist is almost impossible. If the fingers are fully flexed prior to flexing the wrist, the fingers have a tendency to extend when the wrist is flexed. If the wrist is flexed first, full flexion of the fingers to make a tight fist is impossible.

In each of the above activities the finger flexors are actively insufficient because they are attempting to produce a full ROM at all joints over which they cross. The tension-generating capacity of the flexors is diminished because maximum overlap of actin and myosin has occurred, all of the available crossbridges have been formed, and no additional sites for crossbridge formation are available when one tries to fully flex the fingers to make a tight fist.

At the same time that the finger flexors are actively insufficient, the inactive finger extensors are being passively stretched over all of the joints that they cross. The extensors are providing a passive resistance to wrist and finger flexion at the same time that the finger flexors are having difficulty performing the movement (Fig. 3–24). Passive insufficiency of the extensors is responsible for pulling the fingers into slight extension when the wrist is flexed prior to attempting finger flexion. Any muscle that becomes actively insufficient through excessive shortening is likely to simultaneously create a passive insufficiency in the opposite muscle group. The one exception to this relationship between opposing muscles is when active insufficiency occurs through overlengthening rather than overshortening. In this situation a two-joint muscle on one side of a joint is in an excessively lengthened position and cannot generate tension (active insufficiency) because of insufficient overlap of actin and myosin. The muscle on the opposite side of the joint instead of being lengthened is passively shortened and slack and thus does not become passively insufficient (Fig. 3–25).

The combination of excessive lengthening and attempted shortening is threatening to the integrity of the muscle and such positions are not usually encountered in normal activities of daily living but may be encountered in sports activities.

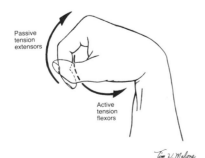

Figure 3–24. Passive and active insufficiency. The individual is attempting to make a tight fist but cannot because the finger flexors are shortened over both the flexed wrist and fingers and have become actively insufficient. The finger extensors are lengthened over both the wrist and the fingers and are passively insufficient.

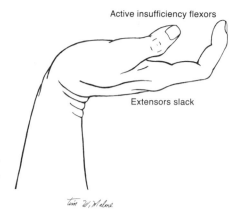

Active insufficiency flexors

Extensors slack

Figure 3-25. The finger flexors have been excessively lengthened and are unable to develop tension and thus are actively insufficient. The extensors are in a slackened position.

Sensory Receptors

Two important sensory receptors, the Golgi tendon organ and the muscle spindle, affect muscle function. The **Golgi tendon organs,** which are located in the tendon at the myotendinous junction, are sensitive to tension and may be activated either by a muscle contraction or by an excessive stretch of the muscle. When the Golgi tendon organs are excited, they send a message to the nervous system to inhibit muscle contraction.

The **muscle spindles,** which consist of 2 to 10 specialized muscle fibers enclosed in a connective-tissue sheath, are interspersed throughout the muscle. These fibers are sensitive to the length and the velocity of lengthening of the muscle fibers, and they send messages to initiate muscle contraction. Once the muscle fiber shortens, the spindles stop sending messages because they are no longer stretched. The muscle spindle is responsible for sending the message to contract when the tendon of a muscle is tapped with a hammer. The quick stretch of the muscle caused by tapping the tendon activates the muscle spindles, and the muscle responds to the spindle message by a brief contraction. This response is called by various names, for example, deep tendon reflex (DTR), muscle spindle reflex (MSR), or simply stretch reflex. Both the Golgi tendon organs and the muscle spindles help to protect the muscle from injury by monitoring changes in muscle length.

The presence of the stretch reflex is beneficial for preventing muscle injury but presents a problem for treatment programs and/or fitness programs in which stretching of a muscle is desirable for improving flexibility and restoring a full range of joint motion. Muscle contraction or reflex activation of motor units during intentional stretching of a muscle creates a resistance to the stretching procedure and makes stretching more difficult and possibly ineffective. Methods of stretching that may prevent reflex activity and motor unit activation during stretching are being investigated.[26,27] The noncontractile components of a muscle also provide resistance to stretching and need to be considered when a muscle-stretching program is implemented.

Joint receptors may have an influence on muscle activity through their signals to the central nervous system. Swelling of the joint capsule and noxious stimuli such as pinching of the capsule will cause reflex inhibition of muscles.

Example: Nocioceptors and other receptors in and around the knee joint can have flexor excitatory and extensor inhibitory action. Even a small joint effusion which is undetectable to the naked eye can cause inhibition.[28-30]

The effects of the sensory receptors on muscle activity adds the involuntary control element of muscle function to the factors previously discussed. A review of the recent literature related to motor control or "movement science" is beyond the scope of this text, but some aspects of motor control will be presented in Chapter 13.

Effects of Immobilization, Injury, and Aging
Immobilization

Immobilization affects both muscle structure and function. The effects of immobilization are dependent on immobilization position (lengthened or shortened), percentage of fiber types within the muscle, and length of the immobilization period. Studies focusing on single muscle fibers and on whole muscles have found that immobilization in a shortened position produces the following structural changes:

- Decrease in the number of sarcomeres[31–33]
- Increase in sarcomere length[31,33,34]
- Increase in the amount of perimysium[34]
- Thickening of endomysium[34]
- Increase in ratio of collagen concentration
- Increase in ratio of connective tissue to muscle fiber tissue
- Loss of weight and atrophy[33,35,36]

Changes in function that result from immobilization in the shortened position reflect the structural changes. The decrease in the number of sarcomeres coupled with an increase in the length of sarcomeres brings muscle to a length wherein it is capable of developing maximal tension in the immobilized position. If sarcomeres were not lost, the muscle would be actively insufficient (maximum overlap of actin and myosin). The loss of sarcomeres displaces the length-tension curve of the muscle so that the maximum tension generated corresponds to the immobilized position. Therefore the muscle is able to generate maximal tension in the shortened position. Although this altered capacity for developing tension may be beneficial while the muscle is immobilized in the shortened position, the muscle will not be able to function effectively at the joint it crosses immediately following cessation of the immobilization. Furthermore, the overall tension-generating capacity of the muscle is decreased. The increase in connective tissue in relation to muscle fiber tissue results in increased stiffness to passive stretch. The reason for increased fatigability of muscle following immobilization in the shortened position is still being investigated.

Muscles immobilized in the lengthened position exhibit fewer structural and functional changes than muscles immobilized in the shortened position. The primary structural changes are an increase in the number and decrease in the length of sarcomeres, and muscle hypertrophy that may be followed by atrophy.[31,34,36] The primary functional changes in muscles immobilized in a lengthened rather than in a shortened position are an increase in maximum tension-generating capacity and displacement of the length-tension curve close to the longer immobilized position. Passive tension in the muscle approximates that of the muscle prior to immobilization.[33]

In addition to the structural and functional changes that are described in the preceding paragraphs, some evidence suggests that FG fibers may be affected (atrophy) more than SO fibers by immobilization.[37] Also, recovery of a muscle from the effects of immobilization in the shortened position (development of preimmobilization maximal tension-generating capacity, relocation of the length-tension curve, and extensibility) may take a relatively long time.

Injury

Overuse may cause injury to tendons, ligaments, bursae, nerves, cartilage, and muscle. The common etiology of these injuries is repetitive trauma that does not allow for complete repair of the tissue. The additive effects of repetitive forces lead to microtrauma, which in turn triggers the inflammatory process and results in swelling. The tissue most commonly affected by overuse injuries is the musculotendinous unit. Tendons can fatigue with repetitive submaximal loading and are most likely to be injured when tension is applied rapidly and obliquely and the muscle group is stretched by external stimuli. Bursae may become inflamed with resultant effusion and thickening of the bursal wall as a result of repetitive trauma. Nerves can be subjected to compression injuries by muscle hypertrophy, decreased flexibility, and altered joint mechanics.[38]

Injuries to muscles may occur as a result of even a single bout of eccentric exercise. After 30 to 40 minutes of eccentric exercise (walking downhill), significant and sustained reductions in maximal voluntary contractions occur. Also a loss of coordination, postexercise muscle soreness (PEMS), or delayed-onset muscle soreness (DOMS), swelling, and a twofold increase in muscle stiffness have been reported. The PEMS or DOMS reaches a peak after 3 to 4 days postexercise.[17,38–40] PEMS or DOMS occurs in muscles performing eccentric exercise but not in muscles performing concentric exercise.[41] The search for a cause of PEMS or DOMS is still under investigation. It is known to be related to the forces experienced by muscles and may be a result of mechanical strain in the muscle fibers and/or in their associated connective tissues. Morphologic evidence shows deformation called Z-line streaming and other focal lesions, following eccentric activity that induces soreness. Biomechanical and histochemical studies have demonstrated evidence for collagen breakdown and for other connective-tissue changes.[17]

Aging

The effect of aging on skeletal muscle is an increase in the concentration of connective tissue within the muscle belly, that is, endomysium and perimysium. The increase in connective tissue is correlated with an increase in muscle stiffness.[42] In animals, resistance exercise does not appear to have any deleterious effects on aging muscles and causes an increase in the size of muscle fibers. However, a more limited response to resistance training occurs in the elderly than in the young.[43]

SUMMARY

There are a great many factors that affect the function of the muscles. A large number of the basic elements of muscle structure and their relationships to muscle function have been presented in this chapter. The interrelationships between structure and function in muscles are complex and often indistinguishable. Muscles are more adaptable than the joints that they serve and are more complex. Artificial joints have been designed and used to replace human joints, but it is as yet impossible to design a structure that can be used to replace a human muscle.

At the present time our knowledge of muscle contraction at the sarcomere level is still incomplete. However, it is known that the structural characteristics of muscle fibers change in response to functional demands. Exercises performed against high resistance with low number of repetitions (weight lifting) will cause the fast-twitch muscle fibers to increase in size.[44] Because the amount of tension a muscle can produce is determined by the total cross-sectional area of active muscle fibers, the increase in muscle fiber size results in an increase in the tension output of the muscle. Changes in a muscle fiber's oxidative capacity have been found after endurance training exercises,[45] but no evidence has been found to suggest that the contractile speed of muscle fibers can be changed in training.[46]

All skeletal muscles adhere to the general principles of structure and function that have been presented in this chapter. The muscles produce motion at the joints as well as providing joint stability. The structure and function of specific muscles and the relationship of the muscles to specific joints will be presented in the following chapters. The role that muscles play in supporting the body in the erect standing posture and in moving the body (walking) will be explored in the last two chapters of this book.

REFERENCES

1. Williams, PL, Warwick, R, Dyson, M, and Bannister, LH (eds): Gray's Anatomy, ed 37. Churchill Livingstone, London, 1989.
2. Johnson, MA, Pogar, J, Weightman, D, and Appleton, D: Data on the distribution of fibre types in thirty-six human muscles: An autopsy study. J Neurol Sci 18:111, 1973.
3. Netter, FH: The Ciba Collection of Medical Illustrations, Vol 8. Ciba-Geigy Corp., New Jersey, 1987.
4. Nordin, M and Frankel, VH: Basic Biomechanics of the Musculoskeletal System, ed 2. Lea & Febiger, Philadelphia, 1989.
5. Gowitzke, BA and Milner, M: Scientific Basis of Human Movement, ed 3. Williams & Wilkins, Baltimore, 1988.
6. Van Zuylen, EJ, Gielen, AM, and Van Der Gon, JJD: Coordination and inhomogenous activation of human arm muscles during isometric torques. J Neurophys 60:1523–1548, 1988.
7. Rosse, C and Clawson, DK: The Musculoskeletal System in Health and Disease. Harper & Row, Hagerstown, Maryland, 1980.
8. Taylor, RG, et al: Fast and slow skeletal muscles: Contractility evaluated by paired stimuli in mice. Arch Phys Med 61:151, 1980.
9. Saltin, B, et al: Fiber types and metabolic potentials of skeletal muscles in sedentary man and endurance runners. Ann NY Acad Sci 301:3, 1977.
10. Eyzaguirre, C and Fidone, S: Physiology of the Nervous System, ed 2. Yearbook Medical Publishers, Chicago, 1975.
11. Colliander, EB, Dudley, GA, and Tesch, PA: Skeletal muscle fiber type composition and performance during repeated bouts of maximal concentric contractions. Eur J Appl Physiol 58:81–86, 1988.
12. Purslow, PP: Strain-induced reorientation of an intramuscular connective tissue network: Implications for passive muscle elasticity. J Biomech 22:21–31, 1989.
13. Cavanaugh, PR: On "muscle action" vs "muscle contraction." J Biomech 21:69, 1988.
14. Parnianpour, M, Nordin, M, Kahanovitz, N, and Frankel, V: The triaxial coupling of torque generation of trunk muscles during isometric exertions and the effect of fatiguing isoinertial moments on the motor output and movement patterns. Spine 13:982–990, 1988.
15. Knapik, JJ, et al: Muscle groups through a range of joint motion. Phys Ther 63:938–947, 1983.
16. Singh, M and Karpovich, PV: Isotonic and isometric forces of forearm flexors and extensors. J Appl Physiol 21:1435–1437, 1966.
17. Stauber, WT: Eccentric action muscles: physiology, injury and adaptation. Ex Sport Sci Rev 17:157–185, 1989.
18. Griffin, JW: Differences in elbow flexion torque measured concentrically, eccentrically, and isometrically. Phys Ther 67:1205–1208, 1987.
19. Hislop, H and Perrine, JJ: The isokinetic exercise concept. Phys Ther 47:114–117, 1967.
20. Murray, P et al: Strength of isometric and isokinetic contractions. Phys Ther 60:4, 1980.
21. Knapik, JL and Ramos, ML: Isokinetic and isometric torque relationships in the human body. Arch Phys Med Rehabil 61:64, 1980.
22. Kumar, S: Isometric and isokinetic back and arm lifting strengths: Device and measurement. J Biomech 21:35–44, 1988.
23. Ivey, FM, et al: Isokinetic testing of shoulder strength: Normal values. Arch Phys Med Rehabil 66:384–386, 1985.
24. Schmid, H and Spring, H: Muscular imbalance in skiers. Manual Med 2:23–26, 1985.
25. Yamashita, N: EMG activities in mono- and bi-articular thigh muscles in combined hip and knee extension. Eur J Appl Physiol 58:274–277, 1988.

26. Guissard, N, Duchateau, J, and Hainaut, K: Muscle stretching and motorneuron excitability. Eur J Appl Physiol 58:47–52, 1988.
27. Entyre, BR and Abraham, LD: Antagonist muscle activity during stretching: A paradox revisited. Med Sci Sports Ex 20:285–289, 1988.
28. Young, A, Stokes, M, Iles, JF, and Phil, D: Effects of joint pathology on muscle. Clin Orthop Rel Res 219:21–26, 1987.
29. Spencer, J, Hayes, KC, and Alexander, IJ: Knee joint effusion and quadriceps reflex inhibition in man. Arch Phys Med Rehabil 65:171–177, 1984.
30. Stokes, M and Young, A: The contribution of reflex inhibition to arthogenous muscle weakness. Clin Sci 67:7–14, 1984.
31. Tabary, JC et al: Physiological and structural changes in the cats soleus muscle due to immobilization at different lengths by plaster casts. J Physiol 224:231–244, 1987.
32. Wills, et al: Effects of immobilization on human skeletal muscle. Orthop Rev XI(11):57–64, 1982.
33. William, PE and Goldspink, G: Changes in sarcomere length and physiological properties in immobilized muscle. J Anat 127:459–468, 1978.
34. Williams, PE and Goldspink, G: Connective tissue changes in immobilized muscle. 138:343–350, 1984.
35. Witzman, FA: Soleus muscle atrophy induced by cast immobilization: lack of effect by anabolic steroids. Arch Phys Med Rehabil 69:81–85, 1988.
36. Booth, F: Physiologic and biomechanical effects of immobilization on muscle. Clin Orthop Rel Res 219:15–20, 1986.
37. Vaughan, VG: Effect of upper limb immobilization on isometric muscle strength, movement time and triphasic electromyographic characteristics. Phys Ther 69:36–80, 1989.
38. Herring, SA and Nilson, KL: Introduction to overuse injuries. Clin Sports Med 6:225–239, 1987.
39. Sargent, AJ and Dolan, P: Human muscle function following prolonged eccentric exercise. Eur J Appl Physiol 56:704–711, 1987.
40. Howell, JN, et al: An electromyographic study of elbow motion during postexercise muscle soreness. J Appl Physiol 1713–1718, 1985.
41. Buroker, KC, and Schwane, JA: Phys Sports Med 17:65–83, 1989.
42. Alnaqeeb, MA, Alzaid, NS and Goldspink, G: Connective tissue changes and physical properties of developing and aging skeletal muscle. J Anat 139:677–689, 1984.
43. Brown, M: Resistance exercise effects on aging skeletal muscle in rats. Phys Ther 69:46–53, 1989.
44. Edgerton, VR: Mammalian muscle fiber types and their adaptability. Am Zool 18:113, 1978.
45. Costill, DL, et al: Adaptations in skeletal muscle following stretch training. J Appl Physiol 46:1, 1979.
46. Gollnick, PD, et al: The muscle biopsy: Still a research tool. Phys Sports Med 8:1, 1980.

STUDY QUESTIONS

1. Describe the contractile and noncontractile elements of a muscle.

2. Explain what happens at the sarcomere level when a muscle contracts.

3. Identify the antagonists in each of the following motions: abduction at the shoulder, flexion at the shoulder, and abduction at the hip.

4. Describe reverse action in the following muscles: triceps, biceps, gluteus medius, iliopsoas, and hamstrings. Give examples of activities in which reverse action of these muscle would occur.

5. Diagram the changes in the MA of the biceps brachii muscle from full elbow extension to full flexion. Explain how these changes will affect the function of the muscle.

6. Identify the muscles that are involved in lowering oneself into an armchair by using one's arms. Is the muscle contraction eccentric or concentric? Is the muscle acting in reverse action? Please explain your answer.

7. Give examples of the following: a tonic muscle, a spurt muscle, and a two-joint muscle.

8. Describe the position of the upper extremity in which each of the following muscles would be actively insufficient: biceps brachii, triceps brachii, and

flexor digitorum profundus. Describe the position in which the same muscles would be passively insufficient.

9. Explain how a motor unit composed of SO fibers differs from a motor unit composed of FG fibers.

10. List the factors that affect muscle function and explain how each factor affects muscle function.

11. Explain how isokinetic exercise differs from other types of exercise such as weight lifting and isometrics.

12. Explain isoinertial exercise.

13. Describe the effects of immobilization on muscles.

Chapter 4

■ ■ ■

The Vertebral Column

■

OBJECTIVES
Following the study of this chapter the reader should be able to:

Describe
1. The curves of the vertebral column using appropriate terminology.
2. The articulations of the vertebral column.

3. The major ligaments of the vertebral column.
4. The structural components of typical regional vertebrae.
5. The structural components of atypical regional vertebrae.
6. The structure of the intervertebral disk.
7. Motions of the vertebral column.
8. Lumbar-pelvic rhythm.
9. The neutral zone.
10. Thoraco lumbar fascia.

Identify
1. Structures that provide stability for the column.
2. Muscles of the vertebral column and their specific functions.
3. Ligaments that limit specific motions (flexion-extension, lateral flexion, and rotation).
4. Forces acting on the vertebral column during specific motions.

Explain
1. The relationship between the intervertebral and zygapophyseal joints during motions of the vertebral column.
2. The role of the intervertebral disk in stability and mobility.
3. How stability of the vertebral column is maintained.

Analyze
1. The effects of deficits using the model presented in this chapter.
2. The effects of an increased lumbosacral angle on the pelvis and lumbar vertebral column.

GENERAL STRUCTURE AND FUNCTION

The interrelationship between structure and function in the human body is clearly illustrated in a study of the vertebral column. The design of the structural elements and of the systems of linkages uniting the elements allows the column to fulfill a variety of functions. The column is able to provide a base of support for the head and internal organs; a stable base for the attachment of ligaments, bones, and muscles of the extremities, rib cage, and the pelvis; a link between the upper and the lower extremities; and mobility for the trunk. In addition, the column protects the spinal cord. Some of these functions require structural stability, whereas others require mobility. The structural requirements for stability frequently are opposite to the requirements for mobility; therefore, a structure that is capable of meeting both functions is complex. Each of the many separate but interdependent components of the vertebral column is designed to contribute to the overall function of the total unit, as well as to perform specific tasks.

Structure

The vertebral column is composed of 33 short bones called vertebrae and 23 intervertebral disks. The column is divided into the following five regions: cervical, thoracic, lumbar, sacral, and coccygeal (Fig. 4–1). The vertebrae adhere to a common basic structural design but show regional variations in size and configuration. The vertebrae increase in size from the cervical to the lumbar region and decrease in size from the sacral to coccygeal region (Fig. 4–2). Twenty-four of the vertebrae in the adult are distinct entities. Seven vertebrae are located in the cervical region,

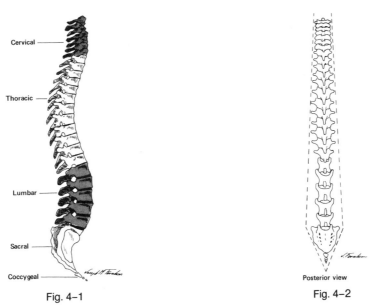

Fig. 4-1   Fig. 4-2

Figure 4-1. Five distinct regions of the vertebral column.

Figure 4-2. A cephalocaudal increase in the size of the vertebrae from the cervical to the lumbar region. The vertebrae show a decrease in size from the first sacral vertebra to the last coccygeal vertebrae.

twelve in the thoracic region, and five in the lumbar region. Five of the remaining nine vertebrae are fused to form the sacrum, while the remaining four form the coccygeal vertebrae. Structural abnormalities in the lower lumbar, sacral, and coccygeal regions are a common finding.[1]

Primary and Secondary Curves

When the vertebral column is viewed from the posterior aspect, all regions together present a single vertical line that bisects the trunk (Fig. 4-3). When the column is viewed from the side, a number of curves that vary with age are evident. The vertebral column of a baby at birth exhibits one long curve that is convex posteriorly. However, when the column of an adult is viewed from the side, four distinct anterior-posterior curves are evident (Fig. 4-4). The two curves (thoracic and sacral) that retain the original posterior convexity throughout life are called **primary curves**, while the two curves (cervical and lumbar) that show a reversal of the original posterior convexity are called **secondary curves**. Curves that have a posterior convexity (anterior concavity) are referred to as **kyphotic curves**, while curves that have a posterior concavity (anterior convexity) are called **lordotic curves**. The secondary or lordotic curves develop as a result of the accommodation of the skeleton to the upright posture. The superior secondary curve in the cervical region develops as the infant begins to hold his or her head up against gravity. The inferior secondary curve in the lumbar region develops as the infant begins to walk and hold his or her trunk upright. These curves continue to develop until growth stops somewhere between the ages of 12 and 17. The curves are interdependent, and if the head is to remain balanced over the sacrum, the region between the head and the pelvis behaves as if it were part of a closed kinematic chain. Changes in the

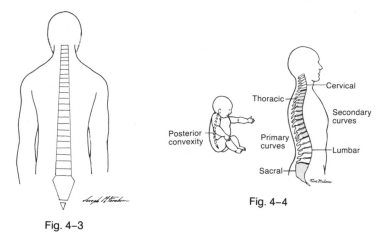

Fig. 4–3

Fig. 4–4

Figure 4–3. From the posterior aspect, the vertebral column appears as a vertical line that divides the trunk into two symmetric parts.

Figure 4–4. Primary and secondary curves. The shaded areas represent the primary curves.

position of any one segment will result in changes in position of adjacent superior or inferior segments.

Articulations

There are two main types of articulations found in the vertebral column: **cartilaginous** joints between the vertebral bodies and the interposed disks; and **diarthrodial**, or **synovial**, joints between the superior articular processes (facets) of one vertebra and the inferior articular processes of an adjacent vertebra above. The joints between the articular processes are called the **zygapophyseal** (apophyseal or facet) joints. All of the facet joints, except for the joint between the first two cervical vertebrae, are plane synovial joints. Synovial joints also are present where the vertebral column articulates with the ribs (Chapter 5) and with the skull. The sacroiliac joints, which are part synovial and part fibrous, are found where the vertebrae articulate with the pelvis.

Generally motion between any two vertebrae is extremely limited and consists of a small amount of gliding (translation) and rotation. According to White and Punjabi,[9] one vertebra can move in relation to an adjacent vertebra in six different directions (three translations and three rotations) along and around three axes (Fig. 4–5). The compound effects of these small amounts of translation and rotation at a series of vertebrae produce a large range of motion for the column as a whole. The motions available to the column as a whole may be likened to that of a joint with 3° of freedom, permitting flexion and extension, lateral flexion, and rotation. However, motions in the vertebral column often are coupled motions. **Coupling** is defined as the consistent association of one motion about an axis with another motion around a different axis. The motion being produced by an external load is termed the **main motion** and all accompanying motions are termed **coupled motions.** For example, when the lumbar spine is axially rotated it bends in the frontal and sagittal planes (axial rotation is coupled with lateral flexion and forward flexion).[9] Another example of a coupled motion is lateral flexion, which is accompanied

Figure 4–5. Translations and rotations of one vertebra in relation to an adjacent vertebra. a. Side-to-side translation (gliding) occurs in the frontal plane. b. Superior and inferior translation (axial distraction and compression) take place vertically. c. Anterior/posterior translation occurs in the sagittal plane. d. Side-to-side rotation (tilting) in a frontal plane occurs around an anterior/posterior axis. e. Rotation occurs in the transverse plane around a vertical axis. f. Anterior/posterior rotation (tilting) occurs in the sagittal plane around a frontal axis.

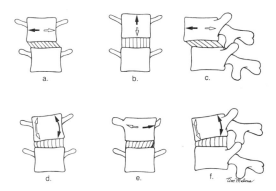

by axial rotation and forward flexion.[2–6,9] Coupling patterns as well as the types and amounts of motion that are available differ from region to region and are dependent on the spinal posture and curves, orientation of the articulating facets, fluidity, elasticity, and thickness of the intervertebral disks and extensibility of the muscles, ligaments, and joint capsules. Coupled motion patterns for left lateral flexion are presented in Table 4–1.

Ligaments and Muscles

The ligamentous system of the vertebral column is extensive and consists of two parts: the **intrasegmental system**, which binds together individual or adjacent vertebrae; and the **intersegmental system**, which binds a number of vertebrae into a unit. The integrity of these two systems is necessary for the stability of the column in motion or at rest; however, total support of the column requires muscular assistance. There are many muscles that contribute to stability as well as provide mobility for the column. The simplest classification of the vertebral muscles is on the basis of function, in which case there are forward flexors, lateral flexors, rotators, and extensors. Generally the forward flexors are located anteriorly, the extensors posteriorly, and the lateral flexors and rotators, on either side of the vertebral column. In addition to the muscles and ligaments, the thoracolumbar fascia has been identified as playing an important role in the stability of the vertebral column.[5,6]

Table 4–1. COUPLING PATTERNS[9]

Spinal Segment	Spinal Motion	Coupling Motion
Midcervical (C-2–C-5) Lower cervical (C-5–T-1) Upper thoracic (T-1–T-4)	Left lateral flexion	Rotation to the left (spinous processes move to the right) and forward flexion of vertebrae
Midthoracic (T-4–T-8) Lower thoracic (T-8–L-1)	Left lateral flexion	Rotation to either left or right (varies with individual) and forward flexion of vertebrae
Upper lumbar (L-1–L4) Lower lumbar (L-4–L-5) Lumbosacral (L-5–S-1)	Left lateral flexion	Rotation to the right and forward flexion of vertebrae

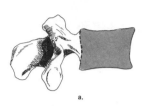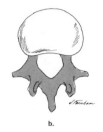

a.   b.

Figure 4-6. a. The anterior portion of a
vertebra is called the *vertebral body.* b.
The posterior portion of a vertebra is
called the vertebral arch.

A Typical Vertebra

The structure of a typical vertebra consists of two major parts: an anterior
cylindrically shaped structure called the **vertebral body** (Fig. 4–6a), and a posterior
irregularly shaped structure called the **vertebral arch** (Fig. 4–6b). The vertebral
body is composed of a block of trabecular or spongy bone, which is covered by a
layer of cortical bone. The cortical covering of the superior and inferior surfaces,
or plateaus, is thickened around the rim where the epiphyseal plates are located
and in the center by a layer of hyaline cartilage called the **cartilaginous end plate**
(Fig. 4–7).[7] Vertical, oblique, and horizontal trabecular systems that correspond to
the stresses placed on the bodies are found within the spongy bone (Fig. 4–8). The
vertical systems help to sustain the body weight and resist compression forces (Fig.
4–9). The other trabecular systems help to resist shearing forces. When the trabecular systems are viewed together (Fig. 4–8), an area of weakness is evident in
the anterior portion of the body, while areas of strength are demonstrable where
the trabeculae cross each other.[8] The area of weakness is a potential site for collapse
of the vertebrae (compression fracture).

The vertebral arch is a more complex structure than the body, since it has many
projections, including four articular and three nonarticular processes (Fig. 4–10).
The three nonarticular processes, two transverse and one spinous, provide sites for
the attachment of ligaments and muscles. The transverse processes divide the arch
into anterior and posterior portions. The portion of the arch anterior to the transverse processes consists of the pedicles, which attach the arch to the upper posterior wall of the vertebral body. The portion of the arch posterior to the transverse
processes consists of the laminae. The posterior portion of the laminae that is
located between the superior and inferior articular processes on each side is called
the **pars articularis.** The laminae join to form the peak of the arch and continue
posteriorly to form the spinous process (Fig. 4–11). The trabecular systems of the
vertebral arch extend into the vertebral body and the area where the transverse and
articular processes (facets) arise is reinforced by many crossing trabeculae.[8] (Fig.
4–8).

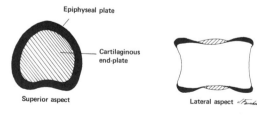

Epiphyseal plate

Cartilaginous
end-plate

Superior aspect

Lateral aspect

Figure 4-7. The cartilaginous end-plates
and epiphyseal plates are located on the
superior and inferior vertebral plateaus.

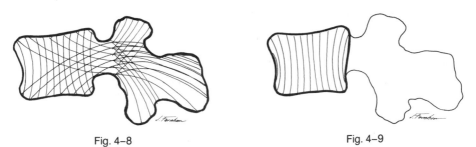

Fig. 4–8   Fig. 4–9

Figure 4–8. Schematic representation of the internal architecture of a vertebra. The various trabeculae are arranged along the lines of force transmission.

Figure 4–9. The vertical trabeculae of the vertebral bodies are arranged to resist compressive loading.

The Intervertebral Disk

The intervertebral disks, which make up about 20 to 33 percent of the length of the vertebral column, increase in size from the cervical to the lumbar region.[9] The disk thickness varies from approximately 3 mm in the cervical region to about 9 mm in the lumbar region.[8] While the disks are largest in the lumbar region and smallest in the cervical region, the ratio between disk thickness and vertebral body height is greatest in the cervical and lumbar regions and least in the thoracic region.[8] The greater the ratio, the greater the mobility, and therefore the cervical and lumbar regions have greater mobility than the thoracic region.

Disk Structure

The disk is composed of two parts: a central portion called the **nucleus pulposus** and a peripheral portion called the **annulus fibrosus** (Fig. 4–12). The composition of the nucleus and annulus are similar in that they both are composed of water, collagen, and proteoglycans (PGs). However, the relative proportions of these substances and the types of collagen that are present differ in the two parts of the disk. Fluid and PG concentrations are highest in the nucleus and lowest in the outer annulus. Conversely, collagen concentrations are highest in the outer annulus and lowest in the nucleus pulposus.

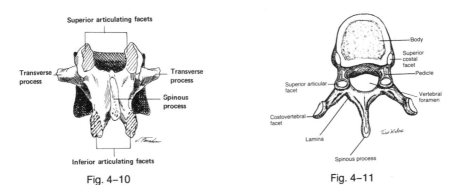

Fig. 4–10   Fig. 4–11

Figure 4–10. When a vertebra is viewed from the posterior aspect, seven projections are evident.

Figure 4–11. Superior view of a thoracic vertebra.

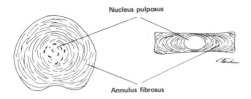

Figure 4-12. The nucleus pulposus is totally enclosed by the fibers of the annulus fibrosus.

Proteoglycan Content

The nucleus pulposus in healthy young adults is composed of a matrix that contains the following glycosaminoglycans: chondroitin 6-sulfate, chondroitin 4-sulfate, keratan sulfate, and hyaluronate. When glycosaminoglycans are linked to proteins, they form large molecules (PGs). The PG molecules have the capacity to attract and retain water. This ability is due to the concentration of chondroitin 4-sulfate within the molecule. If the concentration of chondroitin 4-sulfate is high, the disk will have a high fluid-attracting and fluid-maintaining capacity.

Fluid Content

The fluid content and composition of the disk varies with aging. In a newborn baby, the fluid content of the nucleus pulposus accounts for approximately 88 percent of the weight.[9,10] In a person at the age of 77, the fluid content may account for only 65 percent of the weight of the nucleus pulposus.[10-12] The composition (PG/collagen ratio) of the disk and the applied load affect the fluid content. Disks with a high PG/collagen ratio (young disks) will hold more fluid than older degenerated disks that have a low PG/collagen ratio. Under equal loads the percentage of fluid will be greatest in disks with the highest PG ratio. The fluid content of any disk may vary considerably during the day because the disk can lose a considerable amount of its fluid content when it is loaded continuously for a number of hours.

Collagen Content

Seven different types of collagen have been identified in intervertebral disks; however, type I and type II collagen predominate.[12] Type I collagen typically is found in tissues that are designed to resist tensile forces, that is, skin, tendon, and bone. Type II collagen is present in high concentrations in tissues that are required to resist compression.[6] The total amount of collagen in the disk increases steadily from about 6 to 25 percent of the dry weight in the center of the nucleus pulposus to 70 percent in the outer annulus.[12] The types of collagen that are present also vary between the nucleus and the annulus and also in different portions of the annulus. The nucleus contains primarily type II collagen,[12-14] which constitutes about 15 to 20 percent of the dry weight of the nucleus in the healthy young adult.[6] The annulus, which is subjected to both tensile and compression loading, contains both types of collagen, but type I collagen predominates.[6]

Maturation, aging, and disease as well as normal and abnormal stresses on the disk may affect the distribution and relative proportions of type I and type II collagen in the annulus. In adolescents and young adults, the concentration of type I collagen is greatest in the outer portion of the annulus and smallest next to the nucleus pulposus. The reverse is true for type II collagen. The concentration of type II collagen is smallest in the outer portion of the annulus and progressively increases toward the nucleus pulposus.

During maturation and aging changes have been found in the relative concentrations of type I and type II collagen in the outer anterior and posterior portions of the annulus fibrosus in lumbar disks. A greater proportion of type II relative to

Figure 4–13. Schematic representation of an intervertebral disk showing arrangement of lamellae in annulus fibrosus. a. The collagen fibers in any two adjacent concentric bands or sheets (lamellae) are oriented in opposite directions. Anterior portions of the lamellae have been removed to show the orientation of the collagen fibers.

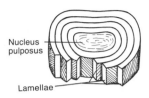

type I collagen has been found in the outer annulus portions of the anterior aspect of the disk, while the reverse was found for the posterior aspect of the disk. Changes also have been found in the distribution of the two types of collagen in annuli fibrosi located at the apices of abnormal curvatures of the vertebral column.

> *Example:* Disks of patients with an abnormal curvature of the spine show a difference in the distribution of types I and II collagen on opposite sides of the disk. The sides of the disks on the concave side of the curvature (which is subject to compression) have more type I collagen than the disks on the convex side of the curve.[12,14] The sides of disks on the convex side of the curve (which are subject to tension) show a decrease in the proportion of type I collagen in the outer portion of the annulus.[14]

These changes are the reverse of what one might expect (sides of disks subject to compression would contain more type II collagen, while sides of disks subject to tension would contain more type I collagen). The reason why these changes occur has not been found. Investigators have hypothesized that the changes may represent remodeling of the collagen in the disk in response to altered mechanical loading.[12,14] An alternative explanation is that the changes in the distribution of collagen preceded the development of a curvature and may have led to the formation of the abnormal curvature.

The collagen fibers of the annulus fibrosus are arranged in sheets called **lamellae** (Fig. 4–13). The lamellae are arranged in concentric rings that totally enclose the nucleus and keep it under constant pressure. Collagen fibers in adjacent rings are oriented in opposite directions at 120° to each other.[9] The annulus fibers are attached to the cartilaginous end plates on the inferior and superior vertebral plateaus of adjacent vertebrae and to the epiphyseal ring region by Sharpey's fibers.

Innervation and Nutrition

In the cervical and lumbar regions, the outer third of the annulus is innervated by branches from the vertebral and sinuvertebral nerves.[15] The sinuvertebral nerve also innervates the peridiskal connective tissue, and specific ligaments associated with the vertebral column. Neither blood vessels[16] nor nerves[17] have been found to penetrate as far as the nucleus. Nutrition of the disk is thought to occur by diffusion through the cartilaginous end plate.[12]

Vertebral End Plates

The vertebral end plates are thin layers of cartilage that cover the superior and inferior surfaces of the vertebral bodies. Sometimes, the vertebral end plate is considered to be part of the disk.[6] The chemical composition of the end plate is similar to that found in the intervertebral disk (PGs, collagen, and water). However, the disk contains more water.[18] The center of the end plate, like the center of the disk, has a higher fluid and PG content and lower collagen content than the peripheral areas.[18] The collagen fibers in the end plate are arranged horizontally to withstand the swelling pressure of the nucleus pulposus, which occurs when the nucleus

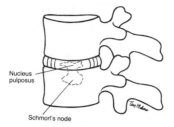

Nucleus pulposus

Schmorl's node

Figure 4–14. A schematic drawing of a Schmorl's node, in which nuclear material is extruded into the vertebral body.

imbibes water and/or is subject to axial compression. In the very young (0 to 6 months) the end plates are covered by multiple small holes and depressions that are thought to represent blood vessel markings.[19] The holes and depressions disappear by the second year of life, and ridges and sulci begin to develop at the periphery of the end plates, especially in the lumbar and lower thoracic regions. The ridges and sulci elaborate up until the age of 18 and give the vertebrae a toothlike appearance. The ridges and sulci disappear when ossification of the end plate occurs. Edelson[19] has suggested that the ridges and sulci may provide translational stability for the end plates in areas of the vertebral column in which the ribs and uncinate processes are not present to provide stability. The uncinate processes (joints of Luschka), which are found only in the cervical spine, prevent posterior translation of one cervical vertebrae on another and limit lateral bending. Following maturation at around 20 to 25 years, osteophytes begin to appear at the periphery of the end plate plateau predominantly in areas of greatest stress in the column (C-4 to C-5, T-9 to T-10, and L-3 to L-4).[19]

By the age of 50 years and with continued aging, osteolytic patches begin to appear in the end plates. These osteolytic patches are found most commonly in the cervical region and least commonly in the thoracic region.[19] Also intrusions of disk tissue into the end plate (Schmorl's nodes) may occur in areas of congenital weakness[20] (Fig. 4–14). These nodes weaken the end plate and may be a precursor to collapse of the end plate and disk degeneration. The nodes adversely affect the ability of the end plate to resist the swelling pressure of the nucleus.[18] If the disk ruptures and herniates, disk material may protrude through the end plate to make contact with the trabecular bone of the vertebral body, and sclerosis of the trabecular bone may occur in areas of contact.[20]

Articulations and Ligaments

The superior and inferior vertebral plateaus of the adjacent vertebral bodies and the interposed disks compose the **intervertebral joints.** These are cartilaginous joints of the symphysis type. The zygapophyseal joints, sometimes referred to simply as **apophyseal** or **facet joints,** are composed of the articulations between the

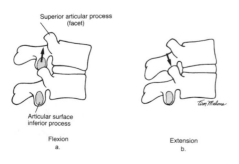

Superior articular process (facet)

Articular surface inferior process

Flexion
a.

Extension
b.

Figure 4–15. Zygapophyseal joints. a. The inferior articular process of the superior vertebra articulates with the superior process of the inferior vertebra. The shaded area on the superior vertebra represents the articular surface of the facet joint as the superior vertebra tilts anteriorly during forward flexion of the vertebral column. b. The articular surfaces are obscured as the superior vertebra tilts posteriorly during extension of the vertebral column. Note also the approximation of the spinous processes and narrowing of foramen that occurs during extension.

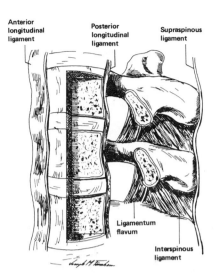

Figure 4–16. The anterior and posterior longitudinal ligaments are located on the anterior and posterior aspects of the vertebral body, respectively. The ligament flavum runs from lamina to lamina on the posterior aspect of the vertebral canal.

right and left superior articulating facets of a vertebra and the right and left inferior facets of the adjacent superior vertebra (Fig. 4–15). The four zygapophyseal joints are plane diarthrodial joints. In the lumbar region, rudimentary fibrous invaginations of the dorsal and ventral capsule have been identified as associated with the joints.[21] These invaginations may be involved in protecting articular surfaces that are exposed during flexion and extension of the vertebral column. The capsular ligaments of the zygapophyseal joints in the lumbar spine play a dominant role in resisting flexion of the intervertebral joints.[22–24] The zygapophyseal joints protect the disks from shear forces in the lumbar region by resisting a large proportion of the shear.[22,23]

There are six main ligaments that are associated with the intervertebral and zygapophyseal joints. They are the anterior and posterior longitudinal ligaments; the ligamentum flavum; and the interspinous, intertransverse, and supraspinous ligaments (Figs. 4–16 and 4–17).

The anterior and posterior longitudinal ligaments, which are part of the intersegmental ligamentous system, are associated with the intervertebral joints. The anterior longitudinal ligament (ALL) runs as a dense band along the anterior and lateral surfaces of the vertebral bodies from the sacrum to the second cervical vertebra. An extension of the ligament from C-2 to the occiput is called the anterior atlanto-axial ligament. The superficial fibers of the ALL are long and bridge several vertebrae, while the deep fibers are short and run between single pairs of vertebrae. The deep fibers also blend with the fibers of the annulus fibrosus, reinforcing the anterolateral portion of the disks and the anterior joint aspects. The ten-

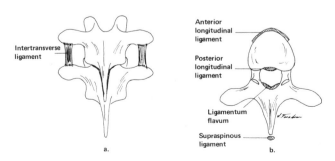

Figure 4–17. a. The intertransverse ligament connects the transverse processes. b. The relative positions of the other ligaments are shown in a superior view of the vertebra.

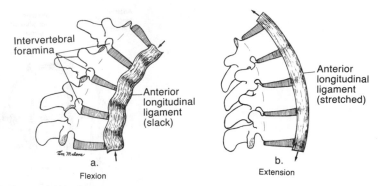

Figure 4–18. Anterior longitudinal ligament (ALL). a. The ALL is slack in forward flexion of the vertebral column. b. In extension of the vertebral column, the ligament is stretched.

sile strength of the ALL is greatest at the high cervical, lower thoracic, and lumbar regions. However, when tested in axial tension, the ligament demonstrates its greatest tensile strength (676 N) in the lumbar area.[25] The ligament is compressed in flexion (Fig. 4–18a) and is stretched in extension (Fig. 4–18b). The ligament may become slack in the neutral position of the spine when the normal height of the disks is reduced, such as might occur when the nucleus pulposus is destroyed and/or degenerated.[26] The ALL is reported to be twice as strong as the posterior longitudinal ligament.[25]

The posterior longitudinal ligament (PLL) runs within the vertebral canal along the posterior surfaces of the vertebral bodies from the second cervical vertebra to the sacrum. Short fibers attach the ligament to the posterior aspect of the intervertebral disk. Superiorly the ligament continues to the occiput, becoming the tectorial membrane at C-2. Inferiorly the ligament narrows, providing little support for the intervertebral joints in the lumbar region. The PLL's resistance to axial tension in the lumbar area is only ⅙ of that of the ALL (160 N compared to 676 N).[25] The PLL is stretched in flexion (Fig. 4–19a) where maximal strain in the ligament occurs and is slack in extension (Fig. 4–19b). However, if the axis of motion changes, as it does when the nucleus pulposus is destroyed either experimentally or by degenerative processes, and the axis moves posteriorly, the ligament may be stretched in extension.[26]

Three of the remaining ligaments, the ligamentum flavum, the interspinous, and the intertransverse ligaments belong to the intrasegmental system, while the supraspinous ligament belongs to the intersegmental system. The ligamentum flavum is a thick, elastic ligament, which is located on the posterior surface of the vertebral canal. The fibers of the ligament run within the canal from the second cer-

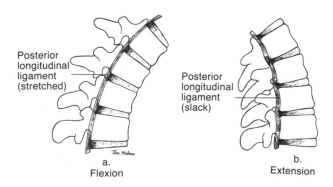

Figure 4–19. Posterior longitudinal ligament (PLL). a. The PLL is stretched during forward flexion of the vertebral column. b. The ligament is slack and may be compressed during extension.

vical vertebra to the sacrum, connecting laminae of adjacent vertebrae. Some fibers extend laterally to cover the articular capsules of the zygapophyseal joints. The ligamentum flavum is strongest in the lower thoracic and weakest in the midcervical region.[25] Although the highest strain in this ligament occurs during flexion when the ligament is stretched,[6,25] this ligament is under constant tension even when the spine is in a neutral position. The ligamentous tension creates a continuous compressive force on the disks, which causes the intradiskal pressure to rise. The raised pressure in the disk makes the disk stiffer and thus more able to provide support for the spine in the neutral position.[27]

The interspinous ligament, which is well developed only in the lumbar area, runs from the base of one spinous process to another. The ligament is stretched in forward flexion and is slack in extension. Its fibers resist the separation of the spinous processes that occurs during flexion. The ligament has a tensile strength of only 24 to 185 N and thus is potentially weaker in tensile strength than the ALL, PLL, and the ligamentum flavum.[25]

The supraspinous ligament runs along the tips of the spinous processes from the seventh cervical vertebra to the sacrum. The fibers of the ligament become indistinct in the lumbar area where they merge with the insertions of the lumbar muscles. In the cervical region the ligament becomes the **ligamentum nuchae.** The supraspinous ligament, like the interspinous ligament, is stretched in flexion and its fibers resist separation of the spinous processes during forward flexion. During hyperflexion the supraspinous and the interspinous are maximally stretched and are the first of the posterior ligaments to fail.[22]

The paired intertransverse ligaments, like the interspinous ligament, are well developed only in the lumbar area. This group of paired ligaments are alternately stretched and compressed during lateral bending. The ligaments on the right side are stretched during lateral bending to the left while the ligaments on the left side are slack and compressed during this motion. The ligaments on the right offer resistance to bending to the left. Conversely, the ligaments on the left side are stretched during lateral bending to the right and offer resistance to this motion.

The zygapophyseal joint capsules assist the ligaments in providing limitation to motion and stability for the vertebral column. However, the exact role that the capsules play and their strength in comparison to ligamentous supporting structures is still under investigation. The capsules are strongest in the thoracolumbar region and at the cervicothoracic junction.[25] These areas of the spine are the sites where the spinal configuration changes from a kyphotic to lordotic curve and from a lordotic to kyphotic curve, respectively. Therefore the potential exists for excessive stress in these areas. The joint capsules, like the supraspinous and interspinous ligaments, are vulnerable to hyperflexion, especially in the lumbar region. It has been suggested that the joint capsules in the lumbar region provide more restraint to forward flexion than any of the posterior ligaments because they fail after the supraspinous and interspinous ligaments when the spine is hyperflexed.[28] During axial loading the upper portion of the capsule is stretched because the height of the disk is reduced and the upper articular processes slide down on the lower processes. The superior portion of the capsule also is stretched during extension as one vertebra slides over another.[26] See Table 4–2 for a summary of the ligaments and their functions.

Function

Stability

The stiffness of the vertebral column is the column's ability to resist an applied load. Stiffness can be represented graphically by the slope of the stress-strain curve.[6] The steeper the slope of the curve, the stiffer the structure. The complexity

Table 4–2. LIGAMENTS OF THE VERTEBRAL COLUMN

Ligaments	Function	Region
Anterior longitudinal	Limits extension and reinforces anterior portion of annulus fibrosus	Axis to sacrum well-developed in lumbar and thoracic
Anterior atlantoaxial (continuation of anterior longitudinal ligament)	Limits extension	Axis and atlas
Posterior longitudinal	Limits flexion and reinforces posterior portion of annulus fibrosus	Axis to sacrum
Ligamentum flavum	Limits flexion, especially in lumbar area	Axis to sacrum
Ligamentum nuchae (continuation of the supraspinous ligaments)	Limits flexion	Cervical region
Supraspinous	Limits flexion	Thoracic and lumbar
Posterior atlantoaxial (continuation of ligamentum flavum)	Limits flexion	Atlas and axis
Tectorial membrane (continuation of posterior longitudinal ligament)	Limits flexion	Axis to occipital bone
Interspinous	Limits flexion	Primarily lumbar
Intertransverse	Limits lateral flexion	Primarily lumbar
Alar	Limits rotation of head to same side and lateral flexion to opposite side	Axis to skull

of the column has made accurate determinations of both the stiffness of the column as a whole and the contributions of various structures to stiffness very difficult to obtain.

Researchers investigating the spine have used small segments of the column, **motion segments,** to determine stiffness. A motion segment consists of two adjacent vertebrae and the intervening soft tissues.[27,29] By applying a specified load to a motion segment, an investigator can determine the stiffness of that particular segment. The sequential removal of ligaments, joint capsules, and portions of the disk followed by repeated measurements of stiffness yields information on how the stiffness of the segment is affected by the removal of the particular structure. Mathematical modeling, and the assessment of hysteresis and creep, have also been investigated in attempts to assess the stiffness of the spine.[30-36] Recently, Panjabi[9,37] has used the size of the **neutral zone** to provide a clinical measure of spinal stability.

The neutral zone is the range of motion through which the spine can be displaced from a neutral position to the point at which elastic deformation begins when a small load is applied. In a stress-strain curve the neutral zone would be represented by the toe region of the curve. Panjabi[37] has suggested that the existence of a large neutral zone indicates instability.

Instability of the vertebral column can be considered as a lack of stiffness, and an unstable structure is one that is not in an optimal state of equilibrium.[38] When the vertebral column is unstable, it exhibits a greater than normal (abnormal) range of motion.

The vertebral column is subjected to axial compression, tension, bending, torsion, and shear stress not only during normal functional activities but also at rest.[39] The column's ability to resist these loads varies among spinal regions and is dependent on the type, duration, and rate of loading; the person's age and posture; the condition and properties of the various structural elements (vertebral bodies, joints, disks, muscles, joint capsules and ligaments); and the integrity of the nervous system.[40]

Axial Compression

Axial compression (force acting through the long axis of the spine at right angles to the disks) occurs due to the force of gravity and forces produced by the ligaments and muscular contractions. Most of the compressive force is resisted by the disks and vertebral bodies, but the arches share some of the load in certain postures and during specific motions. The compressive load is transmitted from the superior end plate to the inferior end plate through the trabecular bone of the vertebral body and the cortical shell. The cancellous body contributes 25 to 55 percent of the strength of a lumbar vertebra under the age of 40 years and the cortical bone carries the remainder. After age 40, the cortical bone carries a greater proportion of the load as the trabecular bone's compressive strength and stiffness decrease with decreasing bone density.[33] Depending on the posture and region of the spine, the zygapophyseal joints carry some of the load in compression 0 to 33 percent. The spinous processes also may share some of the load when the spine is in hyperextension.

The nucleus pulposus acts as a ball of fluid that can be deformed by a compression force. The pressure created in the nucleus actually is greater than the force of the applied load.[41] When a weight is applied to the nucleus pulposus from above, the nucleus loses height as it exhibits a swelling pressure and tries to expand outward toward the annulus and the end plates (Fig. 4–20). As the nucleus attempts to distribute the pressure in all directions, stress is created in the annulus and central compression loading occurs on the vertebral end plates. The force of the nucleus on the annulus and the annulus on the nucleus form an interaction pair. Normally, the annulus and the end plates are able to provide sufficient resistance to the swelling pressure in the nucleus to reach and maintain a state of equilibrium. The pressure exerted on the end plates is transmitted to the superior and inferior

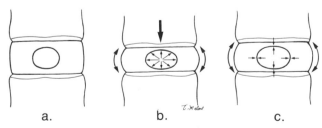

Figure 4–20. Compression of an intervertebral disk. a. In this schematic representation of a disk, the nucleus is shown as a round ball in the middle of the annulus fibrosus. b. Under compressive loading, the pressure is exerted in all directions as the nucleus attempts to expand. Tension in the annulus fibrosus rises as a result of the nuclear pressure. c. A force equal in magnitude but opposite in direction is exerted by the annulus on the nucleus, which restrains radial expansion of the nucleus and establishes equilibrium. The nuclear pressure is transmitted by the annulus to the end-plates.

vertebral bodies. The annulus fibrosus is under tensile stress and thus is able to better resist the compressive load. The disks and trabecular bone are able to undergo a greater amount of deformation without failure than the cartilaginous end plates or cortical bone when subjected to axial compression. The end plates are able to undergo the least deformation and therefore will be the first to fail (fracture) under high compressive loading. The disks will be the last to fail (rupture).

When the disks are subjected to a constant load by forces that are not large enough to cause permanent damage, the disks exhibit creep. Under sustained compressive loading such as incurred in the upright posture, the rise in the swelling pressure causes fluid to be expressed from the nucleus pulposus and the annulus fibrosus. The amount of fluid expressed from the disk depends both on the size of the load and the duration of its application. The expressed fluid is absorbed through microscopic pores in the cartilaginous end plate. When the compressive forces on the disks are decreased in the recumbent posture or absent in weightlessness, the disk imbibes fluid back from the vertebral body.[31] The recovery of fluid that returns the disk to its original state explains why a person getting up from bed is taller in the morning than in the evening. It also explains why an astronaut returning from weightlessness of space is taller on his return than on his departure. In the elderly the amount of creep that occurs is greater than in the young and the recovery from creep and hysteresis is slower.[31] The loss of height that occurs as people grow older is due to the fact that the nucleus loses a large proportion of its fluid-imbibing capacity with aging.

Bending

Bending causes both compression and tension on the structures of the spine. In forward flexion the anterior structures (anterior portion of the disk, anterior ligaments, and muscles) are subjected to compression, while the posterior structures are subjected to tension. The resistance offered to the tensile forces by collagen fibers in the posterior outer annulus fibrosus, zygapophyseal joint capsules and posterior ligaments help to limit extremes of motion and hence provide stability in flexion. The creep that occurs when the vertebral column is subjected to sustained loading in the fully flexed or fully extended postures, such as gardening and painting the ceiling, leads to an increase in the range of motion beyond normal limits. The elongation of the supporting structures that is caused by creep leads to a lack of stability and places the vertebral structures at risk of injury.

In extension the posterior structures generally are either unloaded or subjected to compression, while the anterior structures are subjected to tension. However, the reverse may occur if the axis of rotation for a particular region has been displaced due to abnormalities in structure.[42] Generally, resistance to extension is provided by the anterior outer fibers of the annulus fibrosus, zygapophyseal joint capsules, passive tension in the anterior longitudinal ligament, and possibly by contact of the spinous processes. In lateral bending, the ipsilateral side of the disk is compressed, that is, in right lateral bending the right side of the disk is compressed while the outer fibers of the left side of the disk are stretched. Therefore, the contralateral fibers of the outer annulus fibrosus and the contralateral intertransverse ligament help to provide stability during lateral bending by resisting extremes of motion.

Torsion

Torsional forces are created during axial rotation that occurs as a part of the coupled motions that take place in the spine. The torsional stiffness of the upper thoracic region T-1 to T-6 is similar in stiffness in flexion and lateral bending, but torsional stiffness increases from T-7/T-8 to L-3/L-4. The highest torsional stiffness is found at the thoracolumbar junction. Torsional stiffness is provided by the outer layers of both the vertebral bodies and intervertebral disks and by the orientation of the facets.[43] The outer shell of cortical bone reinforces the trabecular bone and provides resistance to torsion.[43] When the disk is subjected to torsion, one half of the annulus fibers resist clockwise rotations, while the other half resist counterclockwise rotations. It has been suggested that the annulus fibrosus may be the most effective structure in the lumbar region for resisting torsion.[44] However, the risk of rupture of the disk fibers is increased when torsion, heavy axial compression, and bending are combined.[34]

Shear

Shear acts on the midplane of the disk and tends to cause each vertebra to move anteriorly, posteriorly, or from side to side in relation to the inferior vertebra. In the lumbar spine the zygapophyseal joints resist some of the shear force while the disks resist the remainder. When the load is sustained, the disks exhibit creep and the zygapophyseal joints may have to resist all of the shear force.

Mobility

Motions at the intervertebral and zygapophyseal joints are interdependent. The amount of motion that is available is determined primarily by the size of the disks, while the direction of the motion is determined primarily by the orientation of the facets. The motion that takes place between the vertebral bodies at the intervertebral joints is similar to what occurs when a rubber ball is placed between two blocks of wood. The blocks may be tilted or rotated in any direction and may glide if the ball rolls. If the size of the ball is increased or decreased, the amount of tilting possible is increased or decreased, respectively.

Example: If the height of the disk is large in comparison to either the anterior-posterior or medial-lateral diameter of the disk (small round ball), the amount of sagittal or frontal plane motion available to the vertebral segment is large. If the diameter of the disk (large flatter ball) is larger than or equal to the height, motion of the vertebral segment is more restricted.

The motions of flexion and extension occur as a result of the tilting and gliding of a superior vertebra over the inferior vertebra. The nucleus pulposus acts like a pivot but, unlike a ball, is able to undergo greater distortion since it behaves as a fluid.

Regardless of the magnitude of motion created by the ratio of disk height to width, a gliding motion occurs at the zygapophyseal joints as the vertebral body tilts (rotates) over the disk at the intervertebral joint. The orientation of the zygapophyseal joint surfaces, which varies from region to region, determines the direction of the tilting (rotation) and gliding (translation) within a particular region. If the superior and inferior joint surfaces of three adjacent vertebrae lie in the sagittal plane, the motions of flexion and extension are facilitated (Fig. 4–21a). On the

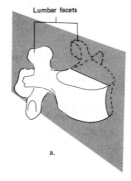
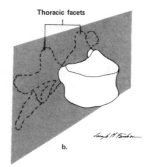

Figure 4-21. a. Sagittal plane orientation of the lumbar zygapophyseal facets favors the motions of flexion and extension. b. Coronal plane orientation of the thoracic zygapophyseal facets favors lateral flexion.

Lumbar facets

Thoracic facets

a.

b.

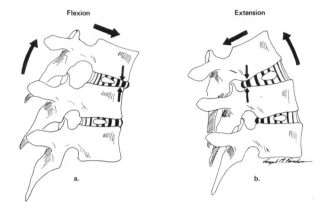

Flexion

Extension

a.

b.

Figure 4-22. a. The superior verte-bra tilts and glides anteriorly over the adjacent vertebra below during flexion. The anterior tilting and glid-ing cause compression and bulging of the anterior annulus fibrosus and stretching of the posterior annulus. b. In extension the superior vertebra tilts and glides posteriorly over the vertebra below. The anterior annu-lus fibers are stretched and the pos-terior portion of the disk bulges posteriorly.

other hand, if the joint surfaces are placed in the frontal plane the predominant motion that is allowed is that of lateral flexion (Fig. 4-21b).

Flexion

In vertebral flexion the anterior tilting and gliding of the superior vertebra causes a widening of the intervertebral foramen and a separation of the spinous proc-esses (Fig. 4-22a). Although the amount of tilting is partly dependent on the size of the disks, tension in the supraspinous and interspinous ligaments resists separa-tion of the spinous processes and thus limits the extent of flexion. Passive tension in the zygapophyseal joint capsules, ligamentum flavum, posterior longitudinal liga-ment, posterior annulus, and the back extensors also imposes controls on excessive flexion. Tension in the posterior ligaments can be produced by contractions of the hip extensors pulling downward on the pelvis. Tension in the thoracolumbar fascia produced by contractions of the transversus abdominis also can limit flexion by the pull of the fascia on the spinous processes in the lumbar area. The disks influence flexion because the anterior portion of the annulus fibrosus is compressed and bulges anteriorly during flexion while the posterior portion is stretched. In exten-sion the reverse situation occurs.

Extension

In extension the intervertebral foramen is narrowed and the spinous processes move closer together (Fig. 4-22b). The amount of motion available in extension, in addition to being limited by the size of the disks, is limited by bony contact of the spinous processes, tension in the zygapophyseal joint capsules, anterior fibers of the annulus, anterior trunk muscles, and the anterior longitudinal ligament. In general, there are many more ligaments that limit flexion than there are ligaments that limit extension. The only ligament that limits extension is the anterior longi-tudinal ligament. The lack of ligamentous restraints may be due in part to the fact that there are bony limitations to extension while there are no similar bony limita-tions for flexion.

Lateral Flexion

In lateral flexion the superior vertebra tilts and rotates over the adjacent ver-tebra below (Fig. 4-23). The annulus fibrosus is compressed on the concavity of the curve and stretched on the convexity of the curve. Passive tension in the annu-lus fibers, intertransverse ligaments, and anterior and posterior trunk muscles on the convexity of the curve limit lateral flexion. The rotation that accompanies lat-eral flexion differs slightly from region to region because of the orientation of the facets.

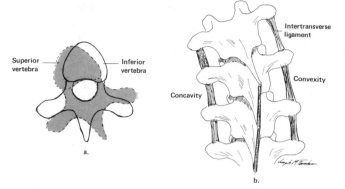

Figure 4-23. a. The superior vertebra tilts laterally and rotates over the adjacent vertebra below during lateral flexion. b. Lateral flexion and rotation of the vertebra are limited by tension in the intertransverse ligament on the convexity of the curve.

All intervertebral and zygapophyseal joint motion that occurs between the vertebrae from the second cervical to the fifth lumbar vertebra adheres to the general descriptions that have been presented. Regional variations in the structure, function, and musculature of the column will be covered in the following sections. Regional variations in the structure of the vertebrae are summarized in Table 4-3.

Table 4-3. REGIONAL VARIATIONS IN VERTEBRAL STRUCTURE

	Cervical Vertebrae	Thoracic Vertebrae	Lumbar Vertebrae
Body	Small 　Transverse diameter greater than anterior-posterior diameter and height Uncinate processes on posterolateral superior and inferior surfaces	Equal transverse and anterior-posterior diameter Anterior height greater than posterior Demifacets for articulation with the ribs on the posterolateral corner of the vertebral plateau in front of the inferior vertebral notch	Massive 　Transverse diameter greater than anterior-posterior diameter and height
Arches 　Spinous process	Short, slender, and extend horizontally Bifid	Slope inferiorly and overlap spinous process of adajcent vertebrae	Broad, thick, and extend horizontally
Transverse process	Foramen for vertebral artery, vein, and venous plexus Gutter for spinal nerves	Thickened ends with facets for articulation with costal tubercle	Long, slender, and horizontal
Facets	Superior zygapophyseal facets face superiorly and medially Inferior facets face inferiorly and laterally	Superior zygapophyseal facets face superiorly and laterally Inferior facets face anteriorly and medially	Superior zygapophyseal facets are concave and face medially and posteriorly Inferior facets are convex and face anteriorly and laterally

REGIONAL STRUCTURE AND FUNCTION

The complexity of a structure that must fulfill many functions is reflected in the design of its component parts. Structural variations that are evident in the first cervical vertebra and the thoracic, fifth lumbar, and sacral vertebrae represent adaptations necessary for joining the vertebral column to adjacent structures. Differences in vertebral structure are also apparent at the cervicothoracic, thoracolumbar, and lumbosacral junctions, where a transition must be made between one type of vertebral structure and another. The vertebrae located at regional junctions are called **transitional vertebrae** and they usually possess characteristics common to two regions. The cephalocaudal increase in the size of the vertebral bodies reflects the increased proportion of body weight that must be supported by the lower thoracic and lumbar vertebral bodies. Fusion of the sacral vertebrae into a rigid segment reflects the need for a firm base of support for the column. In addition to the above variations, there are a large number of minor alterations in structure that occur throughout the column. However, only the major variations will be presented.

Structure of the Cervical Region

The first two cervical vertebrae C-1 and C-2, or respectively the atlas and axis, are atypical vertebrae. The seventh cervical vertebra is a transitional vertebra and therefore has characteristics of two regions. The rest of the cervical vertebrae, C-3 to C-6, conform to the general basic structural design of all vertebrae with the following regional variations.

TYPICAL CERVICAL VERTEBRAE (FIG. 4–24)

Body	The body is small, with a transverse diameter greater than anterior-posterior diameter and height. The body supports uncinate processes on the posterolateral superior and inferior surfaces.
Arches	
Spinous processes	The spinous processes are short, slender, and extend horizontally. The tip of the spinous process is bifid (split into two portions).
Transverse processes	There is a foramen for the vertebral artery, vein, and venous plexus. Also there is a groove for the spinal nerves.
Facets	The superior facets face superiorly and medially, while the inferior facets face inferiorly and laterally.

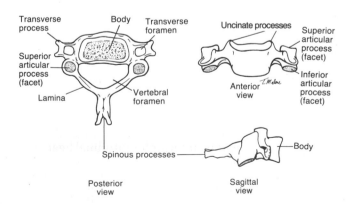

Figure 4–24. The body of a typical cervical vertebra is small and supports uncinate processes on the posterolateral superior and inferior surfaces.

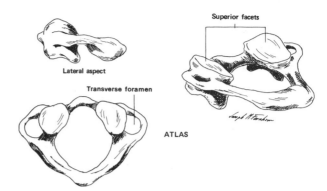

Figure 4-25. The atlas is a markedly atypical vertebra. It lacks a body and a spinous process.

Atlantoaxial Complex

The first two cervical vertebrae are markedly atypical vertebrae. The atlas (C-1) is different from other vertebrae in that it has no body or spinous process and is shaped like a ring (Fig. 4–25). It has four articulating facets, two superior and two inferior, on the thickened lateral portions of the ring. The superior facets are slightly concave and are designed to articulate with the slightly convex surface of the occipital bone. The inferior facets are slightly convex, flattened, and directed inferiorly for articulation with the superior facets of the axis (C-2). The atlas also possesses a facet on the internal surface of the anterior arch for articulation with the dens (odontoid process) of the axis. The axis is atypical in that the anterior portion of the body extends inferiorly and a vertical projection called the dens arises from the superior surface of the body (Fig. 4–26). The dens has an anterior facet for articulation with the atlas and a posterior groove for articulation with the transverse (cruciform) ligament. The arch of the axis has inferior and superior facets for articulation with the adjacent inferior vertebra and the atlas, respectively. The spinous process of the axis is elongated and the superior facets face upward and laterally.

The atlanto-occipital joint is a plane synovial joint and is composed of the two concave superior facets of the atlas that articulate with the two convex occipital condyles of the skull. The atlantoaxial joint is composed of three separate joints: the median atlantoaxial (atlanto-odontoid) joint between the dens and the atlas and two lateral joints between the superior facets of the axis and the inferior facets of the atlas (Fig. 4–27). The median joint is a synovial trochoid (pivot) joint in which the dens of the axis rotates in an osteoligamentous ring formed by the anterior arch of the atlas and the transverse (cruciform) ligament. The two lateral joints are plane synovial joints.

Ligaments

Besides the ligaments mentioned earlier in the chapter, there are a number of ligaments that are specific to the cervical region. Many of these ligaments attach to the axis, atlas, or skull and serve to reinforce the articulations of the upper two vertebrae. Four of the ligaments are continuations of the longitudinal tract system, while the two remaining ligaments are specific to the cervical area. The posterior atlantoaxial, the anterior atlantoaxial, the tectorial membrane, and the ligamentum nuchae are continuations of the ligamentum flavum, anterior longitudinal ligament, posterior longitudinal ligament, and the supraspinous ligament, respectively. The transverse atlantal ligament stretches across the ring of the atlas and divides the ring into a large posterior section for the spinal cord and a small anterior space for

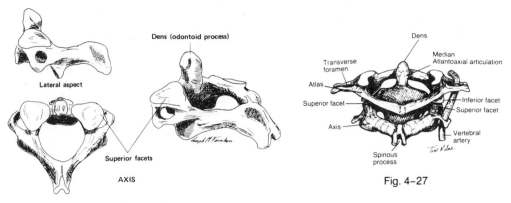

Fig. 4-26

Fig. 4-27

Figure 4-26. The dens (odontoid process) arises from the anterior portion of the body of the axis. The superior articulating facets are located on either side of the dens.

Figure 4-27. Atlantoaxial articulation. The median atlantoaxial articulation is shown with the posterior portion (transverse ligament) removed to show the dens and the anterior arch of the atlas. The two lateral atlantoaxial joints between the superior facets of the axis and the inferior facets of the atlas can be seen on either side of the median atlantoaxial joint.

the dens. The transverse ligament has a thin layer of articular cartilage on its anterior surface for articulation with the dens. Longitudinal fibers of transverse ligament extend superiorly to attach to the occipital bone and inferior fibers descend to the posterior portion of the axis. The transverse ligament and its vertical bands are sometimes referred to as the **atlantal cruciform ligament** (Fig. 4-28). The transverse portion of the ligament holds the dens and the atlas in close approximation and therefore plays a critical role in maintaining stability at the median atlantoaxial joint. Although the transverse ligament serves as an articular surface for the dens, its primary function is to prevent anterior displacement of C-1 on C-2. The superior and inferior vertical bands of the transverse ligament provide some assistance in providing stability.

The two alar ligaments are also specific to the cervical region. These paired ligaments arise from either side of the dens and attach to the medial aspects of the occipital bone. These ligaments are relaxed in extension and taut in flexion. The right ligament limits rotation of the head and neck to the left and the left ligament limits head and neck rotation to the right. The right upper and left lower portions of the alar ligaments limit left lateral flexion of the head and neck.[9] These ligaments also help to prevent distraction of C-1 on C-2.

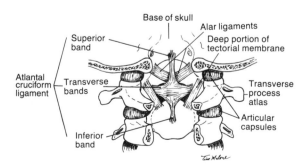

Figure 4-28. The transverse (atlantal cruciform) ligament. This is a posterior view of the vertebral column in which the posterior portion of the vertebrae (spinous processes and portion of the arches) have been removed to show the atlantal cruciform and alar ligaments.

Function of the Cervical Region
Stability

The cervical region differs from the thoracic and lumbar regions in that the cervical region bears less weight and is generally more mobile. Although the cervical region demonstrates the most flexibility of any of the regions of the vertebral column, stability of the cervical region, especially of the atlanto-occipital and atlantoaxial joints, is essential for support of the head and protection of the spinal cord and vertebral arteries. The design of the atlas is such that it provides more free space for the spinal cord than any other vertebra. The extra space helps to ensure that the spinal cord is not impinged on during motion. The bony configuration of the atlanto-occipital articulation confers some stability, but the application of small loads produces large rotations across the occipito-atlantoaxial complex[45,46] and also across the lower cervical spine.[45] The neutral zone across the occipital-atlantoaxial complex has been estimated to be 50 percent larger than in the lower cervical spine.[45] The existence of a large neutral zone implies that the ligaments and joint capsules are lax and that the muscles play an important role in providing stability for the occipito-atlantoaxial complex.[45] The muscles responsible for providing stability are the multifidus, interspinalis, semispinalis capitis, and semispinalis cervicis.[47] If these muscles are cut during surgery or torn during injury, the stability of the complex will be severely compromised.

No disks are present at either the atlanto-occipital or atlantoaxial articulations. Therefore, the weight of the head (compressive load) must be transferred directly through the atlanto-occipital joint to the articular facets of the axis. These forces are then transferred through the pedicles and laminae of the axis to the inferior surface of the body and to the two inferior articular processes. Subsequently the forces are transferred to the adjacent inferior disk. The laminae of the axis are large, which reflects the adaptation in structure that is necessary to transmit these compressive loads.

The loads imposed on the cervical region vary with the position of the head and body and are minimal in a well-supported reclining body posture. The loads appear to be carried both by the bodies and disks and by the zygapophyseal facets, with about one third carried by the body and disk and one third by each facet.[48] Compressive loads are relatively low during erect stance and sitting postures and high during the end ranges of flexion and extension.[27] Cervical motion segments tested in bending and axial torsion exhibit lower stiffness than lumbar motion segments but exhibit similar stiffness in compression.[49] Stability of the remainder of the cervical column is provided for by the same structures that were presented on pages 135–143 and which are summarized in Table 4–4. The joint capsules in the cervical region are lax and therefore provide less restriction to motion than in the thoracic and lumbar regions. The uncinate processes provide additional stability by reinforcing the posterolateral aspects of the disks and limiting flexion.

Mobility

The motions of flexion and extension, lateral flexion, and rotation are permitted in the cervical region. The range of motion (ROM) in lateral flexion and rotation is greater in the cervical region than in any other region. The largest range of rotation occurs between C-1 and C-2. Lateral flexion below the level of C-2 is coupled and is accompanied by rotation. Similarly, rotation initiates lateral flexion owing to the configuration of the articulating facets.

It is generally agreed that the atlanto-occipital joint permits primarily a nodding motion of the head (flexion and extension in the sagittal plane around a coronal axis);[50-52] however, some axial rotation and lateral flexion is possible.[46] There is less

Table 4–4. MOBILITY AND STABILITY OF THE VERTEBRAL COLUMN

Region	FACTORS AFFECTING MOBILITY AND STABILITY	
	Bony Factors	Soft Tissue Factors
Cervical Atlanto-occipital	Bony contact of the anterior ring of the foramen magnum limits flexion.	Tectorial membrane limits flexion.
Atlantoaxial		Tectorial membrane limits flexion and extension. Posterior atlantoaxial ligament limits flexion. Anterior atlantoaxial ligament limits extension. Transverse ligament prevents anterior dislocation of C-1 on C-2. Alar ligaments limit rotation and lateral flexion of the head to the opposite side.
C-2–C-7	Uncinate processes prevent posterior translation of the vertebral bodies and limit lateral flexion.	The annulus fibrous provides stability by limiting the amount of vertebral tilting. Great height and small diameter of disks permit large range of sagittal and frontal plane motion. The posterior longitudinal, ligamentum nuchae, and ligamentum flavum ligaments limit flexion. The anterior longitudinal ligament limits extension.
	Hyperextension limited by contact of spinous processes.	Zygapophyseal capsular ligaments are lax and permit a large range of motion.

agreement about the ROM. The combined ROM for flexion-extension reportedly ranges from 10 to 30°.[45,50–52] Flexion at the atlanto-occipital joint is limited by osseous contact of the anterior ring of the foramen magnum on the dens. Extension is checked by the tectorial membrane. Motion at the atlantoaxial joint includes flexion, extension, lateral flexion, and rotation. Approximately 50 percent of the total rotation of the cervical region occurs at the median atlantoaxial joint. This occurs before rotation in the rest of the cervical region. The atlas pivots about 45° to either side or a total of about 90°. Rotation at the atlantoaxial joint is limited by the alar ligaments and flexion and extension are limited by the tectorial membrane.

The only uncoupled motion that exists in the cervical spine from C-2 to C-7 is that of flexion and extension in the sagittal plane. The site of maximum motion in flexion and extension occurs between C-4 and C-6. In lateral flexion coupled with rotation the spinous processes of the vertebral bodies (C-2 to C-7) move toward the convexity of the lateral flexion curve or opposite to the rotation of the vertebral body. For example, in left lateral flexion, the spinous processes move to the right. The amount of rotation that occurs with lateral flexion decreases from C-2 to C-7.[9] Generally, motion is limited by tension in the ligaments, annulus fibrosus, and facet orientation. Lateral flexion is limited by the uncinate processes, while hyperextension is limited by bony contact of the spinous processes. The height in relation to the diameter of the disks also plays an important role in determining the amount of motion available in the cervical spine. The height is large in comparison to the anterior-posterior and transverse diameters of the cervical disks. Therefore a large amount of flexion, extension, and lateral flexion may occur at each segment, especially in the young when there is a large amount of water in the disks. The laxity of the zygapophyseal joint capsules also promotes mobility.

Muscles

The muscles that flex the head on the neck are the rectus capitis anterior and rectus capitis lateralis. The sternocleidomastoid muscle will flex the head and the neck when acting bilaterally. Flexion of the lower cervical region is produced by the action of the longus colli, longus capitis, rectus capitis anterior, and rectus capitis lateralis. The scalene muscles acting bilaterally may either flex the neck on the thorax or elevate the upper ribs when the cervical spine is stabilized.[53]

Lateral flexion of the head and neck is brought about by unilateral contractions of the scalene muscles, sacrospinous and splenius capitis. A unilateral contraction of the sternocleidomastoid causes a combined motion of lateral flexion and contralateral rotation.

The primary extensors of the head and neck are numerous, and most of these muscles are also capable of producing rotation. Extensors that produce rotation to the opposite side are the multifidus, rotators, and semispinalis. The extensor muscles that cause rotation to the same side are the oblique capitis and erector spinae.

Structure of the Thoracic Region

The majority of the thoracic vertebrae adhere to the basic structural design of all vertebrae except for some minor variations. The 1st and 12th thoracic vertebrae are transitional vertebrae.

TYPICAL THORACIC VERTEBRAE (FIG. 4–29)

Body	The body has equal transverse and anterior-posterior diameters. The anterior height of the vertebra is greater than the posterior height. There are **demifacets** for articulation with the ribs on the posterolateral corners of the vertebral plateaus and adjacent to the vertebral notch.
Arches	
Spinous processes	The spinous processes slope inferiorly and overlap the spinous process of the adjacent inferior vertebra.
Transverse processes	The transverse processes have thickened ends for articulation with the costal tubercles.
Facets	The superior zygapophyseal facets face superiorly and laterally. The inferior zygapophyseal facets face anteriorly and medially.

Each thoracic vertebra articulates with a pair of ribs by way of two joints: the costovertebral and the costotransverse joints. These joints will be discussed in the next chapter. The ligaments associated with the thoracic region are the same as those described previously for the vertebral column, except that the ligamentum flavum and anterior longitudinal ligaments are thicker in the thoracic region than in the cervical region. The zygapophyseal facet joint capsules are less lax than in the cervical region.

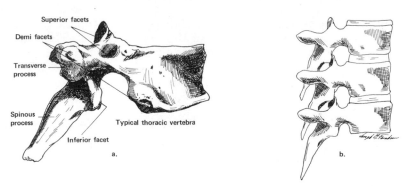

Figure 4-29. a. Lateral view of the thoracic vertebra shows the superior and inferior facets of the zygapophyseal joints and the demifacets for articulation with the ribs. b. Overlapping of spinous processes in thoracic region.

Function of the Thoracic Region
Stability and Mobility

The thoracic region is less flexible and more stable than the cervical region due to the limitations imposed by structural elements such as the rib cage, spinous processes, zygapophyseal joint capsules, the ligamentum flavum, and the dimensions of the vertebral bodies. All motions are possible in the thoracic region, but the range of flexion and extension is extremely limited in the upper thoracic region, T-1 to T-6, where the facets lie in the frontal plane. In the lower part of the region, T-9 to T-12, the facets lie more in the sagittal plane, allowing an increased amount of flexion and extension. Lateral flexion is free in the upper thoracic region, and increases in the lower region. Rotation, which also is free in the upper thoracic region, decreases caudally. While lateral flexion is always coupled with some rotation, the amount of accompanying rotation decreases in the lower part of the region due to the change in orientation of the facets. In the upper part of the thoracic region, rotation is accompanied by movement of the spinous process toward the convexity of the curve, while rotation in the lower region may be accompanied by rotation of the spinous process towards the concavity of the curve. However, the direction of the rotation may vary among individuals.[9]

Flexion in the thoracic region is limited by tension in the posterior longitudinal ligament, the ligamenta flava, the interspinous ligament, and the capsules of the zygapophyseal joints. Extension of the thoracic region is limited by contact of the spinous processes, zygapophyseal facets, size of the disks, and tension in the anterior longitudinal ligament, joint capsules, and abdominal muscles. Lateral flexion is restricted by impact of the facets on the concavity of the lateral-flexion curve and by limitations imposed by the rib cage. Rotation in the thoracic region also is limited by the rib cage. When a thoracic vertebra rotates, the motion is accompanied by distortion of the associated rib pair (Fig. 4-30). The posterior portion of the rib on the side to which the vertebral body rotates becomes more convex, while the anterior portion of the rib becomes flattened. The reverse occurs for the rib on the side opposite to the vertebral rotation. The amount of rotation that is possible is dependent on the ability of the ribs to undergo distortion and the amount of motion available in the costovertebral and costotransverse joints. As a person ages, the costal cartilages ossify and allow less distortion. This results in a reduction in the amount of rotation that is available with aging.

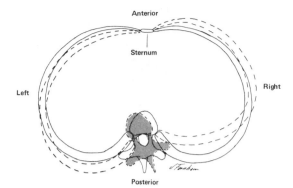

Figure 4-30. Rotation of a thoracic vertebral body to the left produces a distortion
of the associated rib pair that is convex
posteriorly on the left and convex anteriorly on the right.

Muscles

The muscles that produce movement of the thoracic region also produce
motion at the lumbar region and therefore the anterior, lateral, and posterior trunk
muscles will be discussed following the sections on the lumbar and sacral regions.
The other muscles specific to the thoracic region are muscles of respiration. These
muscles will be discussed in chapter 5.

Structure of the Lumbar Region

The first four lumbar vertebrae are similarly structured (Fig. 4–31). The fifth
lumbar vertebra has structural adaptations for articulation with the sacrum.

TYPICAL LUMBAR VERTEBRAE

Body
: The body of the typical lumbar vertebra is massive with a transverse diameter that is greater than anterior diameter and height.

Arches
  Spinous
  process
: The spinous process is broad, thick, and extends horizontally.

  Laminae
: The cortical bone in the pars interarticularis, which is that portion
of the lamina that lies at the junction of the vertically oriented
lamina and horizontal projecting pedicle is thickened.

Transverse
process
: The transverse process is long, slender, and extends horizontally.

Facets
: The superior zygapophyseal facets are concave and face medially
and posteriorly. The inferior facets are convex and face anteriorly and laterally. Both facets lie in the sagittal plane. The superior facets contain mamillary processes on their posterior edges.

The fifth lumbar vertebra is a transitional vertebra and differs from the rest of
the lumbar vertebrae in that it has a wedge-shaped body. The anterior portion of
the body is of greater height than the posterior portion. The superior diskal surface
area of L-5 is about 5 percent greater than the areas of disks at L-3 and L-4 while
the inferior diskal surface area of L-5 is smaller than the diskal surface area at other

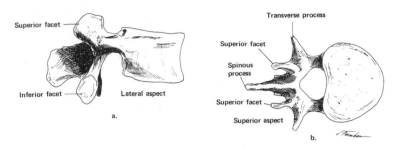

Figure 4–31. a. A lateral view of a typical lumbar vertebra shows the large body and facets. b. A superior view of a typical lumbar vertebra shows transverse and spinous processes.

lumber levels.[54] Also the spinous process is small and the transverse processes are large and directed superiorly and posteriorly. The inferior facets of this vertebra are adapted for articulation with the sacrum by being widely spaced.

The lumbosacral articulation is formed by the fifth lumbar vertebra and first sacral segment. The first sacral segment, which is inclined slightly anteriorly and inferiorly, forms an angle with the horizontal called the **lumbosacral angle** (Fig. 4–32). The size of the angle varies with the position of the pelvis and affects the lumbar curvature. An increase in this angle will result in an increase in the anterior convexity of the lumbar curve and will increase the amount of shearing stress at the lumbosacral joint.

The ligaments associated with the lumbar region are the same ligaments described previously, except for the iliolumbar ligaments and the thoracolumbar fascia. The iliolumbar ligaments consist of the following five bands: anterior, posterior, superior, inferior, and vertical. These bands extend from the transverse processes of L-4 and L-5 to attach on the iliac crests. The ligaments as a whole are very strong and play a significant role in stabilizing the fifth lumbar vertebra (preventing the vertebra from anterior displacement).[55]

The thoracolumbar fascia consists of three layers (anterior, middle, and posterior) that arise from the transverse and spinous processes and blend with other tissues. The fascia completely surrounds the muscles of the lumbar spine (Fig. 4–33). The anterior layer of the thoracolumbar fascia is derived from the fascia of the quadratus lumborum muscle. The middle layer is not as well defined, but is thought

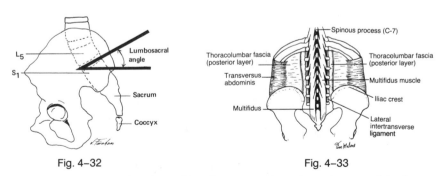

Fig. 4–32   Fig. 4–33

Figure 4–32. The lumbosacral angle is formed by a line drawn along the inclination of the first sacral segment and a horizontal line.

Figure 4–33. The thoracolumbar fascia. The anterior and middle layers of the fascia have been partially removed to show the posterior layer fusing with the transversus abdominis muscle.

to lie posterior to the quadratus lumborum muscle. It attaches medially to the tips of the transverse processes. The posterior layer consists of two laminae, one with fibers oriented caudomedially and the other with fibers oriented caudolaterally. The two laminae of the posterior layer fuse with the transversus abdominis muscle and thus provide an indirect attachment for the transversus abdominis muscle to the lumbar spinous processes.[55] Gracovetsky has designated the anterior layer of the thoracolumbar fascia as the "passive part" and the posterior layer as the "active part."[56] According to Gracovetsky, the passive part serves to transmit tension from a contraction of the hip extensors to the spinous processes. The active portion is activated by a contraction of the transversus abdominis muscle, which tightens the fascia. The tension in the fascia transmits longitudinal tension to the tips of the spinous processes of L-1 to L-4 and may help the spinal extensor muscles to resist an applied load.[56]

Function of the Lumbar Region
Stability

One of the primary functions of the lumbar region is to provide support for the weight of the upper part of the body in static as well as in dynamic situations. The increased size of the lumbar vertebral bodies and disks in comparison to their counterparts in the other regions helps the lumbar structures to support the additional weight. The compressive load that must be sustained by the lumbar structures is altered by changes in the lumbar curvature or arrangement of body segments. Changes in the position of body segments will change the location of the body's center of gravity and thus change the forces acting on the lumbar spine. In the normal standing posture the line of gravity passes through the combined axis for the lumbar vertebrae and therefore no net torque exists. Any deviation of the line of gravity will lead to torque production. Muscle contraction creates additional compression on the vertebrae and the potential for excessive torsional and shear stresses.

> *Example:* In a situation in which a person is standing holding an object in his or her right hand with arm extended, two bending moments or torques will be created: a forward flexion bending moment and a right lateral flexion moment. The muscles of the lumbar and trunk regions will have to contract to exert an opposing moment (countertorque) to maintain equilibrium of the spine in the static upright position and prevent motion of the trunk in the direction of the external moments.[57]

In a closed kinematic chain situation, repositioning of one segment will influence adjacent segments and may affect lumbar stability. For instance, an anterior tilt of the pelvis will cause a change in the inclination of the first sacral segment. The change in inclination will result in an increase in the lordotic curvature of the lumbar region. An increase in the anterior convexity of the lumbar curve will increase the shearing forces acting on the lumbar vertebrae and increase the likelihood of anterior displacement of the fifth lumbar vertebra on the first sacral segment. The increase in compression forces on the posterior structures that accompanies an increase in the lumbar curvature may damage the spinous processes, zygapophyseal joints, and posterior longitudinal ligament. The anterior longitudinal ligament, which is well developed in the lumbar area, will be stretched to support the anterior position of the lumbar vertebrae and disks. The abdominal muscles will contract to exert an upward pull on the pelvis, which posteriorly tilts the pelvis. But when these muscles contract, they create a compressive force on the

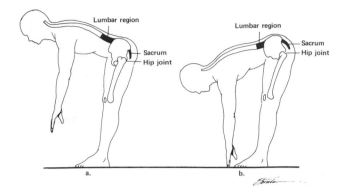

Figure 4-34. Lumbar-pelvic rhythm (a) the lumbar spine flexes and (b) the pelvis rotates anteriorly in the sagittal plane.

structures of the lumbar region. The annulus fibrosus, joint capsules, and configuration of the facets also will help to provide stability.

Mobility

The lumbar facets from L-1 to L-4 lie in the sagittal plane and favor flexion and extension but limit lateral flexion and rotation. Flexion of the lumbar spine is more limited than extension and, normally, it is not possible to flex the lumbar region to form a kyphotic curve. The amount of flexion varies at each interspace of the lumbar vertebrae, but most of the flexion takes place at the lumbosacral joint. A specific instance of coordinated activity of the lumbar flexion and anterior tilting of the pelvis in the sagittal plane is called **lumbar-pelvic rhythm.** The activity of bending over to touch one's toes with knees straight is dependent on lumbar-pelvic rhythm.[59] The first part of bending forward consists of lumbar flexion (Fig. 4–34a). This is followed by anterior tilting of the pelvis at the hip joints (Fig. 4–34b). The integration of motion of the pelvis with motion of the vertebral column not only increases the ROM available to the total column but also reduces the amount of flexibility required of the lumbar region. The contribution to motion from multiple areas to produce a larger ROM is similar to what is found at the shoulder in scapulohumeral rhythm. A restriction of motion at either the lumbar spine or at the hip joints may disturb the rhythm and prevent a person from reaching his or her toes. Restriction of motion at one segment also may result in hypermobility of the unrestricted segment. A return to the erect posture requires posterior tilting of the pelvis at the hips followed by extension of the lumbar spine. During flexion and extension the greatest mobility of the spine occurs between L-4 and S-1, which is also the area that must support the most weight. In an erect standing posture, pressures on the lower disks are much greater than the weight of the body, and these pressures increase with movement and muscle contraction.

Lateral flexion and rotation of the lumbar vertebrae are most free in the upper lumbar region and progressively diminish in the lower region. Rotation of the vertebrae in the upper area is accompanied by movement of the spinous process toward the concavity of the curve, which is similar to rotation in the lower thoracic region. Little or no lateral flexion is possible at the lumbosacral joint because of the orientation of the facets. Table 4–5 provides a stability-mobility summary for the thoracic and lumbar regions.

Structure of the Sacral Region

Five sacral vertebrae are fused to form the triangular or wedge-shaped structure that is called the sacrum. The base of the triangle, which is formed by the first sacral vertebra, supports two articular facets that face posteriorly for articulation

Table 4–5. MOBILITY AND STABILITY OF THE VERTEBRAL COLUMN

	FACTORS AFFECTING MOBILITY AND STABILITY	
Region	Bony Factors	Soft Tissue Factors
Thoracic	Rib cage. Costovertebral and costotransverse joints.	Anterior longitudinal ligament limits extension.
	Bony contact of spinous processes limits extension.	Posterior longitudinal ligament limits flexion and reinforces posterior aspect of disk.
	The orientation of the facets limits flexion and extension but permits lateral flexion and rotation.	The ligamentum flavum, supraspinous, and interspinous ligaments also limit flexion.
T-12–L-1	Facet orientation limits rotation.	The capsular ligaments of the zygapophyseal joints are short and taut and limit zygapophyseal joint motion. The intertransverse ligament limits lateral flexion.
Lumbar L-1–L-5	Facet joint orientation limits rotation and lateral flexion.	Taut zygapophyseal joint capsules limit motion and provide stability. The intertransverse and iliolumbar ligaments limit lateral flexion. The interspinous and ligamentum flavum ligaments limit flexion. The anterior longitudinal ligament limits extension and the posterior longitudinal ligament limits flexion.

with the inferior facets of the fifth lumbar vertebra. The apex of the triangle, formed by the fifth sacral vertebra, has a small facet for articulation with the coccyx. Differences in the structure among vertebrae in this region may be due to anomalies such as asymmetric or incomplete development of the vertebral arches.

Articulations and Ligaments

The two sacroiliac joints consist of the articulations between the left and right articular surfaces on the sacrum (which are formed by fused portions of the first, second, and third sacral segments) and the left and right iliac bones (Fig. 4–35).

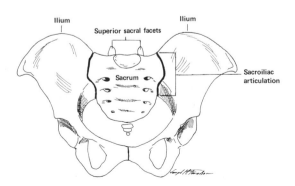

Figure 4–35. The sacroiliac joints consist of the articulations between the first three sacral segments and the two ilia of the pelvis.

The sacroiliac joints are unique in that both the structure and function of these joints change significantly from birth through adulthood. The sacroiliac joint has all of the typical features of a synovial joint[61] except that the articular cartilage on the iliac side of the joint is fibrocartilage while the cartilage covering the opposing joint surface (sacral) is hyaline cartilage.[62] Therefore, the joint is sometimes described as a part synovial and part fibrous joint.[63]

The sacroiliac joint surfaces in early childhood are smooth and gliding motions are possible in all directions, which is typical of a synovial plane joint.[62] However, after puberty, the joint surfaces change their configuration and motion is restricted to the anterior-posterior movement of the sacrum on the ilium or ilium on sacrum (flexion or rotation and extension or counterrotation). Flexion or nutation involves movement of the anterior tip of sacral promontory of the sacrum anteriorly and inferiorly while the coccyx moves posteriorly in relation to the ilium. Extension or counternutation refers to the opposite movement in which the anterior tip of the sacral promontory moves posteriorly and superiorly while the coccyx moves anteriorly in relation to the ileum. The change in position of the sacrum during flexion and extension affects the diameter of the pelvic brim and pelvic outlet. During flexion (nutation) the anterior-posterior diameter of the pelvic brim is reduced and the anterior-posterior diameter of the pelvic outlet is increased. During extension the reverse situation occurs. The anterior-posterior diameter of the pelvic brim is increased and the diameter of the pelvic outlet is decreased.[8] Accurate descriptions of the sacroiliac joints and the motions that occur at these joints have been difficult to obtain because the planes of the joint surfaces are oblique to the angle of an x-ray beam used to make a standard anterior-posterior radiograph of the pelvis.[64]

Articulating Surfaces on the Sacrum

The articulating surfaces on the sacrum are auricular-(C-)shaped,[62] and are located on the sides of the fused sacral vertebrae lateral to the sacral foramina. The portions of the sacrum on which the articulating surfaces are found are broad and irregular and covered with hyaline cartilage. In the adult a central groove develops that extends the length of the articulating surfaces.[62,65]

Articulating Surfaces on the Ilia

The articular surfaces on the ilia are also C-shaped and are located posteriorly. In the first decade of life, the iliac joint surfaces are smooth, flat, and covered with fibrocartilage. Following puberty, the joint surfaces develop a central ridge that extends the length of the articulating surface and corresponds to the grooves on the sacral articulating surfaces.[62,65]

Ligaments

Four groups of ligaments are associated with the sacroiliac joints: the iliolumbar ligaments, the sacroiliac ligaments, the sacrospinous ligaments, and the sacrotuberous ligaments. The iliolumbar ligaments have been described previously in this chapter. The sacroiliac ligaments extend from the iliac crests to attach to the tubercles of the first four sacral vertebrae. The sacrospinous ligaments connect the ischial spines to the lateral borders of the sacrum and coccyx. The sacrotuberous ligaments connect the ischial tuberosities to the posterior spines at the ilia and the lateral sacrum and coccyx (Fig. 4–36). The sacrospinous ligament forms the inferior border of the greater sciatic notch while the sacrotuberous ligament forms the inferior border of the lesser sciatic notch.[8,66]

Symphysis Pubis

The symphysis pubis is a cartilaginous joint located between the two ends of the pubic bones. The end of each pubic bone is covered with a layer of articular

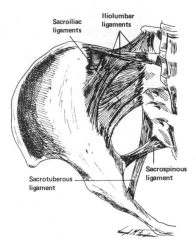

Figure 4–36. The sacroiliac and iliolumbar ligaments reinforce the sacroiliac and lumbosacral articulations, respectively. The sacrospinous ligament forms the inferior border of the greater sciatic notch and the sacrotuberous ligament forms the interior border of the lesser sciatic notch.

cartilage and the joint is formed by a fibrocartilaginous disk that joins the hyaline cartilage-covered ends of the bones. The disk has a thin central cleft,[8] which in women may extend throughout the length of the disk.[4] The three ligaments that are associated with the joint are the superior pubic ligament, the inferior pubic ligament, and the posterior ligament.[8] The superior ligament is a thick and dense fibrous band that attaches to the pubic crests and tubercles and helps to support the superior aspect of the joint. The inferior ligament arches from the inferior rami on one side of the joint to the inferior portion of the rami on the other side and thus reinforces the inferior aspect of the joint. The posterior ligament consists of a fibrous membrane that is continuous with the periosteum of the pubic bones.[8] The anterior portion of the joint is reinforced by aponeurotic expansions from a number of muscles that cross the joint (Fig 4–37). Kapandji describes the muscle expansions as forming an anterior ligament consisting of expansions of the transversus abdominis, rectus abdominis, internal obliquus abdominis, and the adductor longus.[8]

Function of the Sacral Region
Stability and Mobility

Stability of the sacroiliac joints is extremely important because these joints must support a large portion of the body weight. In normal erect posture the weight of head, arms, and trunk (HAT) is transmitted through the fifth lumbar vertebra and lumbosacral disk to the first sacral segment. The force of the body weight tends to

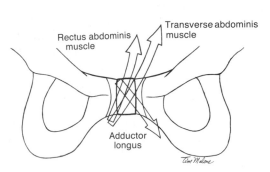

Figure 4–37. The aponeurotic extensions of the muscles crossing the anterior aspect of the symphysis pubis.

separate the sacrum from the ilia and tends to push the first sacral segment into flexion (nutation). The sacroiliac ligaments form the main bond that keeps the ilia and the sacrum in close approximation. Tension developed in the sacrotuberous, sacrospinous, and anterior sacroiliac ligaments counteract the downward and forward movement of the sacrum.

The sacroiliac joints permit a small amount of motion that varies among individuals. The sacroiliac joints are linked to the symphysis pubis in a closed kinematic chain, and therefore any motion occurring at the symphysis pubis is accompanied by motion at the sacroiliac joints and vice versa.[67] During pregnancy the ligaments supporting both of these joints become softened. Consequently, the joints become more mobile and less stable and the likelihood of injury to these joints is increased. The combination of loosened posterior ligaments and an anterior weight shift caused by a heavy uterus may allow excessive movement of the ilia on the sacrum and result in stretching of the sacroiliac joint capsules.

The sacroiliac joints and symphysis pubis are closely linked functionally to the hip and intervertebral joints and therefore affect and are affected by movements of the trunk and lower extremities. For example, weight shifting from one leg to another is accompanied by motion at the sacroiliac joints. Fusions of the lower lumbar vertebrae have been found to cause compensatory increases in motion at the sacroiliac joints.[68] Shearing forces are created at the symphysis pubis during the single-leg-support phase of walking as a result of lateral pelvic tilting. In a normal situation, the joint is capable of resisting the shearing forces and no appreciable motion occurs. If, however, the joint is dislocated, the pelvis becomes unstable during gait with increased stress on the sacroiliac and hip joints as well as the vertebral column. The joints of the pelvis are linked to the hip and vertebral column in non-weight-bearing as well as in weight-bearing postures. Hip flexion in a back-lying position tilts the ilia posteriorly in relation to sacrum. This pelvic motion causes nutation at the sacroiliac joints, which increases the diameter of the pelvic outlet. During the process of birth the increase in the diameter of the pelvic outlet facilitates delivery of the fetal head. Counternutation is brought about by hip extension in the supine position and enlarges the pelvic brim. Therefore a hip-extended position is favored early in the birthing process to facilitate the descent of the fetal head into the pelvis, whereas the hip-flexed position is used during delivery.[8]

MUSCLES OF THE VERTEBRAL COLUMN

Flexors

Muscles that flex the trunk are located anteriorly and laterally with attachments on the ribs, sternum, and pelvis. These muscles act indirectly on the vertebral column by exerting a pull on the adjacent structures. Isometric or concentric contractions of the flexor muscles cause compression forces on the vertebral column. If the pelvis and ribs are free to move, a shortening contraction of the flexors will pull these structures closer together and as a consequence flex the total spine, as in a situp. If the ribs are fixed, a shortening contraction of the rectus abdominis muscle will exert an upward pull on the pelvis. The resulting posterior rotation of the pelvis in the sagittal plane (posterior pelvic tilt) will produce flexion of the lumbar spine.

Forward flexion of the trunk from the erect standing posture does not require any action of the trunk flexors because the gravitational force will pull the trunk forward. However, any activity that involves pushing, pulling, or lifting will initiate an immediate isometric contraction of the flexors in order to stabilize the ribs and pelvis and indirectly the vertebral column.

The flexor muscles are not active during normal erect standing. However, they are considered essential for balancing the pull of the back extensor and the hip flexor muscles in dynamic situations and for keeping the pelvis in a normal position. When either the back extensor or hip flexor muscles act unopposed, they cause an anterior tilting of the pelvis in the sagittal plane (anterior pelvic tilt) and an increase in the lumbar curve. In any motion that involves flexion of the trunk against gravity, the abdominal muscles will be required to produce the flexion motion. Also, the abdominal muscles perform the function of protecting and supporting the viscera. During pregnancy, especially in the second and third trimesters and immediately postpartum, the rectus abdominis muscle may separate or relax along the linea alba (diastasis recti abdominis). This condition may have adverse effects on the ability of the flexors to function as static or dynamic stabilizers and movers of the vertebral column.[69]

The psoas major muscle has been described as a flexor, stabilizer, and extensor of the lumbar spine.[70] When the spine is in the flexed position, the fibers cross anterior to the axis of rotation of the lumbar intervertebral joints and thus, the muscle has a flexion moment and a concentric contraction can produce flexion. When the lumbar spine is extended, most of the fibers are posterior to the axis of rotation and produce an extension moment. During the activity of lifting the psoas is active in stabilization of the lumbar spine.[70]

Extensors

Muscles that extend the vertebral column are located posteriorly (Fig. 4–38). The sacrospinalis muscle (erector spinae), which consists of three divisions (lateral, medial, and intermediate), represents the largest portion of the posterior musculature. This group of muscles extend from the sacrum to the occipital portion of the skull attaching to the transverse and spinous processes of all vertebrae and the angles of the ribs by various divisions. The most lateral division that attaches to the ribs is called the **iliocostalis group** and consists of the iliocostalis cervicis, thoracis, and lumborum (Fig. 4–39). The most medial division of the erector spinae muscle attaches to the spinous processes and is called the **spinalis group,** which includes the spinalis cervicis and thoracis. The intermediate division that attaches

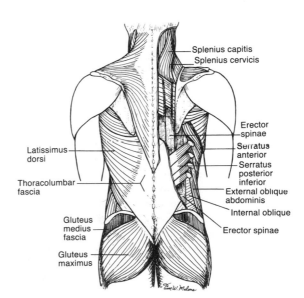

Figure 4-38. Posterior back muscles. The superficial muscles have been removed on the right side to show the erector spinae. The anterior layer of the thoracolumbar fascia is intact on the left side of the back.

Splenius capitis
Splenius cervicis
Erector spinae
Serratus anterior
Serratus posterior inferior
External oblique abdominis
Internal oblique
Erector spinae
Latissimus dorsi
Thoracolumbar fascia
Gluteus medius fascia
Gluteus maximus

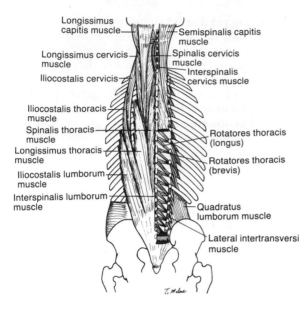

Longissimus
capitis muscle

Semispinalis capitis
muscle

Longissimus cervicis
muscle

Spinalis cervicis
muscle

Iliocostalis cervicis

Interspinalis
cervics muscle

Iliocostalis thoracis
muscle

Spinalis thoracis
muscle

Rotatores thoracis
(longus)

Longissimus thoracis
muscle

Rotatores thoracis
(brevis)

Iliocostalis lumborum
muscle

Interspinalis lumborum
muscle

Quadratus
lumborum muscle

Lateral intertransversi
muscle

Figure 4–39. Erector spinae and deep back muscles. The erector spinae muscle has been removed from the right side of the back to show the deep back muscles.

to the transverse processes is called the **longissimus**. The longissimus consists of the longissimus capitus, cervicis, thoracis, and lumborum. Four other muscle groups lie deep to the erector spinae: semispinalis, multifidus, rotatores, and the interspinales and intertransversarii (Fig. 4–39). These muscles attach to the transverse, spinous, or articular processes of the vertebrae.

All of the posterior trunk muscles can produce extension of the spine and can increase the lumbar curve. Conversely, a contraction of the flexor muscles decreases the lumbar curve. Intermittent activity of the back extensors is necessary to maintain stability of the column in erect standing. The extensor muscles also are responsible for controlling forward flexion of the vertebral column in the standing position. The gravitational moment will produce forward flexion, but the extent and rate of flexion is controlled partially by eccentric contractions of the extensors and partially by the thoracolumbar fascia and posterior ligamentous system. The extensors are active until approximately two thirds of maximal flexion has been attained, at which point they become electrically silent.[6,55,56] According to Gracovetsky,[56] control of flexion becomes the responsibility of the passive elastic response of the thoracolumbar fascia and posterior ligamentous system. The posterior ligaments (supraspinous and interspinous ligaments) have longer moment arms than the extensor muscles and thus have a mechanical advantage over the extensors. The longissimus thoracis and iliocostalis thoracis control movement of the thorax on the lumbar spine, while the multifidus, longissimus lumborum, and iliocostalis lumborum control flexion of the lumbar spine. The latter two muscles also help to prevent anterior translation of the lumbar vertebrae that may accompany flexion.

The extensors lie parallel to the vertebral column and thus, like the abdominals, exert a compression force on the column during contractions. The moment arm of the lumbar extensors is considerably decreased when the trunk is in a forward flexed position and increased when the lumbar lordotic curvature is increased.[71,72]

Role of Flexors and Extensors in Lifting

The disadvantageous position of the extensors in the forward flexed position is one reason why lifting in this position is discouraged. However, the primary reason for not lifting in the forward flexed position is because of the increase in intradiskal pressures in the lumbar area.[73] The critical factors in lifting in the flexed posture appear to be the distance of the object to be lifted from the body[6] and the velocity of the lift. The farther away the load is from the body, the greater the gravitational moment acting on the vertebral column. This requires greater muscle activity to perform the lift, and consequently creates greater pressure in the disk. The higher the velocity of the lift, the greater the amount of weight that can be lifted, but the higher the load on the lumbar disks.

The prevalence of back problems in the general population and the difficulties of resolving these problems has generated a great deal of research both to explain the mechanisms involved in lifting and to determine the best method of lifting so that back injuries can be prevented. One theory that has been used in the past to explain how the forces needed to perform a heavy lift are generated and the compressive forces on the disks are reduced is the intra-abdominal pressure (IAP) theory. The IAP theory was first proposed by Bartelink in 1957, who postulated that a contraction of the abdominal muscles (transversus abdominis and internal oblique muscles) in the presence of a closed glottis raises the intra-abdominal pressure, supporting the thorax and thus assisting the back muscles in raising the weight.[74] This theory has been challenged on the basis of calculations that demonstrate that to provide the required upward force on the thorax, the intra-abdominal pressure would have to exceed systolic aortic blood pressure. Furthermore, it has been calculated that the force of the contraction necessary to generate this pressure exceeds the maximum possible hoop tension of the abdominal muscles.

The questions raised about the viability of the IAP theory have led investigators to develop new theories to explain how lifting is accomplished. Gracovetsky[56] has proposed a theory that includes intra-abdominal pressure as a component, but he has ascribed a different role for intra-abdominal pressure than was proposed in the IAP theory. He suggests that the back extensor muscles are assisted in lifting a large weight by extension moments generated by passive tension in the posterior ligamentous system and passive and active tension in the thoracolumbar fascia (TLF). According to Gracovetsky, tension is created by forward flexion of the trunk, which stretches the posterior ligamentous system, including the zygapophyseal joint capsules, posterior longitudinal, supraspinous, and ligamentum flavum, and interspinous ligaments. Contractions of the hip extensors in a closed kinematic chain exert a posterior tilting force on the pelvis, which places tension along the posterior ligamentous system.

Passive tension in the TLF also is created by forward flexion of the spine and posterior tilting of the pelvis. Active tension in the TLF is created by a contraction of the latissimus dorsi, internal oblique muscles, and transversus abdominis muscles. Contractions of the transversus abdominis and internal oblique muscles are thought to be responsible for increasing the IAP in the presence of a closed glottis.[70] The IAP adds tension to the TLF and thus increases the force of the extension moment that the TLF can generate. The TLF requires an appropriate amount of IAP to function properly and a degree of spinal flexion to function efficiently.[56] When the IAP is low, tension on the TLF is reduced and consequently the extension moment generated is decreased.

Gracovetsky's theory, which includes a role for the central nervous system (CNS), has not been proven, but the theory provides a direction for future research efforts. According to the theory, the CNS acts as a controlling computer that monitors the amount of stress acting at each intervertebral joint and adjusts the amount of stress to minimize and equalize stress in order to protect the joint structures. Stress minimization can be accomplished through reduction in muscle contraction, switching to ligamentous support, switching the relative contributions made by the muscles and ligaments, changing the posture of the spine, or by aborting the lift.[56]

The use of ligamentous rather than muscle support also conserves energy. Ladin and associates[57] found that muscles appear to function to conserve energy in that they have periods of activity and inactivity in response to an applied load. Furthermore, they found that muscle responses to the application of an external load can be predicted. Ladin and associates have called the loading threshold required for a muscle or group of muscles to respond as the **switching curve.**

Changes in the velocity of the lift also have been proposed as a means of increasing the ability of a person to lift a large load. When lifting is performed slowly or when a weight is held continuously and the ligamentous system and TLF are providing most of the force for the activity, the ligaments are subjected to creep. The ligamentous elongation during creep imposes a limit on the maximum weight that can be lifted slowly because the ligaments must remain taut to balance the external moments. Only about one quarter of the amount of weight can be lifted at slow speed in comparison to the amount that can be lifted at a high speed.[56] However, during fast-speed lifting higher moments may occur at the L-5 to S-1 levels than during slower-speed lifting.[75]

Rotators and Lateral Flexors

Rotation of the trunk is usually coupled with some degree of lateral flexion. Anterior muscles that produce rotation and lateral flexion are the **external** and **internal oblique abdominals.** Rotation of trunk to the left requires a simultaneous contraction of the right external oblique and the left internal oblique. Rotation of the trunk to the right requires a simultaneous contraction of the left external oblique and the right internal oblique. The posterior muscles that rotate the trunk may be divided into two groups: muscles that produce rotation and lateral flexion to the *same side* and muscles that produce rotation to the opposite side. Rotation and lateral flexion to the same side are a function of the iliocostalis, longissimus, spinalis muscles, quadratus lumborum, and serratus posterior superior. Muscles that produce rotation to the *opposite side* are the semispinalis thoracis, multifidus, rotators, and transverse thoracis. The lateral flexors of the trunk are the quadratus lumborum and the iliopsoas. When either of these muscles contracts unilaterally, it causes lateral flexion of the trunk ipsilaterally if the pelvis and femur are fixed. If lateral flexion occurs when the person is standing erect, the force of gravity will continue the lateral flexion movement and the contralateral muscles will be required to balance the gravitational moment and control the movement by contracting eccentrically. A unilateral contraction of the quadratus lumborum will "hike the hip" or laterally tilt the pelvis in the frontal plane if the pelvis is free to move, but will laterally flex the trunk if the pelvis is fixed. The psoas muscle, when acting bilaterally in a closed kinematic chain with the femurs fixed, will pull the lumbar spine anteriorly and thus extend the lumbar spine and increase the lordosis. The psoas muscle also acts to flex the hip when either the femur or the pelvis and lumbar vertebrae are fixed. See the Appendix to Chapter 4 for a complete listing of the muscles of the vertebral column.

Muscles of the Pelvic Floor

Structure

Although the levator ani and coccygeus muscles neither play a major supporting role for the vertebral column nor produce movement of the column, these muscles are mentioned here because of their proximity to the column and possible influence on the linkages that form the pelvis. The levator ani muscles comprise two

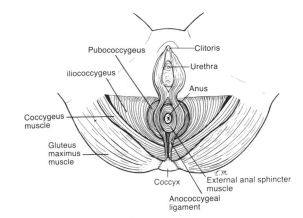

Figure 4–40. Muscles of the pelvic floor.

distinct parts, the iliococcygeus and the pubococcygeus, which help to form the floor of the pelvis and separate the pelvic cavity from the perineum. The muscles arise from the inner surface of the pubis anteriorly and from the obturator membrane and pelvic surface of the ischial spine posteriorly. The muscles insert into the perineal body, sides of the anal canal, and the anococcygeal ligament between the anal canal and the coccyx. The coccygeus muscle arises from the spine of the ischium and attaches to the coccyx and lower portion of the sacrum. The gluteal surface of the muscle blends with the sacrospinous ligament (Fig 4–40).

Function

Voluntary contractions of the levator ani muscles help to constrict the openings in the pelvic floor (urethra and anus) and prevent unwanted micturition and defecation (stress incontinence). Involuntary contractions of these muscles occur reflexly during coughing or holding one's breath when the IAP is raised. In women, these muscles surround the vagina and help to support the uterus. During pregnancy the muscles can be stretched or traumatized and result in stress incontinence whenever the IAP is raised. In men, damage to these muscles may occur following prostate surgery. The coccygeus muscle assists the levator ani in supporting the pelvic viscera and maintaining IAP. It also exerts an anterior pull on the coccyx and thus may contribute to posterior tilting of the pelvis or resistance to anterior pelvic tilting.

GENERAL EFFECTS OF AGING, INJURY, AND DEVELOPMENT DEFICITS

The vertebral column, like other structures in the body, is subject to aging, injury, disease processes, and development deficits. Any one or more of the structural components may be affected by these conditions, but injuries and degenerative diseases are more likely to occur in those areas that are subjected to the greatest stress. Normally the spine is able to withstand large amounts of stress, but when the stresses are unexpected, prolonged, or excessive, the likelihood of injury is increased. Even relatively minor stresses may cause damage when the integrity of the structure has been previously compromised.

Aging

Aging causes changes in the structure of the disks and, consequently, in the function of the disk and related structures. For example, with increasing age the average fluid content of the disk decreases, and as a result the height of the disk decreases. The decrease in disk height brings the vertebrae closer together and alters the relationships of the zygapophyseal joints. The closer proximity of the vertebrae causes more compressive stress to be applied to these joint surfaces. The posterior ligamentous system also is affected by the loss in height and becomes slack. The lack of tension in the ligaments permits more movement in flexion and increases the extent of the neutral zone, and as a result decreases the stability of the spine. These age-related changes in disk structure have been found to affect men and women differently. For example, in a study by Miller and colleagues[76] of 600 disks from the lumbar spine (ages of subjects, 0 to 90 years) disks from men showed signs of degeneration, that is, presence of osteophytes and/or loss of disk height, earlier than disks from women. For men the signs of disk degeneration first occurred in the 11- to 19-year age range, while in women signs of degeneration did not occur until a decade later. Age-related changes may occur in all structures of the vertebral column; however, this text is not designed to provide this type of information.

Development Deficits

The lumbar region in particular is susceptible to injuries and development deficits. Spondylolysis, which is a development defect in the lamina of the vertebra, is common in the lumbar area but may also occur in other regions. As a result of the weakened lamina, the shear forces acting on the lumbar spine may cause a forward slippage of the affected vertebra. This condition is called **spondylolisthesis**, which is described as the slippage of all or part of one vertebra on another ("spondylos" = vertebra and "olisthesis" = slip or slide down an incline). Spondylolisthesis may be due to a wide variety of causes, and although it is common in the lumbar area, it may also occur in other regions (Fig. 4–41). The altered location of the slipped vertebra changes its relationship to adjacent structures and creates excessive stress on associated supporting ligaments and joints. Overstretched ligaments may lead to the lack of stability or **hypermobility** of the segment. Narrowing of the posterior joint space, which occurs with forward slippage of a vertebra, may cause stress to nerve roots. The slipped vertebrae may even create a shear stress on the spinal cord and result in a paralysis. Pain in spondylolisthesis may arise from excessive stress

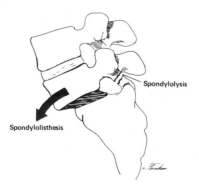

Figure 4–41. Shearing forces on the fifth lumbar vertebra may cause spondylolisthesis when a spondylolysis is present.

on any of the following pain-sensitive structures: anterior and posterior longitudinal ligaments, interspinous ligament, nerve roots, vertebral bodies, zygapophyseal joint capsules, synovial linings, or the muscles.

Model for Determining Effects of Deficits

The hypothetical effects of an injury, disease, or development deficit may be analyzed by taking the following points into consideration:
1. The normal function that the affected structure is designed to serve
2. The stresses that are present during normal situations
3. The anatomic relationship of the structure to adjacent structures
4. The functional relationship of the structure to other structures

Normal Structure and Function

We will consider the posterior longitudinal ligament as an example. (1) The normal function of this ligament is to reinforce the posterior aspect of the intervertebral joints and to limit flexion of the vertebral column. (2) During flexion of the column, the ligament normally is subjected to distraction. Under *normal conditions* the ligament elongates during flexion, provides resistance to flexion, and provides stability at the intervertebral joints. When the spine returns to neutral position, the ligament returns to its normal state. (3) Anatomically, the ligament is adjacent to the spinal cord on one side, and the disks and vertebral bodies on the other. (4) Functionally, the ligament works in conjunction with the supraspinous, interspinous, and ligamentum flavum to limit flexion of the column and provide stability of the joints.

Hypothetical Effects of Injury

If one uses the background knowledge of normal function and structure of the ligament as a foundation, then it is possible to offer a number of hypotheses regarding the effects of ligamentous damage due to injury or disease (Table 4–6). (1) There will be a lack of support for the intervertebral joint, which may lead to excessive tilting of the vertebra during flexion. (2) The excessive tilting combined with lack of support for the posterior annulus may lead to the possibility of tearing the annulus fibers. (3) Under compression forces, the posterior part of the annulus may bulge excessively into the spinal canal as a result of the lack of normal reinforcement provided by the posterior longitudinal ligament. The bulging annulus may cause pressure on the spinal cord and result in disturbed functioning of structures supplied by the spinal cord. (4) Other structures that are functionally related, such as the supraspinous ligaments and the joint capsules of the facet joints, will be under increased stress due to the diminished stability at the intervertebral joints. These structures may be excessively stretched during flexion of the column, resulting in further instability.

Summary

The model for analyzing the effects of alterations in structure, which was presented in the preceding paragraphs, can be used throughout this text. The hypotheses and Table 4–6 represent just a sampling of the type of theoretical reasoning one can apply to determine the effects of injury and disease on any structure. The reader is encouraged to follow this method of analysis, using a variety of structures both in this chapter and in subsequent chapters.

Table 4–6. INJURY—POSTERIOR DISK HERNIATION AT L-4 TO L-5 CAUSED BY RUPTURE OF POSTERIOR ANNULUS FIBROSUS

Normal Intervertebral Disk	Hypothetical Effects of Injury
Normal Function	*Loss of Normal Function*
1. Shock absorption	1. Loss of shock-absorbing capability.
2. Distribution of forces	2. Abnormal concentration of forces on vertebral bodies.
3. Pivot for motion	3. Decreased motion—disturbance of normal tilting action.
4. Stability-joint integrity	4. Decreased stability initially, but subsequent replacement of damaged tissue by fibrocartilaginous tissue.
Normal Stresses	*Abnormal Stresses*
1. Compression	1. Increased compression due to loss of shock absorbing capability and ability to distribute forces. Also initial spasm of surrounding muscles may increase compression.
2. Tension	2. Increased tension on any remaining posterior annulus fibrosus fibers and on posterior longitudinal ligament during flexion.
Anatomic Relationships	*Disrupted Anatomic Relationships*
1. Posterior longitudinal	1. Pressure exerted on the posterior longitudinal ligament when nuclear material extrudes posteriorly.
2. Spinal cord	2. Possible pressure on spinal cord and/or nerve roots due to either nuclear protusion or narrowing of foramen.
Functional Relationships	*Disrupted Functional Relationships*
1. Intervertebral joints	1. The diminished joint space at the intervertebral joint caused by the loss of the hydrostatic pressure in the nucleus may cause compression at the facet joints during extension of the column.
2. Spinous processes	2. Impingement of one spinous process on another may occur in extension as a result of the diminished joint space at the intervertebral joint.

REFERENCES

1. Carmichael, SN and Burkart, SL: Clinical anatomy of the lumbosacral complex. Phys Ther 59:8, 1979.
2. Panjabi, M: Symposium on the lumbar spine 11. Orthop Clin North Am 8:169–179, 1977.
3. Panjabi, M, Yamamoto, I, Oxland, T, and Crisco, J: How does posture affect coupling in the lumbar spine? Spine 14:1002–1011, 1989.
4. Palastanga, N, Field, D, and Soames, R: Anatomy and Human Movement: Structure and Function, Heinemann Medical Books, Halley Court, Jordan Hill, Oxford, 1989.
5. Gracovetsky, S and Farfan, H: The optimum spine. Spine 11:543–573, 1986.
6. Bogduk, N and Twomey, LF: Clinical Anatomy of the Lumbar Spine. Churchill-Livingstone, Edinburgh, 1987.
7. Jensen, GM: Biomechanics of the lumbar intervertebral disk: A review. Phys Ther 60:4, 1980.
8. Kapandji, IA: The Physiology of the Joints 3, ed 2. Churchill-Livingstone, Edinburgh, 1974.
9. White, AA and Panjabi, MM: Clinical Biomechanics of the Spine, ed 2. JB Lippincott, Philadelphia, 1990.

10. Twomey, LT and Furniss, BI: The life cycle of the intervertebral disc: A review. Aust J Physiol 24:4, 1978.
11. Urban, JPG and McMullin, JF: Swelling pressure of the lumbar intervertebral discs: Influence of age spinal level, composition and degeneration. Spine 13:179–186, 1988.
12. Ghosh, P: The Biology of the Intervertebral Disc. Vol. 1, CRC Press, Boca Raton, Florida, 1988.
13. Adam, M and Deyl, Z: Degenerated annulus fibrosus of the intervertebral disk contains collagen type III. Ann Rheum Dis 43:258–263, 1984.
14. Brickley-Parsons, D and Glimcher, MJ: Is the chemistry of collagen in the intervertebral disc an expression of Wolff's law? Spine 9:148–182, 1984.
15. Bogduk, N, Windsor, M, and Inglis, A: The innervation of the cervical intervertebral discs. Spine 13:2–7, 1988.
16. Finneson, B: Low Back Pain, ed 2. JB Lippincott, Philadelphia, 1980.
17. Lamb, DW: The neurology of spinal pain. Phys Ther 59:8, 1979.
18. Roberts, S, Menage, J, and Urban, JPG: Biochemical and structural properties of the cartilage end-plate and its relation to the intervertebral disc. Spine 14:100–170, 1909.
19. Edelson, JG and Nathan, H: Stages in the natural history of the end-plates. Spine 13:21–26, 1988.
20. McFadden, KD and Taylor, JR: End-plate lesions of the lumbar spine. Spine 14:867–869, 1989.
21. Bogduk, N and Engel, R: The menisci of the lumbar zygapophyseal joints. Spine 9:454–460, 1984.
22. Adams, MA and Hutton, WC: The mechanical function of the lumbar apophyseal joints. Spine 8: 1983.
23. Ghosh, P: The Biology of the Intervertebral Disc. Vol 11, CRC Press, Boca Raton, Florida, 1988.
24. Cyron, BM and Hutton, WC: The tensile strength of the capsular ligaments of the apophyseal joints. J Anat 132:145–150, 1981.
25. Myklebust, JB et al: Tensile strength of spinal ligaments. Spine 13:526–531, 1988.
26. Hedtmann, A et al: Measurements of human spine ligaments during loaded and unloaded motion. Spine 14:175–185, 1989.
27. Nordin, M and Frankel, VH: Basic Biomechanics of the Musculoskeletal System, ed 2. Lea & Febiger, Philadelphia, 1989.
28. Twomey, LT and Taylor, JR: Sagittal movements of the human vertebral column: A qualitative study of the role of the posterior vertebral elements. Arch Phys Med Rehabil 64:322–324, 1983.
29. Adams, MA, Dolan, P, and Hutton, WC: The lumbar spine in backward bending. Spine 13:1019–1026, 1988.
30. Twomey, L and Taylor, J: Flexion creep deformation and hysteresis in the lumbar vertebral column. Spine 7:116–122, 1982.
31. Twomey, LT, Taylor, JR, and Oliver, MJ: Sustained flexion loading, rapid extension loading of the lumbar spine, and the physical therapy of related injuries. Physiother Pract 4:129–138, 1988.
32. Hansson, TH, Keller, TS, and Punjabi, M: A study of the compressive properties of lumbar vertebral trabeculae: Effects of tissue characteristics. Spine 12: 1987.
33. Keller, TS et al: Regional variations in the compressive properties of lumbar vertebral trabeculae: Effects of disc degeneration. Spine 14: 1989.
34. Shirazi-Adl, A: Strain in fibers of a lumbar disc. Spine 14:98–103, 1989.
35. Adams, MA and Hutton, WC: Gradual disc prolapse. Spine 10:524–531, 1985.
36. Panjabi, M et al: Intrinsic disk pressure as a measure of integrity of the lumbar spine. Spine 13:913–917, 1988.
37. Panjabi, M et al: Spinal stability and intersegmental muscle forces: A biomechanical model. Spine 14:194–199, 1989.
38. Pope, M and Panjabi, M: Biomechanical definitions of spinal stability. Spine 10:255–256, 1985.
39. Gracovetsky, SA: The resting spine: A conceptual approach to the avoidance of spinal reinjury during rest. Phys Ther 67:549–553, 1987.
40. Parnianpour, M, Nordin, M, Frankel, VH, and Kahanovitz, N: The effect of fatigue on the motor output and pattern of isodynamic trunk movement. Isotechnol Res Abstr, April 1988.
41. Nachemson, AL: The lumbar spine: An orthopedic challenge. Spine 1:1, 1976.
42. Klein, JA and Hukins, DWL: Relocation of the bending axis during flexion-extension of lumbar intervertebral discs and its implications for prolapse. Spine 8: 1983.
43. Klein, JA and Hukins, DWL: Functional differentiation in the spinal column. Eng Med 12:83–85, 1983.
44. Haher, TR et al: Contribution of the three columns of the spine to rotational stability: A biomechanical model. Spine 14:663–669, 1989.
45. Goel, VK et al: Moment-rotation relationships of the ligamentous occipito-atlanto-axial complex. J Biomech 8:673–680, 1988.
46. Panjabi, M et al: Three dimensional movements of the upper cervical spine. Spine 13:726–730, 1988.
47. Nolan, JP and Sherk, HH: Biomechanical evaluation of the extensor musculature of the cervical spine. Spine 13:9–11, 1988.
48. Pal, GP and Sherk, HH: The vertical stability of the cervical spine. 13:447–449, 1988.

49. Maroney, SP, Schultz, AB, Miller, JAA, and Andersson, GBJ: Load-displacement properties of lower cervical spine motion segments. J Biomech 21:769–779, 1988.
50. Kent, BA: Anatomy of the trunk: A review. Part 1. Phys Ther 54:7, 1974.
51. Basmajian: Primary Anatomy, ed 7. Williams & Wilkins, Baltimore, 1976.
52. Cailliet, R: Neck and Arm Pain, ed 2. FA Davis, Philadelphia, 1981.
53. Brunnstrom, S: Clinical Kinesiology, ed 3. FA Davis, Philadelphia, 1972.
54. Columbini, P et al: Estimation of Lumbar Disc Areas.
55. Macintosh, JE and Bogduk, N: The morphology of the lumbar erector spinae. Spine 12:658–668, 1987.
56. Gracovetsky, S: The Spinal Engine. Springer-Verlag, New York, 1988.
57. Ladin, Kurukundi, RM, and DeLuca, CJ: Mechanical recruitment of low-back muscles. Spine 14:927–938, 1989.
58. Parnianpour, M, Nordin, M, Kahanovitz, and Frankel, V: Triaxial Coupling of Torque Generation. Spine 13:982–990, 1988.
59. Cailliet, R: Soft Tissue Pain and Disability. FA Davis, Philadelphia, 1977.
60. Hollinshead, WH: Functional Anatomy of the Limbs and Back. WB Saunders, Philadelphia, 1976.
61. Cox, HH: Sacroiliac subluxation as a cause of backache. Surg Gynecol Obstet 45:637–649, 1927.
62. Bowen, V and Cassidy, JD: Macroscopic and microscopic anatomy of the sacroiliac joint from embryonic life until the eighth decade. Spine 6:620–627, 1981.
63. DonTigny, RL: Function and pathomechanics of the sacroiliac joint: A review. Phys Ther 65:35–44, 1985.
64. Reilly, JP et al: Disorders of the sacroiliac joint in children. J Bone Joint Surg 1:40, 1988.
65. Mierau, DR, Cassidy, J, Hamin, T, and Milne, RA: Sacroiliac joint dysfunction and low back pain in school aged children. J Manipulative Physiol Ther 7:81–84, 1984.
66. Gould, JA and Davies, GJ (eds): Orthopaedics and sports physical therapy. CV Mosby, St. Louis, 1985.
67. Coventry, MB and Taper, EM: Pelvic instability. J Bone Joint Surg 54A:83, 1972.
68. Grieve, GP: The sacro-iliac joint. Physiotherapy 62:8, 1979.
69. Bartelink, DL: The role of abdominal pressure in relieving the pressure on the lumbar intervertebral disc. J Bone Joint Surg 39B:718–725.
70. Sullivan, MS: Back support mechanisms during manual lifting. Phys Ther 69:52–59, 1989.
71. Edgar, M: Pathologies associated with lifting. Physiotherapy 65:8, 1979.
72. Troup, JDG: Biomechanics of the vertebral column. Physiotherapy 65:8, 1979.
73. Nachemson, A: The load on lumbar discs in different positions of the body. Clin Orthop 45:107, 1966.
74. Bissonault, JS and Blaschak, MJ: Incidence of diastasis recti abdominis during the childbearing years. Phys Ther 68:1082–1086, 1988.
75. Buseck, M, et al: Influence of dynamic factors and external loads on the moment at the lumbar spine in lifting. Spine 13:918–920, 1988.
76. Miller, JAA, Schmatz, C, and Schultz, AB: Lumbar disc degeneration: Correlation with age, sex, and spine level in 600 autopsy specimens. Spine 13:173–178, 1988.

STUDY QUESTIONS

1. Which region of the vertebral column is most flexible? Explain why this region has greater flexibility.

2. Describe the relationship between the zygapophyseal joints and the intervertebral joints.

3. What is the zygapophyseal facet orientation in the lumbar region? How does this orientation differ from that of other regions? How does the facet joint orientation in the lumbar region affect motion in that region?

4. How would a tear in the supraspinous ligament affect function at the intervertebral joints? How would this injury affect the interspinous, intertransverse, and posterior longitudinal ligaments? Would any other structures be affected?

5. Which structures would be affected if a person has an increased anterior convexity in the lumbar area? Describe the type of stress that would occur, where it would occur, and how it would affect different structures.

6. Explain how a limitation of motion at the hip joints affects motion at the lumbar spine.

7. Describe the function of the intervertebral disk during motion and in weight bearing.

8. During rotation of the spine, in which area will you find the spinous processes rotating to the opposite side (convexity) of the curve? In which area do they rotate to the same side as the direction of the vertebral body?

9. Identify the factors that limit rotation and lateral flexion in the thoracic spine. Explain how the limitations occur.

10. Which muscles cause extension of the lumbar spine? In which position of the spine are they most effective?

11. Explain how the Valsalva maneuver may help to provide stability for the lumbar spine.

12. Describe the forces that act on the spine during motion and at rest.

13. Explain how "creep" may adversely effect the stability of the vertebral column?

14. Describe how muscles and ligaments interact to provide stability for the vertebral column.

15. What role has been attributed to the thoracolumbar fascia in lifting?

16. How does the position of hip flexion in the supine position assist in childbirth?

APPENDIX

THE MUSCLES OF THE LUMBAR VERTEBRAL COLUMN*

Muscles	Description	Origin and Insertion	Action
Psoas major (PM)	Muscle is layered with fibers originating from higher lumbar levels forming the outer surface of the muscle and those from lower levels buried deeper and sequentially within its substance. Fibers run inferiorly and laterally following the iliacus m. and pelvic brim.	Arises from medial-anterior surface of transverse process, IV disk, margins of vertebral bodies next to disk and to a fibrous arch that connects the upper and lower margins of each lumbar vertebral body. Arch covers the concavity of the lateral surfaces of the vertebral bodies leaving a space between the arch and bone, which transmits lumbar arteries and veins. Inserts at lesser trochanter.	Hip flexion and flexion of lumbar spine with the thigh fixed.
Intertransversarii lateralis (IL)	Two parts: ventrales (ILV) and dorsales (ILD).	ILV connects margins of consecutive transverse processes. ILD connects accessory process to transverse process below.	Thought to act with the QL in lateral flexion of trunk. According to *Gray's Anatomy* the IL also provides posture control.
Quadratus lumborum (QL)	The muscle is complex with many oblique and longitudinally running fibers that connect the lumbar transverse processes, ilium, and also the 12th rib where the majority of fibers are connected.	Arises from L-5 transverse process, the superior and anterior iliolumbar ligaments, and iliac crest lateral to the point of attachment of the iliolumbar ligament. Inserts at inferior border of last rib and L-1–L-4 transverse processes.	Fixes 12th rib during respiration and same side lateral flexion of lumbar spine.
Interspinales (Int)	Short, paired muscles on either side of the interspinous ligament.	Spinous process to spinous process of adjacent lumbar vertebrae.	Lumbar extension and postural control.
Intertransversarii mediales (IM)	Considered "true" back muscles due to their innervation.	Arise from accessory process, the adjoining mamillary processes and mamillo-accessory ligament that connects these processes. Inserts	Feedback for posture control. *Gray's* suggests IM couples with MULT in

THE MUSCLES OF THE LUMBAR VERTEBRAL COLUMN° *CONTINUED*

Muscles	Description	Origin and Insertion	Action
		into superior aspect of the mamillary process of vertebrae below.	lateral flex of trunk. Twomey disagrees.
Multifidus (MULT)	Largest and most medial of the lumbar spine muscles. Lumbar MULT consists of large fascicles arranged segmentally that radiate from spinous processes. Fascicles are arranged in five overlapping groups so each lumbar vertebra gives rise to one group. Each fascicle arises from a common tendon at the caudal tip of each lumbar spinous process. Each fascicle splits caudally to assume separate attachments to mamillary processes, iliac crest and sacrum. Deeper fibers are attached to zygapophyseal joints and protect the joint capsules from being pinched during spinal movements.	Fascicle from base of L-1 spinous process inserts into the L-4 mamillary process, while fascicles of the common tendon insert at mamillary processes of L-5 and S-1 and at the posterior superior iliac spine. Fascicle at base of L-2 inserts on the mamillary process of L-5 while those from the common tendon insert at the S-1 mamillary process, posterior superior iliac spine, and an area just below it. Fascicle from L-3 spinous process inserts into mamillary process of the sacrum, posterior superior iliac spine and lateral edge of third sacral segment. L-4 fascicles insert on the sacrum, medial to the L-3 insertion and lateral to the dorsal sacral foramina. L-5 fascicles insert on an area medial to the dorsal sacral foramina.	Theoretically, the MULT causes lumbar spine extension and opposes or stabilizes against the flexion effect of the abdominal muscles as they produce spinal rotation. In addition, deep fibers attached to the zygapophyseal joints protect the joint capsules from becoming caught inside the joint during spinal movements.
Longissimus lumborum (LONG.L)	L-5 fascicles lie deep to other fascicles that lie progressively more superficial so that the L-1 fascicle is the most superficial. Lumbar component of the longissimus m. *Gray's Anatomy* does not recognize the existence of a lumbar component	Composed of five fascicles. Each one arises from accessory processes and adjacent medial end of the dorsal surfaces of the transverse processes of the lumbar vertebrae. L-1 through L-4 fascicles form tendons that converge and form the intermuscular aponeurosis of the lumbar spine which attaches on the ilium.	Action depends on whether the muscle contracts unilaterally, or bilaterally. Contracting unilaterally, the longissimus produces lateral flexion of the vertebral column.

THE MUSCLES OF THE LUMBAR VERTEBRAL COLUMN° *CONTINUED*

Muscles	Description	Origin and Insertion	Action
	of the longissimus muscle.	L-5 fascicle inserts on the medial posterior superior iliac spine.	Bilateral contraction produces posterior sagittal rotation and posterior translation.
Iliocostalis lumborum (L.ILIO)	*Gray's* and Bogduk & Twomey are not in agreement. *Gray's Anatomy* refers to L.ILIO as one muscle group while Bogduk & Twomey detail the specific fascicles of the L.ILIO and their attachments.	L.ILIO consists of four overlapping fascicles originating from L-1 through L-4 vertebrae and from the tips of their transverse processes over 2 to 3 cm onto the middle layer of the thoracolumbar fascia. L-4 is the deepest and inserts directly into the iliac crest just lateral to the posterior superior iliac spine. It is covered by the fascicle of L-3 which inserts just dorsolaterally of the L-4 insertion on iliac crest. L-2 covers L-3 and L-1 covers L-2 sequentially with iliac crest insertions becoming more dorsal and lateral. *Gray's* cites this muscle originating from the erector spinae and inserting at the inferior borders of the lower 6 or 7 costal angles.	Bilateral contraction causes extension of vertebral column. Unilaterally, contraction causes same side lateral flexion and cooperates with the MULT to oppose the flexion effect of abdominal muscles when they act to rotate the trunk.

°The muscles of the vertebral column have been compiled with descriptions, origins and insertions, and actions based on the work of Bogduk and Twomey,[6] and Williams, PL, Warwick, R, Dyson, M, and Bannister, LH (eds): Gray's Anatomy, ed 37. Churchill Livingstone, London, 1989.

MUSCLES OF THE THORACIC VERTEBRAL COLUMN

Muscles	Description	Origin and Insertion	Action
Longissimus thoracis (T.LONG)	Largest part of the erector spinae consisting of 11–12 pairs of small fascicles.	According to Bogduk and Twomey, the T.LONG arises from the spinous processes of L-3 to S-3 and inserts into the ribs and transverse processes of T-1 and T-12. According to *Gray's Anatomy*, the insertion is specifically on the tips of all thoracic transverse processes and the lower 9 or 10 ribs between their tubercles and angles.	Bilateral contraction indirectly produces extension of the lumbar vertebral column through pull on the erector spinae aponeurosis. Unilateral contraction causes lateral flexion of the thoracic and lumbar regions.
Iliocostalis thoracis (ILIO.T)	According to Bogduk and Twomey, it represents the thoracic component of the muscle traditionally known as the iliocostalis lumborum.	According to Bogduk and Twomey, the muscle arises from angle of lower seven or eight ribs traveling downward and combining with the erector spinae and ultimately attaching to the posterior superior iliac spine. There is no attachment to the lumbar vertebrae. According to *Gray's*, the muscle arises from upper borders of lower costal angles and ascends to the upper borders of upper costal angles and the posterior portion of the C-7 transverse process.	Causes an increase in lordotic curve and aids in derotating the thoracic cage and lumbar spine.
Spinalis thoracis (SPIN)	Medial part of erector spinae, scarcely a separate muscle from the thoracic longissimus.	Muscle arises from three or four tendons from the 11th thoracic to the 2nd lumbar spinous process. The muscles unite and insert via separate tendons to the upper thoracic spinous processes.	Extension of thoracic vertebral column.

MUSCLES OF THE THORACIC VERTEBRAL COLUMN
CONTINUED

Muscles	Description	Origin and Insertion	Action
Semispinalis thoracis (SEMI)		Muscle arises from T-6 through T-10 transverse processes and inserts into the upper four thoracic and possibly the lower two cervical spinous processes.	Extension of thoracic vertebral column.
Rotatores thoracis (ROT)	Eleven pairs deep to the MULT, more prominent in the thoracic region.	This muscle connects the upper posterior part of a transverse process to the lower lateral surface of T-12–T-1 lamina between T12–T1.	Rotation of vertebral column.

MUSCLES OF THE HEAD AND CERVICAL VERTEBRAL COLUMN

Muscles	Description	Origin and Insertion	Action
Trapezius (TRAP)	A flat and triangular muscle that extends over the dorsum of the neck and upper thoracic area between the scapulae.	Medially, the **TRAP** arises from superior nuchal line, external occipital protuberance, ligamentum nuchae, and spinous processes of C-7 through T-12. According to *Gray's*, superior fibers descend, inferior fibers ascend, and intermediate fibers travel horizontally to converge laterally at the shoulder. Superior fibers attach at the posterior border of lateral one third of clavicle, middle fibers attach at the medial acromion and superior lip of scapular spine, inferior fibers are attached at medial end of scapular spine.	Stabilizes scapula during arm movements; with shoulder fixed, **TRAP** may bend the neck and head posterolaterally. Combined with the levator scapula, rhomboids, and serratus anterior the **TRAP** produces various scapular rotations.

MUSCLES OF THE HEAD AND CERVICAL VERTEBRAL COLUMN *CONTINUED*

Muscles	Description	Origin and Insertion	Action
Sternocleidomastoid (SCOM)	According to *Gray's Anatomy*, the muscle forms a large band which runs obliquely forward and downward to the anterolateral aspect of the neck and is easily visible; inferiorly it has two bands, a sternal and a clavicular portion.	The sternal head attaches to the upper anterior surface of the manubrium of the sternum. The clavicular head attaches at the medial one third of the superior surface of the clavicle. Fibers insert at the mastoid process.	Unilateral contraction: rotation contralaterally, lateral flexion ipsilaterally and extension. Bilateral contraction: accentuates cervical lordosis and produces neck extension.
Scalenus (SCAL)	The SCAL consists of three pairs of muscles located on the anterolateral aspect of the cervical vertebral column.	A.SCAL arises from transverse processes of C-3–C-6, M.SCAL from C-2–C-7 transverse processes. P.SCAL C-4–C-6 transverse processes. Insertions are at the first rib (A.SCAL and M.SCAL) and the second rib (P.SCAL).	Bilateral contraction; anterolateral flexion; unilateral contraction: lateral flexion to same side. All muscles elevate ribs when acting in reverse action.
Spinalis cervicis (SPIN)	This muscle is not always present.	The muscle ascends from the lower ligamentum nuchae and C-7 to insert into the spine of the axis.	The spinales muscles are extensors of the vertebral column.
Spinalis capitus (SPIN)	According to *Gray's, Anatomy* the muscle is often indiscernible from the SEMI capitus.		
Semispinalis cervicis (SEMI)	Cervical portion of SEMI.	The muscle attaches at the upper 5 or 6 thoracic transverse processes and inserts superiorly into spines of C-5–C-2.	The SEMI cervicis extends and contralaterally rotates the cervical spine.
Semispinalis capitus (SEMI)	This muscle is deep to SPL and medial to LONG.	The muscle arises from T-7–C-7 transverse processes and inserts into the area between the superior and inferior nuchal lines.	The SEMI capitus extends and contralaterally rotates the head.

MUSCLES OF THE HEAD AND CERVICAL VERTEBRAL COLUMN *CONTINUED*

Muscles	Description	Origin and Insertion	Action
Splenius (SPL)	This muscle is deep to the TRAP and the rhomboids.	Muscle arises from the lower ligamentum nuchae and spines of C-7–T-4 and inserts into the lateral nuchal line.	Retraction of the head and flattens the cervical spine.
Longus colli cervicis (L.COL)	Composed of tendinous slips which have attachments to cervical and thoracic vertebrae.	The muscle arises from the anterior aspects of T-1–T-3 and inserts into the anterolateral surface of the atlas.	Bilateral contraction: flexion of cervical region. Unilateral contraction: lateral flexion and rotation.
Longus capitus (L.CAP)	Composed of tendinous slips.	Muscle arises from transverse processes of C-3–C-6 and inserts into the inferior surface of occipital bone anterior to the foramen magnum.	Flexion of head.
Rectus capitus anterior (RCA)	Consists of short and flat muscles deep to the L.CAP.	Arises from anterior surface and base of transverse process of atlas. Inserts at inferior surface of occipital bone anterior to foramen magnum.	Flexion of head at the atlanto-occipital joint.
Rectus capitus lateralis (RCL)	Overlies the anterior surface of the atlanto-occipital joint.	Arises from the upper surface of transverse process of atlas and inserts into the occipital bone.	Lateral flexion of the head at the atlanto-occipital joint.
Iliocostalis cervicis (ILIO.C)	Medial to the tendons of the ILIO.T.	Arises from T-6–T-3 costal angles and ascends to insert at the posterior tubercles of C-4–C-6 transverse processes.	Aids in extension and lateral flexion of vertebral column.
Longissimus cervicis (LONG.Cerv)	Located medial to the LONG.T.	Arises from T-5–T-1 transverse processes to insert to the posterior tubercles of C-6–C-2.	Aids in extension and lateral flexion of vertebral column.
Longissimus capitus (LONG.Cap)	Located between the LONG.Cerv and SEMI.Cap muscles.	Arises from transverse processes of T-5–C-4 and inserts at the posterior mastoid process.	Aids in head extension and rotation to the same side.

MUSCLES OF THE HEAD AND CERVICAL VERTEBRAL COLUMN *CONTINUED*

Muscles	Description	Origin and Insertion	Action
Suboccipital Muscles			
Rectus capitus posterior major (RCPmajor)	Deep to SEMI capitus.	Arises as a pointed tendon from axial spine and broadens as it goes superior and lateral to insert at the inferior nuchal line.	Head extension and rotation to the same side.
Rectus capitus posterior minor	Deep to RCP major.	Arises as pointed tendon from posterior tubercle of atlas and inserts at inferior nuchal line.	Extension of the head.
Obliquus capitus inferior (OCI)	Deep to SEMI capitus.	Extends laterally from the spine and lamina of axis to the transverse process of atlas.	Same side rotation of the head.
Obliquus capitus superior (OCS)	Deep to SEMI capitus.	From transverse process of atlas going superiorly to insert between superior and inferior nucal lines.	Posterolateral flexion of the head.

Chapter 5

■ ■ ■

The Thorax and Chest Wall

───────────────── ■ ─────────────────

OBJECTIVES

Following the study of this chapter, the reader should be able to:

Describe

1. The articulations of the ribs with the thoracic vertebrae.
2. The articulations of the ribs and clavicle with the manubriosternum.
3. The structure and function of the manubriosternal and interchondral joints.
4. The shape and functions of the diaphragm.
5. The functions of the interosseous intercostals, levatores costarum, parasternals, scalenes, and triangularis sterni muscles.
6. The functions of the accessory muscles of ventilation.
7. The motions of the chest wall and abdomen during ventilation.

Identify

1. Functions of the abdominals in inspiration and expiration.
2. The structures that provide shape, motion, and stability of the chest wall/thorax.

3. Specific muscles of inspiration, expiration, and those that contribute to both functions.
4. Muscles involved in quiet breathing, during exercise, and in standing and supine positions.

Explain
1. The relationship between the parasternal and scalene muscles during ventilation.
2. The relationship between the diaphragm and abdominal muscles during ventilation.

Compare
1. The relationship between abdominal and thoracic pressures during ventilation.
2. Thoracic structure and function differences between an infant and older person.

Analyze
1. The effects of pregnancy as a normal functional alteration of rib cage and diaphragmatic structure and function.
2. How structural scoliosis alters normal structure and function of ventilation.
3. The effects of hyperinflation on the chest wall and on the efficiency of respiratory muscle function.

GENERAL STRUCTURE AND FUNCTION

Optimal ventilation of the lungs is essential to function of the human body. The process of inspiration and expiration is dependent on an extraordinary and complex set of musculoskeletal and kinesiologic interrelationships involved with the structure and function of the chest wall. The ventilatory "bellows," as the entire mechanism is known, is a complex involving some 88 joints and more than 46 muscles.[1] The respiratory muscles are the only skeletal muscles on which the body depends to sustain life and consequently, must function continuously.[2]

The chest wall in humans has three parts: the rib cage, the diaphragm, and the abdomen[2,3] (Fig. 5–1). The coupling and interaction of these three parts during

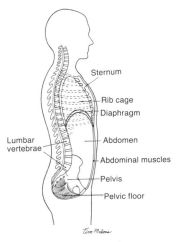

Figure 5–1. The chest wall.

ventilation of the lungs involves a complex set of integrated components that will be examined in detail in this chapter.

Some examples of alterations of normal structure and function will also be presented to illustrate how selected normal functional, musculoskeletal, and systemic conditions can affect the biomechanics of the lungs and chest bellows.

RIB CAGE

The rib cage consists of 12 pairs of ribs, the thoracic vertebrae to which the ribs attach, and the related joints or linkages. The thoracic vertebrae, the base for the rib cage, is the least mobile, and consequently, the most stable section of the vertebral column.[4] The ribs articulate posteriorly with the bodies and transverse processes of the thoracic vertebrae and anteriorly join the sternum via the costal cartilages. The 1st through 7th ribs are known as "true" ribs because their cartilages attach directly to the sternum (Fig. 5–2). The 8th to 10th ribs articulate with the sternum via the costal cartilages above them. The 11th and 12th ribs are "floating" ribs; that is, they have no attachment to the sternum at all.[5] Anteriorly, the sternum provides an osseous protective plate for the heart and is composed of the manubrium, the body, and the xiphoid process.

Articulations of the Rib Cage

The articulations that comprise the chest wall (or thorax) include the manubriosternal (MS), xiphisternal (XS), costovertebral (CV), costotransverse (CT), costochondral (CC), chondrosternal (CS), and interchondral joints (see Fig. 5–2).

Manubriosternal and Xiphisternal Joints

The MS joint is a synchondrosis; that is, a joint with a fibrocartilaginous disk between the hyaline cartilage-covered articulating ends of the manubrium and sternum. The MS joint is so similar to the symphysis pubis of the pelvis that some authors refer to it as the "symphysis sterni."[6] Ossification of the joint occurs in approximately 10 percent of older adults. In approximately one third of adult women and elderly adults, a synovial-lined joint cavity exists secondary to resorption of the central portion of the disk.[6,7] This phenomenon has been linked to involvement of the MS joint in rheumatoid arthritis.[6]

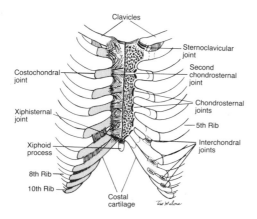

Figure 5–2. An anterior view of the articulations of the rib cage: The shaded areas indicate costal cartilage. The costal cartilages join the ribs at the CC joints. The costal cartilages of the first through the seventh ribs articulate directly with the sternum via the CS joints. The costal cartilages of the eighth through the tenth ribs articulate indirectly with the sternum through the costal cartilages of the adjacent superior rib at the interchondral joints. Interchondral joints may also exist between the fifth through the ninth costal cartilages. The MS joint has a fibrocartilaginous disk between the hyaline cartilage-covered joint surfaces of the manubrium and sternum.

Figure 5-3. A lateral view of the CV joints and ligaments. The three bands of the radiate ligament reinforce the CV joints. The superior and inferior portions of the radiate ligament attach to the capsular ligament (removed) and to the vertebral body. The third part of the ligament attaches to the intervertebral disk. In the middle of the illustration, the CV joint is shown with the radiate ligament and capsule removed to show the intra-articular ligament, which attaches the head of the rib to the annulus.

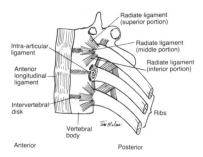

The XS joint is a synchrondrosis articulating the body of the sternum with the xiphoid process. This joint tends to ossify by about 40 to 50 years of age.[4]

Costovertebral Joint

The CV joint is a plane synovial joint formed by the head of the rib and two demifacets found on adjacent thoracic vertebrae. The costal facet on the head of the rib is small, oval-shaped, and slightly convex. Facets on vertebrae T-2 to T-8 are formed by two concave demifacets on the inferior body of one vertebra and on the superior body of the vertebra below it. The vertebral facets have migrated backward onto the pedicle as one looks at CV joint articulations from T-9 to T-12. The articular surface of the CV joint also includes articulation with the interposed intervertebral disk. The articular surface of the head of ribs 2 to 10 fit snugly into the "angle" formed by the vertebral facets and the disk. The 1st, 11th, and 12th ribs articulate with one vertebra only.[4,5,7]

Each of the costal and vertebral articular surfaces are joined by a thin, fibrous capsule and supporting ligaments. The articulation of ribs 2 to 10 includes an **interosseous** or **intra-articular** ligament that lies within the capsule and tethers the head of the rib to the annulus that lies between the demifacets. The intra-articular ligament divides the synovial joint into two cavities.[4,7] The **radiate (or stellate) ligament** (Fig. 5–3) is firmly connected to the anterolateral capsular ligament and has three bands, the superior and inferior, which insert on the vertebral bodies, and the intermediate, which inserts on the intervertebral disk.[4,5,8,9] Posteriorly, the capsular ligament merges with the posterior longitudinal ligament of the vertebral column.

The CV joints that articulate with a single vertebra are more mobile than those joints that articulate with two vertebrae. Motions at all CV joints are rotation and gliding.[9,10]

Costotransverse Joint

The CT joint is formed by the articulation of the costal tubercle of the rib with a costal facet on the transverse processes of the 1st through 10th thoracic vertebrae.[4] The last two or three ribs (floating ribs) have neither articular tubercles nor CT joints. The CT joint is a plane synovial joint. The costal facet is concave and the costal tubercle convex from T-1 to T-6 or T-7 allowing some rotation between segments. Both articular surfaces are flat from T-7 or T-8 where gliding motions predominate.[4,8] The CT joint capsule is strengthened by three major ligaments: the **medial, lateral,** and **superior** CT ligaments.[4,5] An **inferior** ligament is sometimes identified as well. Together they provide stability for the CT joint.

Ribs 1 to 10 articulate with the vertebral column by two synovial joints (the CV and CT joints) and with the manubriosternum. The joints therefore form a closed kinematic chain. The closed-chain relationship means the segments are

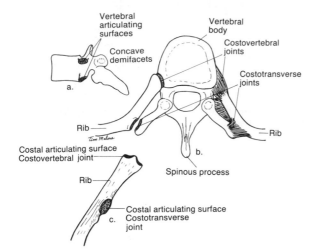

Figure 5-4. CV and CT joints. a. Lateral view of a thoracic vertebra showing the articulating surfaces of the CV joints. b. Posterior-superior view of a vertebra showing paired CV and CT joints. Joint capsules and ligaments have been removed on the left to show the articulating surfaces. c. The articulating surfaces on the rib.

interdependent and motion is more restricted. The 11th and 12th ribs, however, form an open kinematic chain and are less restricted. The alignment of the CV and CT joints (the CV joint of each rib is superior to the CT joint) contributes to the downward sloping of the ribs posteriorly (Fig. 5–4).

Costochondral and Chondrosternal Joints

The first through the seventh ribs articulate with costal cartilages (CC joints) and the cartilages articulate with the manubriosternum (CS joints). The CC joints are synchondroses surrounded by periosteum.[8] The CC joints have no ligamentous support.

The CS joints of ribs two through seven are synovial joints divided by an intra-articular ligament similar to the intra-articular ligament of the CV joint (see Fig. 5–2.)[4,8] Ligamentous support for these joints includes a thin capsule reinforced by anterior and posterior radiate ligaments.[8] Grieve also describes a costoxiphoid ligament of the CS joints.[4] The first rib's costal cartilage is much stiffer than the others and is connected to the manubriosternum by a primary synchondrotic cartilaginous joint. The first rib's CS joint also has no intra-articular ligament.[5] The CS joints may be obliterated with aging.[5]

Interchondral Joints

The 8th through 10th (and sometimes 11th) costal cartilages articulate with the cartilage immediately above them, thus attaching to the sternum by a fused costal cartilage (see Fig. 5–2). The interchondral joints are synovial-like and are supported by a capsular ligament and interchondral ligaments. The interchondral articulations, like the CS joints, tend to become fibrous and fuse in old age.[4,8]

Kinematics of the Ribs and Manubriosternum

The movements of the ribs are an amazing combination of complex geometrics governed by the types and angles of the articulations (especially the CV and CT), the movement of the manubriosternum, and the contribution of the elasticity of the costal cartilages. A controversy exists in the literature regarding the mechanisms and types of motions that are actually occurring for each rib, *especially* regarding the axis of motion at the CV and CT joints.[9,10]

Investigators generally agree regarding the structure and motion of the first rib. As mentioned earlier, the first costal cartilage is stiffer than the rest. Also, the first CS joint is cartilaginous, not synovial, and therefore is firmly attached to the manubrium and permits relatively little movement. The first rib articulates at the CV joint with a single facet, increasing the mobility of the joint. During inspiration, the first rib elevates, moves superiorly and posteriorly at the CV joint, and pushes the rigidly attached manubrium upward. The MS joint also tends to move so that the top of the manubrium moves posteriorly. The first thoracic vertebra may also extend slightly.

The major controversy regarding rib motion centers on the types of motion at the CV articulations and whether or not the rib cage can be deformed during inspiration/expiration. Kapandji and others believe the CV and CT joints are mechanically linked with a *single* axis passing through the center of both joints.[3,4,9] Saumarez argues that the rib is rigid and, therefore, cannot rotate about a single fixed axis but rather moves as successive rotations about a shifting axis.[10] The functional relevance of the controversy is unclear, since the geometry and anatomy of the rib cage result in only restricted rib movement.[4,10]

Using the theory of a common axis for CV and CT joints, motions in ribs 2 to 10 generally occur around axis(es) lying nearly in the frontal plane for the upper ribs and around an axis approaching the sagittal plane for the lowermost ribs (Fig. 5–5). Given the closed kinematic chain formed by the ribs, some motion is likely at the CS joints of these ribs. Motion at the CS joints is limited to slight vertical motion with no rotation.[9] Motion at the interchondral joints of ribs 8 to 10 is a superior-inferior sliding.[8] The resultant movement of the elliptically shaped ribs is an upward and forward motion of the upper ribs called "pump-handle motion." The lower ribs move upward and laterally in what is known as "bucket-handle motion." The intermediate ribs move minimally both forward and laterally (Fig. 5–5).[4,5,9,11] The sternum is also pushed upward and forward by the action of the ribs, increasing the anterior-posterior diameter of the chest wall during inspiration.

Ribs 11 and 12 are free to move in any direction because they have no anterior

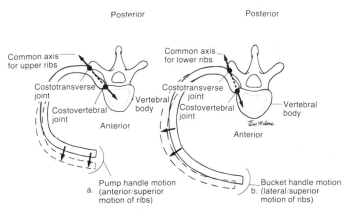

Figure 5–5. Axes and motions. a. The arrow represents the common axis of motion for the upper ribs. It lies nearly in the frontal plane and passes through the centers of the CV and CT joints. The upper ribs move upward and forward in a pump handle motion. b. The axis for the lower ribs lies closer to the sagittal plane. The upward and lateral motion of these ribs is referred to as "bucket handle motion."

articulations and articulate only with a single vertebral facet. During ventilation the quadratus lumborum depresses and fixes these ribs to provide adequate diaphragmatic muscle tension.[4,8]

Muscles Associated with the Rib Cage

The muscles that "power" the ventilatory bellows are generally referred to as the **respiratory muscles** (Fig. 5–6). Any muscle that attaches to the chest wall and many that attach to the shoulder girdle and thoracic vertebrae contribute to ventilation. The number of muscles involved and the extent of their involvement also is related to whether the ventilation is quiet (occurring at rest) or forced (occurring during exercise or in pathologic states).

The respiratory muscles, especially those that participate in normal quiet breathing, are skeletal muscles that differ from other skeletal muscles in three major ways: (1) they must contract rhythmically and intermittently throughout life, (2) the control of these muscles is both voluntary and involuntary, and (3) they must primarily work against elastic (chest wall and lungs) and airway resistive loads, rather than against the gravitational forces encountered by most other skeletal muscles.[2,12]

The respiratory muscles are usually classified as inspiratory or expiratory. The actual complexity of the actions of the respiratory muscles makes this classification somewhat inaccurate and misleading.

Intercostal and Related Muscles

The intercostal muscles connect adjacent ribs and are termed **internal** or **external** intercostals depending on their anatomic orientation. Both sets of intercostal muscles attach to the lower margins of the first 11 ribs and the costal cartilages and upper margins of the ribs and cartilages below. The internal intercostals lie deep to the external intercostals with fibers that run down and posteriorly. The most superficial external intercostal fibers run down and anteriorly in a "hand-in-pocket" pattern.[3,13]

The internal intercostals extend from the sternocostal junctions posteriorly to the angles of the ribs and then form an internal intercostal membrane. The external intercostals extend from the tubercles of the ribs anteriorly to the CC junctions where they form an external intercostal membrane. The intercostal spaces there-

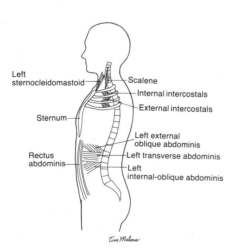

Left
sternocleidomastoid

Scalene

Internal intercostals

External intercostals

Sternum

Left external
oblique abdominis

Rectus
abdominis

Left transverse abdominis

Left
internal-oblique abdominis

Tim Malone

Figure 5–6. Respiratory muscles.

fore have only one muscle layer anteriorly (the internal intercostal) and posteriorly (the external intercostal) with two layers laterally (referred to as the **interosseous** or **lateral** intercostals). The muscle fiber orientation of the internal and external intercostals lie approximately at right angles to each other.[3,13] The interchondral part of the internal intercostals are referred to as the **parasternal muscles.**

The functions of the intercostal muscles has long been a source of debate and controversy. It is widely reported that the external intercostals have an inspiratory action and the internal intercostals have an expiratory function, although the actual biomechanics of these actions is *not* well understood. The parasternals have been conclusively shown to be inspiratory agonists, even during quiet breathing.[3,14] The action of the parasternals appears to be a rotation of the CS junctions, resulting in elevation of the ribs and descent of the sternum.

The kinesiology of the interosseous (lateral) portions of the internal and external intercostals is more complex than originally thought. Hamberger, in 1727, proposed a theory (now debated for centuries) that states that the orientations and insertions of the intercostal muscle fibers determine the actions of these muscles on the ribs.[5] Recent studies using electromyographic (EMG) and electrical stimulation techniques show that both interosseous intercostals have similar effects on the ribs into which they insert, but the ribs are not equally compliant (movable) at all times.[16,17] The ribs move cephalad more easily at low lung volumes and caudally more easily at high lung volumes. Further, this compliance/motion is determined by the apparent "stabilizing" of the upper ribs during inspiration (passive tension is higher in the neck muscles at low lung volumes) and the lower ribs during expiration (passive tension is greater in the abdominal muscles at high lung volumes when the rib cage is above its neutral position).[3,16,17] The lateral (interosseous) intercostals may be more involved with postural motions, especially trunk rotation, than respiratory motions.[3,17]

The **levatores costarum** are paravertebral muscles functionally associated with the intercostal muscles. The levatores costarum lie between C-7 through T-11. Fibers run from the transverse processes of a vertebra to the posterior external surface of the next lower rib between the tubercle and the angle. The levatores costarum muscles assist with inspiration, even during quiet breathing.[8,13]

The **triangularis sterni or transversus thoracis** (or sternocostalis) is a deep and flat layer of muscle deep to the internal intercostals. The triangularis sterni attaches to the posterior surface of the xiphoid process and body of the sternum and the internal surfaces of the costal cartilages of ribs three through seven.[3,13] Although there is no universal agreement, recent studies have shown that this muscle is primarily an expiratory muscle, especially when expiration is active (e.g., talking, coughing, laughing). It appears to always be active during expiration below functional residual capacity (FRC) or below a normal tidal expiration.[3,18] The triangularis sterni muscle has further been shown to be active during quiet expiration in older subjects in standing (it is usually silent in the supine position).[19]

The scalene muscles attach proximally to the transverse processes of C-3 to C-7 and distally to the upper borders of the first rib (scalenus anterior and medius) and second rib (scalenus posterior). The scalenes contribute to quiet inspiration. Two important functions of the scalenes are that they counteract the downward pull of the parasternals on the sternum and expand the upper rib cage in its anterior-posterior dimensions (pump-handle).[14,20]

Accessory Muscles

The muscles that attach to the rib cage, shoulder girdle, or vertebral column, which assist with inspiration in situations of stress (e.g., exercise, disease) but *not* during quiet breathing in normal humans are **accessory muscles** of ventilation.

These muscles are numerous and will not be discussed in detail in this chapter. Many of them are discussed in other chapters in this text in the context of the muscle's primary functions.

A partial list of the most common accessory muscles are the sternocleidomastoids, the pectoralis major, the pectoralis minor, the trapezius, and the subclavius muscles. The ability of the accessory muscles that attach to the shoulder girdle to assist in inspiration has been demonstrated to be decreased during unsupported arm exercise in both normal subjects and subjects with obstructive lung disease. This is because the accessory muscles of inspiration assist in increasing thoracic diameters by moving the rib cage up and outward when the shoulder girdle is fixed (e.g., patient is leaning forward on the hands or forearms).[20]

The quadratus lumborum, which is normally thought of as a lateral flexor of the vertebral column and a hip hiker, is an accessory muscle of expiration. The quadratus lumborum, which attaches to the 12th rib, acts as a stabilizer for the diaphragm as the diaphragm eccentrically contracts during phonation.[8,13,21,22]

DIAPHRAGM

Anatomically the diaphragm consists of two major components. The first component is the central tendon, which is noncontractile and ''boomerang''-shaped (forming the top of the so-called dome of the diaphragm). The central portion of the central tendon fuses with the pericardium. The lateral ''leaflets'' of the central tendon are the domes of the right and left hemidiaphragms (Fig. 5-7). The second component of the diaphragm is a circular margin of radially oriented muscle fibers that attach proximally to the central tendon.[23,24] *Functionally* (and embryologically), the muscular portion of the diaphragm can be thought of as two different muscles: the crural (vertebral or posterior) diaphragm and the costal diaphragm.[25]

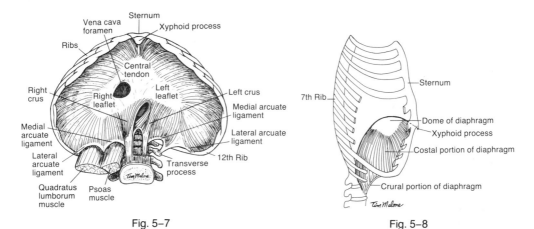

Fig. 5-7　　　　　　　　　　　　　　　　　Fig. 5-8

Figure 5-7. The diaphragm. An inferior view of the diaphragm looking up into the dome from below. The muscles (quadratus lumborum and psoas major) have been removed on the left side of the diagram to show the attachments of the medial and lateral arcuate ligaments to the transverse processes.

Figure 5-8. The diaphragm is shown in a lateral view with the intermediate ribs removed to show the dome of the diaphragm and costal attachments to the inner surfaces and costal cartilages of the lower six ribs and the xyphoid process. (Adapted from Kapandji.[9])

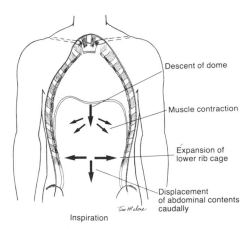

Descent of dome

Muscle contraction

Expansion of
lower rib cage

Displacement
of abdominal contents
caudally

Inspiration

Figure 5–9. An anterior view of the diaphragm
with the ribs removed to show motion of the ribs
and diaphragm during inspiration.

The **crural protion** of the diaphragm inserts on the anterolateral surfaces of the bodies and disks of L-1 to L-3 and on the aponeurotic arcuate ligaments. The medial arcuate ligament arches over the upper anterior part of the psoas muscles and extends from the L-1 or L-2 vertebral body to the transverse process of L-1, L-2, or L-3. The lateral arcuate ligament covers the quadratus lumborum muscles and extends from the transverse process of L-1, L-2, or L-3 to the 12th rib.[23,26] The costal part of the diaphragm attaches by muscular slips to the posterior aspect of the xiphoid process and inner surfaces of the lower six ribs and costal cartilages (Fig. 5–8).[23,24]

The shape of the diaphragm and angle of pull of its fibers are essential for its usual function. The diaphragm is elliptical and domed with its costal fibers running vertically. The portion of the diaphragm that is directly apposed to the inner wall of the lower rib cage is called the **zone of apposition**.[3] When the diaphragm contracts during inspiration, the shortening of the muscle fibers decreases the length of the apposed diaphragm with a *descent* of the dome of the diaphragm. The axial displacement of the dome displaces the abdominal contents caudally and increases intra-abdominal pressure (Fig. 5–9). This increased abdominal pressure acts as a "stabilizer" for the descending central tendon. Continued shortening of the costal fibers then generate an upward pull on the lower ribs that results in an upward and outward (secondary to the bucket-handle phenomenon) motion of the lower rib cage.[2,20,26] The crural portion of the diaphragm has a less direct inspiratory effect on the lower rib cage than the costal portion.[3,25] Indirectly, the action of the crural portion (via its effect on pulling the central tendon downward) increasing intra-abdominal pressure, which is transmitted across the apposed diaphragm to help expand the lower rib cage.[3,25]

ABDOMEN

The four muscles of the abdominal wall that have respiratory functions include, from deep to superficial, the tranversus abdominis, the internal abdominal oblique, the external abdominal oblique, and the rectus abdominis. The **tranversus abdominis** muscle attaches to the inner surface of the posterior aspect of the lower six ribs and runs circumferentially to insert in an anterior aponeurosis forming the rectus sheath. The internal abdominal oblique begins at the iliac crest and the inguinal ligament, running superiorly and medially to attach to the costal margin and the rectus sheath. The external abdominal oblique attaches to the external surfaces of the lower eight ribs with fibers running inferiorly and medially to the iliac crest, inguinal ligament, and the linea alba. The rectus abdominis attaches from the anterior surfaces of the fifth to seventh costal cartilages and the xiphoid process running inferiorly to the pubis.[3,13]

The abdominals have long been considered expiratory muscles as well as trunk flexors and rotators. The major function of the abdominals with respect to ventilation is to assist with forced expiration. The abdominals, however, have a significant role to play in inspiration as well. The abdominals may work on the rib cage during inspiration by passively increasing intra-abdominal pressure, which could push the lower rib cage outward at the zone of apposition. Increased intra-abdominal pressure can also force the diaphragm cranially and exert some passive pull on the costal diaphragm fibers, pulling the lower ribs upward.[3] The major inspiratory influence of the abdominals is most likely the facilitation of the diaphragm by optimizing its length-tension during standing. This is one reason why vital capacity is decreased in the standing position in individuals with paralyzed abdominal muscles. The abdominals tend to be quiet in supine but in standing may contract at the end of expiration in response to increases in ventilation. Activity of the abdominals to enhance ventilation increases during exercise.[3,24]

COORDINATION AND INTEGRATION OF VENTILATORY MOTIONS

The coordination and integration of the skeletal and muscular chest wall components during breathing are not easily understood or measured. More recent studies have used EMG techniques, electrical stimulation, and sophisticated computerized motion analysis techniques. Some analyses and descriptions of chest wall motion and muscular actions published in the 18th century are only now being questioned.[15-17] The complexity of the coordinated actions of the many muscle groups involved even in quiet breathing should be apparent at this point, but the complexities have only been touched on in the above discussions. Many of the respiratory muscles participate in activities other than respiration, including speech, defecation, and the maintenance of posture. A high and complex level of coordination is necessary for these muscles to carry out these alternate activities and breathing at the same time.

Interconnection of the Thorax and Abdomen

One important aspect of the mechanics of ventilation not yet discussed is the hydraulic interconnection between the diaphragm and abdomen. Physiologically, any change in intrathoracic (transpulmonary) pressure must be matched by an opposite change in intra-abdominal pressure. For any given lung (rib cage) volume, a downward displacement of the diaphragm (fibers are shortened) must have an outward displacement of the abdominal wall. Also, if the abdominal wall moves inward (such as when the abdominal muscles are actively contracting), the diaphragm is displaced upward, thereby stretching and lengthening its fibers and optimizing its length-tension. This inverse relationship between thoracic and abdominal pressures during ventilation (often called transdiaphragmatic pressure) is related to Laplace's law, tension = pressure $\times$ radius. This means that a diaphragm that has a smaller radius (and therefore is more domed) will be able to generate a higher transdiaphragmatic pressure and is more effective than a wider, flatter diaphragm.[2,24] Increasing abdominal muscle tension just before inspiring could therefore result in a stronger contraction of the diaphragm and greater potential for higher transdiaphragmatic pressure differences during inspiration.

Normal Sequence of Chest Wall Motions during Breathing

When observing the abdomen and chest wall of a normal, healthy person during quiet breathing, the following sequence of motion is usually apparent. First, the diaphragm contracts and the central tendon moves caudally. Intra-abdominal pressure increases and abdominal contents are displaced such that the anterior epigastric abdominal wall is pushed outward. Once the central tendon is "fixed" or stabilized on the abdominal organs, the appositional, vertical fibers pull the lower ribs upward and outward resulting in lateral movement of the lower chest. Following abdominal expansion with continued inspiration, the parasternals, scalenes, and levatores costarum actively rotate the upper ribs and elevate the manubriosternum, resulting in an outward (anterior) motion of the upper chest. The lateral and anterior motions of the chest can occur simultaneously. Expiration during quiet breathing is passive, with recoil of the elastic components of the lungs and chest wall.

DEVELOPMENTAL ASPECTS OF STRUCTURE AND FUNCTION

Skeletal Changes

The configuration (shape), distensibility (compliance), and mobility of the chest wall changes significantly from infancy to old age. The normal newborn has an extremely compliant chest wall because it is primarily cartilaginous. The cartilaginous ribs allow the distortion necessary for the infant's thorax to travel through the birth canal. Complete ossification of the ribs does not occur for at least several months after birth. With such a compliant rib cage, the infant's chest wall muscles are primarily "stabilizers" to counteract the tendency of the diaphragm to paradoxically pull the lower ribs inward during inspiration.[27] The horizontal alignment of the ribs in the infant (versus elliptical in the older child and adult) changes the costal diaphragm's angle of insertion, increasing the tendency for those fibers to pull the lower ribs inward, thereby decreasing efficiency of ventilation and increasing distortion of the chest wall.[28,30]

With aging, other skeletal changes occur that may affect pulmonary function and ventilation. The costal cartilages ossify, which interferes with their axial rotation.[9] Roughly 70 percent of persons over 75 years have osteoporosis of the ribs and vertebrae.[30] Increased kyphosis is often observed in older individuals, and when accompanied by decreased abdominal tone the kinematics of ventilation would likely be disturbed because the resting position of the diaphragm would be less domed.[9] Chest wall, rib cage, and diaphragm-abdomen compliance have all been found to be significantly reduced with aging. Reduction in diaphragm-abdomen compliance is probably indirectly linked to the decreased rib cage compliance (especially the lower ribs, which are part of the zone of apposition) and some, as yet unexplainable, decrease in abdominal distensibility with aging.[31]

Articular Changes

Many of the articulations of the chest wall undergo fibrosis and fuse in old age. These include the joints of the anterior chest wall, particularly the CC, CS, interchondral, MS, and XS joints. The other chest wall articulations are true synovial joints and are less likely to fuse as a result of aging. The synovial joints can, however, undergo morphologic changes associated with aging and "wear and tear," resulting in reduced mobility. It is important to keep in mind that decreased mobility of the ribs may contribute to decreased mobility of the thoracic vertebrae.

Muscular Changes

Normal newborns have approximately 20 percent slow, oxidative (high-oxidative) muscle fibers in their diaphragms compared to 50 percent in adults. This discrepancy predisposes infants to earlier diaphragmatic fatigue.[27] Aging appears to have little effect on the function of the respiratory muscles when malnutrition and disease are not present.[30]

CHANGES IN NORMAL STRUCTURE AND FUNCTION

This section will briefly present some examples of how normal structure and function of the chest wall are altered by various conditions. The examples represent problems of the musculoskeletal and respiratory systems.

Pregnancy

Pregnancy is not considered to be a pathologic state, but pregnancy results in alteration of every organ system and every tissue of the body of the pregnant woman.[32] The effects of pregnancy on the biomechanics of the chest wall are apparent during the second half of the pregnancy, especially during the last trimester. Progressive uterine distension repositions the diaphragm cephalad, which results in increased chest circumference and decreased functional residual capacity of the lungs.[33,34] Tidal volumes are actually *increased,* but vital capacity is unchanged.

Very few studies have been done to specifically examine the above phenomenon. The information we have to date supports the theory that augmented rib cage volume displacement is likely the result of two factors. One factor is enhanced diaphragmatic contraction secondary to improving its length-tension relationship (lengthened diaphragmatic fibers have increased ability to generate tension). The other factor is the increased area of diaphragmatic opposition to the lower rib cage created by the elevation of the diaphragm. Increased intra-abdominal pressure during enhanced diaphragmatic contraction of the diaphragm transmits an increased force to expand and elevate the lower rib cage.[34]

Scoliosis

Structural scoliosis is an example of a musculoskeletal abnormality that can affect chest wall biomechanics. With this type of scoliosis the musculoskeletal changes that affect ventilation and chest wall biomechanics are related to the severity of the curve. On the concave side of the scoliotic curve the transverse processes of the vertebrae rotate anteriorly, carrying the ribs with them. As the ribs are torsioned anteriorly, they are also approximated, with decreased intercostal spaces (see Fig. 4–30). On the convex side the effects are just opposite to that described above. These changes, if severe enough, will move the axes of rotation of the ribs, will affect the efficiency of the intercostal (and other perhaps other) muscles, and will reduce rib cage compliance. These abnormalities result in a restrictive lung problem that can eventually decrease lung volumes and capacities.[8]

Chronic Obstructive Pulmonary Disease

A major manifestation of chronic obstructive pulmonary disease (COPD) is air-trapping with hyperinflation of the lungs. Hyperinflation affects the bony and muscular components of the chest wall. First, it results in an inflated resting position of the rib cage (the so-called barrel chest), putting many of the inspiratory muscles (parasternals, external intercostals, scalenes, and sternocleidomastoids) in shortened positions and making them much less efficient. The diaphragm is very compromised because the hyperinflation creates a flattening of the diaphragm, thereby decreasing its efficiency. In severe cases a flattened diaphragm results in a paradoxical pulling in of the lower rib cage as the costal fibers are no longer aligned vertically. These problems are compounded by the increased load against which the respiratory muscles must work in COPD if the airways are narrowed by bronchospasm or secretions and if the lung and chest wall compliance is decreased.[20,35]

SUMMARY

In this chapter, a more comprehensive coverage of the structure and function of the joints and muscles of the chest has been provided than was presented in the first edition of this text. Additional information on accessory respiratory muscle structure and function will be presented in chapter 7. Scoliosis will be presented in greater depth in chapter 13. In the following chapter, another area of the body that was not included in the first edition will be presented. The temporomandibular joint, like the chest, has both clinical and functional significance and, therefore, requires an understanding of normal joint structure and function.

REFERENCES

1. Hoppenfeld, S: Physical Examination of the Spine and Extremities. Appleton-Century-Crofts, New York, 1976.
2. Derenne, JPH, Macklem, PT, and Roussos, CH: State of the art: The respiratory muscles: Mechanics, control and pathophysiology. Am Rev Respir Dis 118:119–133, 1978
3. DeTroyer, A and Estenne, M: Functional anatomy of the respiratory muscles. Clin Chest Med 9:175–193, 1988.
4. Grieve, GP: Common Vertebral Joint Problems, ed 2. Churchill Livingstone, New York, 1988, pp 32–39, 110–129.
5. Williams, PL and Warwick, R (eds): Gray's Anatomy, ed 37. Churchill-Livingstone, New York, 1989, pp 496–499, 591–595.
6. Kormano, M: A microradiographic and histological study of the manubriosternal joint in rheumatoid arthritis. Acta Rheum Scand 16:47–59, 1970.
7. Fam, AG and Smythe, HA: Musculoskeletal chestwall pain. Can Med Assoc J 133:379–389, 1985.
8. Schafer, RC: Clinical Biomechanics: Musculoskeletal Actions and Reactions. Williams & Wilkins, Baltimore, 1983, pp 328–374.
9. Kapandji, IA: The Physiology of the Joints, ed 2. Churchill Livingstone, New York, 1974, pp 130–163.
10. Saumarez, RC: An analysis of possible movements of human upper rib cage. J Appl Physiol 60:678–689, 1986.
11. Wilson, TA, et al: Geometry and respiratory displacement of human ribs. J Appl Physiol 62:1872–1877, 1987.
12. Rochester, DF and Braun, NM: The respiratory muscles. Basics Respir Dis 6:1–6, 1978.

13. Gardner, E, Gray, DJ, and O'Rahilly, R: Anatomy: A Regional Study of Human Structure, ed 3. WB Saunders, Philadelphia, 1969, pp 286–297.
14. DeTroyer, A: Actions of the respiratory muscles or how the chest wall moves in upright man. Bull Eur Physiopathol Respir 20:409–413, 1984.
15. Hamberger, GE: De respirationis mechanismo et usu genuino. Jena, Germany, 1749.
16. DeTroyer, A, Kelly, S, and Zin, WA: Mechanical action of the intercostal muscles on the ribs. Science 220:87–88, 1983.
17. DeTroyer, A, et al: Mechanics of intercostal space and actions of external and internal intercostal muscles. J Clin Invest 75:850–857, 1985.
18. DeTroyer, A, et al: Triangularis sterni muscle use in supine humans. J Appl Physiol 62:919–925, 1987
19. Estenne, M, Ninane, V, and DeTroyer, A: Triangularis sterni muscle use during eupnea in humans: Effect of posture. Respir Physiol 74:151–162, 1988.
20. Celli, BR: Respiratory muscle function. Clin Chest Med 7:567–584, 1986
21. Massery, M: Respiratory rehabilitation secondary to neurological deficits: Understanding the deficits. In Frownfelter DL: Chest Physical Therapy and Pulmonary Rehabilitation, ed 2. Year Book, Chicago, 1987, pp 499–528.
22. Basmajian, JV: Muscles Alive: Their Functions by Electromyograph, ed 2. Williams & Wilkins, Baltimore, 1967, p 303.
23. Panicek, DM, et al: The diaphragm: Anatomic, pathologic and radiographic considerations. Radiographics 8:385–425, 1988.
24. Celli, BR: Clinical and physiologic evaluation of respiratory muscle function. Clin Chest Med 10:199–214, 1989.
25. DeTroyer, A, et al: The diaphragm: Two muscles. Science 217:237–238, 1981.
26. Deviri, E, Nathan, H, and Luchansky, E: Medial and lateral arcuate ligaments of the diaphragm: Attachment to the transverse process. Anat Anz 166:63–67, 1988.
27. Davis, GM and Bureau, MA: Pulmonary and chest wall mechanics in the control of respiration in the newborn. Clin Perinatol 14:551–579, 1987.
28. Crane, LD: Physical therapy for the neonate with respiratory dysfunction. In Irwin, S and Tecklin, JS (eds): Cardiopulmonary Physical Therapy, ed 2. CV Mosby, St Louis, 1990, pp 389–415.
29. Muller, NL and Bryan, AC: Chest wall mechanics and respiratory muscles in infants. Pediatr Clin North Am 26:503, 1979.
30. Krumpe, PE, et al: The aging respiratory system. Clin Geriatr Med 1:143–175, 1985.
31. Estenne, M, Yernault, JC, and DeTroyer, A: Rib cage and diaphragm-abdomen compliance in humans: Effects of age and posture. J Appl Physiol 59:1842–1848, 1985.
32. Wilder, E (ed): Obstetrics and gynecologic physical therapy. Clin Phys Ther 20:1–225, 1988.
33. Artal, R and Wiswell, RA (eds): Exercise in Pregnancy. Williams & Wilkins, Baltimore, 1986, pp 147–148.
34. Gilroy, RJ, Mangura, BT, and Lavetes, MH: Ribcage and abdominal volume displacements during breathing in pregnancy. Am Rev Respir Dis 137:668–672, 1988.
35. Sharp, JT: The respiratory muscles in emphysema. Clin Chest Med 4:421–432, 1983.

STUDY QUESTIONS

1. Describe the articulations of the chest wall and thorax, including the CV, CT, CC, CS, interchondral, and MS joints.
2. What is the normal sequence of chest wall motions during breathing? Explain why these motions occur.
3. What is transdiaphragmatic pressure?
4. What is the role of the following during breathing: the diaphragm, the intercostal muscles, and the abdominal muscles?
5. Describe the "accessory" muscles and explain their functions.
6. Compare the action of the paravertebral muscles with that of the scalene muscles.
7. What effect does COPD have on the inspiratory muscles?
8. How does the aging process affect the structure and function of the chest wall?

Chapter 6

■ ■ ■

The Temporomandibular Joint

■

OBJECTIVES

Following the study of this chapter, the reader should be able to:

Describe

1. The articular surfaces of the temporal mandibular (TM) joint.
2. The structure and function of the disk.
3. The structure and functions of the ligaments of the TM joint.
4. The movement available at the TM joint.
5. The motions of the disk and condyle necessary for normal mouth opening and closing.
6. The significance of two articulations of the mandible.
7. The muscular control necessary for normal TM joint motion.
8. The joint structures that provide stability for the TM joint.

9. The characteristics of the cartilage covering the articular surfaces of the mandible and temporal bone.
10. The normal rest position of the mandible.
11. The effect of dentition on the function of the TM joint.
12. The effect of symmetrical and asymmetrical condylar motion on mandibular deviation during mouth opening and closing.

Locate

1. The attachments of selected muscles of the TM joint.

Differentiate

1. Between the motions available in the upper and lower joint compartments of the TM joint.
2. Between structures that have a passive effect on the function of the TM joint and those that have an active or elastic effect on the joint.

Predict

1. The deviations of the jaw that would be present with various internal derangements of the joint structures.

GENERAL FEATURES

The temporomandibular (TM) joint is unique in the body. The mandible is a horseshoe-shaped bone that has an articulation with the temporal bone at each end, giving it two completely separate but solidly connected joints. In addition to the two separate articulations, each TM joint has a disk that separates the joint into two joints, an upper and a lower joint. Therefore, in considering mandibular movement, four distinct joints are affected with any motion.

Functionally the joint is also unique. Few other joints are moved as often as the TM joint. In addition to the motions of eating or chewing, which can create great force within the joint, speaking and swallowing require TM joint motion that is finely controlled and requires little force. The TM joint exhibits a combination of complexity, close-to-continuous use, and capacity for force and finesse that is remarkable.

Each TM joint is formed by the condyle of the mandible, the articular eminence of the temporal bone, and articular disk that is interposed between the two. The joint formed by the condyle and the inferior surface of the disk is a hinge joint (ginglymus). The joint formed by the articular eminence and the superior surface of the disk is a gliding joint (amphiarthrodial). Sicher[1] describes the joint as a hinge joint with movable sockets. The TM joint is a synovial joint, although there is no hyaline cartilage covering the articular surfaces. Rather, the surfaces are covered by dense collagenous tissue that may have some chondroblasts in it, thereby allowing it to be considered fibrocartilage.[1-5] The articular surfaces will be described in more detail later.

STRUCTURE

Articular Surfaces

The mandible is divided into a body and two rami (Fig. 6–1). The angle of the jaw is where the body and the ramus join. The articular portion of the mandible is the anterior portion of the condyle, which is composed of trabecular bone.[2] The

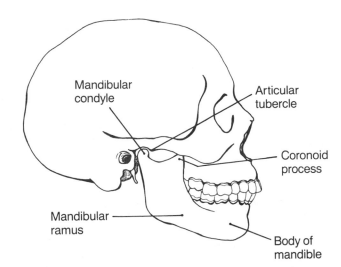

Figure 6-1. Lateral view of cranium and mandible. (Modified from Perry, JF, Rohe, DA, and Garcia, OA: The Kinesiology Workbook. FA Davis Company, Philadelphia, p 162, 1991, with permission.)

condyle is 8 to 10 mm deep (anterior-posterior measurement) and 15 to 20 mm long (medial-lateral measurement) with roughened medial and lateral poles.[1,3,4] The lateral pole of the condyle is almost even with the lateral aspect of the ramus, but the medial pole projects much farther medially than the medial aspect of the ramus.[3] Lines following the axis of medial-lateral poles of each condyle will intersect just anterior to the foramen magnum.[1]

Anterior to the condyle is another projection, the coronoid process. In the closed-mouth position, this process sits under the zygomatic arch, but it can be palpated below the arch when the mouth is open. The coronoid process serves as an attachment for the temporalis muscle.

The condyle of the mandible articulates with the temporal bone in the area of the posterior glenoid spine, glenoid fossa, articular eminence, and articular tubercle (Figs. 6-1 and 6-2). These structures are immediately anterior to the external auditory meatus. Since no other osseous structure is lateral to the spine, the condyle is available for palpation through the external auditory meatus.[6] The glenoid fossa, on superficial inspection, looks like the articular surface for the TM joint. On closer inspection, however, the bone in that area is thin and translucent and not at all appropriate for an articular surface.[1-3] The articular eminence, however, has a major area of trabecular bone and serves as the primary articular surface for the

Figure 6-2. Lateral view of TM joint showing attachments of capsule and disk. (Modified from Perry, JF, Rohe, DA, and Garcia, OA: The Kinesiology Workbook. FA Davis Company, Philadelphia, p 163, 1991, with permission.)

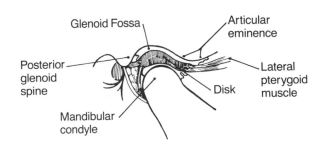

TM joint.[2] The articular tubercle is anterior to the articular eminence and is not an articular surface.

The articular surfaces of the condyle and the articular eminence are covered with dense, avascular collagenous tissue that contains some cartilaginous cells.[1-6] Since some of the cells are cartilaginous, some authors refer to the covering as fibrocartilage.[3,4,6] The greatest amount of fibrocartilage is found on the articular eminence and the anterior, superior aspect of the condyle, which is further evidence supporting the claim that these are the primary areas of articulation.[2,3,6] The presence of fibrocartilage rather than hyaline cartilage is significant because of the ability of fibrocartilage to repair.[3,6] Typically, fibrocartilage is present in areas that are intended to withstand repeated and high-level stress. The directions of the collagen fibers are perpendicular to the surface in the deeper layers of the fibrocartilage and serve to withstand stresses. The fibers near the surface are aligned in a parallel arrangement to facilitate gliding of the joint surfaces.[3] Fibrocartilage also has the capability of remodeling.[6] The ability to remodel has implications for the mechanisms of injury and for potential treatments. Defects in cartilage can be created or corrected through the application of relatively gentle forces over a prolonged period of time. The disk, however, does not possess this capability.[8]

Disk and Capsule

The disk of the TM joint allows the convex surface of the condyle and the convex surface of the articular eminence to remain congruent throughout the motion available to the joint[3,8] (Fig. 6–2). The disk is biconcave, that is, both its superior and inferior surfaces are concave. The disk is firmly attached to the medial and lateral poles of the condyle of the mandible. It is not, however, attached to the joint capsule laterally or medially.[1,3,6,8] This allows the disk to rotate around the condyle quite freely in an anterior-posterior direction. Anteriorly the disk is attached to the joint capsule; fibers of the superior head of the lateral pterygoid muscle also insert into the anterior portion of the disk. Posteriorly the disk has complex attachments, which are collectively called the bilaminar retrodiskal pad. Two bands are attached to the disk posteriorly (Fig. 6–3). The superior band, also called a strata or lamina, is made of elastic fibers that allow it to stretch. The posterior aspect of the superior lamina attaches to the tympanic plate.[3,9] The inferior lamina is inelastic and attaches to the neck of the condyle. Between the two lamina is loose areolar connective tissue rich in arterial and neural supply.[1-4,6,8]

The superior lamina allows the disk to translate forward along the articular eminence during mouth opening. Its elastic properties assist in repositioning the disk as it translates posteriorly during mouth closing. The inferior lamina simply serves as a tether on the disk, limiting forward translation, but does not assist with repositioning the disk during mouth closing.[3,8]

The disk itself is of uneven thickness, varying from 2 mm anteriorly to 3 mm posteriorly to 1 mm in the middle.[3] This arrangement allows the disk to adapt to the bony surfaces with which it articulates, and creates greater congruence of the joint surfaces.[3] The anterior and posterior portions of the disk are vascular and innervated; the middle band, however, is avascular and not innervated.[2,3,8] The lack of vascularity and innervation is consistent with the observation that the middle bank is the force-accepting surface of the disk.

The capsule of the TM joint is not as well defined as many joint capsules. Inferiorly the capsule attaches to the neck of the mandibular condyle. Superiorly it attaches to the articular tubercle anteriorly, the squamotympanic fissure posteriorly, and the circumference of the glenoid fossa laterally and medially. The capsule and the disk are attached to one another anteriorly and posteriorly, but not medi-

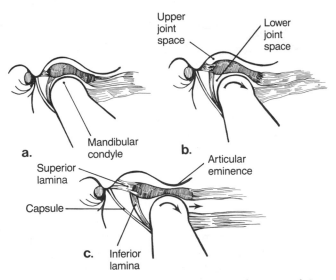

Figure 6-3. Lateral view of TM joint. (Modified from Perry, JF, Rohe, DA, and Garcia, OA: The Kinesiology Workbook. FA Davis Company, Philadelphia, p 163, 1991, with permission.)

ally and laterally. Since there are no medial or lateral attachments of the disk and capsule, translation of the disk within the capsule can occur.[9]

The portion of the capsule above the disk is quite loose, while the portion of the capsule below the disk is tight.[5] The capsule is quite thin and loose in its anterior, medial, and posterior aspects, but the lateral aspect is thicker and stronger, and is reinforced by the TM ligament.[1] The lack of strength of the capsule anteriorly and the incongruence of the bony articular surfaces predisposes the joint to anterior dislocation of the condyle.[1] The capsule is highly vascularized and innervated, which allows it to provide a great deal of information about position and movement.

Joints

The disk divides the TM joint into two separate joint spaces (see Fig. 6–3). The upper joint is the larger of the two. Each joint has its own synovial lining. All of the joint surfaces except the articulating surfaces are covered with synovial membrane. The nutrition of the fibrocartilage covering the joint surfaces and the avascular middle portion of the disk is provided by the synovial fluid. Intermittent pressure on these collagenous structures during joint motion causes the synovial fluid to be pumped in and out of them, providing their nutrition.

Lower Joint

The lower joint of the TM joint, a hinge joint, is formed by the anterior surface of the condyle of the mandible and the inferior surface of the disk. The condyle and disk are firmly attached at the medial and lateral poles of the condyle. These attachments allow free rotation of the disk on the condyle or the condyle under the disk. The axis of this rotation is a line passing through both poles of the condyle. Rotation of the condyle forward results in a relative posterior rotation of the disk. The firm medial and lateral attachments cause the disk and condyle to glide as a unit. There is minimal translation between the two structures.

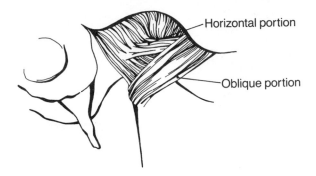

Figure 6–4. Temporomandibular ligament.

Upper Joint

The upper joint of the TM joint, a gliding or amphiarthrodial joint, is formed by the articular eminence of the temporal bone and the superior surface of the disk. There are no direct attachments of the disk to the temporal bone, thereby allowing translation of the disk on the temporal bone.[3] The middle portion of the disk is the articular surface of the disk, and it translates forward with the condyle during mouth opening to maintain contact with the articular eminence.

Ligaments

The primary ligaments of the TM joint are the capsule or capsular ligament, the TM ligament, the stylomandibular ligament, and the sphenomandibular ligament. The TM ligament is a strong ligament that is composed of two parts (Fig. 6–4). The outer oblique portion attaches to the neck of the condyle and the articular tubercle. It serves as a suspensory ligament and limits downward and posterior motion of the mandible as well as rotation of the condyle during mouth opening.[3] The inner portion of the ligament is attached to the lateral pole of the condyle and posterior portion of the disk and to the articular tubercle. Its fibers are almost horizontal and they resist posterior motion of the condyle. This serves to protect the retrodiskal pad.[3] Neither of these bands limits forward translation of the condyle or disk, but they do limit lateral displacement.[9]

The stylomandibular ligament is a band of deep cervical fascia that runs from the styloid process of the temporal bone to the posterior border of the ramus of the mandible (Fig. 6–5). It inserts between the masseter and the medial pterygoid muscles.[3] Its function is apparently controversial with some authors indicating that it limits protrusion of the jaw[3] and others stating that it has no function.[1,10]

The sphenomandibular ligament attaches to the spine of the sphenoid and to

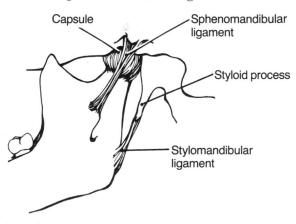

Figure 6–5. Capsule, sphenomandibular, and stylomandibular ligaments.

the middle surface of the ramus of the mandible (see Fig. 6–5). Some authors state that it serves to suspend the mandible[11] and to check the mandible from excessive forward translation.[8] One author, however, states that it has no function.[10]

Motions

The motions available to the TM joint are mouth opening (mandibular depression), mouth closing (mandibular elevation), jutting the chin forward (mandibular protrusion), sliding the teeth backward (mandibular retrusion), and sliding the teeth to either side (lateral deviation of the mandible). These motions are created by combinations of rotation and gliding in the upper and lower joints. The role that various joint structures play in creating these motions is described in the following sections.

The functional movements of the mandible are obtained by combinations of intra-articular movements that are controlled by the delicate interplay of many muscles. The functions that the TM joints support are chewing, talking, and swallowing.[3] For purposes of this chapter, we will only describe the movements of the mandible that occur without resistance (empty-mouth movements).

Mandibular Elevation and Depression

In normally functioning joints, mandibular elevation and depression are relatively symmetrical motions occurring around a coronal axis passing through both condyles. The motion at each TM joint follows a similar pattern. During opening, the first portion of the motion is accomplished by anterior rotation of the condyle on the disk (see Fig. 6–3a and b). This motion occurs in the lower joint between the disk and the condyle and accounts for 11 mm[6,11] to 25 mm of opening.[2] The position at the end of the rotation phase can be described as a posterior rotation of the disk on the condyle or anterior rotation of the condyle on the disk.

The second portion of the opening motion involves translation of the disk-condyle complex anteriorly and inferiorly along the articular eminence (see Fig. 6–3b and c). This occurs in the upper joint between the disk and the articular eminence and accounts for the remainder of the opening. Normal mouth opening is considered to be 40 to 50 mm depending on the author consulted.[2,6,11]

Of that motion, between 11 mm[6,11] and 25 mm[2] is gained from rotation of the condyle in the disk. The remainder is from translation of the disk and condyle along the articular eminence. For a quick and rough, but useful, estimate of function, an individual may use his or her proximal interphalangeal (PIP) joints to assess opening. If two PIP joints can be placed between the central front incisors, the amount of opening is functional. Three PIP joints is considered normal.[2]

Mandibular elevation (mouth closing) is the reverse of these motions. It consists of translation posteriorly and superiorly followed by rotation of the condyle posteriorly on the disk (or the disk anteriorly on the condyle).

Mandibular Protrusion and Retrusion

This motion occurs when all points of the mandible move forward the same amount. No rotation occurs in the TM joint during protrusion. The motion is all translation and occurs in the upper joint. The teeth are separated when protrusion occurs. The condyle and disk together translate anteriorly and inferiorly along the articular eminence. During protrusion, the posterior attachments of the disk (the bilaminar retrodisk tissue) stretch 6 to 9 mm to allow the motion to occur.[3]

Retrusion occurs when all points of the mandible move posteriorly the same amount. This range is limited to 3 mm of translation.[3] The TM ligament limits this

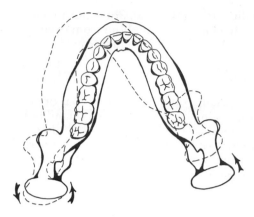

Figure 6-6. Demonstration of lateral deviation of the mandible to the left. (From Perry, JF, Rohe, DA, and Garcia, OA: The Kinesiology Workbook. FA Davis Company, Philadelphia, p 171, 1991, with permission.)

motion as does the soft tissue present in the retrodiskal area between the condyle and the posterior glenoid spine. The ligament limits motions by becoming taut, the retrodiskal tissue limits movement by occupying the space into which the condyle would move during retrusion.

Protrusion should be adequate to allow the upper and lower teeth to touch edge-to-edge. Retrusion is seldom measured, but about 3 mm of motion is available.[6]

Mandibular Lateral Deviation

The mandible can move asymmetrically around either a vertical or an anterior-posterior axis at one of the condyles. In moving around a vertical axis, one condyle spins and the other condyle moves forward[3,5] (Fig. 6–6). For example, deviation to the right would involve the right condyle spinning and the left condyle translating or gliding forward. The result is movement of the center of the mandible (or chin) to the right. Normally, the amount of lateral excursion of the joint is about 8 mm.[2,6]

Another asymmetrical movement is rotation around an anterior-posterior axis.[3] This also involves one condyle spinning, but in this situation the spin is in the frontal plane, whereas the spin discussed above was in the horizontal plane. As one condyle spins, the other condyle depresses. This results in the center of the mandible moving downward and deviating from the midline slightly toward the condyle that is spinning.

These two asymmetrical lateral deviation motions are combined into one complex motion used in chewing and grinding food.[11]

A functional measurement of lateral motion of the mandible involves the use of the width of the two upper central incisors. If the mandible can move the full width of one of the central incisor in each direction, motion is considered normal.[2,6]

FUNCTION

Frequency of Use

The TM joint is one of the most frequently used joints in the body. It is involved in chewing, talking, and swallowing.[3,10,11] Most TM joint movements are empty-mouth movements;[8] that is, they occur with no resistance from food or contact between the upper and lower teeth.

The joint is well designed for this intensive use. The cartilage covering the articular surfaces is designed to tolerate repeated and high-level stress. As discussed earlier, fibrocartilage has greater repair ability than does hyaline cartilage. In addition to a joint structure that supports the high level of usage, the musculature is designed to provide both power and intricate control.[8] Speech requires fine control, while the ability to chew requires great strength. Both are available through the musculature of the TM joint.

Function of the Disk

The biconcave shape of the disk provides three advantages to the TM joint. First, it provides increased congruence of the joint surfaces through a wide range of positions.[3,11] Second, the shape of the disk (thin in the center and wider anteriorly and posteriorly) allows greater flexibility of the disk so that it can conform to the articular surfaces of the condyle and the temporal bone as the condyle first rotates and then translates over the articular eminence.[3,8] Finally, the thick-thin-thick arrangement provides a self-centering mechanism for the disk on the condyle.[3,8] As pressure between the condyle and the articular eminence increases, the disk rotates on the condyle so that the thinnest portion of the disk is between the articulating surfaces. When less pressure is exerted between the joint surfaces and they separate, the disk is free to rotate to place a wider portion of the disk between the surfaces.

Control of the Disk

There are two forms of control over movements of the disk, passive and active. The passive control is exerted by the attachments of the disk to the condyle and the joint capsule. The control exerted by the medial and lateral attachments of the disk to the poles of the condyle has been discussed. These two attachments serve to limit motion of the disk on the condyle to rotation in one plane. The inferior retrodiskal lamina serves to limit the forward motion of the disk during rotation and translation. The attachment of the disk to the condyle through the inferior portion of the anterior capsule limits posterior motion of the disk. Elastic forces are provided by the superior retrodiskal lamina. When the disk is in a forward position, the superior lamina is stretched and can provide tractive force to return the disk to a more posterior position. At rest, however, the lamina is not stretched so there is no tensile force on the disk from the retrodiskal lamina.

The active control of the disk is exerted through the muscular forces (attachments of the superior portion of the lateral pterygoid into the disk) Bell[8] also discusses two newly described muscles that assist with maintaining the disk position. These two muscles are derived from the masseter and are attached to the anterolateral portion of the disk. They help overcome the medial pull of the anteromedially attached lateral pterygoid.

Now let us look at how these forces interact to control the disk during function. During the initial phase of opening, only rotation is occurring and all of the rotation occurs between the disk and the condyle in the lower joint compartment. The medial and lateral attachments of the disk to the condyle limit the motion of the disk on the condyle to the motion of rotation. During the translation phase of opening, the biconcave shape of the disk allows it to follow the condyle; no other force is necessary. The inferior retrodiskal lamina limits forward excursion of the disk. The superior portion of the lateral pterygoid is *not* active during mouth opening.

During mouth closing, the elastic character of the superior retrodiskal lamina applies a posterior tractive force on the disk. In addition the superior portion of the lateral pterygoid applies a force that controls the posterior movement of the disk through an eccentric contraction. Again, the medial and lateral attachments of the disk to the condyle limit the motion to rotation of the disk around the condyle.

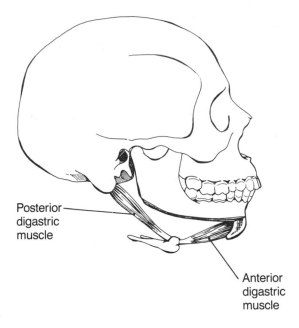

Posterior
digastric
muscle

Anterior
digastric
muscle

Figure 6-7. Digastric muscle. (Modified from Perry, JF, Rohe, DA, and Garcia, OA: The Kinesiology Workbook. FA Davis Company, Philadelphia, p 168, 1991, with permission.)

Muscular Control of the TM Joint

The primary muscle responsible for **mandibular depression** is the digastric muscle (Fig. 6–7).[1,5,8] Some classify the lateral pterygoid muscles as depressors,[5] but Bell[8] cites electromyographic research that the superior portion of the lateral pterygoid is not active during mouth opening, although the lower portion is active. Bell states that the upper and lower portions of the lateral pterygoids function independently. Note that gravity is also a mandibular depressor.

Mandibular elevation is primarily accomplished by the temporalis (Fig. 6–8), masseter (Fig. 6–9), and medial pterygoid muscles (Fig. 6–10).[1,3,6,11] The superior portion of the lateral pterygoid is also active during elevation of the mandible. The purpose of this activity is to rotate the disk anteriorly on the condyle.[3,6,8] This can also be viewed as maintaining the disk in a forward position as the condyle begins to rotate posteriorly. The superior portion of the lateral pterygoid contracts eccentrically to allow the disk-condyle complex to translate upward and posteriorly, and

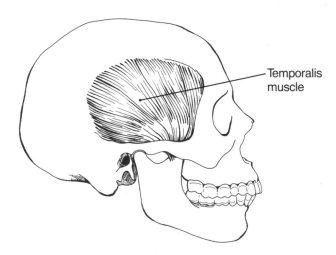

Temporalis
muscle

Figure 6-8. Temporalis muscle. (Modified from Perry, JF, Rohe, DA, and Garcia, OA: The Kinesiology Workbook. FA Davis Company, Philadelphia, p 166, 1991, with permission.)

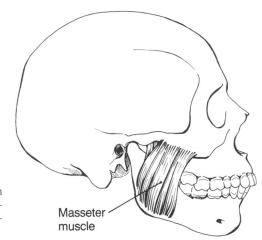

Figure 6-9. Masseter muscle. (Modified from Perry, JF, Rohe, DA, and Garcia, OA: The Kinesiology Workbook. FA Davis Company, Philadelphia, p 165, 1991, with permission.)

Masseter muscle

then maintains the disk in a forward position until the condyle has completed its posterior rotation to return to its normal rest position.

Mandibular protrusion results from bilateral action of the masseter, medial pterygoid,[3,11] and lateral pterygoid muscles.[1,3,11] **Retrusion** or retraction is obtained through the bilateral action of the posterior fibers of the temporalis muscles[5,11] with assistance from the digastric and suprahyoid muscles.[11]

Lateral deviation, or movement of the chin or the center of the mandible away from the midline, is caused by unilateral action of various muscles. Both the medial and lateral pterygoid muscles deviate the mandible to the opposite side.[3,5] The temporalis muscle can deviate the mandible to the same side. An exception to this is the combined action of the lateral pterygoid and temporalis on the same side. These two muscles can function as an effective force couple.[11] For example, the left lateral pterygoid is attached to the medial pole of the condyle and pulls the condyle for-

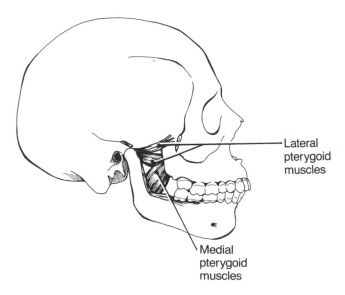

Figure 6-10. Medial and lateral pterygoid muscles. (Modified from Perry, JF, Rohe, DA, and Garcia, OA: The Kinesiology Workbook. FA Davis Company, Philadelphia, p 167, 1991, with permission.)

Lateral pterygoid muscles

Medial pterygoid muscles

ward. The left temporalis is attached to the lateral pole of the condyle and pulls it posteriorly. Together they effectively spin the condyle to create deviation of the mandible to the left. Acting alone, the left lateral pterygoid would tend to deviate the mandible to the right. Since the temporalis is also an elevator of the mandible, this combination of muscular activity is particularly useful in chewing.

Relationship with the Cervical Spine

The cervical spine and the TM joint are intimately connected. Muscles attach the mandible to the cranium, the hyoid bone, and the clavicle. The cervical spine is, in essence, interposed between the proximal and distal attachments of some of the muscles controlling the TM joint. The tension in those muscles has an effect on the position of the mandible. The posture of the head can also affect the rest position of the mandible. A forward shift of the cranium will decrease the freeway space.[8]

Many of the symptoms of TM joint dysfunction are similar to symptoms of cervical spine problems. Many patients with TM joint dysfunction have concurrent cervical spine problems. With the intimate relationship of these two areas, any patient being seen for complaints in one area should have the other examined as well.

Dentition

The teeth are intimately involved in the function of the TM joint. Not only is chewing one of the functions of the TM joint, but the contact of the upper and lower teeth limits motion of the TM joint during empty-mouth movements. The presence and position of the teeth are critical to normal TM joint function. Some basic nomenclature and information about dentition are presented below.

Normal adult dentition includes 32 teeth. For identification they are divided into four quadrants. The only teeth we will refer to by name are the upper and lower central incisors. These are the two central teeth of the maxilla and the two central teeth of the mandible.[7]

When the teeth are in firm approximation, the position is called **maximal intercuspation**[3] or the **occlusal position**.[5] This is not, however, the normal resting position of the mandible. Rather, 1.5 to 5 mm of "freeway" space between the upper and lower teeth is normally maintained.[3,8] This freeway space is particularly important. By maintaining this space, the intra-articular pressure within the joint is decreased, thereby decreasing the stress on the articular structures, and the tissues of the area are able to rest and repair.[3]

DYSFUNCTIONS

Many dysfunctions of the TM joint occur. Some are caused by direct trauma such as motor vehicle accidents or falls; others are the result of years of poor postural or oral habits such as forward head posture or bruxism (teeth grinding). Only two problems will be described here—reciprocal click and osteoarthritis.

Reciprocal Click

A patient who has an anteriorly dislocated disk will have an audible click from the TM joint on opening and a second when the mouth is closing. This is called a **reciprocal click**.[2] In this situation, at rest, the condyle is in contact with the retrodiskal tissue rather than the disk. On mouth opening, the condyle slips under the disk to obtain a normal relationship with the disk. When the condyle slips under the disk, an audible click is often present. Once the condyle is in the proper relationship with the disk, motion continues normally through opening and closing until the condyle again slips out from under the disk, when another click is heard. One would expect a click to signify that the condyle and disk have lost a normal relationship. In the case of an anteriorly dislocated disk, however, the initial click signals regaining a normal relationship. When the click occurs early in opening and late in closing, the amount of anterior displacement of the disk is relatively limited. The later on in the opening that the click occurs, the more severe the dislocation.[2] There is some evidence that the timing of the clicks during opening and closing can determine treatment prognosis.[12]

Osteoarthritis

Hertling and Kessler[11] state that 80 to 90 percent of the population over 60 years of age have some symptoms of osteoarthritis of the TM joint. According to Mahan,[9] osteoarthritis usually occurs unilaterally (unlike rheumatoid arthritis, which is usually bilateral). The primary cause of osteoarthritis is repeated minor trauma to the joint, particularly trauma that creates an impact between the articular surfaces.[9,11] One cause of this is a loss of posterior teeth, which creates an internal joint environment in which simple occlusion of the remaining teeth causes impact between the joint surfaces.[9,11]

The primary symptoms of osteoarthritis are pain on translation of the condyle on the articular eminence with almost pain-free rotation of the condyle, flattening of the condyle and the articular eminence, and narrowing of the joint space. In more advanced stages of the disease, perforation of the disk and lipping around the articular surfaces can also occur.[9,11] Persons with osteoarthritis may limit their mouth opening to the range available without translation (11 to 25 mm). Typically, the symptoms decrease over time, with most pain disappearing after approximately 8 months and apparently normal function (with crepitus, however) returning within 1 to 3 years.[9,13]

REFERENCES

1. Sicher, H: Functional anatomy of the temporomandibular joint. In Sarnat, BG (ed): The Temporomandibular Joint, ed 2. Charles C Thomas, Springfield, Ill, 1964.
2. Rocabado, M: Course notes, 1988.
3. Bourbon, BM: Anatomy and biomechanics of the TMJ. In Kraus, SL: TMJ Disorders: Management of the Craniomandibular Complex. Churchill-Livingstone, New York, 1988.
4. Ermshar, CB: Anatomy and neuroanatomy. In Morgan, DH, House, LR, Wall, WP, and Vamvas, SJ, (eds): Diseases of the Temporomandibular Apparatus: A Multidisciplinary Approach, ed 2. CV Mosby, St Louis, 1982.
5. Warwick, R and Williams, PL: Gray's Anatomy, ed 35. WB Saunders, Philadelphia, 1973.
6. Kraus, SL: Temporomandibular joint. In Saunders, HD (ed): Evaluation, Treatment, and Prevention of Musculoskeletal Disorders, ed 2. Viking Press, New York, 1985.
7. Brand, RW and Isselhard, DE: Anatomy of Orofacial Structures, ed 2. CV Mosby, St Louis, 1982.
8. Bell, WE: Temporomandibular Disorders: Classification, Diagnosis, Management, ed 3. Yearbook Medical Publishers, Chicago, 1990.
9. Mahan, PE: The temporomandibular joint in function and pathofunction. In Solberg, WK and Clark, GT (eds): Temporomandibular Joint Problems: Biologic Diagnosis and Treatment. Quintessence, Chicago, 1980.
10. Helland, MM: Anatomy and function of the temporomandibular joint. J Orthop Sports Phys Ther 1:145–152, 1980.
11. Hertling, D and Kessler, RM: Management of Common Musculoskeletal Disorders: Physical Therapy Principles and Methods, ed 2. JB Lippincott, Philadelphia, 1990.

12. Kirk, WS and Calabrese, DK: Clinical evaluation of physical therapy in the management of internal derangement of the temporomandibular joint. J Oral Maxillofacial Surg 47:113–119, 1989.

13. Nickerson, JW and Boering, G: Natural course of osteoarthritis as it relates to internal derangement of the temporomandibular joint. Oral Maxillofacial Surg Clin North Am 1:1, 1989.

STUDY QUESTIONS

1. Describe the articulating surface of the TM joint.

2. What is the significance of the differing thicknesses and the differing vascularity of the disk?

3. How do the superior and inferior lamina of the retrodiskal area differ?

4. Describe the sequence of motions in the upper and lower joints during mouth opening and closing.

5. What limits posterior motion of the condyle? How is the motion limited?

6. What would be the consequences of having a left TM joint that could not translate?

7. What would be the consequences of having a right disk that could not rotate freely over the condyle?

8. Describe the control of the disk in moving from an open-mouth to a closed-mouth position.

Chapter 7

■ ■ ■

The Shoulder Complex

■

OBJECTIVES

Following the study of this chapter, the reader should be able to:

Define
1. The terminology unique to the shoulder complex.

Describe
1. The articular surfaces of the joints of the complex.
2. The function of the ligaments of each joint.
3. Accessory joint structures and the function of each.

4. Motions and ranges of motion available at each joint and movement of articular surfaces within a joint.
5. The normal mechanism of dynamic stability of the glenohumeral joint, utilizing principles of biomechanics.
6. The normal mechanism of glenohumeral stability in the dependent arm.
7. Scapulohumeral rhythm, including contributions of each joint.
8. The extent of dependent or independent function of each joint in scapulohumeral rhythm.
9. How restrictions in the range of elevation of the arm may occur.
10. One muscular force couple at a given joint and its function.
11. The effect a given muscular deficit may have on shoulder complex function.

Compare
1. The advantage and disadvantages of the coracoacromial arch.
2. The structural stability of these joints, including the tendency toward degenerative changes and derangement.

Draw
1. Action lines of muscles of the shoulder complex and the moment arm for each and resolve each into components.

The upper limbs of the human body constitute slightly less than 10 percent of the total body weight.[1] That small segment of body mass, however, contains one of the principal physical features separating humans from the rest of the animal world: the human hand. The intricate gross and skilled functions performed by the hand are dependent on the mobile yet strong base provided by the shoulder complex.

The shoulder complex is composed of the scapula, clavicle, humerus and the joints that link these bones into a functional entity. These components constitute one half of the weight of the entire upper limb.[2] The shoulder complex is connected to the axial skeleton by a single anatomic joint (the sternoclavicular [SC] joint) and is suspended by muscles that serve as the primary mechanism for securing the shoulder girdle to the rest of the body. The precarious arrangement between the upper extremity and the trunk promotes a wide range of motion for the hand, but conflicts with the need for a stable base of operation and the need to move the hand against large resistances. The contradictory mobility-stability requirements necessitate **dynamic stability,** a concept for which the shoulder complex is the classic example and within which shoulder complex function can best be understood. In essence, dynamic stability results when a segment or set of segments are more dependent on muscles than on joint structures for maintenance of integrity. The resulting wide range of mobility for the complex is based on an arrangement that works well under usual conditions, but is susceptible to problems that may arise from any of the structures that facilitate the mobility-stability compromise.

COMPONENTS OF THE SHOULDER COMPLEX

The shoulder complex consists of the scapula, the clavicle, and the humerus. These segments are responsible for movement of the hand through space. The three segments are controlled by four interdependent linkages: a functional articulation known as the scapulothoracic (ST) joint, the SC joint, the acromioclavicular (AC) joint, and the glenohumeral (GH) joint. A fifth functional articulation is commonly described as part of the complex and is formed by the coracoacromial arch and the head of the humerus. The coracoacromial arch, or so-called suprahumeral joint, plays an important role in shoulder function and dysfunction, but will not be considered as a separate linkage. Each of the articulations and components of the shoulder complex must be examined individually before an understanding of the integrated dynamic function of the shoulder complex can be appreciated.

Scapulothoracic Joint

The ST joint is formed by the articulation of the scapula with the thorax beneath it. It is not a true anatomic joint since it has none of the usual joint characteristics (union by fibrous, cartilaginous, or synovial tissues). Although the scapula is separated from the thorax by layers of interposed muscle, the motion of the scapula on the thorax can best be described in the same way one would describe classically linked bony segments. In describing scapular motions, however, it must be noted that movements of the scapula on the thorax are inescapably associated with related motions of the SC and AC joints. The relationship of these joints is due to the fact that the scapula is attached by its acromion process to the lateral end of the clavicle via the AC joint. The clavicle, in turn, is attached to the axial skeleton at the manubrium of the sternum via the SC joint. Any movement of the scapula on the thorax must result in movement at one or both of these joints. That is, the functional scapulothoracic joint is part of a true closed kinematic chain with the AC and SC joints.

Position and Motions

Normally, the scapula is said to rest at a position on the posterior thorax approximately 2 in from the midline, between the second through seventh ribs[3] (Fig. 7–1). The motions of the scapula from this reference position are given as elevation-depression, abduction-adduction (also known as protraction-retraction), and upward-downward rotation. These motions are classically described as if they could occur independently of each; however, the linkage of the scapula to the AC and SC joints actually prevents such pure motions from occurring. Instead, for example, elevation may be associated with concomitant abduction and upward rotation. An understanding of these motions is developed by defining them one at a time. The more complex scapular function can be best appreciated when integrated function of the shoulder complex is presented.

Elevation and depression of the scapula are translatory motions in which the scapula moves upward (cephalad) or downward (caudally) along the rib cage from its resting position (Fig. 7–2). Abduction and adduction of the scapula are also translatory motions, occurring as the scapula slides along the rib cage away from or toward the vertebral column (Fig. 7–3). Upward and downward rotation are rotatory motions that tilt the glenoid fossa upward or downward, respectively. The same motions can also be described by movement of the inferior angle away from the vertebral column (upward rotation) or movement of the inferior angle toward the vertebral column (downward rotation) (Fig. 7–4).

The scapula has two other motions that are less commonly described but still critical to its movement along the curved rib cage. These motions are **scapular winging** and **scapular tipping**. Scapular winging and tipping (known by a wide variety of other terms) are not usually overt movements of the scapular segment but are joint motions necessary for maintaining close contact of scapula to thorax. Such movements of the scapula can and will occur overtly when the range of motion (ROM) is exhausted or in certain pathologic situations. These movements can best be understood in the context of the AC joint at which they occur and will be discussed in that section.

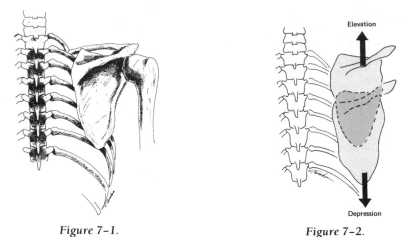

Figure 7–1.   Figure 7–2.

Figure 7–1. Resting position of the scapula on the thorax.

Figure 7–2. Elevation/depression of the scapula at the scapulothoracic joint.

Stability

According to Steindler,[3] the primary force holding the scapula to the thorax is atmospheric pressure. Scapular stability is also provided by the structures that maintain integrity of the linked AC and SC joints. The muscles that attach to both the thorax and scapula maintain contact between these surfaces while producing the movements of the scapula.

The ultimate function of scapular motion is to orient the glenoid for optimal contact with the maneuvering arm, to add range to elevation of the arm, and to provide a stable base for the controlled rolling and sliding of the articular surface of the humeral head. The scapula, with its associated muscles and linkages, performs these mobility and stability functions so well that it must serve as the premier example of dynamic stability in the human body.

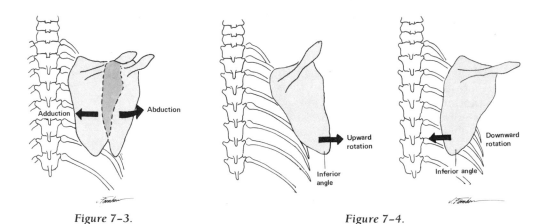

Figure 7–3.   Figure 7–4.

Figure 7–3. Abduction/adduction of the scapulothoracic joint.

Figure 7–4. Upward-downward rotation of the scapula at the scapulothoracic joint.

Sternoclavicular Joint

The SC joint might be considered the "base of operation" for the scapula since, via the interposed clavicle, it is the only structural attachment of the scapula to the rest of the body. Movement of the clavicle at the SC joint inevitably produces movement of the scapula. Similarly, certain scapular motions must produce motion at the SC joint. The SC joint is a plane synovial joint with 3° of freedom of motion. It has a joint capsule, three major ligaments, and a joint disk.

Articulating Surfaces

The SC articulation consists of two saddle-shaped surfaces, one at the sternal end of the clavicle and one at the notch formed by the manubrium of the sternum and first costal cartilage (Fig. 7–5). While tremendous individual differences exist in each of the components of the shoulder complex, the sternal end of the clavicle and the manubrium are invariably incongruent; that is, there is little contact between the articular surfaces at this joint. The superior portion of the medial clavicle does not contact the manubrium at all; instead it serves as the attachment for the joint disk and the interclavicular ligament. At rest, the SC joint space is wedge-shaped (open superiorly).[3] SC joint movement results in changes in the areas of contact among the clavicle, the joint disk, and the manubriocostal cartilage.

Motions

The motions that occur at the SC joint are elevation-depression of the clavicle, protraction-retraction of the clavicle, and rotation of the clavicle. Motions of any lever are always described osteokinematically by the movement of the distal segment of the lever. The horizontal alignment of the clavicle (rather than the vertical alignment of most of the appendicular levers of the skeleton) can sometimes create confusion and impair visualization of the clavicular motions. The motions of elevation-depression and protraction-retraction should be visualized as movements of the lateral end of the clavicle. Clavicular rotation is a rolling motion of the entire clavicle and does not seem to create the same visualization problems that may occur with the other motions.

Elevation and Depression of the Clavicle

The motions of elevation and depression occur between a convex clavicular surface and a concave surface formed by the manubrium and first costal cartilage around an anterior-posterior axis. The SC axis is considered to lie lateral to the joint at the costoclavicular ligament. The shape of the surfaces and the location of the

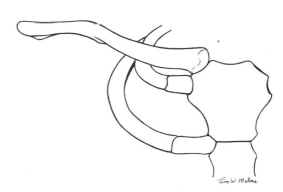

Figure 7–5. The clavicular and sternal segments of the SC joint.

axis indicate arthrokinematically that the convex surface of the clavicle must slide on the concave manubrium and first costal cartilage in a direction opposite to movement of the lateral end of the clavicle. That is, elevation of the clavicle results in downward sliding of the medial clavicular surface on the manubrium and first costal cartilage. The location of the axis so far from the joint accentuates the intra-articular motion. The clavicular range of elevation averages about 45°,[4] while there is about 15° of depression.[3] Elevation and depression of the *clavicle* is invariably associated with elevation and depression of the *scapula* since the scapula is attached to the lateral end of the clavicle. The elevation of the scapula that is associated with clavicular elevation is not a pure motion, but is associated with concomitant upward rotation. The upward rotation component, as shall be presented later, plays a significant role in increasing the range of elevation of the arm.

Protraction and Retraction of the Clavicle

Protraction and retraction of the clavicle occur around a vertical axis that also lies at the costoclavicular ligament. Given the saddle-shape of the joint, however, the configuration of the surfaces has reversed: In this plane the medial end of the clavicle is concave, and the manubrial side of the joint is convex. Arthrokinematically, the clavicular surface will now slide on the manubrium and first costal cartilage in the same direction as the lateral end of the clavicle. That is, protraction of the clavicle is accompanied by anterior sliding of the medial clavicle on the manubrium and first costal cartilage. The ROM averages about 15° protraction and 15° retraction. Protraction and retraction of the *clavicle* are invariably associated with abduction (protraction) and adduction (retraction) of the *scapula* since the scapula is attached to the distal end of the clavicle.

Rotation of the Clavicle

Rotation of the clavicle occurs as a spin between the saddle-shaped surfaces of the clavicle and manubriocostal facet. Unlike many joints that can rotate in either direction from resting position of the joint, the clavicle rotates in only one direction from its resting position. The clavicle rotates posteriorly from neutral, bringing the inferior surface of the clavicle to face anteriorly (Fig. 7–6). From its fully rotated position, the clavicle can rotate anteriorly again to return to neutral. The axis for rotation runs longitudinally through the clavicle, intersecting the joint. The ROM of clavicular rotation averages 30 to 45°.[3,5] Clavicular rotation is associated with scapular rotation, but does not have a direct (or one-to-one) relationship with it. The importance of clavicular rotation to scapular rotation will be presented with integrated function of the shoulder complex.

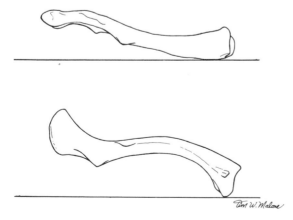

Figure 7–6. a. Neutral position of the clavicle. b. Position of the clavicle after the clavicle has rotated posteriorly (as during flexion and abduction of the upper extremity).

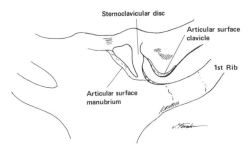

Figure 7-7. SC joint shown with the clavicle elevated to view the articulation of the manubrium, the medial clavicle, and the interposed SC disk.

Sternoclavicular Joint Disk

When two bony articular surfaces are incongruent, as they are at the SC joint, there is frequently an accessory joint structure that contributes to the overall contact of the surfaces. At the SC joint a fibrocartilage **joint disk**, or meniscus, is found interposed between the articular surfaces. The upper portion of the disk is attached to the superior clavicle and lower portion to the manubrium and first costal cartilage, diagonally transecting the SC joint space (Fig. 7-7). In this way the disk actually divides the joint into two separate cavities.[2] The disk acts like a hinge or pivot point during SC motion. In elevation and depression, the medial clavicle pivots on the upper attachment of the disk, with the intra-articular movement occurring primarily between the medial clavicle and a single unit formed by the disk and the manubrial facet. In protraction-retraction, the clavicle and disk now form a single unit that moves on the facet of the manubrium, pivoting around the inferior attachment of the disk.[2] The disk, therefore, is part of the manubrium in elevation-depression and part of the clavicle in protraction-retraction. As the disk switches its participation from one segment to the other during clavicular motions, mobility between the segments is maintained while stability is enhanced. The resultant movement of the clavicle in both elevation-depression and protraction-retraction is a fairly complex set of motions with the actual axis for clavicular movement lying at the level of the costoclavicular ligament while the clavicle intra-articularly pivots about the SC disk.

The SC disk also serves an important stability function by increasing joint congruence and absorbing forces that may be transmitted along the clavicle from its lateral end. While the manubrium and costal cartilage only contact a portion of the medial end of the clavicle, the disk is able to fully contact the medial clavicle, absorbing some of the forces transmitted along the clavicle before these forces arrive on the small manubrial facet. The unique diagonal attachment of the SC disk also checks medial movement of the clavicle on the manubrial facet, a movement with very little in the way of bony checks.

Sternoclavicular Joint Capsule and Ligaments

The SC joint is surrounded by a fairly strong capsule but is dependent on three ligaments for the majority of its support (Fig. 7-8). The anterior and posterior SC ligaments reinforce the capsule. The SC ligaments serve primarily to check anterior and posterior movement of the head of the clavicle. The costoclavicular ligament is a very strong ligament providing substantial stability to the joint while, as previously mentioned, the site of the axis or fulcrum for elevation-depression and protraction-retraction. The ligament is also the main check to clavicular elevation and to superior glide of the clavicle, both of which are strong movements created by the several muscles for which the clavicle is the inferior attachment. The interclavicular ligament checks excessive depression or downward glide of the clavicle,

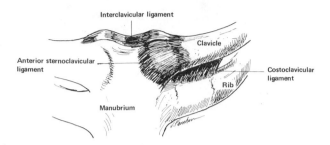

Figure 7–8. SC ligaments.

which is critical to protecting structures like the brachial plexus and subclavian artery, which pass between the clavicle and the first rib below it. In fact, when the clavicle is depressed and the interclavicular ligament and superior capsule are taut, the tension in these structures can support the weight of the upper extremity.[6]

The bony segments of the SC joint, its joint capsule and ligaments, and the SC disk combine to produce a joint that meets its dual functions of mobility and stability well. The SC joint serves its purpose of maintaining contact of the upper limb to the axial skeleton, while both contributing to mobility and withstanding imposed stresses. In spite of its complexity, the SC joint performs its tasks with a minimum of degenerative changes commonly seen at other joints of the shoulder complex.[5,7] Although not considered a congruent joint, its compensatory mechanisms are successful enough that SC dislocations represent only 1 percent of joint dislocations in the body, and when they occur, produce little discomfort or dysfunction.[8]

Acromioclavicular Joint

The AC joint appends the scapula to the clavicle. It is a plane synovial joint with 3° of freedom. It has a joint capsule and two major ligaments; a joint disk may or may not be present. The primary function of the AC joint is to maintain the relationship between the clavicle and the scapula in the early stages of elevation of the upper limb and to allow the scapula additional range of rotation on the thorax in the latter stages of elevation of the limb. **Elevation of the upper extremity** refers to the combination of scapular, clavicular, and humeral motion that occurs when the arm is either raised *forward* or *to the side* (including flexion, abduction, and all the motions in between).

Articulating Surfaces

The AC articulation consists of the small convex facet on the lateral end of the clavicle and a small concave facet on the acromion of the scapula (Fig. 7–9). Given the size and contour of the facets, the AC joint is considered to be incongruent. The inclination of the articulating surfaces varies from individual to individual. Depalma[5] described three joint types in which the angle of inclination of the contacting surfaces varied on the average from 36.1 to 16° from vertical. The closer the surfaces were to the vertical, the more prone the joint was to the wearing effects of shear forces.

Motions

The articular facets of the AC joint are small, afford limited motion, and have a wide range of individual differences. For these reasons studies are inconsistent in identifying the movement and axes of motion for this joint. Morris[4] cites three

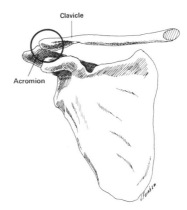

Figure 7-9. AC joint. Articulation between acromion of the scapula and the lateral clavicle is shown.

motions and axes that are understandable and consistent with observations of other authors: scapular rotation, winging, and tipping.[4]

Rotation of the Scapula

Of the three motions of the scapula cited by Morris,[4] the primary movement of the scapula at the AC joint is scapular rotation. It occurs around an anteroposterior (AP) axis lying between the joint and coracoclavicular ligament.[3] Scapular rotation allows the glenoid to tilt upward (upward rotation of the scapula) or downward (downward rotation of the scapula). Rotation occurring at the AC joint, therefore, is identical to and synonymous with rotation occurring at the scapulothoracic joint.

Winging of the Scapula

The remaining two movements of the scapula at the AC joint are less distinct than scapular rotation. The first of these smaller AC joint motions is described as a **winging of the scapula**, which occurs around a vertical axis. There is no consensus on what label to give this movement. However, the term "winging" is almost universally used to describe a pathologic posterior displacement of the vertebral border of the scapula. As is commonly done elsewhere in the human body, we will employ the same term to describe a normal response of nonpathologic magnitude (e.g., normal lordosis versus exaggerated or abnormal lordosis). Winging, therefore, can be used to describe the normal posterior movement of the vertebral border of the scapula (or anterior movement of the glenoid fossa) that must occur to maintain the contact of the scapula with the horizontal curvature of the thorax as the scapula slides around the thorax in abduction and adduction (Fig. 7-10). If abduction of the scapulothoracic joint occurred as a pure translatory movement, the scapula would move directly away from the vertebral column and the glenoid fossa would face laterally. Only the vertebral border of the scapula would remain in contact with the rib cage. In reality, full scapular abduction results in the glenoid fossa facing anteriorly with the full scapula in contact with the rib cage. The scapula has followed the contour of the ribs by rotating about a vertical axis at the AC joint, with the vertebral border of the scapula moving posteriorly and the glenoid fossa moving anteriorly.

Tipping of the Scapula

The second of the smaller AC motions is **tipping of the scapula**. It can be visualized as a movement of the inferior angle posteriorly while the superior border moves anteriorly around a coronal axis. As is true for scapular winging, there is no

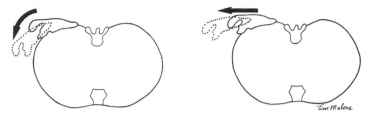

Figure 7–10. a. Abduction of the scapula is normally accompanied by winging of the scapula. Winging keeps the scapula against the ribcage and results in an anterior orientation of the glenoid fossa when abduction is complete. b. If winging did not accompany abduction, only the vertebral border of the scapula would remain on the ribcage, and the glenoid fossa would maintain its lateral orientation.

consensus on how to label this movement. The term "tipping" also comes from a pathologic descriptor of the scapula that is widely (although not universally) used. Scapular tipping, like the motion of winging, occurs to maintain the contact of the scapula with the contour of the rib cage. As the scapula moves upward or downward on the rib cage in elevation or depression, the scapula must adjust its position to maintain full contact with the vertical curvature of the ribs. This adjustment requires posterior movement of the inferior angle of the scapula at the AC joint around a coronal axis (Fig. 7–11). The movement of scapular tipping at the AC joint also occurs during rotation of the clavicle. If the relationship between the clavicle and scapula were fixed (no AC joint), clavicular rotation at the SC joint would carry the scapula along, displacing the inferior angle of the scapula anteriorly into the rib cage. Instead, the AC joint permits the scapula to counterrotate, in effect permitting a posterior movement of the inferior angle of the scapula around a coronal axis during the clavicular rotation.

Given the configuration of the facets of the AC joint, the arthrokinematics of the joint are the same for AC rotation, winging, and tipping. In each instance, the concave acromial articulation will slide on the convex clavicular facet in the same direction as the movement of the scapula.

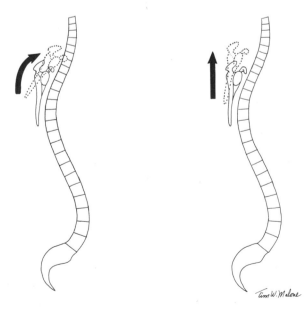

Figure 7–11. a. Elevation of the scapula is accompanied by tipping of the scapula. The inferior angle of the scapula moves posteriorly to accommodate the increased curvature of the ribcage. b. If scapular tipping did not occur, the superior aspect of the scapula would come off the ribcage.

Acromioclavicular Joint Disk

The disk of the AC joint is variable in size and differs among individuals, at various times in the life of the same individual, and between sides of the same individual. Through 2 years of age, the joint is actually a fibrocartilaginous union. Over time a joint space develops, usually maintaining some fibrocartilage within the joint. Degeneration of any remaining disk is usually marked by the fourth decade,[7] with the joint space itself commonly narrowed by the sixth decade.[9]

Acromioclavicular Capsule and Ligaments

The capsule of the AC joint is weak and cannot maintain integrity of the joint without reinforcing ligaments (Fig. 7–12). The superior and inferior AC ligaments assist the capsule in apposing articular surfaces and controlling horizontal joint stability. The fibers of the superior AC ligament are reinforced by aponeurotic fibers of the trapezius and deltoid, making the superior joint support stronger than the inferior.[10]

The coracoclavicular ligament, although not belonging directly to the anatomic structure of the joint, provides much of the AC joint stability and firmly unites the clavicle and scapula. This ligament is divided into a lateral portion, the **trapezoid ligament,** and a medial portion, the **conoid ligament.** The trapezoid ligament is quadrilateral in shape and lies predominantly in the sagittal plane. The conoid ligament is more triangular and lies essentially in the frontal plane (medial and slightly posterior to the trapezoid). The two ligaments are separated by adipose tissue and a large bursa.[6] Although the conoid and trapezoid ligaments each contribute to horizontal stability, they are critical to preventing superior dislocation of the clavicle on the acromion. Horizontal subluxation of the AC joint can occur with the conoid and trapezoid intact (through disruption of the weak capsule and AC ligaments), but vertical dislocation of the joint is almost always accompanied by tearing of the coracoclavicular ligaments.[10] Both portions of the ligament limit rotation of the scapula. The extremely strong bands of the coracoclavicular ligament also assist in the transmission of compression forces from the scapula to the clavicle. A fall to the side on an outstretched hand would tend to translate the scapula medially from the impact of the humerus. As the scapula and its coracoid process attempt to move medially, the trapezoid ligament becomes taut, transferring the force of impact to the clavicle and, ultimately, to the strong SC joint. The most critical role played by the coracoclavicular ligament, as shall be seen later, is in producing the longitudinal rotation of the clavicle necessary for a full ROM in elevation of the upper extremity.

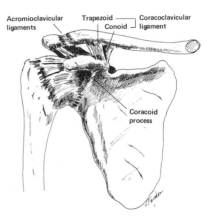

Figure 7–12. AC ligaments.

Unlike the stronger SC joint, the AC joint is extremely susceptible to both trauma and degenerative change. Treatment of sprains, subluxations, and dislocations of this joint occupy a large amount of the literature on the shoulder complex. Controversy centers on description and classification of subluxations and dislocations and on both the nonsurgical and surgical management.[10-12] This relatively unstable joint, however, appears to do reasonably well after injury regardless of whether or not the periarticular structures remain loose and plastic or the joint is overstabilized through some form of internal fixation.

Glenohumeral Joint

The GH joint is a ball-and-socket synovial joint with 3° of freedom. It has a capsule and several associated ligaments and bursae. The articulation is made up of the large head of the humerus and the small glenoid fossa (Fig. 7–13). Since the glenoid fossa of the scapula is the proximal segment of the GH joint, any motions of the scapula (and its interdependent SC and AC linkages) may affect GH joint function. The GH joint has sacrificed congruency to serve the mobility needs of the hand. It is, according to Fenlin,[13] a "sloppy arrangement," susceptible both to degenerative changes and to derangement. The potential for such pathologies varies with the particular joint configuration of the individual; there are a variety of possible deviations between different persons and between the two sides of the same person.

Articulating Surfaces

The glenoid fossa of the scapula serves as the proximal articular surface for this joint. When the scapula is at rest on the thorax, the vertebral border of the scapula is not necessarily vertical, nor does the scapula lie in the true frontal plane. Consequently, the orientation of the shallow concavity of the glenoid may vary according to the resting position of the scapula. Some authors have found that the fossa consistently faces superiorly and anteriorly when a subject is at rest[14]; others have found variations from person to person, with the majority showing the fossa tilting slightly inferiorly.[15-20] The curvature of the surface of the fossa is greater in the frontal plane (length) than in the sagittal plane (width). Steindler[3] gives the GH arcs as 75° in the frontal plane and 50° in the sagittal plane.[3]

The humerus is the distal segment of the GH joint. The humeral head has an articular surface that is invariably larger than that of the proximal segment, forming one third to one half of a sphere.[3,4] As a general rule the head faces medially, supe-

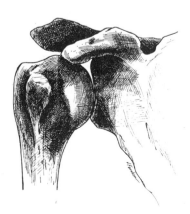

Figure 7–13. GH joint.

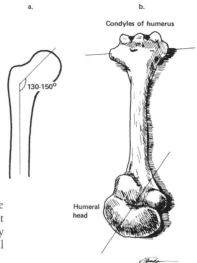

Figure 7–14. Humeral head is angled in two planes. a. The head in the frontal plane is angled 130° to 150° with respect to the shaft. b. The head in the transverse plane is commonly angled posteriorly with respect to an axis through the hemeral condyles (retroversion).

riorly, and posteriorly with respect to the shaft of the humerus and the humeral condyles. An axis through the humeral head and longitudinal axis of the shaft of the humerus may form an angle of 130 to 150° in the frontal plane (Fig. 7–14a).[3,20] This is commonly known as the **angle of inclination** of the humerus. In the transverse plane the axis through the humeral head and an axis through the humeral condyles form an angle that varies far more than other parameters but may be given for illustration as 30° posteriorly (Fig. 7–14b).[20] This angle is known as the **angle of torsion.** The normal posterior torsion of the humeral head may be termed **posterior torsion** or **retrotorsion** of the humerus.

Glenoid Labrum

When the arms hang dependently at the side, the two articular surfaces of the GH joint have little contact. The majority of the time the inferior surface of the humeral head rests only on a small inferior portion of the fossa[3,21,22] (Fig. 7–13). The total available articular surface of the glenoid fossa is somewhat enhanced by an accessory structure known as the **glenoid labrum.** This structure surrounds and is attached to the periphery of the glenoid fossa, providing at least some enhancement to the depth or curvature of the fossa (Fig. 7–15). Although the labrum was tradi-

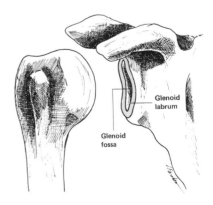

Figure 7–15. The glenoid labrum. As either a fibrocartilage structure or as a redundant capsular fold, the labrum deepens the glenoid fossa.

tionally thought to be synovium-lined fibrocartilage,[17] more recently it has been proposed that it is actually a redundant fold of dense fibrous connective tissue with little fibrocartilage other than at the attachment of the labrum to the periphery of the fossa.[23] This connective tissue fold is continuous with the capsule of the joint superficially and seems to disappear anteriorly with lateral rotation of the humerus.[7]

Glenohumeral Capsule and Ligaments

The entire GH joint in the resting position (arm dependent at the side) is surrounded by a large, loose capsule that is taut superiorly and slack anteriorly and inferiorly (Fig. 7–16). The capsule is twice the size of the humeral head[19] and, when slack, allows more than 1 in of distraction of the head from the glenoid fossa.[7] The relative laxity of the GH capsule is necessary for the large excursion of joint surfaces but provides little stability without the reinforcement of ligaments and muscles. Such reinforcement is weakest inferiorly,[24] but most evident anteriorly since forces on the humerus are much more likely to thrust the humeral head anteriorly than inferiorly. When the humerus is abducted in the frontal plane, the capsule twists on itself and tightens somewhat. The twisting is accentuated when the humerus is concomitantly laterally rotated, making abduction and lateral rotation the close-packed position for the GH joint.[25]

The ligaments that reinforce the GH joint capsule are the GH ligaments and the coracohumeral ligament (Fig. 7–17). The three GH ligaments (superior, middle, and inferior) form a Z on the anterior capsule and may exist as mere capsular thickenings or, as described by Cailliet,[7] as horizontal pleats in the anterior capsule.[7] Each of the GH ligaments becomes taut in, and provides a check to, certain motions of the humerus. All portions tighten on lateral rotation of the humerus and on motions involving anterior glide of the humeral head. The contribution of the GH ligaments and the associated capsule to joint stability is negligible, especially since there is a clear area of weakness between the superior and the middle GH ligaments (known as the foramen of Weitbrecht).

The coracohumeral ligament originates from the coracoid process and blends with the superior capsule and the supraspinatus tendon to insert on the greater

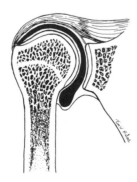

Figure 7–16. GH capsule. When the arm is at rest at the side, the large capsule is taut superiorly and lax inferiorly.

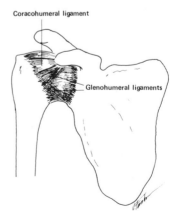

Coracohumeral ligament

Glenohumeral ligaments

Figure 7–17. GH ligaments.

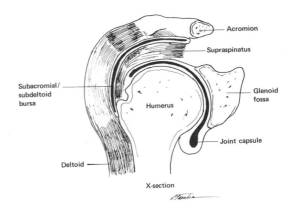

Figure 7–18. Subacromial and subdeltoid bursae permit smooth gliding of the supraspinatus muscle and the humeral head under the deltoid muscle and the acromion process.

tubercle. It checks lateral rotation of the humerus but serves a more important function in passive support of the upper limb against the force of gravity, as will be described shortly.

Bursae

There are several bursae associated with the shoulder complex in general and the GH joint specifically. While all contribute to function, the most important are the subacromial and subdeltoid bursae (Fig. 7–18). These bursae separate the supraspinatus tendon and head of the humerus from the acromion, coracoid process, coracoacromial ligament, and deltoid muscle. The bursae may be separate, but are commonly continuous with each other. Collectively the two are known as the **subacromial bursa.** The subacromial bursa permits smooth gliding between the humerus and supraspinatus tendon and its surrounding structures. Interruption or failure of this gliding mechanism is a common cause of pain and limitation of GH motion.

Coracoacromial Arch

The structures that overlie the subacromial bursa (the acromion and coracoacromial ligament) are together known as the coracoacromial (or suprahumeral) arch (Fig. 7–19). The coracoacromial arch forms an osteoligamentous vault that

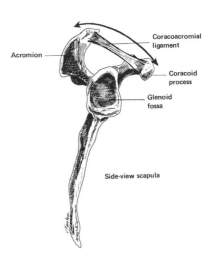

Figure 7–19. Coracoacromial arch. The arch is formed by the coracoid process anteriorly, the acromion posteriorly, and the coracoacromial ligament superiorly.

protects the top of the humeral head and the sensitive muscles, tendons, and bursae that lie above the humeral head from direct trauma from above. Such trauma is quite common and can occur through such simple daily tasks as carrying a heavy bag slung over the shoulder. The arch also prevents the head of the humerus from dislocating superiorly, since an unopposed upward translatory force on the humerus would cause an impact of the head of the humerus on the coracoacromial arch. Paradoxically, the impact of the humeral head into the arch (while beneficially preventing dislocation) would simultaneously cause painful impingement of the structures lying between the humeral head and the arch.

Motions of the Glenohumeral Joint

Osteokinematics

The GH joint is usually described as having 3° of freedom: flexion-extension, abduction-adduction, and medial-lateral rotation. The range of each of these motions occurring solely at the GH joint varies considerably. Steindler[3] cites the range of flexion-extension of the GH joint as 100 to 150°, of which the majority of movement occurs as flexion.[3] The range of medial-lateral rotation of the humerus varies with position. With the arm at the side, medial and lateral rotation may be limited to as little as 50° of combined motion. Abducting the humerus to 90° frees the arc of rotation to 120°.[3] The restricted arc of rotation in the pendant limb is due to the impact of the lesser tubercle on the anterior glenoid fossa with medial rotation and the impact of the greater tubercle on the acromion with lateral rotation. When the arm is abducted, these bony restrictions play little role, so the checks of motion become capsular and muscular.

The maximum range of abduction at the GH joint is the topic of much disagreement. There is consensus, however, that the range of abduction of the humerus in the frontal plane (whether done actively or passively) will be diminished if the humerus is maintained in medial rotation; the humerus will not abduct on the glenoid fossa beyond 60°.[7,24] The restriction is caused by the impingement of the greater tubercle on the acromion. When the humerus is laterally rotated, the greater tubercle will pass under or behind the acromion, and abduction continues unimpeded. There is not the same need for rotation of the humerus for flexion to achieve its full range. Given the forward movement of the humerus in flexion, the greater tubercle automatically slides behind the acromion process.

The ROMs for flexion and abduction of the GH joint (presuming impact of the greater tubercle will not occur) are reported to be anywhere from 90°[3] to 120°,[7,24] or as much as 135°.[26] Adding to the confusion, some studies have examined the range in the traditional frontal (abduction) and sagittal (flexion) planes, while others have investigated elevation in the so-called plane of the scapula. The plane of the scapula lies through the scapula in its resting position and reflects the fact that the scapula does not lie in the frontal plane, but is winged 30 to 45° toward the sagittal plane.[27] The plane of the scapula, therefore, lies approximately midway between the frontal and sagittal planes. When the humerus moves in the plane of the scapula, there is less restriction to motion. The capsule is not twisted and, therefore, is less tense than it is with motion in the frontal plane. Similarly, when the humerus moves in the plane of the scapula, lateral rotation of the humerus is not required to prevent impact of the greater tubercle on the acromion. Elevation at the GH joint in the plane of the scapula is commonly referred to as **abduction in the plane of the scapula** and was found to have a range of 107° by Freedman and Monroe[17] and 112° by Doody and Waterland.[28] These values fall between the 90° and 120° limits cited by Steindler[3] for abduction.

An additional factor that will influence the range of GH abduction is whether the motion is done actively or passively. Cailliet[7] proposes that there is 90° of

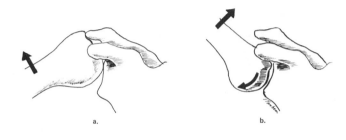

Figure 7–20. a. Abduction of the humerus as a pure rolling of the large humeral head on the small glenoid fossa would cause impaction of the head into the acromion. b. Abduction of the humerus occurring as a combination of rolling and sliding prevents impaction and allows a full ROM.

abduction when the motion is done actively and 120° when performed passively.[7] The restriction of GH abduction to 90° of active motion is quite evident when the scapula is immobilized for some reason,[29] but is more difficult to assess when there is simultaneous scapular movement. Some authors contend that there is only 90° of GH abduction regardless of whether the motion is either performed actively or passively or if there is simultaneous motion of the scapula. There does appear to be consensus among investigators that GH flexion does not vary between active or passive action, but is consistently 120°.

Arthrokinematics

The glenoid fossa and humeral head are incongruent surfaces; the convex humeral head is not parallel to the concave fossa.[7] Given this incongruence, rotation of the joint in any direction cannot take place as a pure spin, but requires that the motions of the humerus be accompanied by a combined rolling and gliding of the head of the humerus on the glenoid fossa in a direction opposite to movement of the shaft of the humerus. For example, abduction of the humerus would cause a superior rolling of the humeral head on the flattened fossa. The large humeral head would soon run out of glenoid surface and the head of the humerus would impact on the overhanging acromion (Fig. 7–20a). If the head of the humerus glides inferiorly while it rolls up the fossa, full ROM can be achieved (Fig. 7–20b). In addition to the inferior gliding required for full abduction, the humeral head may also be required to glide anteriorly or posteriorly. It has been shown, for instance, that the cocking phase of pitching a ball results in a posterior glide of the humeral head on the fossa; the acceleration phase of the throw is accompanied by an anterior glide of the head on the fossa.[30]

It is the function of the muscles that cross the GH joint to (1) move the humerus, (2) provide intra-articular gliding, and (3) maintain apposition of joint surfaces. The last two are necessary for dynamic stability requirements of the GH joint and are effectively performed. In the normal GH joint, impact of the humeral head on the acromion or the coracoacromial ligament does *not* occur.[7,31]

Static Stabilization of the Dependent Arm

Given the incongruence of the GH articular surfaces, the bony surfaces alone cannot maintain joint contact in the dependent position (arm hanging at the side). As the humeral head rests on the fossa, gravity acts on the humerus parallel to the shaft in a downward direction (negative translatory force). This would appear to require a vertical upward pull to restore equilibrium. Such a vertical force could only be supplied by muscles such as the anterior deltoid or the long heads of the biceps brachii and triceps brachii. Basmajian and Bazant[14] and MacConnail and Basmajian[25] have shown that all muscles of the shoulder complex are electrically silent in the relaxed, unloaded limb and even when the limb is tugged vigorously downward. The mechanism of joint stability, therefore, must be passive. As can be

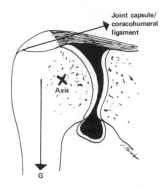

Figure 7–21. Mechanism for stabilization of the dependent arm. When the arm is relaxed at the side, the dislocation effect of gravity (G) is counteracted by the passive tension in the superior joint capsule and the coracohumeral ligament.

seen in Figure 7–21, gravity (G) acting on the humerus is a pure translatory force but lies at a distance from the eccentrically located center of rotation of the humeral head.[3] Given the axis and the line of pull, gravity creates an adduction moment (counterclockwise torque) on the humerus. Gravity must be offset by a force that can apply a torque of equal magnitude in the direction of abduction. Such a force can be applied by the superior joint capsule and the coracohumeral ligament, which are taut when the arm is at the side. Since the superior capsule and coracohumeral ligament insert on the greater tubercle, the moment arm (MA) of this passive force is nearly twice that of the more centrally located force of gravity. The upward translatory component of the action line of the superior capsule and coracohumeral ligament will offset the downward translatory component of gravity. The rotatory component of the passive force is also toward the joint, compressing joint surfaces.

The passive tension in the capsule and coracohumeral ligament is sufficient in most cases to counteract the effect of gravity. When the limb is loaded, the adduction moment (gravity plus the load) may require more force than can be safely provided by these structures. Basmajian and Bazant[14] showed that at such times the supraspinatus muscle contracts.[14] The action line of the supraspinatus is virtually identical to that of the superior capsule and coracohumeral ligament, since these share a common distal attachment. The supraspinatus may be assisted by activity of the posterior deltoid when additional contractile force is necessary.

Whenever the passive force of the capsule and coracohumeral ligament is inadequate, the active tension of the supraspinatus is recruited. In fact, the role of the supraspinatus may be more critical than its electromyographic activity indicates. Although *not* contracting when the unloaded arm hangs at the side, paralysis or dysfunction in the supraspinatus may lead to gradual inferior subluxation of the GH joint. Without the reinforcing passive tension of the intact supraspinatus muscle, the sustained load on the superior capsule and coracohumeral ligament apparently causes these structures to creep and results in a loss of joint stability. GH subluxation is commonly encountered in patients with diminished rotator cuff function due to stroke. Evidence, then, leads to the conclusion that static stabilization of the dependent arm is due to passive tension in the coracohumeral ligament, the superior capsule, and the inactive supraspinatus muscle. Additional reinforcement can be provided by active contraction of the supraspinatus.

Dynamic Stability of the Glenohumeral Joint

Prime Movers

It is generally accepted that the deltoid and supraspinatus muscles serve as the prime movers for GH abduction. The anterior deltoid is also considered the prime

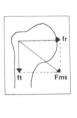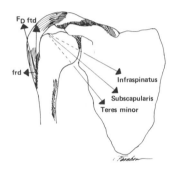

Figure 7–22. Force couple of the deltoid muscle (F_D) and the muscles of the musculotendinous cuff. The infraspinatus, subscapularis, and teres minor muscles together (Fms) have a negative translatory component (ft, inset) that nearly offsets the positive translatory component (ftd) of the deltoid force.

mover in GH flexion. Both abduction and flexion are elevation activities with many biomechanical similarities. Although the segments of the deltoid that participate will vary with role and function,[32,33] examination of the resultant action lines of the deltoid muscle and of the supraspinatus muscle in abduction can be used to highlight the stability needs of the GH joint in either elevation activity. Figure 7–22 shows the action line of the deltoid muscle with the arm at the side (the action line of the three segments of the deltoid acting together coincide with the fibers of the middle deltoid). When the muscle action line (F_D) is resolved into its translatory (ftd) and rotatory components (frd), the translatory component is by far the larger. That is, the majority of the force of contraction of the deltoid causes the humeral head to translate superiorly; only a small proportion of force is causing rotation (abduction) of the humerus. The forces of the deltoid constitute an example in which a translatory force applied in the direction of the joint is *not* a stabilizing influence. The articular surface of the humerus is not in line with the shaft of the humerus; therefore, a force parallel to the bone creates a dislocating rather than a stabilizing effect. The superior (positive) translatory force of the deltoid, if unopposed, would cause impaction of the head into coracoacromial arch before much abduction had occurred. Once the inferiorly directed inferior force of the coracoacromial arch has been introduced, rotation of the humeral head could, theoretically, continue against the leverage provided by the acromion, but pain from impinged structures would prevent much motion. The inferior translatory pull of gravity cannot offset f_{td}, since the resultant force of the deltoid must exceed that of gravity before any rotation can occur. Another force or set of forces must be introduced. This is a major function of the muscles of the rotator, or musculotendinous, cuff.

Rotator Cuff

The supraspinatus, infraspinatus, teres minor, and subscapularis muscles compose the rotator or musculotendinous cuff. These muscles are considered to be part of a "cuff" because the inserting tendons of each muscle of the cuff blend with and reinforce the GH capsule. More importantly, all have action lines that significantly contribute to the dynamic stability of the GH joint. The action lines of the infraspinatus, teres minor, and subscapularis are shown in Figure 7–22. When the force of any one (or all three taken together) is resolved into its components (inset), it can be seen that the rotatory force (fr) not only tends to cause at least some rotation of the humerus, but f_r also compresses the head into the glenoid fossa. Now we have an example of a rotatory component creating joint stabilization! This is due, once again, to the fact that the articular surface of the humerus lies nearly perpendicular to the shaft.

Although the muscles of the rotator cuff are important GH joint compressors, equally (or perhaps more) critical to the stabilizing function of cuff muscles is the negative (inferior) translatory pull (ft) of the muscles. The sum of the three negative

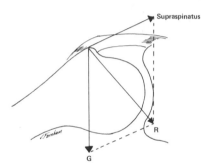

Figure 7–23. Gravity acts as a stabilizing synergist to the supraspinatus muscle. Activity of the supraspinatus and gravity (G) produce a resultant force (R) that abducts the humerus and causes the downward sliding of articular surfaces necessary for a full ROM.

translatory components of the rotator cuff nearly offsets the positive translatory force of the deltoid muscle. The translatory component of the rotator cuff muscle force reduces the shear between the humeral head and the glenoid fossa. The resultant superior translatory force or shear on the humeral head (deltoid force plus rotator cuff force) is greatest when the GH joint is at about 45° of elevation; at about 60° of GH motion, the shear and compressive forces are equal. By the time that the GH joint has nearly completed its range, the shear force is negligible.[6]

In addition to their stabilizing role, the teres minor and infraspinatus muscles, unlike the subscapularis, contribute to abduction by providing the lateral rotation necessary to prevent the greater tubercle from impacting the acromion.

The action of the deltoid along with the combined actions of the infraspinatus, teres minor, and subscapularis form a **force couple.** In a force couple the divergent pulls of the forces create a pure rotation. In this case, the divergent pulls create an *almost* perfect spinning of the humeral head around a fixed axis of rotation.

Although the supraspinatus muscle is also part of the rotator cuff, the action line of the supraspinatus muscle has a positive (superior) translatory component, rather than the negative (inferior) component found in the other muscles of the cuff. Given its line of pull, the supraspinatus is of no use in offsetting the upward dislocating action of the deltoid. The supraspinatus still is very effective as a stabilizer of the GH joint since, like the other cuff muscles, its rotatory component generates a strong compressive force. Unlike the other cuff muscles, the rotatory component of the supraspinatus has a large enough MA so that it is a significant abductor. As an abductor it is capable by itself not only of producing a full range of abduction but also of stabilizing the joint with the assistance of gravity.[25] Gravity acts as a stabilizing synergist to the supraspinatus by offsetting the small upward translatory pull of the muscle (Fig. 7–23). Saha[26] refers to the supraspinatus as a steering muscle.[27] A steering muscle causes a changeover of surfaces within the joint, usually by gliding, and directs the articular surfaces to the appropriate points of contact. Gravity and the supraspinatus act as vertical steerers; the resultant of the two forces causes an inferior gliding of the humeral head during abduction of the shaft, allowing full articulation of the surfaces and preventing superior displacement. The other muscles of the rotator cuff also are steerers, according to Saha, although at some points in the range their importance is in horizontal steering; that is, posterior and anterior gliding of the humeral head. The subscapularis is credited with being able to posteriorly steer the humeral head, thus offsetting anterior dislocating forces.[15]

Costs of Dynamic Stability

The requirements for dynamic stability of the GH joint vary somewhat throughout the range. Anterior stability is considered to be a function of the subscapularis muscle, while the infraspinatus and teres minor muscles protect the

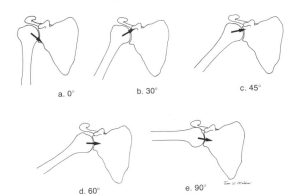

a. 0°   b. 30°   c. 45°

d. 60°   e. 90°

Figure 7–24. When the GH joint is in neutral position (a), the joint reaction force is on the inferior glenoid fossa. As abduction is initiated (b), the joint reaction force shifts to the superior fossa but begins the return to the central portion of the fossa almost immediately, where it remains for the remainder of the GH motion (c–e).

humeral head posteriorly. Some common stability requirements exist regardless of the plane in which the humerus is moved. Equilibrium at any point in the range is a function of (1) the force of the prime mover(s), (2) the force of gravity, and (3) the force of the compressors and steerers. Inman, Saunders, and Abbott[29] appropriately add a fourth point, the force of friction and the joint reaction force. Any shear force within the GH joint creates some friction across its joint surfaces. More importantly, all compressive forces that snug the head into the glenoid fossa must be opposed by an equal force from the glenoid fossa in the opposite direction (joint reaction force). When all equilibrium factors are intact and operating normally, the head of the humerus rotates on a relatively fixed center of rotation, with the only significant excursion of the axis occurring in the early range of elevation[18] (Fig. 7–24).

Although little superior gliding of the humeral head occurs during normal elevation activities, there are substantial changes that occur in the pressure within the subacromial bursa as the humerus moves through its ROM. These pressures are related to both arm position and load, with greater pressures in the bursa evidenced as the arms are loaded and maintained in an elevated position.[34] The elevated subacromial bursal pressures and the coincident poor vascularization of the supraspinatus tendon that lies beneath the bursa may be responsible for the degenerative changes seen in the supraspinatus tendon with increasing age. Degeneration of the supraspinatus tendon results in occurrence of tendon tears with increasingly minor trauma as one ages.[35] This may explain the high incidence of shoulder pain that occurs with increasing age. The pain may also be due to the narrowed AC joint spaces found with increasing age.[36] Whether complaints of pain are due to tendon trauma or to narrowed joint spaces, symptomatic and asymptomatic rotator cuff tears are seen in almost all people over the age of 70, with the supraspinatus likely to show lesions before the other tendons of the cuff.[37] The frequency of supraspinatus lesions may be due not only to its limited vascularity but also to its role both in moving the arm and in sustaining the weight of the dependent upper extremity. Unlike the other GH muscles, which contract during their primary activity and then relax, the supraspinatus is either contracting (in flexion, abduction, and the loaded dependent arm) or passively tensed (unloaded dependent arm) in almost all positions of the GH joint. Rotator cuff lesions typically produce pain with shoulder motions between 60 and 120° of combined GH and scapular abduction. This range constitutes what is known as the **painful arc.** Pain due to AC degeneration is more typically found when the arm is raised beyond the painful arc.[38]

Any disruption of the synergistic action of the four equilibrium factors for dynamic GH joint stability may lead to varying centers of rotation of the humeral head and to excessive excursion of the humeral head on the glenoid fossa. Such

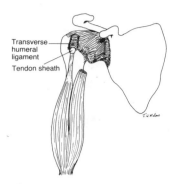

Transverse humeral ligament

Tendon sheath

Figure 7–25. The long head of the biceps brachii passes through a fibro-osseous tunnel formed by the bicipital groove and the transverse humeral ligament. It is protected within the tunnel by a tendon sheath.

mechanical deviations may result in degenerative changes in other structures of the joint besides the rotator cuff and to subluxation of the joint. A predisposition to actual dislocation of the articular surfaces (usually dislocation of the humeral head) exists when the individual structural variations that occur are in the direction of: (1) anterior tilt of the glenoid fossa; (2) excessive retrotorsion of the humeral head; or (3) weakened horizontal steerers (rotator cuff).[26]

Biceps Mechanism

The long head of the biceps runs superiorly from the anterior shaft of the humerus through the bicipital groove between the greater and lesser tubercles to attach to the supraglenoid tubercle above the glenoid fossa. It enters the joint capsule through an opening between the supraspinatus and subscapularis muscles, penetrating the capsule but not the synovium (Fig. 7–25). Within the bicipital groove, the biceps tendon is enveloped by a tendon sheath and tethered there by the transverse humeral ligament. The long head of the biceps, because of its position, is sometimes considered to be part of the reinforcing cuff of the GH joint. The biceps muscle is capable of contributing to the force of flexion and can, if the humerus is laterally rotated, contribute to the force of abduction.[29]

Its relevance at the shoulder, however, may have more to do with dysfunction than with function. That is, its contribution to normal GH motion has less impact than its contribution to shoulder problems. Whenever the humerus elevates in flexion and abduction, whether the biceps is actively contributing to the motion or whether it is passive, the tendon of the biceps must slide within the groove and under the transverse humeral ligament. If the bicipital tendon sheath is worn or inflamed, the gliding mechanism may be interrupted and pain produced. A tear in the transverse humeral ligament may result in the biceps tendon popping in and out of the bicipital groove with rotation of the humerus, a potentially wearing and painful microtrauma. Since the long head of the biceps, like the supraspinatus tendon, is poorly vascularized,[38] it is subject to some of the same degenerative changes and the same trauma seen in the more active muscles of the rotator cuff.

INTEGRATED FUNCTION OF THE SHOULDER COMPLEX

The shoulder complex acts in a coordinated fashion to provide the smoothest and greatest ROM possible to the upper limb. Motion available to the GH joint alone would not account for the full range of elevation (abduction or flexion) available to the humerus. The remainder of the range is contributed by the scapulothoracic joint through its SC and AC linkages. Under normal and unconstrained conditions, each joint makes its contribution not only in a fairly consistent manner, but following a pattern of *concomitant* and *coordinated* movement known as scapulohumeral rhythm.

Scapulohumeral Rhythm
Scapulothoracic and Glenohumeral Contributions

The scapulothoracic joint contributes to both flexion and abduction (elevation) of the humerus by upwardly rotating the glenoid fossa 60° from its resting position. If the humerus were fixed to the fossa, this alone would result in 60° of elevation of the humerus. The humerus is not fixed, of course, but can move independently on the glenoid fossa. The GH joint contributes 120° of flexion and anywhere from 90 to 120° of abduction (depending on individual structural variations and on one's philosophy of available GH abduction). The combination of scapular and humeral movement results in what is commonly held to be a maximum range of elevation to 180° (Fig. 7–26) and in an *overall* ratio of 2° of GH to 1° of scapulothoracic motion.[3] As already noted, there is disagreement over the available range of GH abduction. Not surprisingly, then, there is disagreement over the ratio of scapular to humeral movement. When less maximum range is cited for the GH joint, the ratio may be close to 3° of GH movement to 2° of scapulothoracic.[17,28]

The purpose of scapulohumeral rhythm is threefold: (1) distributing the motion between two joints permits a large ROM with less compromise of stability than would occur if the same range occurred at one joint; (2) maintaining the glenoid fossa in an optimal position to receive the head of the humerus increases joint congruency while decreasing shear forces; and (3) permitting muscles acting on the humerus to maintain a good length-tension relationship minimizes or prevents active insufficiency of the GH muscles.

During the initial 60° of flexion or the initial 30° of abduction of the humerus, an inconsistent amount and type of scapular motion takes place relative to GH motion. During this period the scapula seeks a position of stability in relation to the humerus (setting phase).[29,32] In this early phase, motion occurs primarily at the GH joint, although stressing the arm may increase the scapular contribution.[28,31] With increasing range, the scapula increases its contribution, approaching a 1:1 ratio with GH movement; in the latter part of the range, the GH joint again increases its contribution.[17,26,28] Poppen and Walker[18] found the GH to scapulothoracic ratio to

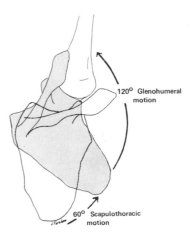

120° Glenohumeral motion

60° Scapulothoracic motion

Figure 7–26. Elevation of the arm through a full ROM normally requires 60° of scapulothoracic motion and 120° of GH motion. These motions occur concomitantly, not sequentially.

be 5:4 between 24° and maximum elevation in the plane of the scapula. They note, however, that the absolute angles achieved at each joint yield an overall ratio of 2° of GH motion for each 1° of scapulothoracic motion.[18] While the ratio of scapulo-humeral rhythm is both individualistic and nonlinear, one can conceptualize the concerted movement of the humerus and scapula as being 2° of GH motion for every 1° of scapulothoracic motion.

Sternoclavicular and Acromioclavicular Contributions

Scapulohumeral rhythm involves the concerted action of the SC and AC joints, as well as the scapulothoracic and GH joints. This is true because the scapulothoracic joint is part of a closed kinematic chain, and movement of the scapula can only occur through movement at one or both of these joints. The 60° arc of upward rotation through which the scapula moves can be attributed to SC and AC motion produced by the force couple of the trapezius and serratus anterior muscles. These two muscles are the *only* muscles capable of upwardly rotating the scapula.

Phase One

The upper and lower portions of the **trapezius muscle** combine with the upper and lower portions of the serratus anterior muscle (Fig. 7–27) to produce an upward rotatory force on the scapula. Although this motion would most likely occur at the AC joint, tension in the conoid and trapezoid (coracoclavicular) ligaments prevents this AC movement. Upward rotation of the scapula at the AC joint would result in movement of the coracoid process of the scapula inferiorly. Since the coracoid process is tied to the clavicle by the coracoclavicular ligament, the movement of the scapula is prevented. The upward rotatory force on the scapula continues, however, as the trapezius and serratus contract; the muscles produce movement at the next available joint—the SC joint. The pull of the muscles on the scapula (and the direct pull of the upper trapezius on the lateral clavicle) force the clavicle to elevate. The clavicular elevation at the SC joint carries the scapula through 30° of upward rotation as the scapula rides on the lateral end of the rising clavicle. The scapulothoracic motion occurs around an axis that appears to intersect the base of the spine of the scapula and the SC joint[32] (Fig. 7–28a). Elevation of the clavicle is checked when the costoclavicular ligament becomes taut.

Through the initial upward rotation of the scapula that occurs with clavicular elevation at the SC joint, the AC joint maintains a relatively fixed relationship between the scapula and clavicle. There is no upward rotation *at the AC joint.* Although the coracoclavicular ligaments prevent upward rotation of the scapula on

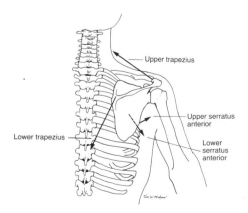

Figure 7–27. The action lines of the upper trapezius, lower trapezius, upper serratus anterior, and lower serratus anterior combine to produce almost pure upward rotation of the scapula.

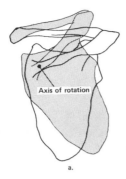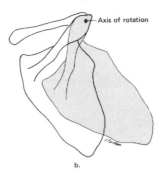

Figure 7–28. a. The first 30° of scapulothoracic motion. Elevation of the clavicle at the SC joint swings the scapula through an arc of motion that has an axis at the base of the spine of the scapula. b. The last 30° of scapulothoracic motion. Clavicular rotation flips the lateral end of the clavicle up. The remaining scapular rotation has an axis at the AC joint.

the clavicle, they permit some tipping and about 10° of winging of the scapula, which maintains the scapula against the changing contour of the rib cage.[33] When the costoclavicular ligament becomes taut and clavicular elevation ends, the scapula will have upwardly rotated through an arc of 30°. Since the scapulothoracic motion occurs *concurrently* with GH motion, the GH joint will have simultaneously elevated about 60° (using an overall 2:1 ratio). The arm will have been raised 90 to 100 ° from the side (30° ST + 60° GH).[29,32]

Phase Two

As the trapezius and the serratus anterior continue to generate an upward rotatory force on the scapula, scapular movement is still restrained by the conoid and trapezoid ligaments at the AC joint and now by the costoclavicular ligament at the SC joint. With no other available motion to dissipate the upward rotatory force being created by the trapezius and serratus muscles, the coracoid process of the scapula pulls downward, tugging on the coracoclavicular ligament and carrying the posteriorly located conoid tubercle of the clavicle downward as well.[33] The resulting motion is rotation of the clavicle around its longitudinal axis. The clavicular rotation will flip the lateral end of the crank-shaped clavicle up without causing further elevation at the SC joint. The scapula, attached to the lateral end of the clavicle, will be carried through an additional 30° of upward rotation around an AP axis through the AC joint (Fig. 7–28b). The AC joint also undergoes a maximum of 20° of tipping[31] and 40° of winging[18] as the scapula finds its final position on the ribcage.

If 180° is accepted as the maximal range of flexion and abduction of the humerus, raising the arm *to the horizontal* involves 60° of GH motion and 30° of scapulothoracic motion, with the scapular movement produced by clavicular elevation at the SC joint. Raising the arm *from the horizontal to vertical position* involves an additional 60° of GH movement (with lateral rotation needed for abduction in the frontal plane) and 30° of scapular movement produced by clavicular rotation and AC motion (Fig. 7–29). For the clavicle to rotate about its longitudinal axis, both the SC and the AC joints need to be free to move.

The sequence of events that are part of scapulohumeral rhythm occur regardless of the plane in which the arm is elevated. That is, although the range may vary somewhat, the component events are similar whether the motion is performed as flexion, as abduction, or in the plane of the scapula. One difference already noted is that abduction in the frontal plan requires concomitant lateral rotation of the

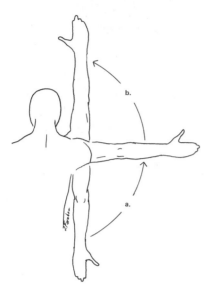

Figure 7–29. a. Bringing the arm from the side to the horizontal requires concomitant GH motion and SC elevation. b. Arm horizontal to overhead position requires concomitant GH motion and SC-AC rotation.

humerus to permit full GH range. There is also another difference between performance of sagittal plane and frontal plane elevation. Although the scapula must upwardly rotate in both instances, flexion requires simultaneous abduction of the scapula. Abduction of the scapula brings the glenoid fossa forward, keeping the fossa in line with the shaft of the humerus. If this did not occur, the head of the humerus would be unprotected posteriorly; posterior dislocations could occur with relatively little force. In abduction in the frontal plane, the scapula tends to remain in its resting position or slightly adducted.

Structural Dysfunction

Completion of the range of elevation of the arm is dependent on the ability of GH, scapulothoracic, SC, and AC joints each to make the needed contribution. Disruption of movement in any of the participating joints will result in a loss of ROM. Once restrictions to function are introduced, however, the concept of scapulohumeral rhythm is no longer relevant. A reduction in GH joint range will *not* result in a proportional decrease in scapulothoracic range. The ratio of movement is no longer pertinent since the body will automatically recruit any and all remaining motion at other joints.

If motion at the GH joint is restricted by pain or disease, the total range available to the humerus will be reduced. Whatever portion of the motion remains at the GH joint will still be accompanied by the full 60° of humeral elevation available from scapulothoracic motion. For example, restriction of the humerus in a position of medial rotation will limit GH abduction to a maximum of 60°. This GH range will combine with 60° of scapulothoracic motion to give a total available range of 120° when the GH joint is maintained in the medially rotated position. Consider also that fusion of the SC joint will eliminate scapular movement. Since both clavicular elevation and rotation take place through the SC joint, SC fusion will eliminate both components of scapular rotation. Movement for the humerus will be available only at the GH joint, with the amount of motion available influenced by whether the motion is performed actively or passively (see the discussion of deltoid function). If the coracoclavicular ligament is disrupted, clavicular rotation will be lost; the range of elevation will be dependent on clavicular elevation and active GH motion.

Muscles of Elevation

Perry[39] describes elevation and depression as the two primary patterns of shoulder complex function. Elevation activities are described as those requiring muscles to overcome or control the weight of the limb and its load, and usually involves components of GH flexion and/or abduction, and scapular upward rotation. The completion of normal elevation is dependent not only freedom of movement and integrity of the joints involved but also on the appropriate strength and function of the muscles producing and controlling movement. A closer look at the activity of these muscles should enhance an understanding of normal function, as well as contribute to an understanding of the deficits seen in pathologic situations.

Deltoid Muscle

The deltoid is at resting length (optimal length-tension) when the arm is at the side. When at resting length, the deltoid's angle of pull will result in a predominance of superior translatory pull on the humerus with an active contraction. With appropriate counteraction of the translatory pull, the rotatory components of the anterior and middle deltoid are effective primary movers for flexion and abduction, respectively. When the humerus is in the plane of the scapula, the anterior and middle deltoid are optimally aligned to produce elevation of the humerus.[18] The action line of the posterior deltoid has too small an MA (and too small a rotatory component) to contribute effectively to abduction; it serves primarily as a joint compressor.[21,33] As the humerus elevates, the translatory component of the deltoid as a whole increases joint compression (and diminishes its superior dislocating influence); the rotatory component must counteract the increasing torque of gravity. Electromyographic activity in the deltoid shows gradually increasing activity in the deltoid, peaking at 90° of humeral abduction and plateauing for the remainder of the motion (Saha[26] found a peak at 120° with dropoff to moderate activity at 180°). The peak activity in flexion does not occur until later in the range and there is less total activity.[26,29] Although the MA of the deltoid gets larger as the humerus elevates[22] and the torque of gravity diminishes above the horizontal, the activity level of the deltoid continues to increase at a high level. The deltoid's shortening fibers are approaching active insufficiency. As a result of the loss of tension due to extreme shortening, a greater number of motor units must be recruited to maintain even equivalent force output. The multipennate structure and considerable cross section of the deltoid help compensate for the relatively small MA, low mechanical advantage, and less-than-optimal length-tension.

Maintenance of appropriate length-tension of the deltoid is strongly dependent on simultaneous scapular movement. When the scapula is restricted, the deltoid can only achieve and barely maintain 90° of GH abduction (whether the supraspinatus is available for assistance or not).[3,24,29] The synergy that occurs between the scapular upward rotators and the deltoid is further discussed in the section on trapezius and serratus anterior function. Deltoid activity is additionally dependent on an intact rotator cuff. With complete derangement of the cuff, deltoid activity results in a shrug of the shoulder rather than in abduction of the humerus. Stimulation of the axillary nerve (innervating the deltoid and teres minor alone) produces approximately 40° of abduction.[40] Partial tears in or partial paralysis of the cuff will weaken the rotation produced by the deltoid.[13]

Supraspinatus Muscle

The supraspinatus muscle is an abductor of the humerus. Like the deltoid muscle, it functions in both flexion and abduction of the humerus. Its role, according to

MacConnail and Basmajian,[25] is quantitative rather than specialized. The pattern of activity of the supraspinatus is essentially the same as that found in the deltoid.[29] The MA of the supraspinatus is fairly constant throughout the ROM and is larger than that of the deltoid for the first 60° of shoulder abduction.[21] When the deltoid is paralyzed, the supraspinatus alone can bring the arm through most if not all of the GH range, but the motion will be quite weak. With a suprascapular nerve block that paralyzes the supraspinatus and the infraspinatus, the strength of elevation in the plane of the scapula is reduced by 35 percent at 0° and by 60 to 80 percent at 150°.[41] The secondary functions of the supraspinatus are to compress the GH joint, to act as a vertical steerer for the humeral head, and to assist in maintaining the stability of the dependent arm. With isolated complete paralysis of the supraspinatus muscle, some loss of abduction force is evident, but most of its functions can be performed by remaining musculature. Isolated paralysis of the supraspinatus is unusual, however, since its innervation is the same as the infraspinatus and related to that of the teres minor. Most commonly, lesions of the rotator cuff muscles occur together, producing a more extensive deficit than seen with paralysis of the supraspinatus alone.

Infraspinatus, Teres Minor, and Subscapularis Muscles

When Inman, Saunders, and Abbott[29] assessed the combined actions of the infraspinatus, teres minor, and subscapularis muscles, electromyographic activity indicated a nearly linear rise in action potentials from 0 to 115° elevation. Activity dropped slightly between 115 and 180°. Total activity in flexion was slightly greater than that in abduction. In abduction an early peak in activity of these muscles appeared at 70° of elevation. Steindler[3] hypothesized that the early peak was a response to the need for depression (downward sliding) of the humeral head, while the latter peak at 115° was a result of increased activity of these muscles in producing lateral rotation of the humerus. The medial rotatory function of the subscapularis acts to steer the head of the humerus horizontally, while continuing with the other cuff muscles to compress and stabilize the joint.[26]

Upper and Lower Trapezius and Serratus Anterior Muscles

The upper trapezius and upper serratus anterior muscles form one segment of a force couple that drives the scapula in elevation of the arm. These two muscle segments, along with the levator scapula muscle, also support the shoulder girdle against the downward pull of gravity. Although support of the scapula in the pendant limb in many individuals is passive, loading the limb will produce activity in these muscles.[25,29] The second segment of the force couple is formed by the lower trapezius and lower serratus anterior muscles. When activity of the upper and lower trapezius and serratus anterior muscles was monitored electromyographically during humeral elevation, the curves were found to be similar and complementary. Activity in the trapezius rises linearly to 180° in abduction, with more undulating activity in flexion. The serratus anterior shows a linear increase in action potentials to 180° in flexion, with undulating activity in abduction.[29] Saha[26] found the upper and lower trapezius activity peaked and plateaued before the end of the range, with some decrease in activity at maximal elevation.

In abduction of the arm, the force of the trapezius seems more critical to the production of upward rotation of the scapula than the force of the serratus anterior. When the trapezius is intact and the serratus anterior is paralyzed, abduction of the arm can occur through its full range although it is *weakened*. When the trapezius is paralyzed (even though the serratus anterior may be intact), abduction of the arm

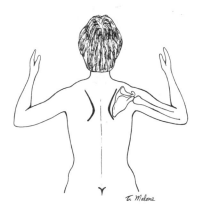

Figure 7–30. With paralysis of the trapezii, attempted abduction of the arms causes the deltoids to have a downward rotatory effect on the scapulae. Although 90° of GH joint motion can be achieved, the position of the scapulae results in the arms only being raised from the side 60° or less.

is both weakened and *limited in range* to 75°.[3] This is only slightly better than the range that can be obtained when neither of the upward rotators of the scapula are present.[40] The remaining range occurs exclusively at the GH joint. Without the trapezius (with or without the serratus anterior), the scapula rests in a downwardly rotated position due to the unopposed effect of gravity on the scapula. When abduction is attempted, the middle and posterior fibers of the activated deltoid (originating on the acromion and spine of the scapula) increase the downward rotatory pull on the scapula. Although the deltoid can still achieve the 90° of GH motion attributed to it when the scapula has been immobilized, the 90° occurs on a downwardly rotated scapula; the net effect is that the arm will only raise from the side about 60 to 75° (Fig. 7–30).

Although the trapezius seems to be the more critical of the two upward rotators in abduction of the arm, the reverse set of circumstances occurs with flexion of the arm. In flexion the anterior orientation of the scapula is important in that this can be produced only by the serratus anterior. If the serratus anterior is intact, trapezius paralysis results in loss of *force* of shoulder flexion but there is no range deficit. If the serratus anterior is paralyzed (even in the presence of a functioning trapezius), flexion will be both *diminished in strength* and *limited in range* to 130 or 140° of flexion. When the scapular adduction component of the trapezius is unopposed by the serratus anterior, the trapezius is unable to upwardly rotate the scapula more than 20° of its potential 60°.[3]

Although the serratus anterior and trapezius are the prime movers for scapulothoracic upward rotation, these muscles serve an equally important function as stabilizing synergists for the deltoid acting at the GH joint. All muscles pull on both attachments (origin and insertion) equally. When both ends are free to move, the lighter end will usually move first. In most instances, the lighter of the two segments of the joint is the distal segment. Rather uniquely, the lighter segment of the GH joint is the proximal scapular segment. If the deltoid acted on its lighter proximal segment rather than its heavier humerus (with its appended forearm and hand), the scapula would rotate downward before the humerus would lift. The deltoid muscle would become actively insufficient before much humeral elevation was produced. The trapezius and serratus anterior muscles, as upward scapular rotators, prevent the undesirable movement of the proximal segment during deltoid contraction. They maintain optimal length-tension in the deltoid and permit the deltoid to carry its heavier distal lever through full ROM. Thus, the role of the scapular force couple is agonistic to scapular movement and synergistic to GH movement. The trapezius and serratus anterior produce desired scapular upward rotation, while preventing undesired movement by the deltoid as it elevates the GH joint.

Middle Trapezius and Rhomboid Muscles

The middle portion of the trapezius and the rhomboid major and minor muscles are all active in elevation of the humerus, especially in abduction. These muscles serve a critical function as stabilizing synergists to the muscles that rotate the scapula. They contract eccentrically to control the change in position of the scapula produced by the upper and lower trapezius and the serratus anterior. Paralysis of these muscles causes disruption of the normal scapulohumeral rhythm and may result in diminished ROM.[3]

Muscles of Depression

Depression is the second of the two primary patterns of shoulder complex function. It involves the *forceful* downward movement of the arm in relation to the trunk, or the forceful movement of the trunk upward in relation to the fixed arm.[39] In this pattern the scapula tends to rotate downward and adduct during the humeral motion, but there is not a consistent scapulohumeral rhythm.

Latissimus Dorsi and Pectoral Muscles

The latissimus dorsi muscle on the free limb serves an important function in adduction and medial rotation of the humerus, as well as in extension of the humerus. Through its attachment to the scapula and by continued action on the humerus, it also adducts and depresses the scapula. When the hand is fixed in weight bearing, the latissimus dorsi muscle will pull its caudal attachment on the pelvis toward its cephalad attachment on the scapula and humerus. This results in lifting the body up as in a seated pushup. When the hands are weight-bearing on the handles of a pair of crutches, a contraction of the latissimus will unweight the feet as the trunk rises beneath the fixed scapula; this will allow the legs to swing forward through the crutches.

Some studies have found the latissimus dorsi muscle to be active in abduction and flexion of the arm.[21,26] Its activity may contribute to joint stability, since it causes compression of the GH joint when the arm is above the horizontal.

The clavicular portion of the pectoralis major can assist the deltoid in flexion of the GH joint but the sternal and abdominal portions are primary depressors of the shoulder complex. The combined action of the pectoralis major's sternal and abdominal portions parallels that of the latissimus dorsi, although the pectoralis is anterior to the GH joint rather than posteriorly as is the latissimus. In activities involving weight-bearing on the hands, both the pectoralis major and the latissimus can depress the shoulder complex, while anterior-posterior movement of the humerus and abduction/adduction of the scapula are neutralized. The depressor function of these muscles is further assisted by the pectoralis minor muscle, which acts directly on the scapula to depress and rotate it downward.

Teres Major and Rhomboid Muscles

The teres major muscle, like the latissimus dorsi, serves to adduct, medially rotate, and extend the humerus. Uniquely, however, it is active only in *static* positions of the humerus. Inman, Saunders, and Abbot[29] found the teres major active in maintained abduction of the humerus; its activity increased with increased loading of the limb, peaking at 90° of abduction. Its role in the activity was not hypothesized.

Function of the teres major muscle is strongly dependent on activity of the rhomboid muscles. Since the teres major muscle originates on the scapula and attaches to the humerus, its lighter segment is its proximal one. The proximal

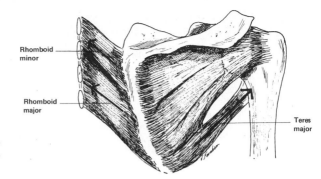

Figure 7–31. Scapular synergy of the teres major and rhomboid muscles. The rhomboid major and rhomboid minor offset the unwanted scapular upward rotation force produced by teres major activity.

attachment of the teres major must be stabilized to permit it to act effectively on the humerus. Unopposed, the teres major would upwardly rotate the scapula. The rhomboid muscles, as downward rotators of the scapula, not only offset the undesired force of the teres major but also contribute to the overall pattern of depression of the shoulder complex (Fig. 7–31).

We have laid the foundation for understanding more distal upper extremity joint function by exploring the intricate dynamic stability of the shoulder complex. The remainder of the upper extremity is dependent on maintenance of the dual roles of stability and mobility. Whereas function in the hand, for instance, can continue on a limited basis with loss of shoulder mobility, loss of shoulder stability can render the remaining function in the hand unusable. We will next explore the elbow as the intermediary between the shoulder and the hand.

REFERENCES

1. Leveau, B (ed): Williams and Lissner's Biomechanics of Human Motion. WB Saunders, Philadelphia, 1977.
2. Dempster, WT: Mechanics of shoulder movement. Arch Phys Med Rehabil 45:49, 1965.
3. Steindler, A: Kinesiology of the Human Body. Charles C Thomas, Springfield, Ill., 1955.
4. Morris, J: Joints of the shoulder girdle. Aust J Physiother 24, June, 1978.
5. Depalma, AF: Degenerative Changes in Sternoclavicular and Acromioclavicular Joints in Various Decades. Charles C Thomas, Springfield, Ill., 1957.
6. Sarrafian, SK: Gross and functional anatomy of the shoulder. Clin Orthop 173:11–18, 1983.
7. Cailliet, R: Shoulder Pain, ed 2. FA Davis, Philadelphia, 1981.
8. Sadr, B and Swann, M: Spontaneous dislocation of the sterno-clavicular joint. Acta Orthop Scand 50:269–274, 1979.
9. Petersson, CJ: Degeneration of the acromio-clavicular joint. Acta Orthop Scand 54:434, 1983.
10. Post, M: Current concepts in the diagnosis and management of acromioclavicular dislocations. Clin Orthop 200:234–247, 1985.
11. MacDonald, PB, Alexander, MJ, Frejuk, J, and Johnson, GE: Comprehensive functional analysis of shoulders following complete acromioclavicular separation. Am J Sports Med 16:475–480, 1988.
12. Bargen, JH, Erlanger, S, and Dick, HM: Biomechanics and comparison of two operative methods of treatment of complete acromioclavicular separation. Clin Orthop 130:267–272, 1978.
13. Fenlin, JM: Total glenohumeral joint replacement. Orthop Clin North Am 6:565, 1975.
14. Basmajian, JV and Bazant, FJ: Factors preventing downward dislocation of the adducted shoulder. J Bone Joint Surg [Am] 41:1182, 1959.
15. Saha, AK: Recurrent anterior dislocation of the shoulder: A new concept. Academic Publications, Calcutta, 1969.
16. Saha, AK: Dynamic stability of the glenohumeral joint. Acta Orthop Scand 42:490, 1971.
17. Freedman, L and Monroe, RR: Abduction of the arm in the scapular plane: Scapular and glenohumeral movements. J Bone Joint Surg [Am] 48:150, 1966.
18. Poppen, NK and Walker, PS: Normal and abnormal motion of the shoulder. J Bone Joint Surg [Am] 58:195, 1976.
19. Rothma, RH, Marvel, JP, Jr, and Heppenstall, RB: Anatomic considerations in the glenohumeral joint. Orthop Clin North Am 6:341, 1975.

20. Kapandji, IA: Physiology of the Joints. E&S Livingstone, London, 1970.
21. Poppen, NK and Walker, PS: Forces at the glenohumeral joint in abduction. Clin Orthop 135:165, 1978.
22. Walker, PS and Poppen, NK: Biomechanics of the shoulder joint during abduction on the plane of the scapula. Bull Hosp Joint Dis Orthop Inst 38:107, 1977.
23. Moseley, HF and Overgaarde, KB: The anterior capsule mechanism in recurrent dislocation of the shoulder. Morphological and clinical studies with special references to the glenoid labrum and glenohumeral ligaments. J Bone Joint Surg [Br]44:913,1962.
24. Lucas, DB: Biomechanics of the shoulder joint. Arch Surg 107:425, 1973.
25. MacConnail, MA and Basmajian, JV: Muscles and Movement: A Basis for Human Kinesiology. Williams and Wilkins, Baltimore, 1969.
26. Saha, AK: Theory of Shoulder Mechanism: Descriptive and Applied. Charles C Thomas, Springfield, Ill., 1961.
27. Johnston, TB: The movements of the shoulder joint: A plea for the use of "plane of the scapula" as the plane of reference for movements occurring at the humeroscapular joint. Br J Surg 25:252, 1937.
28. Doody, SG and Waterland, JC: Shoulder movements during abduction in the scapular plane. Arch Phys Med Rehabil 51:595, 1970.
29. Inman, VT, Saunders, JB, and Abbott, LC: Observations of function of the shoulder joint. J Bone Joint Surg [Br] 26:1,1944.
30. Howell, SM, Galinat, BJ, Renzi, AJ, and Marone, PJ: Normal and abnormal mechanics of the glenohumeral joint in the horizontal plane. J Bone Joint Surg [Am] 70:227–232, 1988.
31. Saha, AK: The Classic: Mechanism of shoulder movements and a plea for the recognition of "zero position" of the glenohumeral joint. Clin Orthop 173:3–9, 1983.
32. Dvir, Z and Berme, N: The shoulder complex in elevation of the arm: A mechanism approach. J Biomech 1:219, 1978.
33. DeDuca, CJ and Forrest, WJ: Force analysis of individual muscles acting simultaneously on the shoulder joint during isometric abduction. J Biomech 6:385, 1973.
34. Sigholm, G, Styf, J, Korner, L, and Herberts, P: Pressure recording in the subacromial bursa. J Orthop Res 6:123–128, 1988.
35. Ozaki, J, Fujimoto, S, Nakagawa, Y, Masuhara, K, and Tamai, S: Tears of the rotator cuff of the shoulder associated with pathological changes in the acromion. J Bone Joint Surg [Am] 70:1224–1230, 1988.
36. Petersson, CJ and Redlund-Johnell, I: The subacromial space in normal shoulder radiographs. Acta Orthop Scand 55:57–58, 1984.
37. Gschwend, N, Ivosevic-Radovanovic, D, and Patte, D: Rotator cuff tear—relationship between clinical and anatomopathological findings. Arch Orthop Trauma Surg 107:7–15, 1988.
38. Kessel, L and Watson, M: The painful arc sydrome: Clinical classification as a guide to management. J Bone Joint Surg [Br] 59:166–1172, May 1977.
39. Perry, J: Normal upper extremity kinesiology. Phys Ther 58:265, March 1978.
40. Celli, L, Balli, A, deLuise, G, and Rovesta, C: Some new aspects of the functional anatomy of the shoulder. Ital J Orthop Traumat 11:83, 1985.
41. Colachis, SC and Strohm, BR: Effects of suprascapular and axillary nerve block on muscle force in the upper extremity. Arch Phys Med Rehabil 52:22–29, 1971.
42. Lehmkuhl, LD and Smith, LK: Brunnstrom's Clinical Kinesiology, ed 4. FA Davis, Philadelphia, 1983.

STUDY QUESTIONS

1. Identify the intra-articular motions of the SC for elevation-depressions and protraction-retraction.

2. What is the role of the costoclavicular and interclavicular ligaments at the SC joint?

3. Discuss the relevance of the sternoclavicular disk to SC joint congruency and joint motion.

4. Identify the scapular movements that take place at the AC joint.

5. Discuss the relevance of the coracoclavicular ligament and the AC joint disk to the AC joint.

6. Discuss the configuration of the humerus and the glenoid fossa as they relate to GH joint stability. What role do the glenoid labrum and joint capsule play in joint stability?

7. What is the most frequent direction of GH dislocation? Why is this true?

8. Compare the relative stability and tendency toward degenerative changes in the GH, AC, and SC ligaments.

9. What are the advantages of the coracohumeral arch? What are the disadvantages?

10. What intra-articular motions must occur at the GH joint for full ROM to take place? What is the normal range? What range will be available to the joint if the humerus is not able to laterally rotate?

11. What muscle is the prime mover in shoulder GH flexion and abduction? What synergy is necessary for normal function of this muscle? Why?

12. Why is the supraspinatus able to abduct the shoulder without additional muscular synergy?

13. What accounts for the static stability of the GH joint when the arm is at the side? What happens if you excessively load the hanging (dependent) limb?

14. Identify five factors that play a role in the dynamic stability of the GH joint in either flexion or abduction of that joint (both flexion and abduction are elevation functions).

15. What is the total ROM available to the humerus in elevation? How is this full range achieved?

16. How does the shape of the clavicle contribute to elevation of the arm?

17. What muscles are necessary to produce the normal scapular and humeral movements in elevation of the arm?

18. If the scapulothoracic joint were fused in neutral position, what range of elevation would still be available to the upper extremity actively?

19. What is the most common traumatic problem at the AC joint? What deficits is a person with this disability likely to encounter?

20. What are the consequences of a rupture of the coracoclavicular ligaments?

21. If the GH joint were immobilized by osteoarthritis, what range of elevation would be available to the upper extremity?

22. If isolated paralysis of the supraspinatus were to occur, what would be the likely functional deficit?

23. If the muscles of the rotator cuff were paralyzed, what would be the effect when abduction of the arm was attempted?

24. When there is paralysis of the trapezius and the serratus anterior, what is the functional deficit when abduction of the arm is attempted?

25. If the deltoid alone were paralyzed, what would happen with attempted abduction of the arm? With attempted flexion of the arm?

26. What is the role of the rhomboids in elevation of the arm?

27. What differences would you see in attempted abduction if the trapezius alone were paralyzed compared with paralysis of both the trapezius and the serratus anterior?

28. What muscular synergy does the teres major require to perform its function?

29. Describe why electromyographic activity of the deltoid in normal abduction shows a gradual rise in activity to between 90 and 120°, with a plateau thereafter.

30. Which of the joints of the shoulder complex is most likely to undergo degenerative changes over time? Which is least likely?

Chapter 8

■ ■ ■

The Elbow Complex

■

OBJECTIVES
Following the study of this chapter, the reader should be able to:

Describe
1. All of the articulating surfaces associated with each of the following joints: humeroulnar, humeroradial, superior, and inferior radioulnar.
2. The ligaments associated with all the joints of the elbow complex.

Identify
1. Axes of motion for supination and pronation and flexion and extension.
2. The degrees of freedom associated with each of the joints of the elbow complex.

3. Structures limiting the range of motion in flexion and extension.
4. Structures that create the carrying angle.
5. Structures limiting motion in supination and pronation.

Compare
1. The translatory and rotatory components of the brachioradialis and brachialis at all points in the range of motion.
2. The moment arms of the flexors at any point in the range of motion.
3. Muscle activity of the extensors in a closed kinematic chain with activity in an open kinematic chain.
4. The role of the pronator teres with the role of the pronator quadratus.
5. The role of the biceps with that of the brachialis.
6. The resistance of the elbow joint to longitudinal tensile forces with its resistance to compressive forces.
7. The features of a classic tennis elbow with the features of cubital tunnel syndrome.
8. The role and structure of the annular ligament with the role and structure of the articular discs.
9. The role of the medial collateral ligament with the role of the lateral collateral ligament at 90° of flexion and in full extension.
10. The etiology of avascular necrosis of the capitulum.

GENERAL FEATURES

The joints and muscles of the elbow complex are designed to serve the hand. They provide mobility for the hand in space by apparent shortening and lengthening of the upper extremity. This function allows the hand to be brought close to the face for eating or placed at a distance from the body equal to the length of the upper extremity. Rotation at the elbow complex provides additional mobility for the hand. In conjunction with providing mobility for the hand, the elbow complex structures also provide stability for skilled or forceful movements of the hand that are necessary when performing activities using tools or implements. Many of the 15 muscles that cross the elbow complex[1] also act either at the wrist or shoulder, and therefore the wrist and shoulder are linked with the elbow in enhancing function of the hand.

The joints that compose the elbow complex are the elbow joints (humeroulnar and humeroradial), and the superior and inferior radioulnar joints. The apparent shortening and lengthening of the upper extremity take place at the elbow joints, which are formed by the distal end of the humerus and the proximal ends of the radius and ulna. The elbow joints, which allow the motions of flexion and extension, are considered as one compound joint that functions as a hinge joint and therefore as a uniaxial diarthrodial joint with 1° of freedom of motion. Flexion and extension take place in the sagittal plane around a coronal axis. Two major ligaments and five muscles are associated with the elbow joint. Three of the muscles are flexors that cross the anterior aspect of the joint. The other two muscles are extensors that cross the posterior aspect of the joint.

The superior and inferior radioulnar joints function as one joint, and therefore the two joints acting together to produce rotation of the forearm have 1° of freedom of motion. These joints are diarthrodial uniaxial joints of the pivot (trochoid) type. Rotation (supination and pronation) takes place in the transverse plane around a longitudinal axis. Six ligaments and four muscles are associated with these joints. Two muscles are for supination and two are for pronation. The elbow and superior radioulnar joints are enclosed in a single joint capsule but constitute distinct articulations.

STRUCTURE: HUMEROULNAR AND HUMERORADIAL JOINTS

Articulating Surfaces on the Humerus

The articulating surfaces on the humerus are the hourglass-shaped **trochlea** and the spherically shaped **capitulum** (Fig. 8–1). These structures, which are covered with articular cartilage, are located between the two epicondyles on the distal end of the humerus. The trochlea, which forms part of the humeroulnar articulation, lies on the anterior medial aspect of the distal humerus. The **trochlear groove** spirals obliquely around the trochlea and divides the trochlea into a medial and lateral portion. The medial portion of the trochlea projects distally more than the lateral portion, and the entire trochlea is set at an angle and slightly anterior to the shaft of the humerus. Individual variations in the obliquity of the trochlear groove affect the direction of movement of the forearm during flexion and extension. The trochlea has an asymmetrical sellar surface that is concave transversely and convex anteroposteriorly.[2] An indentation located just above the trochlea is called the **coronoid fossa.** The capitulum, which is part of the humeroradial articulation, is located on the anterior lateral surface of the distal humerus. The capitulum, like the trochlea, lies anterior to the shaft of the humerus. A groove, called the **capitulotrochlear groove,** separates the capitulum from the trochlea. An indentation located just above the capitulum is called the **radial fossa.** Posteriorly, the distal humerus is indented by a deep fossa called the **olecranon fossa** (Fig. 8–2).

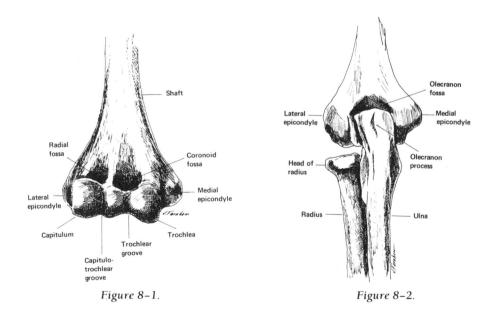

Figure 8–1.

Figure 8–2.

Figure 8–1. Articulating surfaces on the anterior aspect of the right distal humerus.

Figure 8–2. Posterior aspect of a left elbow joint in the extended position. The olecranon fossa is shown partially obscured by the olecranon process of the ulna. In the fully extended position, there is no contact between the radial head and the capitulum.

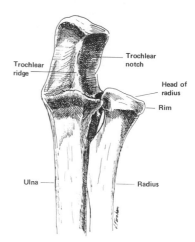

Figure 8–3. An anterior view of the articulating surfaces on the radius and ulna. Drawing depicts a left forearm.

Articulating Surfaces on the Radius and Ulna

The articulating surfaces of the ulna and radius correspond to the humeral articulating surfaces (Fig. 8–3). The ulnar articulating surface of the humeroulnar joint is a semicircular-shaped concave surface called the **trochlear notch.** The proximal portion of the notch is divided into two unequal parts by the trochlear ridge. The ridge corresponds to the trochlear groove on the humerus. The radial articulating surface of the humeroradial joint is composed of the proximal end of the radius (head of the radius). Its slightly cup-shaped concave surface is surrounded by a rim. The concavity of the radial head corresponds to the convex surface of the capitulum and the radial head's convex rim fits into the capitulotrochlear groove.

Articulation

Articulation between the ulna and humerus at the humeroulnar joint occurs primarily as a sliding motion of the ulna on the trochlea. In extension, sliding continues until the olecranon process enters the olecranon fossa (Fig. 8–4a). In flexion, the trochlear ridge of the ulna slides along the trochlear groove until the coronoid process reaches the floor of the coronoid fossa in full flexion[3] (Fig. 8–4b). Articu-

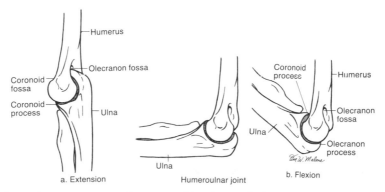

Figure 8–4. Schematic representation of motions of the ulna on the humerus at the humeroulnar joint. a. In extension, the olecranon process enters the olecranon fossa. b. In flexion, the coronoid process reaches the coronoid fossa.

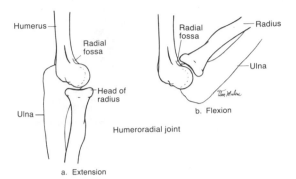

Figure 8–5. Schematic representation of motions of the radius at the humeroradial joint. a. In full extension, there is no contact between the capitulum and the radial head. b. During flexion, the rim of the radius slides in the capitulotrochlear groove and in full flexion reaches the radial fossa on the humerus.

lation between the radial head and the capitulum at the humeroradial joint involves sliding of the concave radial head over the convex surface of the capitulum. In full extension no contact occurs between the articulating surfaces (Fig. 8–5a). In flexion, the rim of the radial head slides in the capitulotrochlear groove and enters the radial fossa in full flexion[4] (Fig. 8–5b). Although sliding of one joint surface over another is the predominant motion at both the humeroulnar and humeroradial joints, London[3] has suggested that a rolling may occur during the last 5 to 10° of flexion and extension.

Joint Capsule

The humeroulnar and humeroradial joints and the superior radioulnar joint are enclosed in a single joint capsule. Anteriorly the capsule is attached to the humerus just above the coronoid and radial fossae, and to the ulna on the margin of the coronoid process. The capsule's attachment to the radius blends with the ligaments of the radioulnar articulation. Posteriorly the capsule is attached to the humerus along the upper edge of the olecranon fossa. The capsule is fairly large, loose, and weak anteriorly and posteriorly, but its sides are reinforced by ligaments. Fat pads are located between the capsule and the synovial membrane adjacent to the olecranon, coronoid, and radial fossae.[2,5]

Ligaments

Most hinge joints in the body have collateral ligaments, and the elbow is no exception. Collateral ligaments are located on the medial and lateral sides of hinge joints to provide medial-lateral stability to the joint and to keep joint surfaces in apposition. The two main ligaments associated with the elbow joints are the medial (ulnar) and lateral (radial) collateral ligaments.

The medial collateral is a triangular-shaped ligament consisting of three parts, anterior, oblique, and posterior (Fig. 8–6a). The anterior part, which is well-defined, extends from the anterior aspect of the medial epicondyle of the humerus to the ulnar coronoid process. The anterior portion of the ulnar collateral is considered to be the primary stabilizer of the elbow to valgus stress in the range of elbow flexion from 20 to 120° of flexion.[6,7] The posterior part of the medial collateral ligament is not as distinct as the anterior medial collateral and sometimes its fibers blend with the fibers of the medial portion of the joint capsule. The fibers of the posterior ligament extend from the posterior aspect of the medial epicondyle of the humerus to attach to the ulnar coronoid and olecranon processes. The posterior medial collateral ligament plays a less significant role than the anterior medial collateral in valgus stability of the elbow.[6,7] The oblique fibers of the medial collateral

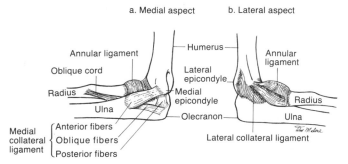

Figure 8–6. Collateral ligaments of the elbow. a. The medial collateral ligament on the medial aspect of the elbow. b. The lateral collateral ligament on the lateral aspect of the elbow.

extend between the olecranon and ulnar coronoid processes. This portion of the ligament assists in providing valgus stability and helps to keep the joint surfaces in approximation.

The lateral collateral ligament is a fan-shaped ligament that extends from the inferior lateral epicondyle of the humerus and attaches to the annular ligament (ligament encircling the head of the radius) and to the olecranon process (Fig. 8–6b). The lateral collateral ligament provides reinforcement for the humeroradial articulation, offers some protection against varus stress in some positions of the elbow, and assists in providing resistance to distraction of the joint surfaces.[8] The lateral collateral is considered to be weaker than the medial collateral and apparently plays a lesser role than the medial collateral in providing reinforcement for the elbow joint.

Muscles

The muscles associated with the elbow joints consist of three flexors and two extensors. The major flexors of the elbow are the brachialis, the biceps brachii, and the brachioradialis. The brachialis muscle originates on the anterior surface of the lower portion of the humeral shaft and attaches to the coronoid process of the ulna. The brachialis has a large physiologic cross-sectional area and a large volume. The biceps brachii arises from two heads, a short and long head. The short head arises from the coracoid process of the scapula and the long head arises from the scapula's supraglenoid tubercle. The muscle fibers arising from the two heads unite in the middle of the upper arm to form the prominent muscle bulk of the upper arm. Muscle fibers from both heads insert by way of the strong tendon on the tuberosity of the radius while other fibers insert into the bicipital aponeurosis that extends medially to blend with the fascia of the forearm that lies over the flexors of the forearm.[9] The brachioradialis muscle originates on the lateral supracondylar ridge of the humerus and inserts into the lower end of the radius just proximal to the styloid process. The brachioradialis has long muscle fibers and a relatively small physiologic cross-sectional area.[1]

The two extensors of the elbow are the triceps and the anconeus. The triceps has three heads, a long, medial, and lateral. Two of the heads, the medial and lateral, arise from the humerus, while the long head arises from the infraglenoid tubercle of the scapula. The three heads insert via a common tendon into the olecranon process. The anconeus is a small triangular muscle that originates from the lateral epicondyle of the humerus and inserts into both the olecranon process and the adjacent posterior surface of the ulna.[9]

FUNCTION: HUMEROULNAR AND HUMERORADIAL JOINTS

Axis of Motion

The axis for flexion and extension is relatively fixed and passes through the center of the trochlea and capitulum bisecting the longitudinal axis of the shaft of the humerus[3,10–12] (Fig. 8–7). When the upper extremity is in the anatomic position, the long axis of the humerus and the long axis of the forearm form an acute angle medially when they meet at the elbow. The angulation is due to the configuration of the articulating surfaces and results in abduction of the forearm in relation to the humerus (Fig. 8–8). This angle is called the **carrying angle** and is slightly greater in women than men. The average angle in men is about 5°, whereas in women it is about 10 to 15°.[13] An increase in the carrying angle is considered to be abnormal, especially if it occurs unilaterally. When the angle is increased beyond the average it is called **cubitus valgus**.

Normally, the carrying angle disappears when the forearm is pronated with the elbow in extension, and in full flexion.[4] The configuration of the trochlear groove determines the pathway of the forearm during passive flexion and extension. In the most common configuration of the groove, the ulna is guided progressively medially from extension to flexion so that in full flexion, the forearm comes to rest in the same plane as the humerus[4] (Fig. 8–9a). In extension, the forearm moves laterally until it reaches a position slightly lateral to the axis of the humerus in full extension. Variations in the direction of the groove will alter the pathway of the forearm so that when the elbow is passively flexed, the forearm will come to rest either medial[4,12] (Fig. 8–9b) or lateral (Fig. 8–9c) to the humerus[4] in full flexion.

Different methods of measuring the carrying angle have led to conflicting con-

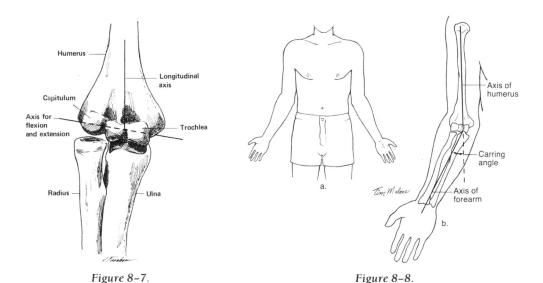

Figure 8–7.　　　　　　　　　　　　　　　　*Figure 8–8.*

Figure 8–7. The axis of motion for flexion and extension. The axis of motion is centered in the middle of the trochlea on a line that intersects the longitudinal (anatomic) axis of the humerus.

Figure 8–8. The carrying angle of the elbow. a. The forearm lies slightly lateral to the humerus when the elbow is fully extended in the anatomic position. b. The long axis of the humerus and the long axis of the forearm form the carrying angle.

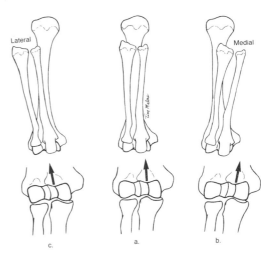

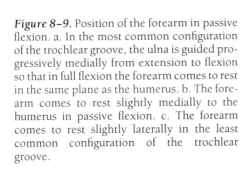

Figure 8–9. Position of the forearm in passive flexion. a. In the most common configuration of the trochlear groove, the ulna is guided progressively medially from extension to flexion so that in full flexion the forearm comes to rest in the same plane as the humerus. b. The forearm comes to rest slightly medially to the humerus in passive flexion. c. The forearm comes to rest slightly laterally in the least common configuration of the trochlear groove.

clusions regarding the carrying angle.[11] For example, London[3] who used the axis of the forearm and a perpendicular to the axis of rotation of the elbow joint rather than the axes of the humerus and forearm to measure the carrying angle concluded that the carrying angle did not change during the range of elbow flexion/extension.

Range of Motion

A number of factors determine the amount of motion that is available at the elbow joint, These factors include the type of motion (active or passive), the position of the forearm, and the position of the shoulder. The range of active flexion at the elbow is usually less than the range of passive motion, because the bulk of the contracting flexors on the anterior surface of the humerus interferes with the approximation of the forearm with the humerus. Active flexion of the elbow with the forearm supinated ranges from about 135 to 145°, while the range for passive flexion is between 150 and 160°.[4] The position of the forearm also affects the flexion range of motion (ROM). When the forearm is in either pronation or midway between supination and pronation, the ROM is less than it is when the forearm is supinated. The position of the shoulder may affect the ROM available to the elbow, because muscles that cross both the shoulder and elbow may become passively insufficient. Passive tension in the triceps may limit elbow flexion when the shoulder is simultaneously moved into full flexion. Passive tension created in the long head of the biceps by shoulder hyperextension may limit full elbow extension.

Other factors that limit the ROM and help to provide stability for the elbow are the configuration of the joint surfaces, the ligaments, and joint capsule. The elbow has inherent articular stability at the extremes of extension and flexion.[5,6] In full extension the elbow joints are in a close-packed position. In this position bony contact of the olecranon process in the olecranon fossa limits the end of the extension range and the configuration of the joint structures helps to provide valgus and varus stability. The bony components, medial collateral ligament, and anterior joint capsule contribute equally to resist valgus stress in full extension.[8] The bony components provide one half of the resistance to varus stress in full extension while the lateral collateral and joint capsule provide the other half of the resistance.[8] Resistance to joint distraction in the extended postion is entirely provided for by soft-tissue structures. The anterior portion of the joint capsule provides the majority of the resistance to anterior displacement of the humerus while the medial and lateral collateral ligaments contribute only slightly.[7,8]

Table 8–1. STABILITY SUMMARY: ELBOW

The Percentage° of Contribution to Resist Applied Valgus, Varus, and Distraction Stresses with the Elbow at 0 and 90°.

	VALGUS†, %		VARUS‡, %		DISTRACTION, §%	
	0°	90°	0°	90°	0°	90°
Medial collateral ligament	31	54	—	—	6	78
Lateral collateral ligament	—	—	14	9	5¶	10¶
Joint capsule	38¶	10¶	31¶	13¶	85	8
Osseous	31	33	55	75	—	—

°Mean percent of four specimens.
†3° of valgus stress.
‡3° of varus stress.
§2.5 mm of distraction.
¶In conjunction with resistance of soft tissue.
Source: Adapted from Morrey, BF and An, KN,[8] pp 315–318.

Approximation of the coronoid process with the coronoid fossa and of the radius in the radial fossa limits extremes of flexion. In 90° of flexion the anterior part of the medial collateral ligament provides the primary resistance to both distraction and valgus stress. If the anterior portion of the medial collateral ligament becomes lax through overstretching, medial instability will result when the elbow is in flexed positions. Also, the carrying angle will increase. The majority of the resistance to varus stress when the elbow is flexed to 90° is provided by the osseous structures of the joint and only a slight amount by the lateral collateral ligament and the joint capsule. The anterior joint capsule contributes only slightly to varus/valgus stability and provides little resistance to distraction when the elbow is flexed.[78] A summary of the contributions of joint surfaces to stability is presented in Table 8–1.

Co-contractions of the flexor and extensor muscles of the elbow, wrist, and hand help to provide stability for the elbow during forceful motions of the wrist and fingers in activities in which the arms are used to support the body weight. During pulling activities such as when one grasps and attempts to pull a fixed rod toward the body, the elbow joints are compressed by the contractions of the forearm muscles of the wrist and hand.[14]

Muscle Action

Flexors

The role that the three flexor muscles play in motion at the elbow is determined by a number of factors, including the location of the muscles, position of the elbow and adjacent joints, position of the forearm, the magnitude of the applied load, the type of muscle contraction, and the speed of motion. The brachialis is inserted close to the joint axis and therefore is considered to be a spurt or mobility muscle. The moment arm (MA) of the brachialis is greatest at slightly more than 100° of elbow flexion,[2] and therefore, its ability to produce torque is greatest at that particular elbow position. The brachialis is inserted on the ulna, and therefore the brachialis is unaffected by changes in the forearm position brought about by rotation of the radius. Studies of muscle activity at the elbow have been performed using electromyography (EMG). This technique is used to monitor the electrical activity that is produced by the firing of motor units. Using EMG it is possible to determine the number of motor units that are firing in a particular muscle during a specific muscle contraction. According to EMG studies, the brachialis muscle works in flexion of the elbow in all positions of the forearm, with and without resistance. It also is active in all types of contractions (isometric, concentric, and eccentric) during slow and fast motions.[15] As a one-joint muscle, the brachialis is not affected by the position of the shoulder.

The biceps brachii is also considered to be a spurt or mobility muscle because of its insertion close to the joint axis. The MA of the biceps is largest between 80 and 100° of elbow flexion, and therefore the biceps is capable of producing its greatest torque in this range. The MA of the biceps is rather small when the elbow is in full extension and most of the muscle force is translatory (joint compression). Therefore, when the elbow is fully extended the biceps is less effective as an elbow flexor than when the elbow is flexed to 90°. When the elbow is flexed beyond 100°, the translatory component of the muscle force is directed away from the elbow joint and therefore acts as a distracting or dislocating force as shown in Figure 1–55 in chapter 1. The biceps is active during unresisted elbow flexion with the forearm supinated or midway between supination and pronation, and in concentric and eccentric contractions but *not* when the forearm is pronated. Whenever the magnitude of the resistance is great, the biceps is active in all positions of the forearm. The biceps also is active in quick extension during an eccentric contraction when there is an applied load.[15] The biceps may become actively insufficient when full flexion of the elbow is attempted with the shoulder in full flexion.

The brachioradialis is inserted at a distance from the joint axis and therefore can be classified as a shunt muscle. During muscle contraction the larger component of muscle force goes toward compression of the joint surfaces and hence toward stability (Fig. 1–56, chapter 1). The brachioradialis shows no electrical activity during eccentric flexor activity when the motion is performed slowly with the forearm supinated. Also the brachioradialis shows no activity during slow, unresisted, concentric elbow flexion. However, when the speed of the motion is increased, the brachioradialis shows moderate activity if a load is applied and the forearm is in either a position midway between supination and pronation or in full pronation.[15] The brachioradialis is unaffected by the position of the shoulder.

The brachialis has a large strength potential (large physiologic cross-sectional area) and a large work capacity (volume).[1] The long head of the biceps has the largest volume among the flexors, but both the biceps and the brachioradialis have relatively small physiologic cross sections because these muscles have longer muscle fibers than the brachialis.[1]

Extensors

The effectiveness of the triceps as a whole is affected by changes in the position of the elbow but not by changes in position of the forearm. The long head of the triceps crosses two joints; therefore, activity of the long head is affected by changing shoulder joint positions. The long head becomes actively insufficient when full elbow extension is attempted with the shoulder in hyperextension. In this instance the muscle is shortened over both the elbow and shoulder simultaneously.

The medial and lateral heads of the triceps are not affected by the position of the shoulder. The medial head is active in unresisted elbow extension against gravity or with gravity eliminated.[15] All three heads are active when heavy resistance is given to extension or when quick extension of the elbow is attempted in the gravity-assisted position. The maximum isometric torque that the triceps can generate is at

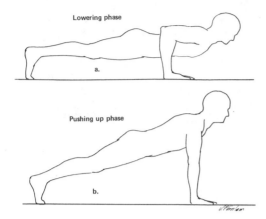

Figure 8–10. Action of the triceps in a push-up. a. The triceps muscle works eccentrically in reverse action to control elbow flexion during the lowering phase of a push-up. b. The triceps works concentrically in reverse action to produce the elbow extension that raises the body in a push up.

an elbow position of 90° of elbow flexion.[16,17] However, the total amount of extensor torque generated at 90° has been found to vary with the position of the shoulder and the body.[18] The triceps is active in extension and flexion of the elbow when it acts in a closed kinematic chain such as in a push-up (Fig. 8–10). It also is active during activities requiring stabilization of the elbow. For example, it acts as a synergist to prevent flexion of the elbow when the biceps is acting as a supinator. The other extensor of the elbow, the anconeus, assists in elbow extension and apparently acts as a stabilizer during supination and pronation.

Synergistic actions of elbow flexor and extensor muscles have been investigated during isometric contractions over a range of torques including varus, valgus, flexion, and extension.[19] Some muscle pairs like the brachialis and brachioradialis and the anconeus and medial head of the triceps brachii are coactivated in a similar manner for all torques. However, the synergistic patterns of other muscles at the elbow complex are complex and vary with torque direction. For example, the brachialis and the long head of the biceps work synergistically during isometric contractions only from 0 to 45° of flexion.

STRUCTURE: SUPERIOR AND INFERIOR RADIOULNAR JOINTS

Superior Radioulnar Joint

The articulating surfaces include the ulnar radial notch, the annular ligament, the capitulum of the humerus, and the head of the radius. The radial notch is lcoated on the lateral aspect of the proximal ulna directly below the trochlear notch (Fig. 8–11a). The surface of the radial notch is concave and covered with articular cartilage. A circular ligament called the **annular ligament** is attached to the anterior and posterior edges of the notch. The ligament is lined with articular cartilage, which is continuous with the cartilage lining of the radial notch. The annular ligament encircles the rim of the radial head, which is also covered with articular cartilage (Fig. 8–11b). The capitulum and the surface of the head of the radius have already been discussed under the elbow joint and therefore will not be discussed here.

Inferior (Distal) Radioulnar Joint

The articulating surfaces include the ulnar notch of the radius, the articular disk, and the head of the ulna (Fig. 8–12). The ulnar notch of the radius is located

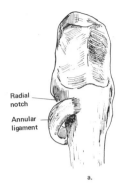
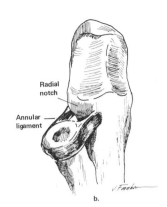

Figure 8-11. The annular ligament. a. Attachments of the annular ligament. b. The head of the radius has been pulled away from its normal position adjacent to the radial notch to show how the ligament partially surrounds the radial head.

at the distal end of the radius along the interosseous border. The notch is concave and the articular disk is attached along its inferior edge. The disk is shaped like a triangle with its base at the radial ulnar notch and its apex attached to the styloid process of the ulna. The disk has been described as resembling a shelf that has its medial border embedded in a wedge of vascular connective tissue containing fine ligamentous bands joining the disk to the ulna and articular capsule. The ventral and lateral borders of the disk have ligamentous attachments to the radius and inner surface of the joint capsule.[20] The proximal surface of the disk articulates with the ulnar head while the distal surface is part of the radiocarpal joint.[2] The ulnar head is convex and is covered with articular cartilage beneath the disk.[21,22]

Articulation

The superior and inferior radioulnar joints are mechanically linked; therefore, motion at one joint is always accompanied by motion at the other joint. The distal radioulnar joint is also considered to be functionally linked to the wrist in that loads are transmitted through the distal radioulnar joint from the hand to the radius and ulna. The radius carries approximately 80 percent of the load and the ulna 20 percent of the load.[22] Pronation of the forearm occurs as a result of the radius crossing over the ulna at the superior radioulnar joint. During pronation and supination the rim of the head of the radius spins within the osteoligamentous enclosure formed by the radial notch and the annular ligament. At the same time the surface of the head spins on the capitulum of the humerus. At the inferior radioulnar joint the concave surface of the ulnar notch of the radius slides over the ulnar head, while the disk follows the radius by twisting at its apex and sweeping along the ulnar head.

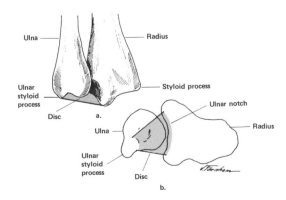

Figure 8-12. The inferior radioulnar joint of a left forearm. a. An anterior view of the inferior radioulnar joint shows the disk in its normal position in a supinated left forearm. b. An inferior view of the disk shows how the disk covers the inferior aspect of the distal ulna and separates the ulna from the articulation at the wrist.

Ligaments

The three ligaments associated with the superior radioulnar joint are the annular and quadrate ligaments and the oblique cord (see Figs. 8–6 and 8–13a). The annular ligament is a strong band that forms four fifths of a ring that encircles the radial head. The inner surface of the ligament that serves as a joint surface is covered with cartilage. The proximal border of the annular ligament blends with the joint capsule, and the lateral aspect is reinforced by fibers from the lateral collateral ligament.[2] The quadrate ligament, which extends from the inferior edge of the ulna's radial notch to the neck of the radius, reinforces the inferior aspect of the joint capsule and helps to maintain the radial head in apposition to the radial notch.[5] The quadrate ligament also limits the spin of the radial head in supination and pronation. The oblique cord is a flat fascial band that extends from an attachment just inferior to the radial notch on the ulna to just below the bicipital tuberosity on the radius.

The inferior radioulnar joint is reinforced anteriorly and posteriorly by the anterior and posterior radioulnar ligaments.[4] The anterior radioulnar ligament is a short ligament that attaches just above the ulnar head and extends to just above the ulnar notch (Fig. 8–13a). The posterior radioulnar ligament is also a short ligament that crosses from the posterior aspect of the head of the ulna to attach on the posterior aspect of the ulnar notch (Fig. 8–13b). The interosseous membrane is a broad collaginous sheet that runs between the radius and ulna. Its fibers run distally and medially from the radius to the ulna. It provides stability for both the superior and inferior radioulnar joints. The interosseous membrane not only binds the joints together, but when it is under tension also provides for the transmission of forces from the hand and distal end of the radius to the ulna. The fibers of the interosseous membrane are under tension when the forearm is in a neutral position (midway between supination and pronation) and the fibers are relaxed in both supinated and pronated positions.[2] The disk also serves to provide stability for the inferior radioulnar joint by binding the distal radius and ulna together.

Muscles

The muscles associated with the radioulnar joints are the pronator teres, pronator quadratus, biceps brachii, and the supinator. The pronator teres arises from the medial epicondyle of the humerus and the coronoid process on the ulna and

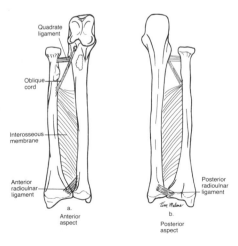

Figure 8–13. Structures that reinforce the superior and inferior radioulnar joints. The radius and ulna have been pulled apart and the annular ligament has been removed to show the quadrate ligament. a. The superior radioulnar joint is reinforced by the quadrate ligament, which helps to bind the radius to the ulna and limits extremes of rotation of the radial head. The anterior aspect of the inferior radioulnar joint is reinforced by the anterior radioulnar ligament. b. The posterior aspect of the inferior radioulnar joint is reinforced by the posterior radioulnar ligament. The articular disk also helps to bind the radius to the ulna.

inserts on the lateral side of the radius at its greatest convexity. The pronator quadratus is located at the distal end of the forearm. It arises from the ulna and crosses the interosseous membrane anteriorly to insert on the radius. The biceps brachii has been discussed previously. The supinator is a short and broad muscle that arises from the lateral epicondyle of the humerus and the lateral aspect of the ulna. It crosses the posterior aspect of the interosseous membrane to insert into the radius just medial and inferior to the bicipital tuberosity.

FUNCTION: RADIOULNAR JOINTS

Axis of Motion

The axis of motion for pronation and supination is a longitudinal axis extending from the radial head to the ulnar head[4] (Fig. 8–14). In supination the radius and ulna lie parallel to one another, while in pronation, the radius crosses over the ulna (Fig. 8–15). There is very little motion of the ulna during pronation and supination. Motion of the proximal ulna is negligible. Motion of the distal ulna is of less magnitude and opposite in direction to motion of the radius.[22] The ulnar head moves distally and dorsally in pronation and proximally and medially in supination. Therefore, the ulnar head glides in the ulnar notch of the radius from a dorsal distal position in pronation to a volar proximal position in full supination.[21]

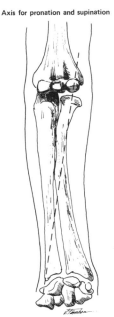

Axis for pronation and supination

Figure 8–14.

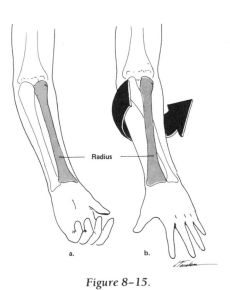

Radius

a.   b.

Figure 8–15.

Figure 8–14. Axis of motion for supination and pronation is illustrated by the dashed line on the drawing of a left upper extremity.

Figure 8–15. Supination and pronation. a. The radius and ulna are parallel to each other in the supinated position of the forearm. b. In the pronated position, the radius crosses over the ulna. Drawing shows a left upper extremity.

Range of Motion

A total ROM of 150° has been ascribed to the radioulnar joints.[2,21] The ROM of pronation and supination is assessed with the elbow in 90° of flexion. This position stabilizes the humerus so that radioulnar joint rotation may be distinguished from rotation that is occurring at the shoulder joint. When the elbow is fully extended, active supination and pronation occur in conjunction with shoulder rotation. Limitation of pronation in the extended position may be caused by passive tension in the biceps brachii. Pronation in all positions is limited by bony approximation of the radius and ulna and by tension in the posterior radioulnar ligament and the posterior fibers of the medial collateral ligament of the elbow.[7] Supination is limited by passive tension in the anterior radioulnar ligament and the oblique cord. The quadrate ligament limits spin of the radial head in both pronation and supination, and the annular ligament helps to maintain stability of the superior radioulnar joint by holding the radius in close approximation to the radial notch.

Muscle Action

The pronators produce pronation by exerting a pull on the radius, which causes its shaft and distal end to turn over the ulna (Fig. 8–16a and b). The pronator teres contributes some of its force toward stabilization of the superior radioulnar joint. The translatory component of the force produced by the pronator teres helps to maintain contact of the radial head with the capitulum. The pronator quadratus, a one-joint muscle, is unaffected by changing positions at the elbow. The pronator quadratus is active in unresisted and resisted pronation and in slow or fast prona-

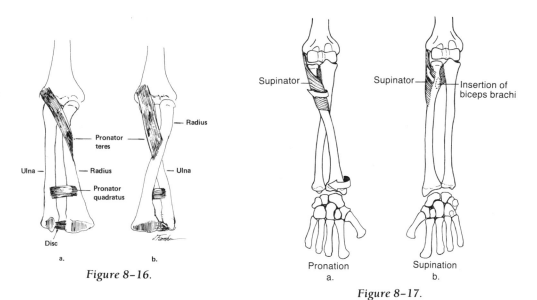

Figure 8–16.

Figure 8–17.

Figure 8–16. Pronation of the forearm. The pronator teres and pronator quadratus produce pronation by pulling the radius over the ulna. Drawing shows a left forearm: (a) in the supinated position; and (b) in the pronated position.

Figure 8–17. Supination of the right forearm. a. In the pronated position the supinator muscle wraps around the proximal radius. A contraction of the supinator and/or the biceps pulls the radius over the ulna. b. The supinator muscle and the insertion site of the biceps are shown in the supinated position.

tion. The pronator quadratus also acts to maintain compression of the distal radio-ulnar joint.[14]

The supinators like the pronators act by pulling the shaft and distal end of the radius over the ulna (Fig. 8–17a and b). The supinator acting alone produces unresisted slow supination in all positions of the elbow or forearm. The supinator acting alone also provides unresisted fast supination when the elbow is extended. However, activity of the biceps is always evident when supination is performed against resistance and during fast supination when the elbow is flexed to 90°. Activity of the biceps is most evident when using a screwdriver to drive a screw into the wood. The anconeus muscle is active in supination and pronation. A stabilization role has been suggested to explain this activity.

The strength of the flexors versus the extensors as determined by isometric testing of these muscle groups at 90° of elbow flexion shows that the elbow flexors are stronger than the elbow extensors and that the supinators are stronger than the pronators.[23]

MOBILITY AND STABILITY: ELBOW COMPLEX

Functional Activities

The joints and muscles of the elbow complex are used in almost all activities of daily living such as dressing, eating, carrying, and lifting. They are also used in tasks such as splitting firewood, hammering nails, and playing tennis. Most of the activities of daily living require a combination of motion at both the elbow and radioulnar joints. A total of about 100° of elbow flexion and 100° of supination/pronation is sufficient to accomplish simple tasks such as eating, brushing one's hair, brushing one's teeth, and dressing. The range of elbow flexion required is from about 30° of flexion to 130° of flexion. Fifty degrees of pronation and 55° of supination are necessary to allow the hand to function normally in these activities. For example, about 40° of elbow flexion, 40° of pronation, and 20° of supination are required to use a telephone.[24] Therefore, mobility of the complex is necessary for normal functioning in most areas of activity.

The design of the radioulnar joints enhances the mobility of the hand. Pronation and supination of the forearm, when the elbow is flexed at 90°, rotate the hands so that the palm faces either superiorly or inferiorly. The mobility afforded the hand is achieved at the expense of stability because the movable forearm is unable to provide a stable base for attachment of the wrist and hand muscles. Therefore, many of the muscles that act on the wirst and hand are attached on the distal end of the humerus rather than on the forearm. The flexors of the wrist and fingers include the flexor carpi radialis, flexor carpi ulnaris, palmaris longus, flexor digitorum superficialis, and flexor digitorum profundus.

All of these muscles, except the flexor digitorum profundus, originate on the medial epicondyle of the humerus. The flexor digitorum profundus originates on the proximal ulna. The extensors of the wrist and fingers, which originate in the region of the lateral epicondyle of the humerus, include the extensor carpi radialis longus, extensor carpi radialis brevis, extensor carpi ulnaris, and extensor digitorum.

Relationship to the Hand and Wrist

The location of the hand and wrist muscles at the elbow and the fact that these muscles cross the elbow create close structural and functional relationships between the elbow and wrist-hand complexes. The elbow complex affects and is affected by the wrist and hand muscles, and therefore these muscles need to be considered in any discourse on the elbow complex. Anatomically, the hand and wrist muscles help to reinforce the elbow joint capsule and therefore contribute to stability of the elbow complex. During muscular contractions the wrist and hand muscles produce compression of the articulating surfaces at the elbow and may contribute to the torque production of the elbow muscles although their primary action is at the wrist and hand. Amis, Dowson, and Wright,[14] who investigated the effect of tensile loads on the forearm during a pulling activity, predicted that both the humeroradial and humeroulnar articulations will be subjected to compressive forces during these activities and that the medial collateral ligament will be heavily loaded. Andersson[25] found that during a pulling task, the flexors, at an elbow position of 90° of flexion, exerted a flexor force of 6000 N.[25]

Changes in position of the elbow or forearm affect the length of the multijoint hand and wrist muscles and therefore affect the function of these muscles at the wrist and hand. Stabilization of the elbow through co-contraction of the elbow muscles is employed to maintain the elbow in a desirable position for optimal functioning of the hand and wrist muscles. If the elbow is allowed to fully flex and the wrist is not stabilized (prevented from flexing), the finger flexors become actively insufficient when a tight fist position is attempted. At the other extreme of elbow motion (elbow extension) passive insufficiency of the finger extensors prevents a full range of simultaneous finger and wrist flexion.

Relationship to the Head, Neck, and Shoulder

Head and neck positions have been found to affect the elbow flexor torque production in healthy young adults.[26] Clinically, it has been observed that restrictions of motion in the neck and shoulder also may affect function of the elbow. Limitations of shoulder joint internal rotation may cause a subject to excessively pronate his or her arm during throwing or racquet swinging activities. Excessive supination of the forearm may be used to compensate for a lack of external rotation at the shoulder.[27]

EFFECTS OF IMMOBILIZATION AND INJURY

Like other joints in the body, the joints and muscles of the elbow complex are subjected to the effects of immobilization and injury. Immobilization of the elbow in 90° of flexion has been found to result in a significant decrease in flexor strength but no significant decrease in extensor strength.[28] This decrease in flexor strength could potentially disrupt normal joint function because the relative strength of the flexors versus the extensors (according to isometric testing at 90 degrees of elbow flexion) shows that the flexors are stronger than the extensors.[23] If the flexors become weaker than the extensors, the flexors might not be able to counteract the pull of the extensors and the joint might hyperextend during extension. Also the flexor muscles' contributions to joint compression would be decreased, and therefore joint stability would be diminished.

Injuries to the elbow are fairly frequent and these injuries usually disrupt normal function. Therefore, an understanding of the mechanisms of these injuries and their relation to elbow joint structures is necessary for determining the effects of the injuries on joint function.

Resistance to Longitudinal Compression Forces

Resistance to longitudinal compression forces at the elbow is provided for mainly by the contact of bony components; therefore, excessive compression forces

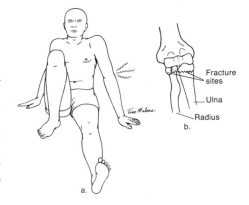

Figure 8–18. A fall on the hand with the elbow in a close-packed position may involve transmission of forces through the bones of the forearm to the elbow. a. Transmission of forces from the hand to the elbow may occur through either or both the radius and ulna. b. Impact of the radial head on the capitulum may cause either a fracture of the radial head and/or neck. A fracture of the coronoid and/or olecranon processes may result from forces transmitted through the ulna.

at the elbow often result in bony failure. Falling on the hand when the elbow is in a close-packed position may result in the transmission of forces through the bones of the forearm to the elbow (Fig. 8–18a). If the forces are transmitted through the radius, a fracture of the radial head may result from impact of the radial head on the capitulum (Fig. 8–18b). If the force from the fall is transmitted to the ulna, a fracture of either the coronoid or olecranon processes may occur from impact of the ulna on the humerus (Fig. 8–18b). If neither the radius nor the ulna fractures and absorbs the excessive force, then the force may be transmitted to the humerus and may result in a fracture of the supracondylar area.

Resistance to Distraction Forces

Resistance of the joints of the elbow complex to longitudinal traction is provided for by ligaments and muscles. A distraction force of sufficient magnitude exerted on the radius may cause the radius to slip out of the annular ligament. Small children are particularly susceptible to this type of injury, because the radial head is not fully developed. Lifting a small child up into the air by one or both hands or yanking a child by the hand is the usual causative mechanism and therefore the injury is referred to as **nursemaid's elbow** (Fig. 8–19a and b).

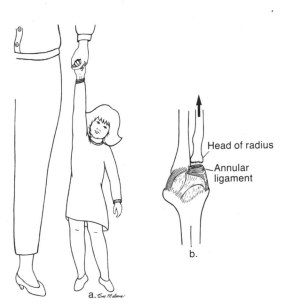

Figure 8–19. Nursemaid's elbow. a. A pull on the hand creates tensile forces at the elbow. b. The radial head is shown being pulled out of the annular ligament.

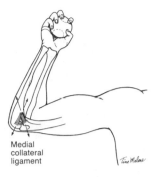

Figure 8–20. Stretching of the medial collateral ligament during throwing.

Resistance to Medial-Lateral Forces

Resistance to medial and lateral stresses at the elbow is provided by the medial and lateral collateral ligaments, articular configuration, and the joint capsule. If either one of the ligaments is overstretched, one aspect of the joint will be subjected to abnormal tensile stresses and the other to abnormal compressive forces. The medial collateral ligament is subjected to tensile stress during the backswing portion of throwing a ball (Fig. 8–20). If the stress on the medial collateral ligament is repetitive, such as in baseball pitching, the ligament may become lax and unable to reinforce the medial aspect of the joint. The resulting medial instability may cause an increase in the normal carrying angle and compression of the radial head on the capitulum. If the abnormal compression forces on the articular cartilage are prolonged, these forces may cause interference with the blood supply of the cartilage and result in avascular necrosis. Repetitive microtrauma from valgus stress can cause epiphyseal plate fractures of the proximal radius.[29]

Tennis Elbow

The current interest in racquet ball, paddle tennis, squash, and tennis, predisposes large numbers of the population to the possibility of elbow injuries. The use of a racquet greatly increases the length of the forearm lever (resistance arm) and subjects the elbow complex structures to great stresses. The classic tennis elbow (epicondylitis of the lateral epicondyle) is caused by repeated forceful contractions of the wrist extensors, primarily the extensor carpi radialis brevis. The tensile stress created at the origin of the extensor carpi radialis brevis may cause microscopic tears that lead to inflammation of the lateral epicondyle.[30,31] Repeated tensile stress on the inelastic tendon may result in microscopic tears at the musculotendinous junction and result in tendinitis. Medial tendinitis or medial epicondylitis may be caused by forceful repetitive contractions of the pronator teres, flexor carpi radialis, and occasionally by the flexor carpi ulnaris. These muscles are involved in the tennis serve when the combined motion of elbow extension, pronation, and wrist flexion is employed. One method of treatment for tennis elbow uses bracing in the form of a nonelastic muscle support. Theoretically, the brace is able to diminish the potential force that a muscle generates by preventing full muscle expansion during a contraction.[32]

Other injuries to the elbow complex that may occur as a result of muscular contraction include nerve compression and bony fracture or dislocation. Repetitive forceful contraction of the flexor carpi ulnaris may compress the ulnar nerve as it passes through the cubital tunnel between the medial epicondyle of the humerus and the olecranon process of the ulna (Fig. 8–21). The resulting injury, called **cubital tunnel syndrome,** results in impaired motion of the thumb and fourth and fifth digits.[33] Sudden forceful contractions of the biceps brachii when the forearm is supinated and flexed at 90° may cause either a fracture or dislocation of the radius.

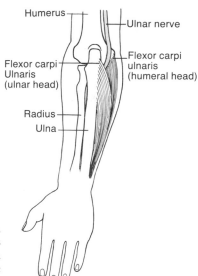

Figure 8–21. Location of the ulnar nerve as it passes through the cubital tunnel. A contraction of the flexor carpi ulnaris muscle can cause compression of the ulnar nerve between the two heads of the muscle, which are located on either side of the ulnar nerve at the elbow.

The interrelationship between the elbow complex and the wrist and hand complex makes normal functioning of the elbow of vital importance. If elbow function is impaired, function of the hand also may be impaired. For example, if the elbow cannot be flexed, it is impossible for the hand to bring food to the mouth. The fact that many important vascular and neural structures that supply the hand are closely associated with the elbow makes it important to prevent excessive stress and to protect the elbow from injury. Supracondylar humeral fractures, which are common in the skeletally immature individual, may injure either the radial or median nerves or the brachial artery.[29] If the radial nerve is injured at the level of the epicondyle, the wrist extensors, supinator, thumb, and finger extensors will be affected. If the median nerve is injured at the level of the elbow, the pronators, flexor carpi radialis, finger flexors, thenar muscles, and lumbricales will be affected. In the following chapter the reader will learn the specific functions of the hand muscles and will be better able to appreciate the significance of injury to some of the muscles.

SUMMARY

Some of the interrelationships between the structure and function of elbow, shoulder, wrist, and hand have been introduced in this chapter. Muscles that have their primary actions at the wrist and hand also cross the elbow and contribute to its stability and function, whereas the stability and ROM at the shoulder and elbow helps to enhance the function of the wrist and hand. Compensations at the elbow complex often are necessary when the ROM is limited at the shoulder or wrist. New relationships for the joints and muscles of the upper extremity will be introduced in the detailed study of the wrist and hand that follows in the next chapter.

REFERENCES

1. An, KN et al: Muscles across the elbow joint: A biomechanical analysis. J Biomech 14:659–669, 1981.
2. Williams, PL, Warwick, R, Dyson, M, and Bannister, L: Gray's Anatomy, ed 37. Churchill-Livingstone, London, 1989.

3. London, JT: Kinematics of the elbow. J Bone Joint Surg [Am] 63a:529–535, 1981.
4. Kapandji, IA: The physiology of the Joints, Vol I. E&S Livingstone, Edinburgh and London, 1970.
5. Palastanga, N, Field, D, and Soames, R: Anatomy and Human Movement: Structure and Function. Heinemann Medical Books, Oxford, 1989.
6. Sojberg, JO, Oveson, J, and Nielsen, S: Experimental elbow instability after transection of the medial collateral ligament. Clin Orthop Rel Res 218:186–190, 1987.
7. Hotchkiss, RN and Weiland, AJ: Valgus stability of the elbow. J Orthop Res 5:372–377, 1987.
8. Morrey, BF and An, KN: Articular and ligamentous contributions to the stability of the elbow joint. Am J Sports Med 11:315–318, 1983.
9. Netter, F: The Ciba Collection of Medical Illustrations, Vol 8. Musculoskeletal System Part 1. Ciba-Geigy, Summit, NJ, 1987.
10. Youm, Y et al: Biomechanical analysis of forearm pronation-supination and elbow flexion-extension. J Biomech 12:245, 1979.
11. Deland, JT, Garg, A, and Walker, PS: Biomechanical basis for elbow hinge-distractor design. Clin Orthop Rel Res 215:303–312, 1987.
12. Morrey, BF and Chao, YS: Passive motion of the elbow joint. J Bone Joint Surg [Am] 58a:501–508, 1976.
13. Hoppenfeld, S: Physical examination of the spine and extremities. Appleton-Century-Crofts, New York, 1976.
14. Amis, AA, Dowson, D, and Wright, V: Elbow joint force predictions for some strenuous isometric actions. J Biomech 13:765–775, 1980.
15. Basmajian, JV: Muscles Alive, ed 4. Williams & Wilkins, Baltimore, 1978.
16. Provins, KA and Salter, N: Maximum torque exerted about the elbow joint. J Appl Physiol 7:393–398, 1955.
17. Currier, DP: Maximal isometric tension of elbow extensors at varied positions. Phys Ther 52:1043–1049, 1972.
18. Bohannon, RW: Shoulder position influences elbow extension force in healthy individuals. JOSPT 12:111–114, 1990.
19. Buchanan, TS, Almdale, DPJ, Lewis, JL, and Rymer, WZ: Characteristics of synergistic relations during isometric contractions of human elbow muscles. J Neurophysiol 56:1225–1241, 1986.
20. Mohiuddin, A and Zanjua, MZ: Form and function of radioulnar joint articular disc. The Hand 14(2):61–66, 1982.
21. Palmer, AK: The distal radioulnar joint. Orthop Clin North Am 15:321–335, 1984.
22. Palmer, AK and Werner, FW: Biomechanics of the distal radioulnar joint. Orthop Clin North Am 27–35, 1984.
23. Askew, LJ et al: Isometric elbow strength in normal individuals. Clin Orthop Rel Res 222:261–266, 1987.
24. Morrey, BF, Askew, LJ, An, KN, and Chao, EY: A biomechanical study of normal functional elbow motion. J Bone Joint Surg 63a:872–876, 1981.
25. Andersson, GBJ and Schultz, AB: Transmission of moments across the elbow joint and the lumbar spine J Biomech 12:747–755, 1979.
26. Deutsch, H et al: Effect of head-neck position on elbow flexor muscle torque production. Phys Ther 67:517–521, 1987.
27. Dilorenzo, CE, Parkes, JC, and Chmelar, RD: The importance of shoulder and cervical dysfunction in the etiology and treatment of athletic elbow injuries. Journal of Orthopedic and Sports Physical Therapy 11:402–409, 1990.
28. Vaughan, VC: Effects of upper limb immobilization on isometric muscle strength, movement time, and triphasic electromyographic characteristics. Phys Ther 69:119–129, 1989.
29. Ireland, ML and Andrews, JR: Shoulder and elbow injuries in the young athlete. Clin Sports Med 7:473–494, 1988.
30. Priest, JD, Braden, J, and Gerberich, SG: The elbow and tennis. Part 1: An analysis of players with and without pain. Phys Sports Med 8:4, 1980.
31. La Freniere, JG: "Tennis elbow" evaluation, treatment and prevention. Phys Ther 59:6, 1979.
32. Nirschi, RP and Sobel, J: Conservative treatment of tennis elbow. Phys Sports Med 9:6, 1981.
33. Craven, PR and Green, DP: Cubital tunnel syndrome. J Bone Joint Surg 62(A):986, 1980.

STUDY QUESTIONS

1. Name and locate all of the articulating surfaces of the joints of the elbow complex. Describe the method of articulation at each joint including axes of motion and degrees of freedom.

2. Explain the stabilizing function of the brachioradialis by diagraming the translatory and rotatory components at different joint angles.

3. Explain why active elbow flexion is more limited than passive flexion. Which structures limit extension?

4. Describe the "carrying angle" and explain why it is present.

5. Which structures limit supination and pronation?

6. If slow pronation of the forearm is attempted without resistance, which muscle will be used?

7. How does the structure and function of the annular ligament differ from that of the medial collateral ligament?

8. Describe the activity of the biceps brachii during a chin-up.

9. What is the mechanism of injury in tennis elbow?

10. Which position of the elbow is most stable? Why?

11. Compare the biceps brachii with the brachialis on the basis of structure and function.

12. Describe the mechanism of injury involved in cubital tunnel syndrome.

Chapter 9

■ ■ ■

The Wrist and Hand Complex

■

OBJECTIVES
Following the study of this chapter, the reader should be able to:

Define
1. The terminology unique to the wrist and hand complexes.

Describe
1. The articular surfaces of the joints of the wrist and hand complexes.
2. The ligaments of the joints of the wrist and hand, including the functional significance of each.

3. Accessory joint structures found in the wrist and hand complexes, including the function of each.
4. Types of movements and ranges of motion of the radiocarpal joint, the midcarpal joint, and the total wrist complex.
5. The sequence of joint activity occurring from full wrist flexion to extension, including the role of the scaphoid; the sequence of joint activity in radial and ulnar deviation from neutral.
6. The role of the wrist musculature in producing wrist motion.
7. Motions and ranges available to joints of the hand complex.
8. The gliding mechanisms of the extrinsic finger flexors.
9. The structure of the extensor mechanism, including the muscles and ligaments that compose it.
10. How metacarpophalangeal extension occurs, including the muscles that produce and control it.
11. How flexion and extension of the proximal interphalangeal joint occur, including the muscular and ligamentous forces that produce and control these motions.
12. How flexion and extension of the distal interphalangeal joint occur, including the muscular and ligamentous forces that produce and control these motions.
13. The role of the wrist in optimizing length-tension in the extrinsic hand muscles.
14. The activity of reposition, including the muscles that perform it.
15. The functional position of the wrist and hand.

Differentiate Between
1. The role of the interossei and lumbrical muscles at the metacarpophalangeal and interphalangeal joints.
2. The muscles used in cylindrical grip to those active in spherical grip, hook grip, and lateral prehension.
3. The muscles that are active in pad-to-pad, tip-to-tip, and pad-to-side prehension.

Compare
1. The activity of muscles of the thumb (in opposition of the thumb to the index finger) with the activity of those active in opposition to the little finger.
2. The characteristics of power grip with those of precision handling.
3. The most easily disrupted form of precision handling with the form of precision handling that may be used by someone without any active hand musculature; what are the prerequisites for each?

The human hand may well surpass all body parts except the brain as a topic of universal interest. The human hand has been characterized as a symbol of power,[1] as an extension of intellect,[2] and as the seat of the will.[3] The symbiotic relationship of the mind and hand is exemplified by sociologists' claim that although the brain is responsible for the design of civilization, the hand is responsible for its formation. Although the hand cannot function without the brain to control it, likewise the encapsulated brain needs the hand as a primary tool of expression. The entire upper limb is subservient to the hand. Any loss of function in the upper limb, regardless of the segment, ultimately translates into diminished function of its distal portion. It is the significance of this potential loss that has led to detailed study of the finely balanced intricacies of the normal upper limb and hand.

THE WRIST COMPLEX

Each joint proximal to the wrist complex (carpus) serves to broaden the placement of the hand in space and to increase the degrees of freedom available to the hand. The shoulder serves as a dynamic base of support, the elbow allows the hand to approach or extend away from the body, and the forearm adjusts the approach of the hand to an object. The carpus, unlike the more proximal joints, serves placement of the hand in space to only a minor degree. The wrist has little functional redundancy with the more proximal joints of the upper limb; lost function in the wrist cannot be replaced by compensatory movements of the shoulder, elbow, or forearm. The major contribution of the wrist complex seems to be to control length-tension relationships in the multiarticular hand muscles and to permit fine adjustment of grip. These functions are performed by two compound joints: the radiocarpal and the midcarpal joints, referred to collectively as the **wrist complex.** The intricacy and variability of the interarticular and intra-articular relationships within the wrist complex are such that the wrist has received a large amount of attention with consensus on very few points. The intent here is to describe the wrist complex in such a way that its general structure is clear and its function understood in terms of optimizing length-tension in the hand muscles and adjusting grip. An attempt will also be made to identify the movements of the carpals that may lead to pathologic function.

The wrist complex as a whole is biaxial, with motions of flexion/extension (volar flexion-dorsiflexion) around a coronal axis, and radial deviation-ulnar deviation (abduction/adduction) around an AP axis. The ranges of motion (ROMs) of the entire complex are generally considered to be 85° of flexion, 70 to 80° of extension, 20 to 25° of radial deviation, and 30 to 35° of ulnar deviation. The ranges are contributed in various proportions by the compound radiocarpal and midcarpal joints. Gilford, Bolton, and Lambrinudi[4] proposed that the two-joint, rather than single-joint, system of the wrist complex: (1) permitted large ROMs with less exposed articular surface and a tighter joint capsule, (2) had less tendency for structural pinch at extremes of ranges, and (3) allowed for flatter multiple-joint surfaces that are more capable of withstanding imposed pressures.

Structural Components of the Wrist Complex

Radiocarpal Joint Structure

The radiocarpal joint is formed by the radius and radioulnar disk proximally and by the scaphoid, lunate, and triquetrum distally (Fig. 9–1). The proximal radiocarpal joint surface has a single continuous biconcave curvature, which is long and shallow side to side (frontal plane), and shorter and sharper anteroposteriorly (sagittal plane). The proximal joint surface is composed of (1) the lateral radial facet, which articulates with the scaphoid; (2) the medial radial facet, which articulates with the lunate; and (3) the inferior surface of the radioulnar disk, which articulates predominantly with the triquetrum (Fig. 9–2). The radioulnar disk accounts for just over 11 percent of the articular surface, while the medial and lateral radial facets contribute 43 and 46 percent, respectively.[5]

The radioulnar disk, or triangular fibrocartilage, participates both as part of the distal radioulnar joint and as part of the radiocarpal joint. The fibrocartilage disk arises from the ulnar notch of the radius. Mohiuddin and Janjua[6] propose that the fibrocartilage portion does not extend as far as the ulnar styloid, but is continuous with a wedge of connective tissue that thickens medially, envelops the ulnar styloid and attaches to the articular capsule (Fig. 9–2). They propose that the connective tissue wedge is more compressible than fibrocartilage, thus permitting greater

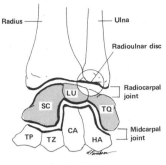

Fig. 9–1

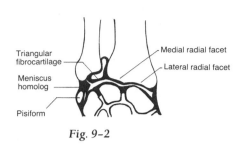

Fig. 9–2

Figure 9–1. Wrist complex. The radiocarpal joint is composed of the radius and the radioulnar disk, with the scaphoid (SC), lunate (LU), and the triquetrum (TQ). The midcarpal joint is composed of the scaphoid, lunate, and triquetrum with the trapezium (TP), the trapezoid (TZ), the capitate (CA), and the hamate (HA).

Figure 9–2. The proximal surface of the radiocarpal joint is formed by the medial and lateral facets of the distal radius and by the triangular fibrocartilage or radioulnar disk. The radioulnar disk and meniscus homolog are together part of the triangular fibrocartilage complex.

extremes of wrist ROM. The disk and its connective tissue wedge do not end with the medial capsular attachment, but continue on distally to attach to the triquetrum, hamate, and the base of the fifth metacarpal with fibers from the ulnar collateral ligaments, radioulnar ligaments, and the sheath of the extensor carpi ulnaris. Palmer and Werner[7] have coined the term "triangular fibrocartilage complex (TFCC)" to refer to the disk, the connective tissue wedge, and the various fibrous attachments. They also refer to the connective tissue wedge as the "meniscus homolog." Since the connections of the radioulnar disk suspend the disk off the head of the ulna (preventing articulation of the lunate or triquetrum with the ulna), the ulna is not considered part of the radiocarpal articulation. The head of the ulna can actually be removed without impairing wrist function.[8] As a whole, the compound proximal radiocarpal joint surface is oblique, angled slightly volarly and ulnarly. The angulation is due primarily to the articular margins of the radius, which extend farther on the dorsal and radial sides of the bone.

The scaphoid, lunate, and triquetrum compose the proximal carpal row and combine to form the distal radiocarpal joint surface. These bones are connected by numerous interosseous ligaments that, like the carpals themselves, are cartilage-covered proximally. The proximal carpal row and ligaments together present a single biconvex joint surface that, unlike a rigid segment, can change shape somewhat to accommodate to the demands of space between the forearm and hand.[9] The pisiform, anatomically part of the proximal row, does not participate in the radiocarpal articulation. It functions entirely as a sesamoid bone, presumably to increase the moment arm (MA) of the flexor carpi ulnaris that passes over it. The curvature of the distal radiocarpal joint surface is sharper than the proximal surface in both directions, making the joint somewhat incongruent. This incongruence and the angulation of the proximal joint surface result in a greater range of flexion than extension, and in greater ulnar deviation than radial deviation. The total range of flexion/extension is greater than the total range of radial-ulnar deviation.[10,11]

The radiocarpal joint is enclosed by a strong but somewhat loose capsule and is reinforced by capsular and intracapsular ligaments. Since most ligaments that cross the radiocarpal joint also contribute to stability at the midcarpal joint, all the ligaments will be presented together after introduction of the midcarpal joint. Similarly, the muscles of the radiocarpal joint also function at the midcarpal joint. In fact, the radiocarpal joint is not crossed by any muscles that act on the radiocarpal joint alone. The flexor carpi ulnaris is the only muscle that attaches to any of the bones of the proximal carpal row. It merely binds its tendon to the pisiform as it passes on to the hamate and fifth metacarpal. Motions occurring at the radiocarpal joint are a result of forces applied by the abundant passive ligamentous structures and by muscles that are attached to the distal carpal row and metacarpals. Consequently, movements of the radiocarpal and midcarpal joints must be examined together.

Midcarpal Joint Structure

The midcarpal joint is the articulation between the scaphoid, lunate, and triquetrum proximally and the trapezium, trapezoid, capitate, and hamate distally (distal carpal row) (Fig. 9–1). The midcarpal joint is a functional rather than anatomic unit. It does not form a single uninterrupted articular surface nor does it have its own capsule as does the radiocarpal joint. However, it is anatomically separate from the radiocarpal joint and has a capsule that is continuous with each intercarpal articulation (Fig. 9–3). The lack of anatomic continuity may be attributed to the dual role played by the scaphoid, which functions both as part of the proximal carpal row in some portions of wrist motion and as part of the distal carpal row in other portions of wrist motion. Ignoring the complicated interplay of the scaphoid, the midcarpal joint surfaces appear to have a reciprocally concave-convex configuration. Given that configuration, it has been oversimplified and described as a hinge joint with only 1° of freedom: flexion/extension.[12] Most investigators describe it as a condyloid joint with 2° of freedom, adding varying amounts of radial and ulnar deviation to flexion/extension. The excursions permitted by the articular surfaces of the midcarpal joint generally favor the range of extension over flexion and radial deviation over ulnar deviation.[13] This predominance of motion is the opposite of what is found at the radiocarpal joint.[14]

Ligaments of the Wrist Complex

The midcarpal joint receives some stability from the common intercarpal joint capsule but primarily depends on support from numerous ligaments, many of which it shares with the radiocarpal joint. The wrist ligaments not only provide support for the radiocarpal and midcarpal joints but also contribute to motion of the wrist complex by application of passive forces. The remaining intercarpal and interosseous ligaments serve to bind the carpals together but may contribute less to carpal motion.[15]

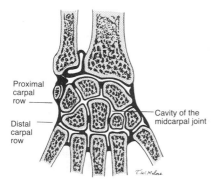

Proximal carpal row

Distal carpal row

Cavity of the midcarpal joint

Figure 9–3. The midcarpal joint formed by the articulation of the bones of the proximal and distal carpal rows is anatomically separated from the radiocarpal joint by interosseous ligaments.

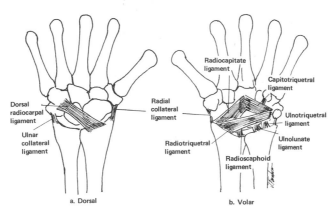

Figure 9–4. a. Dorsal ligaments of the wrist complex. b. Volar ligaments of the wrist complex, including the volar radiocarpal ligament (radiocapitate, radiotriquetral, and radioscaphoid) and the ulnocarpal ligament (ulnolunate, ulnotriquetral, and capitotriquetral).

Dorsally, the only major wrist ligament is the dorsal radiocarpal ligament (Fig. 9–4a). Passing from the radial styloid process to the lunate and triquetrum, the dorsal radiocarpal ligament exists as a thickening of the capsule. Its primary function appears to be to maintain contact between the lunate and the radius. It is looser than its volar counterpart, although it does become taut in full wrist flexion.[15]

The ulnar and radial collateral ligaments, like the dorsal radiocarpal, are capsular ligaments (Figs. 9–4a and b). The collaterals, however, provide significant passive control of radiocarpal motion in the frontal plane. The radial collateral ligament originates on the radius and passes to the scaphoid, trapezium, and first metacarpal. The ulnar collateral ligament originates on the ulna and inserts on the pisiform and triquetrum. Mayfield[16] has found that the radial collateral fails under lower tension loads than the ulnar collateral.

On the volar surface of the wrist complex, one finds the volar radiocarpal ligament. The volar radiocarpal ligament is the most important ligament for wrist complex motion and stability. It consists of three distinct ligaments that are intracapsular and are named for their proximal and distal attachments. They are the radiocapitate, radiotriquetral, and radioscaphoid ligaments (Fig. 9–4b). The radiotriquetral ligament is the strongest, stiffest, and most distinct of the three.[16] It attaches to and supports the lunate. The radiocapitate and radioscaphoid ligaments are less classically defined but play a significant part in wrist motion and stability by checking movement of joint surfaces and maintaining joint integrity.

The ulnocarpal ligament arises from the triangular fibrocartilage complex. It has bands attaching to the lunate (ulnolunate) and either directly to the capitate (ulnocapitate) or indirectly via the ulnotriquetral and capitotriquetral ligaments (Fig. 9–4b). Additional intercarpal ligaments that have been described include the scaphotrapezial ligaments complex[17] and the scapholunate interosseous ligament.[18] These intercarpal ligaments prevent diastasis (separation) of the carpals to which they attach, a problem that would result in wrist instability.

Function of the Wrist Complex

Movements of the Radiocarpal and Midcarpal Joints

Movement at the radiocarpal joint is predominantly a gliding of the proximal carpal row on the radius and the radioulnar disk. Because the convex surface of the carpals moves on the concave surface of the distal radius and radioulnar disk, the glide of the proximal carpal row occurs in a direction opposite to movement of the hand. For example, in wrist flexion the carpals slide dorsally on the radius and disk, while in wrist ulnar deviation the carpals slide radially. The major ligaments and, in some instances, bony structure check each slide at the end of the radiocarpal range. In radial deviation the ulnar slide of the proximal carpal row is checked when the ulnar collateral and, to a lesser extent, the ulnocarpal ligament become taut. The scaphoid may also impact the radial styloid process. Since the ligaments and bony configuration are the only checks to radiocarpal range, individual variations can alter the permissible range to a considerable extent.

Motions at the radiocarpal joint are caused by a rather unique combination of active and passive forces. Because there are no muscular forces applied directly to the bones of the proximal carpal row, the proximal carpals serve as a mechanical link between the radius and the distal carpals to which the muscular forces are actually applied. Gilford and associates[4] suggest that the proximal carpal row is an **intercalated segment**, a relatively unattached middle segment of a three-segment linkage. When compressive forces are applied across an intercalated segment, the middle segment tends to collapse and move in the wrong direction. The proximal carpal row would crumple and move in the wrong direction when compressive muscular forces were applied across the radiocarpal and midcarpal joints. Some type of stabilizing mechanism is required to normalize combined midcarpal-radiocarpal motion and prevent collapse of the middle proximal carpal row between the distal radius and the distal carpal row. The stabilization mechanism would seem to be the intercarpal bridge provided by the scaphoid and the wrist complex ligaments.

Flexion/Extension of the Wrist

When the wrist is moved toward extension from a fully flexed position, the following sequence of events gives an understanding of the relative motions of the various segments and of their interdependence. The conceptual framework presented, however, is oversimplified and ignores some of the simultaneous interactions that occur among the key carpal bones. Extension is initiated at the distal carpal row with the capitate at its center.[19] Midcarpal motion occurs with the distal carpals (capitate, hamate, trapezium, and trapezoid) gliding on the relatively fixed proximal bones (scaphoid, lunate, and triquetrum). Although the surface configurations of the midcarpal joint vary from carpal to carpal, arthokinematic movement appears to occur most commonly as a glide of the distal carpal row in the same direction as motion of the hand. When neutral wrist complex position is reached (long axis of the third metacarpal in line with the long axis of the forearm), the ligaments spanning the capitate and scaphoid draw the capitate and scaphoid together into a close-packed position. Continued extensor force now moves the combined unit of the distal carpal row and the scaphoid on the relatively fixed lunate and triquetrum. At approximately 45° of hyperextension of the wrist complex, the scapholunate interosseous ligaments bring the scaphoid and lunate into close-packed position. This unites all the carpals and causes them to function as a single unit. Wrist complex extension is completed as the proximal articular surface of the carpals move as a solid unit on the radius and radioulnar disk. All ligaments become taut as full extension is reached and the entire wrist complex is close-packed.[12,15,20,21] The scaphoid, through mediation of the wrist ligaments, participates at different times in scaphoid-capitate, scaphoid-lunate, or radioscaphoid motion (Fig. 9–5). Crumpling of the proximal carpal row (intercalated segment) is prevented and full ROM is achieved. Wrist motion from full extension to full flexion occurs in the reverse sequence.

Since the projected component sequence of wrist flexion and extension is dependent on the individually varied ligamentous checks and locking of intercarpal

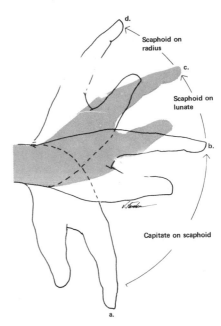

Figure 9-5. Wrist complex extension. From a to b movement occurs entirely at the midcarpal joint. From b to c the scaphoid locks onto the distal carpal row to move on the stable lunate and triquetrum. From c to d all the carpals lock and motion occurs entirely at the radiocarpal joint.

surfaces, it is understandable that studies on wrist kinematics have drawn different conclusions on the degrees of movement contributed by each component. Sarafian, Melamed, and Goshgarian[11] presented the results of 14 studies, each providing varied evidence that the midcarpal contribution to wrist extension was either greater than, equal to, or less than the radiocarpal contribution. There appeared to be consensus only that the predominance of one joint's contribution over the other (where found) in extension reversed in wrist flexion. What appears to be a redundancy in function at the midcarpal and radiocarpal joints ensures maintenance of the minimum ROM required for activities of daily living. Brumfield and Champoux[22] found that a series of hand activities necessary for independence required a functional motion of 10° of flexion and 35° of extension.

Radial/Ulnar Deviation of the Wrist

Studies of radial and ulnar deviation of the wrist complex show considerably more consistency than do studies of flexion and extension. In radial deviation from neutral position of the wrist complex, the distal carpal row moves radially on the proximal row until ligamentous and bony structures lock the rows together. Continued force toward radial deviation would appear to cause the carpals as a unit to slide ulnarly on the radius and radioulnar disk, although Youm and associates[13] believe that the radiocarpal motion is negligible. It has been found that both the scaphoid and the lunate flex during radial deviation, while the distal carpal row extends.[5,9,23] It is hypothesized that the opposing extension/flexion motions of the carpal rows that occur with radial deviation permit the carpals to accommodate to the narrowing space between the trapezoid and the radial styloid process. In full radial deviation, both the radiocarpal and midcarpal joints are in close-packed position.[11,13,21,24]

In ulnar deviation from neutral wrist complex position, active and passive forces move the distal carpal row ulnarly until checked by ligaments. The hamate is simultaneously pulled proximally, causing the proximal carpals to spread and slide radially until checked by the radial ligaments. Ulnar deviation occurs at both the midcarpal and radiocarpal joints, although some investigators have found radiocarpal motion to predominate.[13,15,20,21,24] As in radial deviation, the proximal and distal carpal rows again move in opposite directions in the sagittal plane while ulnar deviation occurs. The scaphoid and lunate extend while the capitate, trapezoid, and trapezium flex.[5,9,23]

The ranges of wrist complex radial and ulnar deviation are greatest when the wrist is in neutral flexion/extension. When the wrist is extended and is in close-packed position, the carpals are all locked and very little radial or ulnar deviation is possible. In wrist flexion the joints are loose-packed and the bones splayed. Further movement of the proximal row cannot occur and, as in extreme extension, little radial or ulnar deviation is possible in the fully flexed position.[20]

Muscles of the Wrist Complex

The primary role of the muscles of the wrist complex is to provide a stable base for the hand, while permitting positional adjustments that allow for optimal length-tension in the long finger muscles. Hazelton and associates[2] investigated the peak force that could be exerted at the interphalangeal (IP) joints of the fingers during different wrist positions. The study found the greatest IP flexor force occurs with ulnar deviation of the wrist (neutral flexion/extension), while the least force occurred with wrist flexion (neutral deviation). The muscles of the wrist, however, are not structured merely to optimize the force of finger flexion. If optimizing finger flexor force outweighed other concerns, one might expect the wrist extensors to be stronger than the wrist flexors. Contrary to this, when the work capacity (ability of a muscle to generate force per unit of cross section) of the wrist muscles is assessed, the wrist flexors have more than twice the work capacity of the extensors. Again contrary to expectation in optimizing finger flexor force, the work capacity of the radial deviators slightly exceeds that of the ulnar deviators.[14] The function of the wrist muscles cannot be understood by looking at any one factor or function, but should be assessed by observing the patterns of activity seen in each wrist muscle when monitored electromyographically in various patterns of use against the resistance of gravity and external loads. While an introduction to muscles of the wrist is in order, discussion of the synergies between hand and wrist musculature must be discussed in the context of hand function.

Volar Wrist Musculature

There are six muscles that have tendons crossing the volar aspect of the wrist and, therefore, are capable of creating a wrist flexion movement. These are the palmaris longus (PL), the flexor carpi radialis (FCR), the flexor carpi ulnaris (FCU), the flexor digitorum superficialis (FDS), the flexor digitorum profundus (FDP), and the flexor pollicis longus (FPL). The first three of these muscles are primary wrist muscles. The last three are flexors of the digits with secondary actions at the wrist. All pass under the flexor retinaculum of the wrist, except the PL.

The positions of the FCR and FCU tendons as they cross the wrist complex indicate that the tendons can, respectively, radially deviate and ulnarly deviate the wrist. However, electromyographic evidence has shown that the FCR is not significantly effective as a radial deviator of the wrist in an isolated contraction. Along with the PL, it functions more as a pure wrist flexor.[10] The FCR must contribute something to radial deviation, however, since it is active when pure radial deviation is performed.

The FCU tendon crosses the wrist farther from the axis for wrist radial-ulnar deviation than does the FCR, so it is more effective in its ulnar deviation function than the FCR is in its radial deviation function. The FCU also passes over the pisiform, a sesamoid bone that increases the FCU moment arm for flexion/extension. The FCU is effective in both flexion and ulnar deviation of the wrist complex.

The FDS, FDP, and the FPL are predominantly flexors of the digits. As multi-joint muscles, their capacity to produce effective force at the wrist is dependent on a synergistic stabilizer to prevent full excursion of more distal joints. If these muscles attempt to act over both the wrist and the more distal joints, they will become actively insufficient. The FDS and FDP show varied activity in wrist radial-ulnar deviation as one might anticipate from the central location of the tendons. The superficialis seems to function more consistently as a wrist flexor than does the profundus,[25] which is logical considering the FDP is a longer muscle and crosses more joints. The effect of the FPL on the wrist has received relatively little attention. The position of the tendon suggests the ability to contribute to flexion and radial deviation of the wrist.

Dorsal Wrist Musculature

The dorsum of the wrist complex is crossed by the tendons of nine muscles, all of which pass under the extensor retinaculum. Three are primary wrist muscles; the extensor carpi radialis longus and brevis (ECRL, ECRB) and the extensor carpi ulnaris (ECU). The other six muscles are finger and thumb muscles that may act secondarily on the wrist; these are the extensor digitorum (ED), the extensor indicis (EI), the extensor digiti minimi (EDM), the extensor pollicis longus (EPL), the extensor pollicis brevis (EPB), and the abductor pollicis longus (APL).

The ECRL and ECRB together make up the predominant part of the wrist extensor mass.[26] The ECRB is somewhat smaller than the ECRL, but generally shows more activity during wrist extension activities, given its more central location.[10,27] One study found the ECRB to be active during all grasp-and-release hand activities, except those performed in supination.[28] The ECRL shows increased activity when either radial deviation or support against ulnar deviation is required, or when forceful finger flexion motions are performed.[12,27]

The ECU extends and ulnarly deviates the wrist. It is active not only in wrist extension, but frequently in wrist flexion as well.[27] Backdahl and Carlsoo[25] hypothesized that the ECU activity in wrist flexion adds an additional component of stability to the structurally less stable position of wrist flexion. This is not needed on the radial side of the wrist, which has more developed ligamentous and skeletal control. When the forearm is pronated, the crossing of the radius over the ulna causes a reduction in the MA of the ECU, making it less effective as a wrist extensor.[26,28]

The EDM and the EI insert into the tendons of the ED and, therefore, have a common function with the ED.[29] The EI and EDM are capable of extending the wrist, but wrist extension is credited more to the ED. The ED is a finger extensor muscle but functions also as a pure wrist extensor. There appears to be some reciprocal synergy of the ED with the ECRB in providing wrist extension, since less ECRB activity is seen when the ED is active.[27]

The three extrinsic thumb muscles that cross the wrist do not contribute much to the functioning of the wrist. The APL and the EPB are both capable of radially deviating the wrist and may serve a minor role in that function. More generally, the wrist deviation detracts from the prime action on the thumb and necessitates a synergistic contraction of the ECU to offset unwanted motion. When ulnar deviators are absent, the thumb extrinsics may produce a significant radial deviation deformity at the wrist. Little evidence has been found to indicate that the more centrally located EPL has any notable effect on the wrist.

Wrist Joint Pathology

The scaphoid, lunate, and capitate form the functional unit of the wrist complex and, as already emphasized, are dependent on both an intact ligamentous structure and musculature for stability. The large compressive forces directed across the wrist seem to focus first on the capitate and then pass to the scapholunate junction and to the distal radius and disk.[30] In an experimental model, the radius received 82 percent of the compressive load across the radiocarpal joint while the distal ulna received 18 percent. Removal of the radioulnar disk resulted in a 12 percent reduction in load on the ulna and a concomitant increase in load on the radius.[7] The disk, therefore, helps shift some of the compressive load off the radius and on to the ulna. Removal of the disk can be expected to result in an increase in stress on the radius. Because of the angulation of the distal end of the radius, the compressive forces across the radiocarpal joint do not act in line with the longitudinal axis of the radius, but obliquely to it. The joint reaction force is ulnarly and volarly directed, tending to glide the proximal carpal row in that direction. This tendency is resisted predominantly by the radiocarpal ligaments and by the radioulnar disk.[5,9]

In spite of the stabilizing influence of the scaphoid, the active forces initiated on the distal carpal row result in the proximal carpal row still acting as an intercalated segment between the forearm and the distal carpal row. That is, forces across the wrist that result in movement of the proximal carpal row on the radius and disk will also result in some additional midcarpal movement as well. Ligamentous structures are responsible for offsetting the tendency toward "collapse" of the scaphoid and lunate beneath the capitate. Failure of one or more of the ligaments may result in what is known as **dorsal intercalated segment instability (DISI)**. This is characterized by maintained extension of the lunate, which brings the capitate dorsal to the long axis of the radius. The capitate will flex in an attempt to improve alignment, resulting in a zigzag pattern from the radius to the capitate.[5,9] DISI is the most common form of carpal instability and is usually secondary to instability (dissociation) between the lunate and the scaphoid. Ligamentous injury between these two bones can then progress to create additional problems around the lunate and on to the capitolunate and triquetrolunate joints.[16] The other common form of carpal instability is **volar intercalated segment instability (VISI)**. VISI is usually secondary to ligamentous disruption beween the lunate and the triquetrum.[9]

THE HAND COMPLEX

The hand consists of five digits, or four fingers and a thumb. Each digit has a carpometacarpal (CMC) joint and a metacarpophalangeal (MCP) joint. The fingers each have two IP joints, while the thumb has only one. There are 19 bones distal to the carpals and 19 joints that make up the hand complex. While there are structural similarities between the joints of the fingers and the joints of the thumb, function differs significantly enough that the fingers shall be examined separately from the thumb.

Structure of the Fingers
CMC Joints of the Fingers

The CMC joints of the fingers are composed of the articulations between the distal carpal row and the bases of the second through fifth metacarpals (Fig. 9–6). The second metacarpal articulates primarily with the trapezoid, and secondarily with the trapezium and capitate. It also articulates with the third metacarpal to which it is contiguous. The third metacarpal articulates primarily with the capitate, but the contiguous sides of the second and fourth metacarpals contribute to the joint surface as well. The fourth metacarpal articulates with the capitate and

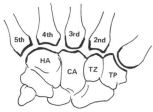

Figure 9-6. Carpometacarpal joints of the fingers. The articulations between the second through fifth metacarpals and the distal carpal row (trapezium = TP; trapezoid = TZ; capitate = CA; hamate = HA).

hamate, as well as with the sides of the contiguous third and fifth metacarpals. Lastly, the fifth metacarpal articulates with the hamate and with the ulnar side of the fourth metacarpal. The second through fourth CMC joints are plane synovial joints with 1° of freedom: flexion/extension. The fifth carpometacarpal joint is a saddle joint with 2° of freedom including not only flexion/extension but also a limited amount of abduction/adduction.[10,31] All finger CMCs are supported by strong transverse and weaker longitudinal ligaments volarly and dorsally. The transverse metacarpal (intermetacarpal) ligament tethers together the heads of the four metacarpals of the fingers, effectively preventing the second through fifth metacarpals from performing any abduction at the CMC joints. It is predominantly the ligamentous structure that controls the total ROM available at each carpometacarpal joint, although some differences in articulations also exist.

The range of CMC motion of the fingers increases from the radial to the ulnar side of the hand. The second and third CMC joints are essentially immobile. The fourth CMC is substantially more mobile than either the second or the third, but less mobile than the fifth. The fifth CMC can simultaneously flex and adduct through a range of 10 to 20°, due largely to the saddle shape of the fifth metacarpal and hamate. The second and third metacarpals essentially form a fixed unit with, respectively, the contiguous trapezoid and capitate bones. Youm and associates[13] investigated the fixed unit theory and found no more than 2° of motion between the metacarpals and their carpals and no detectable motion between the capitate and trapezoid. The stability of the second and third CMCs is a functional adaptation that enhances function of the radial wrist flexors and extensors (FCR, ECRL, and ECRB). These muscles insert on the bases of the second and third metacarpals and would act initially on the CMCs before flexing or extending the midcarpal and radiocarpal joints. With the more distal attachment of the radial wrist muscles to the metacarpals and with immobility of the second and third CMC joints, the muscles obtain an increased lever arm (LA) without the concomitant loss of tension resulting from excessive ROM.

The function of the CMC joints of the fingers is primarily to contribute to the hollowing of the palm. The so-called **palmar arches,** or **palmar cupping,** allow the hand and digits to conform optimally to the shape of the object being held. This maximizes the amount of surface contact, enhancing stability as well as increasing sensory feedback. When the first metacarpal and first CMC joints are included, the arches can easily be visualized as occurring obliquely across the palm and longitudinally down the palm (Fig. 9–7). In fact the immobile second and third CMCs provide a fixed axis about which the mobile first metacarpal and fourth and fifth metacarpals can move.[32] The concavity of the palm of the hand persists even when the hand is fully opened (Fig. 9–8). This consistent concavity is known as the **carpal arch.**

The carpal arch is created partly by the curved shape of the carpals but also by the ligaments that maintain the concavity. The ligaments that maintain the arch are the flexor retinaculum (or transverse carpal ligament) and the intercarpal ligaments that are transversely oriented. The intercarpal ligaments that are longitudinally ori-

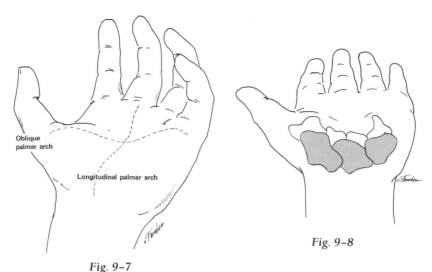

Oblique
palmar arch

Longitudinal palmar arch

Fig. 9–7

Fig. 9–8

Figure 9–7. Oblique and longitudinal palmar arches.

Figure 9–8. Structural arch of the carpal bones. When the hand is fully opened, the structural arch of the carpal bones remains.

ented appear to have little function in supporting the arch.[33] The carpal arch and the flexor retinaculum together are known as the **carpal tunnel.** In addition to the function of the arch in increasing contact of objects within the palm, the carpal tunnel is the pathway and protection for the long digital flexors and for the median nerve. When the flexor retinaculum is surgically sectioned as it might be when the carpal tunnel creates pressure on the median nerve, there is a widening of the arch. The transverse stability of the arch is retained, however, as long as the other transverse intercarpal ligaments remain intact.[33]

Muscles that act over the CMC joints will contribute to palmar cupping (mobility of the palmar arches). Since the muscles acting on the first CMC will be discussed separately, only the radial wrist muscles (FCR, ECRL, and ECRB) and the ECU act uniquely over the remaining CMC joints. Of these, the radial flexors and extensors produce very little CMC motion. Most of the CMC motion available at the fourth and fifth CMCs are produced by musculature that also crosses more distal joints. The FDS and the FDP cross and act on the CMC joints as well as the more distal joints that each crosses. Hollowing of the palm, therefore, accompanies finger flexion, and relative flattening of the palm accompanies finger extension. The carpal arch itself also has some flexibility that can increase the contouring of the palm. Increased arching occurs with activity of the FCU and with intrinsic hand muscles that insert on the transverse carpal ligament.[8,34]

MCP Joints of the Fingers

Each of the four MCP joints of the fingers are composed of the convex metacarpal head proximally and the concave base of the first phalanx distally (Fig. 9–9). The MCP joint is condyloid with 2° of freedom: flexion/extension and abduction/adduction. The large metacarpal head has 180° of articular surface in the sagittal plane, with the predominant portion lying volarly. This is apposed to approximately 20° of articular surface on the phalanx. In the frontal plane the articular surfaces are more congruent. The joint is surrounded by a capsule that is generally considered to be lax in extension and, in conjunction with the poorly mated surfaces,

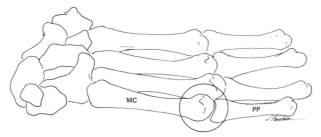

Figure 9-9. Metacarpophalangeal joints of the fingers. A view of the volar surface of the hand shows the articulation between the large head of the metacarpal (MC) and the smaller base of the proximal phalanx (PP).

allows some passive axial rotation of the proximal phalanx in that position. The presence of a volar and two collateral ligaments contributes to joint stability.

The Volar Plate

When there is joint incongruence as there is at the MCP joints, it is common to find an accessory joint structure to enhance joint stability and congruence. The volar plate (volar ligament) at the MCP joint is a unique structure that does more than merely reinforce the joint capsule. It is actually fibrocartilage that is firmly but not rigidly attached to the base of the proximal phalanx. It becomes membranous proximally to blend with the capsule and attach to the metacarpal head (Fig. 9-10). The four volar plates of the MCP joints of the fingers also blend with and are interconnected by the more superficial transverse metacarpal ligament, which passes volarly across the heads of the metacarpals of the four fingers. The plate, therefore, is a multilayered structure consisting of a fibrocartilaginous thickening of the inside of the joint capsule, the joint capsule itself and the more superficially located transverse metacarpal ligament whose transverse fibers blend with the longitudinally oriented fibers of the MCP joint capsule. The inner surface of the volar plate is actually an extension of the distal articular surface of the joint, thereby adding surface to the base of the proximal phalanx. In extension the plate adds to the amount of surface in contact with the large metacarpal head. The plate also helps to restrict the hyperextension that is permitted by the loose capsule. In flexion the flexible attachment of the plate to the phalanx permits the plate to glide proximally along the volar surface of the metacarpal head without restricting motion. This fibrocartilage mechanism also prevents pinching of the long flexor tendons during flexion.

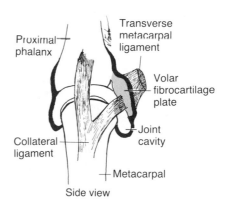

Figure 9-10. Schematic representation of the volar fibrocartilage plate at the metacarpophalangeal joint. The plate is attached to the volar surface of the proximal phalanx, adding to the contact with the metacarpal head below.

The volar plates are stabilized on the volar surface of the metacarpal head by the attachments of the collateral ligaments.[8]

The Collateral Ligaments

The collateral ligaments of the MCP joint are, like the joint capsule, generally considered to be slack in extension, although Shultz and associates[35] did not find this to be true. They concluded that the collateral ligaments provided stability throughout the MCP ROM with parts of the fibers taut at various points in the range. They also proposed that the limitation in MCP abduction/adduction seen in full MCP flexion is due to a change in shape of the head of the metacarpal rather than tension in the collateral ligaments. They found that the metacarpal head assumed a bicondylar shape on its volar surface, so that at about 70° of MCP flexion there was a bony block to abduction/adduction. There is consensus that the full flexion is considered to be the close-packed position for the MCP joint.

Fisher and associates[36] did a series of dissections of fingers seeking an explanation for the relatively small incidence of osteoarthritis in MCP joints as compared to the fairly common changes seen in the distal interphalangeal (DIP) joints and, to a lesser extent, in the proximal interphalangeal (PIP) joints. They confirmed the presence in almost all MCP, PIP, and DIP joints of fibrocartiliginous intra-articular collagenous tissue. That is, they found fibrocartilage that projected into the joints from the inner surface of the extensor hood, from the volar plate, and from the collateral ligaments. The base of these projections contributed to a ring of fibrocartilage firmly attached to the base of the phalanx and merging with the capsule externally. The fibrocartilage projections were most impressive in the MCP joints.[36] These fibrocartilage projections, like the volar plate itself, increase the surface area on the small base of the phalange for contact with the large metacarpal (and phalangeal) heads.

Range of Motion

The total ROM available at the MCP joint varies with each finger. Flexion/extension increases radially to ulnarly, with the index finger having approximately 90° of MCP flexion and the little finger approximately 110°. Hyperextension is fairly consistent between fingers but varies widely among individuals. The range of passive hyperextension has been used as a measure of generalized body flexibility.[37] The range of abduction/adduction is maximal in MCP extension; the index and little finger have more frontal plane mobility than the middle and ring fingers. As previously mentioned, abduction/adduction is most restricted in MCP flexion.

IP Joints of the Fingers

Each of the PIP joints and DIP joints of the fingers are composed of the head of a phalanx and the base of the phalanx distal to it. Each IP joint is a true synovial hinge joint with 1° of freedom (flexion/extension), a joint capsule, a volar fibrocartilage plate, and two collateral ligaments (Fig. 9–11).

The base of each middle and distal phalanx has two shallow concave facets with a central ridge. The distal phalanx sits on the pulley-shaped head of the phalanx proximal to it. The joint structure is similar to that of the MCP joint in that the proximal articular surface is larger than the distal articular surface. There is little pos-

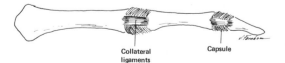

Collateral   Capsule
ligaments

Figure 9–11. Capsule and collateral ligaments of the proximal and distal interphalangeal joints.

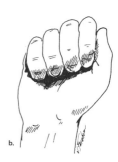

Figure 9–12. a. The grip is tighter ulnarly than radially. b. Pattern of increasing flexion from the radial to the ulnar side of the hand.

terior articular surface and, therefore, little hyperextension. The DIP joint has some passive hyperextension, but the proximal joint has essentially none. Volar plates reinforce each of the joint capsules and enhance stability. The plates at the IP joints are structurally and functionally identical to those at the MCP joint, except that the plates are not connected by a transverse ligament. Fisher and associates[36] found fibrocartilage projections from the extensor mechanism, the volar plate, and the collateral ligaments attached to the bases of the phalanges at both the PIPs and the DIPs, with the structures more obvious at the PIPs. The collateral ligaments of the IP joints are not fully understood but remain taut and provide support in all joint positions.

The total range of flexion/extension available to the index finger is greater at the PIP joint (100 to 110°) than it is at the DIP joint (80°). The range at each IP joint increases ulnarly, with the proximal and distal joints achieving 135° and 90°, respectively, in the little finger.

The increasing flexion/extension ROM from the radial to the ulnar side of the hand is consistent at all joints in the hand. The pattern that results from simultaneous flexion at all joints is shown in Figure 9–12b. The additional range allocated to the more ulnarly located fingers angles them toward the thumb. This facilitates opposition of these fingers with the thumb and produces a grip that is tighter, or has greater closure, on the ulnar side of the hand. Many objects are constructed so that the shape is narrower at the ring and little fingers and widens toward the long and index fingers (Fig. 9–12a).

Finger Musculature

Extrinsic Finger Flexors

Mechanisms of Finger Flexion

There are two muscles originating outside the hand that contribute to finger flexion. These are the FDS and the FDP. The superficialis can flex the MCP joint and the PIP joint. The profundus can flex these joints and the DIP joint as well. Both the profundus and superficialis are dependent on wrist position for maintenance of optimal length-tension. If there is not a counterbalancing extensor force at the wrist, the volarly located forces of the FDP and FDS muscles will cause wrist flexion to occur. As this happens there is a concomitant loss of tension at the more distal joints. Wrist flexion reduces the efficiency of the finger flexors to one fourth of that available in wrist extension.[10] The counterbalancing wrist extensor force is usually supplied by an active wrist extensor such as the ECRB or, in some instances, the ED.

Loss of tension in the FDP and FDS can also be observed in the more ulnarly located fingers during active contraction, because of the greater available range of MCP and IP flexion in those fingers. If the object to be held by the flexors is heavy or requires strong grip, the object may be shaped so that it is wider ulnarly than radially. This limits the MCP/IP flexion and, therefore, limits diminution of FDS/FDP tension. The so-called pistol grip of most tools is an example of modifying an object to optimize hand function.

Optimal function of the FDS and FDP is dependent not only on the wrist musculature but also on intact gliding mechanisms. The gliding mechanism consists of retinaculae, ligaments, bursae, and tendon sheaths that tether the long flexor tendons to the hand while still permitting friction-free excursion of the tendons. The retinaculae and ligaments prevent bowstringing of the tendons, which would result in loss of tension in the contracting muscles. The tendons must be anchored without interfering with their pull or creating frictional forces that would cause degeneration of the tendons over time.

As the FDS and FDP cross the wrist to enter the hand, they must first pass beneath the flexor retinaculum. Friction between the tendons themselves and friction of the tendons on the retinaculum are prevented by the radial and ulnar bursae that encase the flexor tendons at this level. All eight tendons of the profundus and superficialis are invested in a common bursa known as the **ulnar bursa.** The bursa is compartmentalized to prevent friction of tendon on tendon. The flexor pollicus longus that accompanies the FDS and FDP through the flexor retinaculum is encased in its own **radial bursa.** The radial and ulnar bursae contain a synovial-like fluid that minimizes frictional forces. Although the pattern may vary among individuals, the ulnar bursa generally ends just distal to the palmar crease. The tendons of the fifth finger, however, commonly remain encased so that the ulnar bursa becomes continuous with the digital tendon sheath for the fifth finger (Fig. 9–13).

The long flexor tendons of each finger pass through three additional fibro-osseous tunnels formed by what are known as **annular pulleys (or vaginal ligaments)** (Fig. 9–13). The first two annular pulleys lie closely together, with one at the head of the metacarpal and the second at the base of the proximal phalanx. The floor of the first pulley is formed by the transverse metacarpal ligament, while the floor of

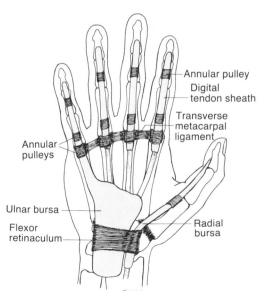

Figure 9–13. The flexor mechanism of the flexor digitorum superficialis and profundus includes the flexor retinaculum, the radial and ulnar bursae, the annular pulleys, and the digital tendon sheaths.

the second is the phalange itself. The third annular pulley lies centrally on middle phalanx, with the floor of the tunnel again formed by the bone. Friction of the FDS and FDP tendons on the annular pulleys is prevented by the digital tendon sheaths, which envelop the tendons from just proximal to the point where the tendons pass into the first annular pulley to the point at which the tendon of the profundus passes through a split in the superficialis tendon. The synovial-like fluid contained in each of the digital tendon sheaths permits gliding of the FDP tendon as it passes volarly through the loop of the FDS as the superficialis is ready to attach to the middle phalanx. Once the FDP has passed distal to the last annular pulley and the attachment of the FDS, the tendon sheath ends, since lubrication of the tendon is no longer needed. Although subject to individual differences, the tendon sheaths of the index, middle, and ring fingers are usually separate from the ulnar bursa, whereas that of the little finger commonly communicates with the bursa.

Any interruption in either the annular pulleys or the digital tendon sheaths can result in substantial impairment of FDS and FDP functioning or in structural deformity. Manske and Lesker[38] found that the third annular pulley appeared to be the most important of the pulleys, but that sectioning of any of them reduced the available flexion range of IP and MCP joints. The gliding mechanism of the fingers is critical not only to proper application of active finger flexion forces but also to the ability of the tendons to undergo passive excursion in finger extension. Trigger finger is one example of the disability that can be created when repetitive trauma to a flexor tendon results in the formation of nodules on the tendon and thickening of the annular pulley. Finger flexion may be prevented completely or the finger may be unable to re-extend.[8]

Finger Flexor Function

The FDP is generally accepted to be the more active than the FDS in finger flexion activities. The insertion of the FDP on the distal phalanx makes it capable of full hand closure without active assistance from other hand musculature. However, when hand closure is attempted by a person with long-standing paralysis of the intrinsic hand musculature, the fingertips roll into the palm and catch against the metacarpal head.[39] Grasp is ineffective. Apparently, the passive viscoelastic constraints imposed by the intact intrinsic musculature at the IP joints are necessary for normal function.

The FDS functions alone in finger flexion only when flexion of the DIP joint is not required. When simultaneous PIP and DIP flexion are required, the FDS acts as a reserve muscle. It joins the FDP by increasing its activity as increased flexor force is needed, or when finger flexion with wrist flexion is desired.[25,29,40] Interestingly, the FDS connection to the little finger is commonly smaller and functionally inferior. Baker and associates[41] found that 34 percent of the persons they studied could not achieve full isolated PIP flexion using the FDS to the fifth digit. When the FDS is inadequate, the function of the intact FDP would be particularly important.

Extrinsic Finger Extensors

The extrinsic finger extensors are the ED, the EI, and the EDM. As each of these muscles passes from the forearm to the hand, it passes beneath the extensor retinaculum. At that point each tendon is encased within its own tendon sheath to prevent friction between tendons and friction on the retinaculum. There are no commonly shared bursae dorsally. The tendon sheaths are also not as variable or as extensive as the sheaths associated with the flexor tendons and there are no annular pulleys. The ED inserts into each finger on the middle phalanx by a central tendon and on the distal phalanx by a terminal tendon formed by fibers from the ED and the finger intrinsics. The EI and EDM insert on the ED tendons of the index and little fingers, respectively. Given the attachments of the EI and EDM to the ED tendon, these muscles appear to add independence of action to the first and fourth fingers, rather than additional strength or additional actions. Since the EI and EDM share innervation, insertion, and function with the ED, discussion of the ED from this point should be assumed to include consideration of both the EI and the EDM.

The ED is the only muscle capable of extending the MCP joints of the fingers. It is also a wrist extensor by continued action. The ability of the ED to produce IP extension is the subject of considerable divergence of opinion. The source of the complexity lies in structure and function of the extensor mechanism.

The Extensor Mechanism

The extensor mechanism of the fingers consists of the extensor expansion (extensor hood or dorsal aponeurosis) and the associated musculature: the ED; the dorsal and volar interossei (DI, VI); and the lumbrical muscles. The structure of the extensor mechanism of each finger (Fig. 9–14) is made up of the ED tendon, its connective tissue expansion, and fibers from the tendons of the interossei and the lumbrical of each finger. The ED tendon passes under the extensor retinaculum of the wrist, protected at that level by a synovial sheath. The tendon continues distally without its sheath and flattens into an aponeurotic hood just distal to the MCP joint. The flattened hood is joined by tendon fibers from the interossei muscles that arise from the sides of the metacarpals. Some interosseous fibers go deep to insert directly into the proximal phalanx. Other fibers join with and become part of the hood that wraps around the proximal phalanx. The hood splits into three segments just proximal to the head of the proximal phalanx. Of the three segments, a central tendon crosses the PIP joint and inserts on the base of the middle phalanx, while two lateral bands pass to either side of the central tendon, cross the proximal joint, and reunite to insert in a single terminal tendon on the distal phalanx. The central tendon and both lateral bands receive contributing fibers from the interossei, with the radial lateral band also receiving the lumbrical tendon. The lateral bands are interconnected dorsally by a triangular band of superficial fibers known as the tri-

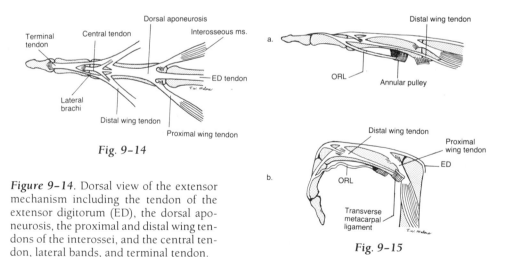

Fig. 9–14

Figure 9–14. Dorsal view of the extensor mechanism including the tendon of the extensor digitorum (ED), the dorsal aponeurosis, the proximal and distal wing tendons of the interossei, and the central tendon, lateral bands, and terminal tendon.

Fig. 9–15

Figure 9–15. a. Each of the oblique retinacular ligaments (ORL) arises from the annular pulley on the proximal phalanx and passes distally to a lateral band of the extensor mechanism. The ORL lies volar to the axis of the proximal interphalangeal (PIP) joint and dorsal to the axis of the distal interphalangeal (DIP) axis through its attachment to the lateral band. b. Full flexion of the PIP joint releases tension in the lateral bands and the ORL. Consequently, the terminal tendon is relaxed and the DIP cannot be actively extended.

angular, or **transverse retinacular, ligament.** The dorsal hood is also attached to the transverse metacarpal ligament, which helps prevent bowstringing of the extensor tendon.[8] With the addition of the oblique retinacular ligament, the structure of the extensor mechanism is complete (Fig. 9–15a). The oblique retinacular ligaments (ORL) arise from the sides of the first phalanx and from the digital tendon sheath that encases the long finger flexors. The ORLs continue distally as slender bands to insert on the distal portion of the lateral bands (Fig. 9–15a).[8,10,42]

Function of the extensor mechanism can be understood by looking at both the active and passive elements that compose it and by referencing the relationship of relevant segments to each joint individually.

Influence on MCP Joint Function. At each MCP joint, the ED tendon passes *dorsal* to the MCP joint axis. An active contraction of the muscle creates tension on the hood, pulls the hood proximally over the MCP joint, and extends the proximal phalanx. The other active forces that are part of the extensor mechanism are the dorsal interosseous, volar interosseous, and lumbrical muscles. Each of these muscles passes *volar* to the MCP joint axis and, therefore, creates a flexor force at the joint. When the ED, interossei, and lumbricals all contract simultaneously, the MCP joint will generally extend, with the torque produced by the ED exceeding that of the other muscles. The magnitude and functional significance of the flexor force of the interossei and lumbricals varies and will be detailed during the discussion of these muscles.

Influence on PIP Joint Function. Each PIP joint is crossed dorsally by the central tendon and lateral bands, and volarly by the oblique retinacular ligaments. The ED that is the structural base of the extensor mechanism and the interossei and lumbrical muscles that have attachments to the hood and/or lateral bands are each capable of producing at least some tension in the central tendon and lateral bands. The ED, interossei, and lumbricals are, therefore, each capable of applying an extensor force to the PIP joint. However, there is consensus among investigators that an active contraction of the ED alone cannot produce sufficient tension in the central tendon or lateral bands to overcome the passive forces of the long finger flexors at the IP joints. Consequently, an isolated contraction of the ED will produce **clawing** of the finger: MCP hyperextension with IP flexion produced by passive flexor pull.[8,10,39,43] An active contraction of an interosseous or lumbrical muscle *is* capable of generating the active tension in the extensor mechanism necessary to extend the PIP joint (although the argument will be presented that some additional passive tension in the extensor mechanism may be needed for these muscles to complete the motion). Extension of the PIP joint is always accompanied by simultaneous extension of the DIP joint, as will be further explained.

Influence on DIP Joint Function. The DIP joint is tied to the PIP joint by both active and passive forces in such a way that DIP extension and PIP extension are interdependent. When the PIP is actively extended, the DIP will also extend. Similarly, active DIP extension will always create PIP extension. When a force is applied to the extensor expansion by the interossei or lumbrical muscles, tension is created simultaneously on the more distal portion of the lateral bands and terminal tendon that cross the DIP. The muscular extensor force on the lateral bands is assisted by the passive action of the oblique retinacular ligaments. The ORLs, which pass volar to the PIP joint, become stretched as the PIP is extended by the interossei or lumbrical. The stretch on the ORLs at the proximal joint creates a pull on the ORLs' insertion at the distal lateral bands and, thus, on the terminal tendon that lies dorsal to the DIP. In this way the DIP joint is forced to extend simultaneously with the PIP joint as a result of the combined active and passive forces applied to the lateral bands and the terminal tendon.[8,10,44]

Stack[45] has suggested that the interossei and lumbricals would not be able to generate sufficient tension to cause IP extension if the ED tendon was completely slack. Tension must be present throughout the extensor mechanism for effective action. Tension in the ED tendon and extensor expansion can be maintained either by the contraction of the ED during active MCP extension or by the passive stretch that is imposed on the ED tendon during active MCP flexion. Whether the tension in the ED is active (as in MCP extension) or passive (as in MCP flexion), an active contraction of either an interosseous or lumbrical muscle on the tensed mechanism is necessary to produce complete IP extension. When the intrinsic musculature is paralyzed, the ED may be used to extend the IP joints, but *only* if the MCP joint is *passively maintained in flexion* by some external force. The ability to use the ED to extend the IPs with passively maintained MCP flexion is known as **Bunnell's sign.**[46] With the stretch imposed on the extensor mechanism by the passive MCP flexion, an active contraction of the ED can create the additional tension needed to pull on the central tendon, oblique retinacular ligaments, and, perhaps, the lateral bands. This is not a particularly functional substitution for the missing intrinsic musculature, however, since it requires an assistive device to maintain the passive MCP flexion before the ED can actively extend the IP joints.

Flexion of the DIP joint produces flexion of the PIP joint by a similar combination of active and passive forces that results in simultaneous active extension of these two joints. When the DIP joint is flexed by the FDP, a simultaneous flexor force is applied over the proximal joint that the FDP also crosses. However, the active force might not be sufficient to produce PIP flexion if dorsal restraining forces were not released at the same time. When the DIP begins flexing, the terminal tendon and distal lateral bands are stretched over the dorsal aspect of the DIP joint. The stretch in the lateral bands pulls the extensor hood (from which the lateral bands arise) distally. The distal migration in the extensor hood causes the central tendon of the extensor expansion to relax. The *release of the extensor influence* of the central tendon at the PIP joint is accompanied by the application of a *passive flexor force* at the PIP by the ORLs. The ORLs, like the distal lateral band to which they attach, are stretched as the DIP is actively flexed. The stretch in the ORLs causes a pull on their proximal portion that is located volar to the PIP joint, creating a flexion force at the PIP joint. The combination of active (FDP) and passive forces (release of the central tendon and pull of the ORLs) might still not be sufficient to flex the PIP joint if the lateral bands remained taut on the dorsal aspect of the PIP joint. The bands, however, are permitted to migrate volarly by the elasticity of the interconnecting triangular ligament. Through the combination of active and passive mechanisms, both active and passive DIP flexion ordinarily result in simultaneous PIP flexion.

The normal coupling of DIP and PIP flexion can be overridden by some individuals. That is, some people at some fingers can actively flex a DIP while maintaining the PIP in extension. This "trick" is not really a violation of the coupling of forces, but underscores the strong influence of the ORLs in coupling DIP and PIP function. Active DIP flexion with simultaneous PIP extension requires that the PIP joint be locked in hyperextension first. When the PIP can be sufficiently hyperextended, the ORLs (ordinarily lying just volar to the PIP joint axis) pass *dorsal* to the PIP joint axis. Now, tension in the ORLs produced with active DIP flexion will accentuate PIP extension since the ORLs have been placed dorsal to the PIP joint and function as passive joint extensors. The trick of active DIP flexion and PIP extension serves no functional purpose and can be accomplished only by those individuals and in those fingers where PIP hyperextension is available.

The functional coupling of proximal and distal interphalangeal joint action can be demonstrated by one other PIP-DIP relationship. This relationship again emphasizes the role of the ORL in IP joint function. When the PIP is fully flexed actively by the FDS or passively by some outside force, the DIP cannot be actively extended (although it can be flexed either actively or passively). When the PIP joint is in full flexion, the dorsally located central tendon is stretched. The tensed central tendon pulls the extensor hood (from which the central tendon arises) distally. This distal movement of the hood releases some of the tension in the lateral bands. The lateral bands are further released as they are permitted to separate and move volarly around the flexing PIP joint by the elasticity of the interconnecting triangular ligament. Relaxation of the lateral bands also relaxes the terminal tendon and the ORLs that attach to them. As 90° of PIP flexion is reached, loss of tension in the terminal tendon and ORLs completely eliminates extensor force from the DIP joint.[44] While active DIP flexion can accompany PIP flexion with the addition of a FDP contraction, the DIP cannot be actively extended again once the PIP joint is fully flexed.

The coupled actions of the PIP and DIP joints are summarized as follows:

- Active extension of the PIP joint will normally be accompanied by extension of the DIP joint.
- Active or passive flexion of the DIP joint will normally be accompanied by flexion of the PIP.
- Full flexion of the PIP joint (active or passive) will prevent the DIP joint from being actively extended.

Intrinsic Finger Musculature

Dorsal and Volar Interossei

The interossei (DI and VI) are sets of muscles arising from between the metacarpals and attaching to the bases of the proximal phalanges and/or to the extensor expansion. Since the DI and VI are alike in location and in some of their actions, they have traditionally been characterized by the capability of producing MCP abduction and adduction, respectively. More recently, the detail of attachments of these muscles to the extensor mechanism has permitted further elucidation on their actions.

The interossei muscle fibers join the finger in two locations. Some fibers attach *proximally* to the proximal phalanx and to the extensor hood via what are termed **proximal wing tendons**; some fibers attach more *distally* to the lateral bands and central tendon via **distal wing tendons** (Figs. 9–14 and 9–15). Although individual variations in muscle attachments exist, studies have found some consistency in the point of attachment of the different interossei.[42,45,47] The first dorsal interosseous muscle has the most consistent attachment of its group, inserting entirely into the bony base of the proximal phalanx and the extensor hood via proximal wing tendons with no distal wing tendons present. The three dorsal interossei of the middle and ring fingers have both proximal wing tendons *and* distal wing tendons attaching them to the lateral bands and central tendon. The proximal and distal wing tendons of the dorsal interossei usually arise from separate bellies, with the belly from which the distal wing tendon arises closely resembling that of a volar interosseous muscle. The abductor digiti minimi muscle is, in effect, a dorsal interosseous with only a proximal wing tendon. The volar interossei appear consistently to be muscles with distal wing tendons only and no proximal attachments.

Given the attachments of the interossei, these muscles can be characterized not only as abductors or adductors of the MCP, but also as proximal or distal interossei according to the pattern of attachment. Proximal interossei will have their predominant effect at the MCP joint alone, while the distal interossei will produce their predominant action at the IP joints, with some effect by continued action at the MCP joint.

All of the interossei muscles (regardless of their designation as dorsal, volar, proximal, or distal) pass dorsal to the transverse metacarpal ligament (Fig. 9–15a) but just volar to the coronal MCP joint axis for flexion/extension. All the interossei, therefore, are flexors of the MCP joint. The ability of the interossei to flex the MCP joint, however, will vary somewhat with MCP joint position and with its proximal versus distal attachment.

Role at the MCP Joint in MCP Extension. When the MCP joint is in extension, the MA (and rotatory component) of any of the interossei for MCP flexion is so small that little flexion torque is produced (Fig. 9–15a). However, the muscles have a reasonable MA for abduction or adduction, since the muscles lie much further from the AP axis for MCP joint deviation (abduction or adduction). Consequently, in the MCP extended position, the interossei can be effective abductors or adductors of the MCP without the loss of tension that would occur if the muscles were simultaneously producing MCP flexion. The interossei with proximal wing tendons are better as MCP joint deviators (abductors/adductors) since they act directly on the proximal phalanx. Interossei with distal wing tendons are less effective at the MCP joint since they must act on the MCP joint by continued action. Since all the dorsal interossei (MCP abductors) have proximal wing tendons and the volar interossei (MCP adductors) have only distal wing tendons, MCP abduction is stronger than MCP adduction. The dorsal interossei also have twice the muscle mass of the volar muscles. In a progressive ulnar nerve paralysis, the relatively ineffective MCP adduction component of interossei action is the first to show weakness.

Role at the MCP Joint in MCP Flexion. When the MCP joint is in flexion, the action line of the interossei lies more volar to the MCP joint axis than it does in MCP extension. The action line of the interossei is nearly perpendicular to the moving segment (proximal phalanx) in full MCP flexion (Fig. 9–15b). Consequently, the ability of the interossei muscles to create an MCP flexion torque increases with the amount of MCP flexion obtained. Once the MCP joint begins flexing, the ability of the interossei to produce MCP joint deviation (abduction or adduction) diminishes because the muscles are now producing two motions simultaneously. Loss of tension in the interossei through MCP joint deviation is prevented, however, because the collateral ligaments of the MCP become increasingly taut with MCP flexion. The tension in the collaterals gradually diminishes the range of deviation available; in full MCP flexion, MCP abduction and adduction are completely restricted by tight collateral ligaments, by the shape of the condyles of the metacarpal head, and by the inability of the interossei muscles to shorten enough to produce both flexion and deviation.

Role at the IP Joint in IP Extension. The ability of the interossei to produce IP joint extension is governed by the distal wing tendons of the muscles. To create tension in the extensor mechanism and extend the IP joints, the muscles must have tendons that attach to the central tendon or lateral bands. All the interossei (volar and dorsal) have distal wing tendons except the first dorsal and the abductor digiti minimi. In other words, all interossei have distal attachments except the two outside abductors.

When the MCP joint is extended, the action lines of the distal interossei are ineffective in producing MCP flexion but still capable of extending the IP joints. Since the distal interossei attach directly to the central tendon and lateral bands, their IP extension action is stronger than their MCP deviation action (which is, again, only performed by continued action). When the MCP joint is in flexion, the action lines of the distal interossei migrate volarly but are restricted in their volar excursion by the transverse metacarpal ligament. The transverse metacarpal ligament not only prevents the distal interossei from becoming slack through volar migration, but has a pulley effect on the distal interossei. That is, the IP extensor function of the distal interossei appears to be more effective in MCP flexion than in MCP extension.

The index and little fingers each have only one interosseous muscle with a distal wing tendon (first VI and fourth VI, respectively) since the first dorsal interosseous and the abductor digiti minimi have proximal wing tendons only. The first and fourth fingers, therefore, are weaker in IP extension than are the middle and ring fingers, which each have two interossei with distal attachments (second and third DI, and second VI and fourth DI, respectively).[47]

Overall, in approaching or holding the position of MCP flexion and IP extension, both proximal and distal interossei contribute to the MCP flexion torque. The proximal components are effective MCP flexors, while the distal components are maximally effective as both MCP flexors and IP extensors. The most consistent activity of all the interossei has been seen when the MCPs are being flexed and the IP joints are simultaneously extended.[39,48]

Although the interossei are not effective MCP flexors when the MCP joint is in extension, the interossei do appear to serve an important function in MCP extension (regardless of whether MCP abduction or adduction is required). When simultaneous activity of the ED and FDP (MCP extension with IP flexion) is required, there is a tendency for the fingers to claw; that is, for the proximal phalangeal segment to collapse into MCP hyperextension as the IPs flex.[39] Controlled positioning of the MCP during ED and FDP contraction requires a balancing MCP flexor force to prevent exaggerated action by the ED and collapse of the proximal segment. This force appears to be provided by the passive viscoelastic tension of the interossei. Although no electromyographic activity is recorded in the interossei during iso-

Table 9–1. SUMMARY OF INTEROSSEI MUSCLE ACTION

Muscle	Attachments	Action	
		MCP Extended	MCP Flexed
First Finger			
DI	Proximal only	MCP abduction	MCP flexion
VI	Distal only	IP extension and MCP adduction°	IP extension and MCP flexion°
Second Finger			
DI	Proximal and distal	MCP abduction and IP extension	MCP flexion and IP extension
DI	Proximal and distal	MCP abduction and IP extension	MCP flexion and IP extension
Third Finger			
DI	Proximal and distal	MCP abduction and IP extension	MCP flexion and IP extension
VI	Distal only	IP extension and MCP adduction°	IP extension and MCP flexion°
Fourth Finger			
DI	Proximal only	MCP abduction	MCP flexion
VI	Distal only	IP extension and MCP adduction°	IP extension and MCP flexion°

°Occurs indirectly by continued action.
Note: DI = dorsal interossei; VI = volar interossei; MCP = metacarpophalangeal; IP = interphalangeal.

lated **ED** or combined **ED-FDP** activity, the same extrinsic activity in a hand with long-standing ulnar nerve paralysis produces exaggerated MCP joint hyperextension. Clawing does not occur in the ulnar-nerve-deficient hand until the viscoelastic tension in the interossei has been lost through atrophy. Once such atrophy occurs, the predominance of **ED** tension even in relaxation is evidenced by the MCP hyperextension posture assumed by each finger of the hand at rest.[8,10,14,39] The overwhelming effect of the **ED** in the absence of the interossei is observable in an ulnar nerve-injured hand even in the index and middle fingers that have *all* other MCP flexors intact. The posture of MCP hyperextension of the fingers in the relaxed hand is the characteristic **claw hand deformity** associated with an ulnar nerve deficit.

A summary of actions of the interossei is presented in Table 9–1.

Lumbrical Muscles

The lumbricals are the only muscles in the body that attach to tendons of other muscles both proximally and distally. Each muscle arises from a tendon of the FDP in the palm, passes volar to the transverse metacarpal ligament, and attaches to the lateral band of the extensor mechanism on the radial side (Fig. 9–16a). Like the interossei, the lumbricals cross the MCP joint volarly and the IP joints dorsally. Differences in function in the two muscle groups can be attributed to the lumbricals' more distal insertion on the lateral band, to their profundus tendon origin, and to their great contractile range.

The insertion of the lumbricals on the distal lateral bands makes them consistently effective IP extensors, regardless of MCP position. In MCP extension, the transverse metacarpal ligament prevents the lumbrical muscle from migrating dorsally and losing tension as the IP joints extend. Studies have found the lumbricals to be more frequently active as IP extensors in the MCP extended position than are the interossei.[39,49] When the lumbrical contracts, it pulls not only on its distal insertion but also on its proximal origin. Because the origin of the lumbrical muscle is on a somewhat movable tendon, shortening of the lumbrical muscle not only pulls the lateral bands proximally to extend the IP joints but also pulls the profundus tendon distally. The distal movement of the FDP tendon releases much of the passive flexor force of the profundus at the MCP and IP joints while also applying an active IP extensor force (Fig. 9–16b). Ranney and Wells[46] confirmed this, finding that the IP joints did not extend until the tension within the lumbrical equaled the tension

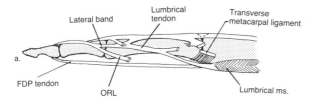

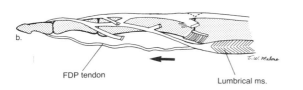

Figure 9–16. a. The lumbrical muscle arises from the tendon of the flexor digitorum profundis (FDP). The lumbrical lies volar to the transverse metacarpal ligament and passes dorsally to attach to the lateral band of the extensor mechanism. b. An active contraction of the lumbrical pulls on both the FDP tendon and the lateral band. The pull on the FDP tendon will move the tendon distally, releasing the passive tension of the tendon at the PIP and DIP.

within the profundus, which permitted the lumbrical to pull the profundus tendon distally. In MCP flexion, the large contractile range of the lumbrical muscles allows them to shorten sufficiently to produce both MCP flexion and IP extension simultaneously. The distal interossei can also exend the IP joints. However, they are less effective as IP extensors in the absence of the lumbricals, since the interossei do not have the same ability to release the passive resistance to IP extension presented by the FDP tendon.

The lumbricals have a greater MA for MCP flexion than the interossei, since the lumbricals lie volar to the interossei. Functionally, however, this component of lumbrical action is relatively weak.[39,47,49,50] This has been attributed to the small cross section of the lumbricals, as compared with the interossei, and to the moving origin of the lumbricals on the profundus. Although the lumbricals cross the MCP joint, a contraction of the lumbricals causes the FDP tendon and the origin of the lumbrical muscle to glide distally. While enhancing the function of IP extension, this same movement minimizes the active force of the lumbrical at the MCP joint.

Given the tendinous origin and insertion of the lumbrical muscles, it should be pointed out that, in the absence of stable attachments, tension in the connecting tendons is critical for lumbrical function. If passive tension were not present in the inactive profundus, an active lumbrical contraction would pull the FDP tendon so far distally that the muscle would become actively insufficient and there would be no effective pull of the lumbrical on the extensor expansion. Similarly, tension (active or passive) in the ED tendon and extensor expansion are necessary if the lumbrical is to fully extend both IP joints.[45]

In summary, the function of the lumbricals is simpler than that of the interossei. The lumbricals are strong extensors of the IP joints, regardless of MCP joint position; they are also relatively weak MCP flexors regardless of MCP joint position. The ability of the lumbricals to extend the IPs appears to be dependent only on intact tension in the extensor mechanism and in the FDP tendons.

Structure of the Thumb
CMC Joint of the Thumb

The CMC joint of the thumb is the articulation between the trapezium and the base of the first metacarpal. Unlike the CMC joints of the fingers, it is a saddle joint with 2° of freedom: flexion/extension and abduction/adduction (Fig. 9–17). The joint also permits some axial rotation, which takes place concurrently with the other motions. The net effect at this joint is a circumduction motion commonly termed **opposition-reposition**. Opposition permits the tip of the thumb to oppose the tips of the fingers. Reposition is the return from opposition.

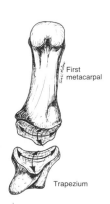

First metacarpal

Trapezium

Figure 9–17. The saddle-shaped portion of the trapezium is concave in the sagittal plane (abduction/adduction) and convex in the frontal plane (flexion/extension). The spherical portion found near the anterior radial tubercle is convex in all directions. The base of the first metacarpal has a reciprocal shape to that of the trapezium.

First CMC Joint Structure

Zancolli and associates[51] have proposed that the first CMC joint surfaces consist not only of the traditionally described saddle-shaped surfaces but also of a spherical portion located near the anterior radial tubercle of the trapezium. The saddle-shaped portion of the trapezium is concave in the sagittal plane (abduction/adduction) and convex in the frontal plane (flexion/extension). The spherical portion is convex in all directions. The base of the first metacarpal has a reciprocal shape to that of the trapezium (Fig. 9–17). Flexion/extension and abduction/adduction take place on the saddle-shaped surfaces, while the axial rotation of the metacarpal that accompanies opposition and reposition take place on the spherical surfaces.[51] Flexion and extension of the joint occur around a somewhat oblique AP axis, while abduction and adduction occur around an oblique coronal axis. The obliquity of the motions occurs because of the inclination of the trapezium. As a consequence, flexion/extension occur nearly parallel to the palm, with abduction and adduction occurring nearly perpendicular to the palm. Cooney and associates[52] have measured the first CMC ranges to average 53° of flexion/extension, 42° of abduction/adduction, and 17° of rotation.

The capsule of the CMC joint is relatively lax, but is reinforced by radial, ulnar, volar, and dorsal ligaments. There is also an intermetacarpal ligament that helps tether the bases of the first and second metacarpals, preventing extremes of radial and dorsal displacement of the base of the first metacarpal.[51,53] Although some investigators hold that the axial rotation seen in the metacarpal during opposition and reposition is a function of incongruence and joint laxity,[10,54] Zancolli and associates[51] theorize that it is a result of the congruence of the spherical surfaces and resultant tensions encountered in the supporting ligaments. The first CMC is close-packed both in extremes of abduction and adduction, with maximal motion available in neutral position.[31]

First CMC Joint Function

It is the unique range and direction of motion at the first CMC that produces opposition of the thumb. Opposition is, sequentially: abduction, flexion, and adduction of the first metacarpal, with simultaneous rotation. These movements change the orientation of the metacarpal, bring the thumb out of the palm, and position the thumb for contact with the fingers. The functional significance of the CMC joint of the thumb and of the movement of opposition can be appreciated when one realizes that use of the thumb against a finger occurs in almost all forms of prehension (handling activities). When the first CMC joint is fused in extension and adduction, opposition cannot occur. Over a period of time, an adaptation of opposition may develop at the trapezioscaphoid joint, which develops a more saddle-shaped configuration.[34] This amazing shift in joint function is an excellent example of the body's ability to replace essential functions whenever possible.

MCP and IP Joints of the Thumb

The MCP joint of the thumb is the articulation between the head of the first metacarpal and the base of its proximal phalanx. It is considered to be a condyloid joint with 2° of freedom: flexion/extension and abduction/adduction. There is an insignificant amount of passive rotation.[34] The main functional contribution of the first MCP joint is to provide additional range to the thumb pad in opposition and to allow the thumb to grasp and contour to objects. The motions at the first MCP joint are more restricted than those at the MCP joints of the fingers. There is only about 50 percent of the flexion available at the fingers, and little if any hyperextension. Abduction and adduction are extremely limited. The metacarpal head is not covered with cartilage dorsally or laterally, and more closely resembles the head of the

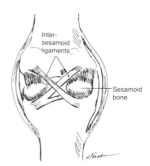

Figure 9–18. Metacarpophalangeal joint of the thumb. Unlike the MCP joints of the fingers, this joint is reinforced volarly by two sesamoid bones and the intersesamoid ligaments that secure them to the joint.

proximal phalanx, minus its central groove. The joint capsule, the reinforcing volar plate, and the collateral ligaments are similar to those of the other MCP joints. Unlike most of the other MCP joints, the first MCP joint is reinforced extracapsularly on its volar surface by two sesamoid bones (Fig. 9–18). These are maintained in position by fibers from the collaterals and by an intersesamoid ligament. Goldberg and Nathan[55] propose that the sesamoids are the result of friction and pressure on the tendons in which they are embedded. They support this by noting that the sesamoids of the first MCP do not appear until around 12 years of age and that sesamoids in some investigations have also been found in as many as 70 percent of fifth MCP joints and 50 percent of second MCPs.

The IP joint of the thumb is the articulation between the head of the proximal phalanx and the base of the distal phalanx. It is structurally and functionally identical to the distal IP joints of the fingers.

Thumb Musculature
Extrinsic Thumb Muscles

There are four extrinsic muscles, one located volarly and three dorsolaterally. The FPL inserts on the distal phalanx and is the correlate of the FDS. At the wrist, the FPL tendon is invested by the radial bursa, which is continuous with its digital tendon sheath. The EPB and APL run a common course from the dorsal forearm, crossing the wrist on its radial aspect to their insertion. The short extensor (EPB) inserts on the proximal phalanx; the long abductor (APL) inserts on the base of the metacarpal. Both muscles radially deviate the wrist slightly and abduct the CMC joint. The EPB also extends the MCP joint. The EPL originates with the previous two dorsal muscles but courses around the dorsal radial tubercle before turning toward the thumb and inserting on the distal phalanx. The EPL creates an extensor force at both the IP and MCP joints. It also extends and adducts the CMC joint of the thumb.

As is true for other extrinsic hand muscles, wrist positioning is an essential factor in providing optimal length-tension for the extrinsic muscles of the thumb. The FPL is ineffective as an IP flexor in wrist flexion. The EPL cannot complete IP extension when the wrist, CMC, and MCP are simultaneously extended. The abductor longus and extensor brevis require the synergy of an ulnar deviator of the wrist to prevent the muscles from creating wrist radial deviation, thus affecting their ability to generate tension over the joints of the thumb.

Intrinsic Thumb Muscles

There are four thenar or intrinsic thumb muscles that primarily take their origin from the carpal bones and the flexor retinaculum. The opponens pollicis (OP) is the only intrinsic thumb muscle to insert on the first metacarpal. Its action line is nearly perpendicular to the long axis of the metacarpal and is applied to the lateral side of the bone. The OP, therefore, is very effective in positioning the metacarpal in an abducted, flexed, and rotated posture. The abductor pollicis brevis (APB), flexor pollicis brevis (FPB), and adductor pollicis (ADP) insert on the proximal phalanx and name their action. The FPB has two heads of insertion. Its larger lateral head inserts with the APB and also applies some abductor force. The FPB crosses the sesamoid bones at the MCP, which increases its MA for flexion. The medial head of the flexor inserts with the adductor and assists in thumb adduction.

The thenar muscles are active in most grasping activities, regardless of the precise position of the thumb as it participates. The OP works together most frequently with the APB and the FPB, although the intensity of the relationship varies. When the thumb is gently brought into contact with any of the other fingers, activity of the OP predominates in the thumb and APB activity exceeds that of the FPB. When opposition to the index finger or middle finger is performed firmly, activity of the FPB exceeds that of the OP. With firm opposition to the ring and little fingers, however, the relationship changes; OP activity increases with firm opposition to the ring finger, equaling activity of the FPB with firm opposition to the little finger.[12] The change in balance of muscle activity with firm opposition and with increasingly ulnar opposition can be accounted for by the increased need for abduction and metacarpal rotation. Increased pressure in opposition additionally appears to bring in activity of the adductor pollicis. The ADP stabilizes the thumb against the opposed finger. In firm opposition to the index and middle fingers, ADP activity exceeds the very minimal activity of the APB. With more ulnarly located position, the increased need for abduction results in simultaneous activity of the abductor and adductor.[12]

Activity of the extrinsic thumb musculature in grasp appears to be partially a function of helping to position the MCP and IP joints. The main function of the extrinsics, however, is in returning the thumb to extension from its position in the palm. **Reposition** is essentially an extrinsic function but has been found to include some OP and abductor brevis activity.[12] This muscular activity would assist in maintaining the thumb in abduction and in maintaining metacarpal rotation, which facilitates the next move of the thumb back into opposition.

The joint structure and musculature of the thumb, the fingers, and the wrist complex have each been examined on an individual basis. Some instances of specific muscle activity have been presented to clarify the potential function of the muscle. A summary of wrist and hand function, however, can only be presented through the assessment of purposeful hand activity. Since the entire upper limb is geared toward execution of movement of the hand, it is appropriate to complete the description of the upper limb by looking at an overview of the wrist and hand in prehension activities.

PREHENSION

Prehension activities of the hand involve the grasping or taking hold of an object between any two surfaces in the hand; the thumb may or may not participate. The number of ways that objects of varying sizes and shapes may be grasped is nearly infinite; however, a broad classification system for grasp has evolved that will permit observations about the coordinated muscular function generally required to produce or maintain a position. Prehension can be categorized as either **power grip** or **precision handling.** Each of these two categories has subgroups that further define the grasp.

Power grip is a forceful act resulting in flexion at all finger joints. When the thumb is used, it acts as a stabilizer to the object held between the fingers and the palm. Precision handling, in contrast, is the skillful placement of an object between fingers, or finger and thumb.[56] The palm is not involved. Landsmeer[57] has suggested that power grip and precision handling can be differentiated on the basis of the dynamic and static phases involved. Power grip is the result of a sequence of (1) opening the hand, (2) positioning the fingers, (3) approaching the fingers to the object, and (4) maintaining a static phase that actually constitutes the grip. This is contrasted to precision handling, which shares the first three steps of the sequence but does not contain a static phase at all. In precision handling the fingers and thumb grasp the object with the intention of manipulating it within the hand; in power grip the object is grasped so that the object can be moved through space by the more proximal joints.

When assessing muscular function during each type of grasp, the synergy of the hand muscles results in almost constant activity of all intrinsic and extrinsic muscles. The task becomes more one of identifying when muscles are *not* working or when the balance of activity between muscles might change. It should also be emphasized that the muscular activity documented by electromyographic studies is very specific to the activity as performed in a given study. Even in studies using similar forms of prehension, variables such as size of object, firmness of grip, timing, and instructions to the subject can cause substantial changes in muscle activity reported. However, as indications of general muscular activity patterns, the studies are useful in the development of a conceptual framework within which hand function can be understood.

Power Grip

The fingers in power grip usually function in concert to clamp onto and hold an object into the palm. The fingers assume a position of sustained flexion that varies in degree with the size, shape, and weight of the object. The palm contours to the object as the palmar arches form around it. The thumb may serve as an addi-

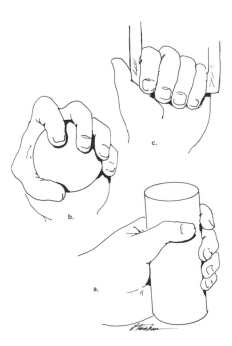

Figure 9–19. Three varieties of power grip: (a) Cylindrical, (b) spherical, and (c) hook.

tional surface to the finger-palm vise by adducting against the object, or it may be removed from the object. Three varieties of power grip studied by Long and associates[40] exemplify the similarities and differences seen in this mode of grasp. These are (1) cylindrical grip, (2) spherical grip, and (3) hook grip (Fig. 9–19). A fourth variety, lateral prehension, will also be considered with power grip.

Cylindrical Grip

Cylindrical grip (Fig. 9–19a) almost exclusively uses flexors to carry the fingers around and maintain grasp on an object. The function in the fingers is performed largely by the FDP, especially in the dynamic closing action. In the static phase the FDS assists when the intensity of the grip requires greater force. Although power grip traditionally has been thought of as an extrinsic activity, recent studies have indicated considerable interosseous muscle activity. The interossei are considered to be functioning as MCP flexors and abductors/adductors. In strong grip the magnitude of force of the interossei in metacarpal flexion has been found to nearly equal that of the extrinsic flexors.[50,56] Because the IP joints are being flexed, the MCP flexion task most likely falls to the proximal (dorsal) interossei. The so-called rotation task of the interossei is to position the MCP joint in the proper abduction or adduction for application of the pad of the finger to the object. The combination of MCP flexion and deviation (abduction/adduction) gives the appearance of rotation of the phalanx. Although the location of the lumbricals indicates a possible contribution to MCP flexion, their lack of electromyographic activity, regardless of strength grip, is consistent with their role as IP extensors.[40]

In cylindrical grip, there is frequently a symmetric ulnar deviation of the second through fifth MCP joints that results from the increased range of flexion available to the more ulnar finger joints. This deviation points the fingers toward the thumb but also tends to produce lateral subluxation forces on the MCP joints and on the tendons of the long flexors at the MCP. The subluxing forces are ordinarily counteracted by the radial collateral ligaments and by the annular pulleys that anchor the long tendons in place. These structures are assisted, especially in isometric functions, by the active or passive tension in the ED. The ED contraction increases joint compression and enhances joint stability.[56] The ED activity sometimes observed in grasping functions may be due to the increased stability needs.

Thumb position in cylindrical grip is the most variable of the digits. The thumb usually comes around the object, then flexes and adducts to close the vise. The FPL and the thenar muscles are all active. As noted previously, the activity of the thenar muscles will vary with the width of the web space, with the CMC rotation required, and with increased pressure or resistance. A distinguishing characteristic of power grip over precision handling is the magnitude of activity of the adductor pollicis. The EPL may be variably active as an MCP stabilizer or as an adductor.

Muscles of the hypothenar eminence (abductor digiti minimi, opponens digiti minimi, and flexor digiti minimi) usually are active in cylindrical grip. The abductor digiti minimi functions as a proximal interosseous muscle to flex and abduct (ulnarly deviate) the fifth MCP joint. The opponens digiti minimi and the flexor digiti minimi are more variable but frequently reflect the amount of abduction and rotation of the first metacarpal. In fact, increased activity of the opponens pollicis automatically results in increased activity of the opponens digiti minimi and flexor digiti minimi.

Cylindrical grip may be accompanied by ulnar deviation of the wrist, since this is the position that optimizes force of the long finger flexors. The least flexion force is generated at these joints in wrist flexion.[2] The heavier an object is, the more likely it is that the wrist will ulnarly deviate. Additionally, a strong contraction of the flexor carpi ulnaris at the wrist will increase tension on the flexor retinaculum. This provides a more stable base for the active hypothenar muscles that originate from that ligament. It is interesting to note that regardless of wrist position, the percent of total IP flexor force allocated to each finger is relatively constant. The ring and little fingers can generate only 70 percent of the flexor force of the index and middle fingers.[2] The ring and little fingers seem to serve as weaker but more mobile assists to the more stable and stronger index and middle fingers.

Spherical Grip

Spherical grip (Fig. 9–19b) is similar in most respects to cylindrical grip. The extrinsic finger and thumb flexors and the thenar muscles follow similar patterns of activity and variability. The main distinction can be made by the greater spread of the fingers to encompass the object. This evokes more interosseous activity than is seen in other forms of power grip.[40] The MCP joints do not deviate in the same direction but tend to abduct. The phalanges are no longer parallel to each other as they commonly are in cylindrical grip. The MCP abductors must be joined by the adductors to stabilize the joints that are in the loose-packed position of semiflexion. Although flexor activity predominates, as it does in all forms of power grip, the extensors do have a role. The extensors not only provide a balancing force for the flexors but are also essential for smooth and controlled opening of the hand and release of the object. Opening, approach, and release are primarily an extensor function, calling in the lumbricals, the ED, and thumb extrinsics.

Hook Grip

Hook grip (Fig. 9–19c) is actually a specialized form of prehension. It is included in power grip because it has more characteristics of power grip than of precision handling. It is a function primarily of the fingers. It may include the palm but never includes the thumb. It can be sustained for prolonged periods of time as anyone who has carried a briefcase or books at his or her side or hung onto a commuter strap on a bus or train can attest. The major muscular activity is provided by the FDP and FDS. The load may be sustained completely by one muscle or the other, or by both muscles in concert. This is dependent on the position of the load relative to the phalanges. If the load is carried more distally so that DIP flexion is mandatory, the FDP must participate. If the load is carried more in the middle of the fingers, the FDS may be sufficient. Some interosseus muscle activity has been demonstrated on electromyography, but its purpose is not fully understood. It may help prevent clawing at the MCP joints, although the activity is not evident in every finger.[40] In hook grip the thumb is held in moderate to full extension by thumb extrinsics.

Lateral Prehension

Lateral prehension is a rather unique form of grasp. Contact occurs between two adjacent fingers. The MCP and IP joints are usually maintained in extension, while the MCP joints simultaneously abduct or adduct. This is the only form of prehension in which the extensor musculature plays a part in the maintenance of the posture; the ED and the lumbricals are active to extend the phalanges. MCP abduction and adduction are performed by the interossei. Lateral prehension is included here as a form of power grip because it involves the static holding of an object that is then moved by the more proximal joints of the upper extremity. Although not a "powerful" grip, neither is lateral prehension used to manipulate objects in the hand. It is generally typified by the holding of a cigarette.

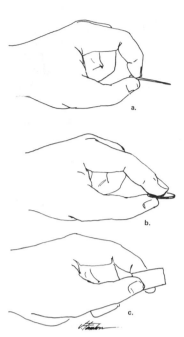

Figure 9–20. Three varieties of precision handling: (a) tip-to-tip prehension, (b) pad-to-pad prehension, and (c) pad-to-side prehension.

Precision Handling

The positions and muscular requirements of precision handling are somewhat more variable than those of power grip and require much finer motor control and intact sensation. The thumb is one jaw of what has been termed a two-jaw chuck; the thumb is generally abducted and rotated from the palm. The second and opposing jaw is formed by either the distal tip, the pad, or the side of a finger. When two fingers oppose the thumb, it is called a three-jaw chuck. The three varieties of precision handling that exemplify this mode are (1) pad-to-pad prehension, (2) tip-to-tip prehension, and (3) pad-to-side prehension (Fig. 9–20). Each tends to be a dynamic function with relatively little static holding.

Pad-to-Pad Prehension

Pad-to-pad prehension (Fig. 9–20b) involves opposition of the pad, or pulp, of the thumb to the pad, or pulp, of the finger. It is in the pad of the distal phalanx of each digit that the greatest concentration of tactile corpuscles are found; 80 percent of precision handling uses this mode of prehension.[1] The finger used in two-jaw chuck is usually the index; in three-jaw chuck the middle finger is added. The MCP and PIP joints of the fingers are partially flexed, with the degree of flexion being dependent on the size of the object being held. The DIP joint may be fully extended or in slight flexion. When DIP is extended, the FDS alone performs the function. When partial DIP flexion is required, the FDP must be activated. Interosseous activity is present both in supplementing MCP flexor force and in providing the MCP abduction or adduction required in object manipulation. In dynamic manipulation the volar and dorsal interossei tend to work reciprocally, rather than in the synergistic co-contraction pattern observed during power grip. In a firmly maintained pad-to-pad pinch, the muscles may again contract.[40]

The thumb in pad-to-pad prehension is held in CMC flexion, abduction, and rotation. The first MCP and IP joints are partially flexed to fully extended. The thenar muscle control is provided by the OP, FPB, and APB, each of which is innervated by the median nerve. Adductor pollicis activity (ulnar nerve) increases with increased pressure of pinch. In ulnar nerve paralysis, loss of adductor pollicis function renders the thumb less stable.

Fine adjustments in the angulation of the DIP joint of the finger and the IP joint of the thumb provide the control for the points of contact on the pads of the digits. In full finger DIP and thumb IP extension, contact occurs on the proximal portion of the distal phalanx. As flexion of the finger DIP and thumb IP joint increases, the contact moves distally toward the nails. The flexion, when required, is provided by the FDP for the finger and by the FPL for the thumb. DIP flexion in the finger is accompanied by a proportional flexion in the PIP joint.

As is found in power grip, the extensor musculature is used for opening the hand to grasp, for release, and for stabilization when necessary. In the thumb, the EPL may be used to maintain the IP joint in extension, when contact is light and on the proximal pad.

Tip-to-Tip Prehension

Although the muscular activity found in tip-to-tip prehension (Fig. 9–20a) is nearly identical to that of pad-to-pad prehension,[40] there is a significant difference. In tip-to-tip prehension the IP joints of the finger and thumb must have the range and available force to create nearly full joint flexion. The MCP joint of the finger must also be ulnarly deviated (or pointed radially) to present the tip to the thumb. In the first finger, the ulnar deviation occurs as MCP adduction. In the remaining fingers, MCP abduction produces ulnar deviation. If the flexion range for the phalanx is not available, or if the active force for IP flexion and MCP ulnar deviation (abduction or adduction) cannot be provided, tip-to-tip prehension cannot be performed. As the most precise form of grasp, it is also the most easily disturbed. It is dependent on the activity of the FDP, the FPL, and the interossei, whereas, pad-to-pad prehension requires only one of these muscles, the interossei.

Pad-to-Side Prehension

Pad-to-side prehension (Fig. 9–20c) differs from the other forms of precision handling only in that the thumb is more adducted and less rotated. The activity level of the FPB increases and that of the OP decreases, as compared with tip-to-tip prehension. Activity of the adductor pollicis also increases over that seen in either tip-to-tip or pad-to-pad prehension.[4] Slight flexion of the distal phalanx of the thumb is required.

Pad-to-side prehension is the least precise of the forms of precision handling; it can actually be performed by a person with paralysis of all hand muscles. The flexor force needed at the MCP and IP joints can be provided by the passive tension created in the extrinsic finger and thumb flexors as they are stretched over an extending wrist. This phenomenon of hand closure during wrist extension, known as tenodesis, accounts for hand function in many persons with quadriplegia. Active control of a wrist extensor muscle is necessary for functional use of tenodesis.

Pad-to-side prehension is the finest grasp that can be accomplished without active hand musculature. The same tenodesis action will produce a cylindrical power grip, if the object can be placed appropriately in the palm. Release of grasp is effected by relaxing the wrist extensors and allowing gravity to flex the wrist. As the wrist flexes, the extrinsic flexors become slack, while the ED and EPL become stretched. The passive tension in the extensors in a dropped wrist is adequate to partially extend both MCP and IP joints.

FUNCTIONAL POSITION OF THE WRIST AND HAND

Although it is difficult to isolate any one joint or function as being singularly important among all those examined, grasp would have to take precedence. There can be little doubt that the hand cannot function either as a manipulator or as a sensory organ unless an object can enter the palmar surface and unless moderate finger flexion and thumb opposition are available to allow sustained contact. Application of either an active muscular or passive tendinous flexor force to the digits requires the wrist to be stabilized in moderate extension and ulnar deviation. Delineation of the so-called functional position of the wrist and hand takes into account these needs and is the position from which optimal function is most likely to occur. It is *not necessarily* the position in which a hand should be immobilized. Position for immobilization depends on the disability.

The functional position is (1) wrist complex in slight extension (20°) and slight ulnar deviation (10°) and (2) fingers moderately flexed at the MCP joint (45°), slightly flexed at the PIP joint (30°), and slightly flexed at the DIP joint[1]. The wrist position optimizes the power of the finger flexors so that hand closure can be accomplished with the least possible effort. It is also the position in which all wrist muscles are under equal tension. With similar considerations for the position of the joints of the digits, the functional position provides the best opportunity for the disabled hand to interact with the brain that controls it.

REFERENCES

1. Harty, M: The hand of man. Phys Ther 51:777, 1974.
2. Hazelton, FT, et al: The influence of wrist position on the force produced by the finger flexors. Biomechanics 8:301, 1975.
3. Simpson, DC: The functioning hand, the human advantage. J R Coll Surg Edinb 21:329, 1976.
4. Gilford, VW, Bolton, RH, and Lambrinudi, C: The mechanism of the wrist joint. Guy's Hosp Rep 92:52, 1943.
5. Linscheid, RL: Kinematic considerations of the wrist. Clin Orthop 202:27–39, 1986.
6. Mohiuddin, A and Janjua, MZ: Form and function of the radioulnar disc. Hand 14:61–62, 1982.
7. Palmer, AK and Werner, FW: Biomechanics of the distal radioulnar joint. Clin Orthop 187:26–35, 1984.
8. Cailliet, R: Hand Pain and Impairment, ed 3. FA Davis, Philadelphia, 1982.
9. Taleisnik, J: Current concepts review: Carpal instability. J Bone Joint Surg 70A:1262–1268, 1988.
10. Kapandji, IA: The Physiology of the Joints: Vol. I, ed 2. E & S Livingstone, Edinburgh and London, 1970.
11. Sarafian, SK, Melamed, J, and Goshgarian, FM: Study of wrist motion in flexion and extension. Clin Orthop 126:153, 1977.
12. MacConaill, MS and Basmajian, JV: Muscles and Movements: A Basis for Human Kinesiology. Williams & Wilkins, Baltimore, 1969.
13. Youm, Y, et al: Kinematics of the wrist. I: An experimental study of radial-ulnar deviation and flexion-extension. J Bone Joint Surg 6A:423, 1978.
14. Steindler, A: Kinesiology of the Human Body under Normal and Pathological Conditions. Charles C Thomas, Springfield, Ill, 1955.
15. Mayfield, JK, Johnson, RP, and Kilcoyne, RF: The ligaments of the human wrist and their functional significance. Anat Rec 186:417, 1976.
16. Mayfield, JK: Wrist ligamentous anatomy and pathogenesis of carpal instability. Orthop Clin North Am 15:209–216, 1984.
17. Drewniany, JJ, Palmer, AK, and Flatt, AE: The scaphotapezial ligament complex: An anatomic and biomechanical study. J Hand Surg 10A:492–498, 1985.
18. Kauer, JM: The interdependence of the carpal articulation chains. Acta Ant 88:481, 1976.
19. Conwell, HE: Injuries to the Wrist. CIBA Pharmaceutical, Summit, NJ, 1970.
20. MacConaill, MD: The mechanical anatomy of the carpus and its bearing on some surgical problems. J Anat 75:166, 1941.
21. Wright, RD: A detailed study of movement of the wrist joint. J Anat 70:137, 1935.

22. Brumfield, RH and Champoux, JA: A biomechanical study of normal functional wrist motion. Clin Orthop 187:23, 1984.
23. Kauer, JM: The mechanism of the carpal joint. Clin Orthop 202:16–26, 1986.
24. Fisk, GR: Carpal instability and the fractured scaphoid. Ann R Coll Surg (Engl)46:63, 1970.
25. Backdahl, M and Carlsoo, S: Distribution of activity in muscles acting on the wrist. Acta Morph Neer Scand 4:136, 1961.
26. Ketchum, LD, et al: The determination of moments for extension of the wrist generated by muscles of the forearm. J Hand Surg 3:105, 1978.
27. Radonjic, F and Long, C: Kinesiology of the wrist. Am J Phys Med 50:57, 1971.
28. Perry, J: Normal upper extremity kinesiology. Phys Ther 58:265, 1978.
29. Boivin, JH, et al: Electromyographic kinesiology of the hand: Muscles driving the index finger. Arch Phys Med Rehabil 50:17, 1969.
30. Volz, RG, Lieb, M, and Benjamin, J: Biomechanics of the wrist. Clin Orthop 149:112–117, 1980.
31. Batmanabane, M and Malathi, S: Movements at the carpometacarpal and metacarpophalangeal joints of the hand and their effect on the dimensions of the articular ends of the metacarpal bones. Anat Rec 213:1002–1010, 1985.
32. Joseph, RB, et al: Chronic sprains of the carpometacarpal joints. J Hand Surg 6:172–180, 1981.
33. Garcia-Elias, M, et al: Stability of the transverse carpal arch: An experimental study. J Hand Surg 14A:277–282, 1989.
34. Kaplan, EB: The participation of the metacarpophalangeal joint of the thumb in the act of opposition. Bull Hosp Joint Dis 27:39, 1966.
35. Shultz, RJ, Storace, A, and Krishnamurthy, S: Metacarpophalangeal joint motion and the role of the collateral ligaments. Intern Orthop (SICOT) 11:149–155, 1987.
36. Fisher, DM, et al: Descriptive anatomy of fibrocartilaginous menisci in the finger joints of the hand. J Orthop Res 3:484–491, 1985.
37. Cailliet, R: Scoliosis: Diagnosis and Management. FA Davis, Philadelphia, 1975.
38. Manske, PR and Lesker, PA: Palmar aponeurosis pulley. J Hand Surg 8:259–263, 1983.
39. Long, C: Intrinsic-extrinsic muscle control of the fingers. J Bone Joint Surg 50A:973, 1968.
40. Long, C, et al: Intrinsic-extrinsic muscle control of the hand in power grip and precision handling. J Bone Joint Surg 52A:853, 1970.
41. Baker, DS, et al: The little finger superficialis - Clinical investigation of its anatomic and functional shortcomings. J Hand Surg 6:374–378, 1981.
42. Salisbury, CR: The interosseous muscles of the hand. J Anat 71:395, 1936.
43. Brand, PW: Paralytic claw hand. J Bone Joint Surg 40B:618, 1958.
44. Landsmeer, JM: The anatomy of the dorsal aponeurosis of the human fingers and its functional significance. Anat Rec 104:31, 1949.
45. Stack, HG: Muscle function in the fingers. J Bone Joint Surg 44B:899, 1962.
46. Ranney, D and Wells, R: Lumbrical muscle function as revealed by a new and physiological approach. Anat Rec 222:110–114, 1988.
47. Eyler, DL and Markee, JE: The anatomy and function of the intrinsic musculature of the fingers. J Bone Joint Surg 36A:1, 1954.
48. Close, JR and Kidd, CC: The functions of the muscles of the thumb, the index and the long fingers. J Bone Joint Surg 51A:1601, 1969.
49. Backhouse, KM and Catton, WT: An experimental study of the functions of the lumbrical muscles in the human hand. J Anat 88:133, 1954.
50. Ketchum, LD, et al: A clinical study of the forces generated by the intrinsic muscles of the index finger and extrinsic flexor and extensor muscles of the hand. J Hand Surg 3:571, 1978.
51. Zancolli, EA, Ziadenberg, C, and Zancolli, E: Biomechanics of the trapeziometacarpal joint. Clin Orthop 220:14–26, 1987.
52. Cooney, WP, et al: The kinesiology of the thumb trapeziometacarpal joint. J Bone Joint Surg 63A:1371–1380, 1981.
53. Pagalidis, T, Kuczynski, K, and Lamb, DW: Ligamentous stability of the base of the thumb. Hand 13:29–35, 1981.
54. Kauer, JMG: Functional anatomy of the carpometacarpal joint of the thumb. Clin Orthop 220:7–13, 1987.
55. Goldberg, I and Nathan, H: Anatomy and pathology of the sesamoid bones. Int Orthop (SICOT) 11:141–147, 1987.
56. Chao, EY, Opgrande, JD, and Axmeare, FE: Three-dimensional force analysis of the finger joints in selected isometric hand functions. J Biomech 9:387, 1976.
57. Landsmeer, JM: Power grip and precision handling. Ann Rheum Dis 22:164, 1962.

STUDY QUESTIONS

1. Name the bones of the wrist complex and describe the compound joints that are formed by these bones.
2. What role does the radioulnar disk play in wrist joint function?

3. What is the total ROM normally available at the wrist complex? How are the motions distributed between the compound joints of the complex?

4. Describe the sequence of joint activity occurring from full wrist flexion to full extension, emphasizing the role of the scaphoid.

5. Compare the ECRL to the ECRB; include the joints crossed, actions produced, and activity levels of each.

6. What is the function of the CMC joints of the fingers? How do the variations in ROM among the four CMC joints of the fingers contribute to function?

7. What role does the transverse metacarpal ligament play at the CMC joint? What role at the MCP joint?

8. Describe the location(s) and function(s) of the volar (or palmar) fibrocartilage plates.

9. What MCP joint position is most prone to injury and why?

10. Compare the joint structure of the MCP joints with that of the IP joints of the fingers. Identify both similarities and differences.

11. Describe the mechanisms, joint motions, and muscles that are necessary for the fingers to gently close into the palm without friction or loss of length-tension.

12. How does the "pistol-grip" design of most tools (larger ulnarly) relate to the MCP ROM and muscular function of the four fingers?

13. When is the FDS active as the primary finger flexor? When does it back up the FDP?

14. What pure wrist position optimizes finger flexion strength? Which wrist position is least effective for grasp?

15. What is an annular pulley, where is it found, and what function(s) does it serve?

16. Identify the bursae of the hand. What are their functions and how are they related to the digital tendon sheaths?

17. What role does the EDC play in active extension of the PIP and DIP joints of the hand?

18. Describe the attachments of the interossei and lumbricals to the extensor mechanism. How do these muscles contribute to IP extension?

19. Why is active DIP flexion normally accompanied by PIP flexion at the same time?

20. Explain why the DIP cannot be actively extended if the PIP is fully flexed.

21. Why will an isolated contraction of the EDC produce flexion of the PIP and DIP joints?

22. What are the primary functions of the "proximal" interossei? How does "distal" interossei function differ?

23. Why is finger extension weaker in the first and fourth fingers?

24. Why does MCP adduction weaken more quickly than abduction in ulnar nerve problems?

25. Which are stronger flexors of the MCP joint, the lumbricals or the interossei?

26. Compare the MCP joint structure of the thumb with the MCP joint structure of the fingers.

27. What does the motion of thumb opposition require in terms of joint function and musculature?

28. What is the primary muscle(s) of reposition?

29. In general, what is the difference between power grip and precision handling? What do they have in common?

30. Cylindrical grip is generally referred to as an extrinsic hand function. Why is this true?

31. What requirement does spherical grip have that differentiates it from cylindrical?

32. Which form of prehension requires only intrinsic musculature?

33. Which forms of prehension do not require the thumb?

34. Why are the interossei needed in precision handling?

35. What requirements does tip-to-tip prehension have that are not necessary for pad-to-pad?

36. What is the finest (most precise) form of prehension that can be accomplished by someone without intact intrinsic or extrinsic hand function?

37. What is the functional position of the wrist and hand? Why is this the optimal resting position when there is no specific hand problem?

38. Why is an ulnar nerve injury called "claw hand"? What deficiency causes the clawing, and in which fingers does it occur?

Chapter 10

■ ■ ■

The Hip Complex

■

OBJECTIVES

Following the study of this chapter, the reader should be able to:

Describe

1. The articulating surfaces of the pelvis and femur.
2. The structure and function of the trabecular systems of the pelvis and femur.
3. The structure and function of the ligaments of the hip joint.
4. The angle of inclination and angle of torsion.
5. The planes and axes of the following pelvic motions and the accompanying motions at the lumbar spine and hip joints: pelvic rotation and anterior, posterior, and lateral tilting of the pelvis.
6. The muscle activity that produces tilting and rotation of the pelvis.

 7. Motions of the femur on the pelvis including planes and axes of motion.
 8. The structure and function of all the muscles associated with the hip joints.
 9. The forces that act on the head of the femur.
10. The position of greatest stability at the hip.
11. The position of greatest articular contact.

Explain
 1. How sagittal and frontal plane equilibrium are maintained in erect bilateral stance.
 2. How frontal plane equilibrium is achieved in unilateral stance.
 3. Three ways to reduce forces acting on the femoral head.
 4. How the function of the two-joint muscles at the hip are affected by changes in the position of the knee and hip.
 5. The functional and structural relationship among the hip, knee, pelvis, and lumbar spine.

Compare
 1. The forces acting on the femoral head in erect bilateral stance with the forces acting on the head in erect unilateral stance.
 2. Coxa valga with coxa vara on the basis of hip joint stability and mobility.
 3. Anteversion with retroversion on the basis of hip stability and mobility.
 4. The motions that occur at the hip, pelvis, and lumbar spine during forward trunk bending with the motions that occur during anterior and posterior tilting of the pelvis in the erect standing position.
 5. The structure and function of the following muscles: flexors and extensors, abductors and adductors, lateral and medial rotators.
 6. The forces across the hip joint produced by using a cane on the same side as hip pain and using a cane on the side opposite hip pain.

UNDERSTANDING THE HIP JOINT

The hip joint, or coxofemoral joint, is formed by the union of the acetabulum of the pelvis and the head of the femur. These two segments form a diarthrodial ball-and-socket joint with 3° of freedom: flexion/extension in the sagittal plane, abduction/adduction in the frontal plane, and medial-lateral rotation in the transverse plane. Although it is tempting to draw an analogy between the hip joint and the shoulder complex, the functional and structural adaptations of each to their respective roles has been so extensive that such comparisons are more of general interest than of functional relevance. The role of the shoulder complex is to provide a wide range of mobility for the hand on a reasonably stable base. The primary function of the hip joint is to support the weight of the head, arms, and trunk (HAT) both in static erect posture and in dynamic postures such as ambulation, running, and stair climbing. The hip joints also provide a pathway for the transmission of forces between the pelvis and the lower extremities.

The hip joint needs to be understood largely in the context of closed kinematic chain function, unlike the shoulder complex, which is designed to operate predominantly in an open chain. Critical issues in both function and dysfunction of the hip joint have to do with weight-bearing stresses and with the interdependence of the hip joints with both proximal and distal segments.

The fact that the hip tends to operate in a closed kinematic chain is not immediately obvious. Although it is clear that the foot (distal end of the chain) is frequently fixed by weight bearing, both the proximal and the distal ends of the chain must be fixed for a closed chain to exist. The proximal end of the chain is, in fact, the head. The head is certainly free to move in space, but most commonly remains

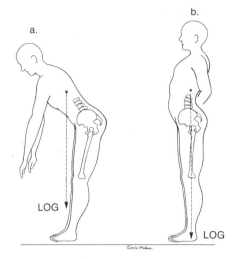

Figure 10–1. a. In an open-chain response to hip flexion, the trunk will be inclined forward. The line of gravity (LOG) must be kept in the base of support by shifting the pelvis slightly posteriorly. b. In a closed-chain response to hip flexion, the lumbar spine is hyperextended to bring the head and trunk back over the center of the base of support.

upright and vertically oriented due to the influence of the tonic labyrinthine and optical righting reflexes that are evident almost immediately at birth[1] and continue to operate throughout life. These reflexes and the learned responses that develop over time result in the head effectively *behaving as if it is fixed in a vertical position,* while it also maintains itself *over the base of support.* When the head is held upright and over the base of support, all the segments between the head and the weight-bearing surface of the feet are part of a closed chain. Compensatory responses of the interposed segments will ensure that the head remains over the base of support and that the entire body does not become unstable. The closed-chain response can be voluntarily overridden, however, since the head is not truly fixed.

A common example of open-chain versus closed-chain function is seen when the hip flexor musculature is tight and the hip joint is maintained in flexion. An open-chain response would result in displacement of the head from vertical and would require some compensatory adjustments to prevent displacement of the line of gravity (LOG) of the body from the base of support (Fig. 10–1a). More commonly, sustained hip flexion in stance is accompanied by compensatory movements of the vertebral column that maintain the head in the upright position and keep the LOG well within the base of support (Fig. 10–1b).

Neither structure nor function of the hip joint can be examined without consideration of the weight-bearing nature of the joint. The weight-bearing function of the hip not only influences the stresses across the joint, but it results in a predominance of closed-chain responses of the joint and its interdependence with the other joints of the lower extremity and spine. The interdependence of the hip joint with the other joints of the lower extremity will continue to be examined as additional joints are presented in subsequent chapters and in the in-depth examination of the composite functions of posture and gait.

STRUCTURE OF THE HIP JOINT

Proximal Articular Surface

The cuplike concave socket of the hip joint is called the **acetabulum** and is located on the lateral aspect of the pelvic bone (innominate or os coxa). There are three bones that form the pelvis: the ilium, the ischium, and the pubis. Each of the

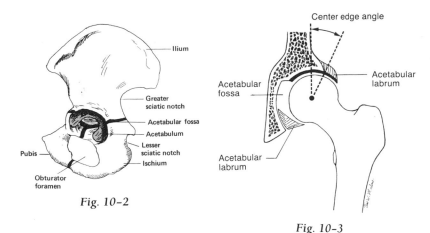

Fig. 10-2

Fig. 10-3

Figure 10-2. The acetabulum is formed by the union of the three bones of the pelvis, with only the upper horseshoe-shaped area being articular.

Figure 10-3. The center edge angle of the acetabulum is formed between a vertical line through the center of the femoral head and a line connecting the center of the femoral head and the bony edge of the acetabulum. The acetabular labrum deepens the acetabulum.

three bones contributes to the structure of the acetabulum (Fig. 10–2). The pubis forms one fifth of the acetabulum, the ischium two fifths, and the ilium the remainder. Until full ossification of the pelvis occurs between 15 and 25 years of age, the separate segments of the acetabulum may remain visible.[2] The acetabulum appears to be a hemisphere, but only its upper margin has a true circular contour,[3] and the roundness of the acetabulum as a whole decreases with age.[4] In actuality, only a horseshoe-shaped portion of the periphery of the acetabulum is covered with hyaline cartilage and articulates with the head of the femur (Fig. 10–2). The inferior aspect of the cartilage-lined portion of the acetabulum (the base of the horseshoe) is interrupted by a deep notch called the **acetabular notch.** The central or deepest portion of the acetabulum, called the **acetabular fossa,** is nonarticular and contains fibroelastic fat covered with synovial membrane. The femoral head does not contact this area (Fig. 10–3).

Center Edge Angle of the Acetabulum

The acetabulum is oriented on the pelvis to face laterally, somewhat inferiorly, and somewhat anteriorly. A line connecting the lateral rim of the acetabulum and the center of the femoral head forms an angle with the vertical known as the **center edge (CE) angle** or the **angle of Wiberg** (Fig. 10–3). The CE angle is, in essence, the amount of inferior tilt of the acetabulum. Using computed tomography, Adna and associates[5] found CE angles in adults to average 38° in men and 35° in women (with ranges in both sexes to be about 22 to 42°). These values are in reasonable agreement with those ascertained by Brinckmann and associates[6] using roentgenograms, although the average values for men and women were the same in their larger sample. The similarity of the CE angle between men and women is somewhat surprising given the increased diameter and more vertical orientation of the sides of the female pelvis.[2] A smaller CE angle (or more vertical orientation) of the acetabulum may result in diminished coverage of the head of the femur and an increased risk of superior dislocation of the head of the femur. Since there is evidence that the CE angle increases with age,[7] the implication is that children have less coverage over the head of the femur and therefore decreased joint stability as compared to adults. In fact, congenital dislocation is more common at the hip joint than at any other joint in the body. This may be due to deficiencies in the superior acetabulum (a diminished CE angle).[2]

Acetabular Anteversion

The magnitude of anterior orientation of the acetabulum may be referred to as the **angle of acetabular anteversion.** Adna and associates[5] found the average value to be 18.5° for men and 21.5° for women,[5] although Kapandji[8] cites larger values of 30 to 40°.[8] Pathologic increases in the angle of acetabular anteversion are associated with decreased joint stability and increased tendency for anterior dislocation of the head of the femur.

Acetabular Labrum

Given the need for stability at the hip joint, it is not surprising to find an accessory joint structure. The entire periphery of the acetabulum is rimmed by a ring of wedge-shaped fibrocartilage called the **acetabular labrum** (see labrum cross section in Fig. 10–3). The acetabular labrum not only deepens the socket but increases the concavity of the acetabulum through its triangular shape and grasps the head of the femur to maintain contact with the acetabulum. The transverse acetabular ligament is part of the labrum and spans the articular gap at the base of the articular horseshoe. The flat fibers of the transverse acetabular ligament cross the acetabular notch, forming the roof of a tunnel through which blood vessels and nerves can enter the hip joint. Although the ligament is considered to be part of the acetabular labrum, unlike the labrum it contains no cartilage cells.[2]

Distal Articular Surface

The head of the femur is a fairly rounded hyaline cartilage-covered surface that may be slightly larger than a true hemisphere or as much as two thirds of a sphere depending on body type.[8] The head of the femur is considered to be circular, unlike the more irregularly shaped acetabulum.[4] The radius of curvature of the femoral head is smaller in women than in men when compared to the dimensions of the pelvis.[6] Just inferior to the most medial point on the femoral head is a small roughened pit called the **fovea** or **fovea capitis** (Fig. 10–4). The fovea is not covered with articular cartilage and is the point at which the ligament of the head of the femur is attached.

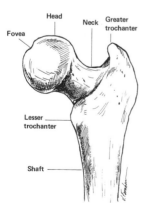

Figure 10–4. An anterior view of the proximal portion of the left femur shows the relationship between the head, neck, and femoral shaft.

The femoral head is attached to the femoral neck; the femoral neck is attached to the shaft of the femur between the greater trochanter and the lesser trochanter (Fig. 10–4). Although the femoral neck is more distinct than the neck of the humerus, it is generally only about 5 cm long.[2] The femoral neck angulates the head in such a way that the head most commonly faces medially, superiorly, and anteriorly. Although the angulation of the femoral head is more consistent across the population than that of the humerus, there are still substantial individual differences and differences from side to side in the same individual.

Angulation of the Femur

There are two angulations made by the head and neck of the femur on the shaft. One angulation (the angle of inclination) occurs in the frontal plane between the axis of the femoral neck and the axis of the femoral shaft. The other angulation (the angle of torsion) occurs in the transverse plane between the axis of the femoral neck and the axis of the femoral condyles. The origin and variability of these angulations can be understood in the context of the embryonic development of the lower limb. In the early stages of fetal development, both upper extremity and lower extremity limb buds project laterally from the body as if in full abduction. During the 7th and 8th weeks of gestational age and prior to full definition of the joints, adduction of the buds begins. At the end of the 8th week, the "fetal position" has been achieved but the upper and lower limbs are no longer positioned similarly. Although the upper limb buds have rotated somewhat laterally (bringing the ventral surface of the limb bud anteriorly), the lower limb buds have rotated medially so that the ventral surface faces posteriorly.[2] The result for the lower limb is critical to function. The knee flexes in the opposite direction from the elbow and the extensor surface of the limb is anteriorly rather than posteriorly located. The head and neck of the femur retain the original position of the limb bud, while the shaft is adducted and medially rotated with respect to the head and neck. The magnitude of adduction and medial rotation is dependent on embryonic growth and, presumably, fetal positioning during the remaining months of uterine life. The development of the angulation of the femur also appears to proceed after birth and through the early years of development.

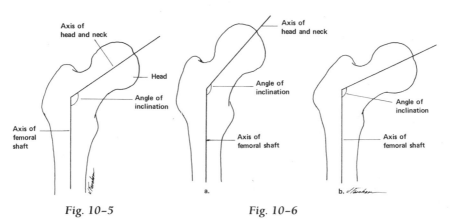

Fig. 10–5   Fig. 10–6

Figure 10–5. The axis of the femoral head and neck form an angle with the axis of the femoral shaft called the *angle of inclination*. This angle, measured medially, is approximately 125° in the adult.
Figure 10–6. Abnormal angles of inclination. a. A pathologic increase in the angle of inclination is called *coxa valga*. b. A pathologic decrease in the angle is called *coxa vara*.

Angle of Inclination of the Femur

The angle of inclination of the femur (Fig. 10–5) in early infancy is about 150° (measured medially). The inclination decreases to an average of 125° in the normal adult and to about 120° in the normal elderly person.[9,10] The angle of inclination varies among individuals and between sexes. In women, the angle is somewhat smaller than it is in men, owing to the greater width of the female pelvis.[2] A pathologic increase in the medial angulation between the neck and shaft is called **coxa valga** (Fig. 10–6a), and a pathologic decrease is called **coxa vara** (Fig. 10–6b).

Angle of Torsion of the Femur

The angle of torsion of the femur can best be viewed by looking down the length of the femur. An axis through the femoral neck will make an angle with an axis through the femoral condyles that reflects the medial rotation of the femoral condyles that occurred during fetal development. However, rather than being seen as medial rotation of the femoral condyles, this twist in the femur is seen as anterior torsion (relative lateral rotation) of the head and neck of the femur (Fig. 10–7). The reversal of perspective from distal to proximal occurs because the femoral condyles normally align themselves so that the axis through the condyles (the knee joint axis) lies in the frontal plane. Consequently, the head and neck of the femur normally face anteriorly with respect to the condyles. The angle of torsion (which may also be known as the angle of anteversion) decreases with age. In the newborn, the angle of torsion is approximately 40°, decreasing substantially in the first 2 years.[11] Svenningsen and associates[7] found a decrease of approximately 1.5° per year until cessation of growth among children with both normal and exaggerated angles of anteversion. In the adult, the angle of torsion is normally around 15°, but may vary from 8 to 30° and, like the angle of inclination, may vary between sexes.[2,8,12]

A pathologic increase in the angle of torsion is called **anteversion** (Fig. 10–8), and a pathologic decrease in the angle or reversal of torsion is known as **retroversion** (Fig. 10–8b). Since the hip joint can only tolerate a limited amount of torsion of the head without jeopardizing congruence, greater degrees of femoral anteversion or retroversion may, in fact, be seen distally at the femoral condyles. Internal

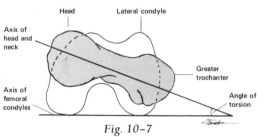

Fig. 10-7

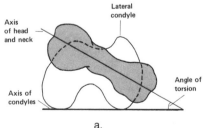

a.

Figure 10-7. A line parallel to the posterior femoral condyles and a line through the head and neck of the femur normally make an angle with each other that averages 15° in the normal adult. The femoral head and neck are twisted anteriorly with respect to the femoral condyles.

Figure 10-8. Abnormal angles of torsion in a right femur. a. A pathologic increase in the angle of torsion is called *anteversion.* b. A pathologic decrease in the normal angle of torsion is called *retroversion.*

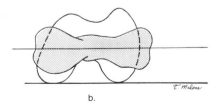

b.

Fig. 10-8

femoral torsion exists when the femoral condyles are turned medially from the frontal plane. Anteversion and internal femoral torsion *are synonymous*. When the anteverted femoral head is twisted anteriorly to a degree where an excessive amount of anterior articular surface is exposed, the entire femur must medially rotate to restore some congruence to the joint. As a result, the hip joint appears normally congruent, but the femoral condyles now face medially (internal femoral torsion). Similarly, retroversion of the femur may manifest itself distally as external femoral torsion when the femoral head is twisted posteriorly enough so that the entire femur must laterally rotate to restore some joint congruence.

Both normal and abnormal angles of inclination and torsion are *properties of the femur and exist independently of the hip joint*. However, abnormalities in the angulations of the femur can substantially alter hip joint stability, the weight-bearing biomechanics of the hip joint, and muscle biomechanics. While some conditions such as femoral anteversion and coxa valga are commonly found together, each may occur independently of the other. Each condition or combination of conditions warrants careful consideration as to the impact on hip joint function *and* function of the joints both proximal and distal to the hip joint. As shall be evident when the knee and foot are presented in subsequent chapters, femoral anteversion (internal femoral torsion) can create substantial dysfunction at both the knee and at the foot. The impact of abnormal angulations in the femur in hip joint function also will be discussed later in this chapter in the section on hip joint pathology.

Articular Congruence of the Hip Joint

The hip joint is commonly cited as being a very congruent joint. Although this is relatively true, the head of the femur is larger than the acetabulum. In the neutral or standing position, the articular surface of the femur remains exposed anteriorly and somewhat superiorly (Fig. 10–9a). Although the superiorly angled femoral head would appear to fit the inferiorly oriented acetabulum, this is not entirely the case. The acetabulum does not fully cover the head superiorly. The anterior torsion of the femoral head is poorly matched to the anterior orientation of the acetabulum, exposing a substantial amount of femoral articular surface anteriorly. Failure of developmental forces to reduce anteversion or valgus angulation of the femur, or to develop the roof of the acetabulum, can result in less congruence and stability of the hip joint in the neutral (standing) position. Articular contact between the femur and the acetabulum can be improved by a combination of flexion, abduction, and slight lateral rotation (Fig. 10–9b). This position (also known as the frog-leg position) corresponds to that assumed by the hip joint in a quadruped position and, according to Kapandji,[8] is the true physiologic position of the hip joint.[8] The combination of hip joint flexion, abduction, and rotation is commonly used for positioning or for immobilization of the hip joint when the goal is to improve joint congruence in conditions such as congenital dislocation of the hip and in Legg-Calvé-Perthes disease.[13]

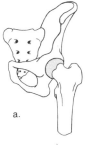 

Figure 10–9. a. In the neutral hip joint, articular cartilage from the head of the femur is exposed anteriorly and to a lesser extent superiorly. b. Maximum articular contact of the head of the femur with the acetabulum is obtained when the femur is flexed, abducted, and laterally rotated slightly.

a.   b.

The hip joint articular surfaces are never fully congruent based on joint position alone. Bullough and associates[4] found under light load conditions that the femoral head did not contact the dome (superior aspect) of the acetabulum. This held true across a wide variety of ages of cadavers and in a variety of positions of the femur, although not as consistently among the elderly. Radin and associates[14] point out that the incongruence of the surfaces under low loads has a functional advantage. When the joint is subjected to high loads (which may reach five times body weight in running), the flattening of articular cartilage and subchondral bone causes maximum surface contact, serving to reduce the force per unit area.[4,14] A congruent fit under low load would lead to incongruence under high load. It should also be remembered that only the periphery of the acetabulum is articular and that the acetabular fossa (deep in the acetabulum) does not contact the femoral head. The fossa may be important in setting up a partial vacuum in the joint, which helps maintain contact between the femoral head and the acetabulum.[8]

Hip Joint Capsule and Ligaments

The articular capsule of the hip is strong and dense. Unlike the weak articular capsule of the shoulder, the hip joint capsule is a substantial contributor to joint stability. The capsule is attached to the entire periphery of the acetabulum. It usually blends with the acetabular labrum, although it extends beyond the acetabulum superiorly.[2] The capsule covers the femoral neck like a sleeve and attaches to the base of the neck. The femoral neck is intracapsular, whereas the greater and lesser trochanters are both extracapsular. The capsule has two sets of fibers: The longitudinal fibers are more superficial and the circular fibers are deeper. The circular fibers form a collar around the femoral neck called the **zona orbicularis**.[2,8] The capsule itself is thickened anterosuperiorly where the predominant stresses occur, while it is relatively thin and loosely attached posteroinferiorly. The capsule permits little or no joint distraction even under strong traction forces.[2] The synovial membrane lines the inside of the capsule.

The anterior portion of the capsule is reinforced by two strong capsular ligaments, whereas the posterior portion is reinforced by one. The ligaments are named by their attachments. The iliofemoral ligament is a fan-shaped ligament that resembles an inverted letter Y (Fig. 10–10). It often is referred to as the **Y ligament of Bigelow.** The apex of the ligament is attached to the anterior inferior iliac spine, while the two arms of the Y fan out to attach along the intertrochanteric line of the femur. The iliofemoral ligament is the strongest ligament at the hip, with all fibers becoming taut during and providing a check to hip hyperextension (extension beyond neutral). The superior fibers of the iliofemoral ligament may become tensed during adduction and its inferior fibers tensed during abduction. The pubo-

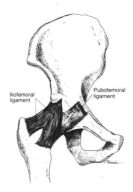

Figure 10–10. An anterior view of the right hip joint shows the two bands of the iliofemoral (Y) ligament and the more inferiorly located pubofemoral ligament.

femoral ligament (Fig. 10–10) is also anteriorly located, arising from the anterior aspect of the pubic ramus and passing to the anterior surface of the intertrochanteric fossa. The fibers of the pubofemoral ligament become taut in hip abduction and in extension. The bands of the iliofemoral and the pubofemoral ligaments form a Z on the anterior capsule similar to that of the glenohumeral ligaments.

The ischiofemoral ligament (Fig. 10–11) attaches to the posterior surface of the acetabular rim and the acetabulum labrum. Some of its fibers spiral around the femoral neck and blend with the fibers of the zona orbicularis. Other fibers are arranged horizontally and attach to the inner surface of the greater trochanter.[2] The spiral fibers tighten during extension but loosen or unwind during hip flexion.

All of the capsular ligaments of the hip are coiled or twisted as they pass from the pelvis to the femur in neutral position.[8] Hip extension or hip hyperextension, which is assumed in the upright posture, further coils and tightens these ligaments, making hip extension the close-packed position of the joint and a position of stability.[2] Full extension of the hip is stable because the ligaments have drawn the joint surfaces together and because further movement into extension is checked by these strong structures. At this point, however, the articular surfaces of the joint are not in optimal contact. The hip joint is one of the few joints where the position of optimal articular contact (combined flexion, abduction, and lateral rotation) is the *loose-packed* rather than close-packed position, since flexion and lateral rotation tend to uncoil the ligaments and make them slack. Under circumstances where the joint surfaces are *neither* maximally congruent *nor* close-packed, the hip joint is at greatest risk for traumatic dislocation. When the hip joint is flexed and adducted (as it is when sitting with the thighs crossed), a strong force up the femoral shaft toward the hip joint may push the femoral head out of the acetabulum.[8,13]

The ligament of the head of the femur (ligamentum teres) is a triangularly shaped band. Its base is on both sides of the acetabular notch where it blends with

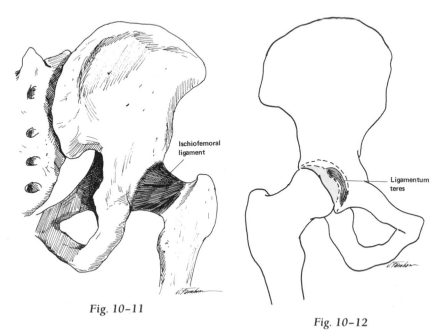

Ischiofemoral ligament

Ligamentum teres

Fig. 10–11

Fig. 10–12

Figure 10–11. A posterior view of a right hip joint shows that the spiral fibers of this ligament are tightened during hyperextension and therefore limit hyperextension.

Figure 10–12. An anterior view of a right hip shows the centrally located ligamentum teres arising from the fovea on the femoral head. The joint capsule and other structures have been removed.

the transverse acetabular ligament and its apex, as already mentioned, is on the fovea of the femur (Fig. 10–12). The ligament reaches the head of the femur by traveling under the transverse acetabular ligament and through the acetabular notch. Given its location within the joint capsule, the ligament is encased in a flattened sleeve of synovial membrane.

The ligament of the head does not appear to play a major role in hip stability, although it may be under tension in combined adduction and semiflexion.[2] Rather, the role of the ligament appears to be to act as a conduit for the blood supply and nerves that travel along the ligament to reach the head of the femur. The medial and lateral circumflex artery (derived from the femoral and deep femoral arteries) supply most of the head and neck of the femur through branches of the extracapsular arterial ring of the femoral neck and the subsynovial intra-articular arterial ring that wind around the femoral neck. However, these vessels cannot cross the cartilaginous epiphysis to supply the femoral head other than on the surface of the bone. Until bony maturation, the femoral head is supplied predominantly through the blood vessels derived from the obturator artery that travel to the fovea of the femur with the ligament of the head.[15] Prior to closure of the epiphysis of the head or to anastomosis of the arterial rings of the neck with the femoral head, disruption of the blood supply through the ligament of the head may lead to avascular necrosis of the head of the femur.

Weight-Bearing Structure of the Hip Joint

The internal architecture of the pelvis and femur reveal the remarkable adaptations that have occurred to accommodate the mechanical stresses and strains created by the transmission of forces between the femur and the pelvis. The trabeculae of bone line up along lines of stress and form systems to meet stress requirements. In the chapter on the vertebral column, we followed the line of weight bearing through the vertebrae of the spinal column to the sacral promontory and on through the sacroiliac joints. The weight-bearing lines of both the pelvis and the femur can be seen by looking at the arrangement of trabeculae (Fig. 10–13). Most of the weight-bearing lines in the pelvis pass to the acetabulum, although some are apparent along the pubic ramus and running to the ischial tuberosities. The stresses along the pubic ramus most likely arise from the function of the pubic bones as tierods preventing separation of the ilia and as compression struts against the medial femoral thrust.[2] The stress lines between the sacroiliac joints and the ischial tuberosities follow the weight-bearing forces in sitting.

The pelvic trabeculae that pass through the acetabulum of the pelvis form two major systems within the femur: the medial trabecular system and the lateral trabecular system. There are also two minor accessory systems of trabeculae. In bilateral static stance, the weight of the HAT is evenly distributed between the left and right hip joints, with the force of half the superimposed body weight traveling through the trabecular systems of each side.

The medial trabecular system arises from the medial cortex of the upper femoral shaft and radiates outward to the cortical bone of the superior aspect of the femoral head (Fig. 10–13). This system of trabeculae is oriented along what is considered to be the vertical rather than oblique weight-bearing force that passes through the hip joint. The region of increased subchondral bone density in the superior acetabulum is the primary weight-bearing surface of the acetabulum.[16,17] The primary weight-bearing surface of the acetabulum can be found on a roetgenogram by drawing a line between the medial and lateral edges of the area of increased density. In the normal hip this line will be horizontal and will lie directly over the center of rotation of the femoral head.[17] As the medial trabecular system continues distally, it coincides with a thickened area of cortical bone on the medial

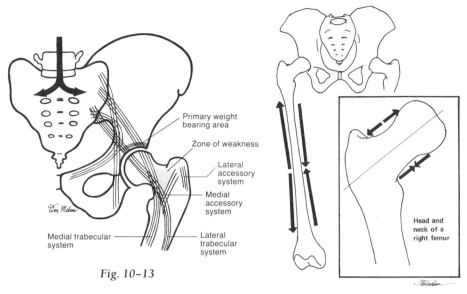

Fig. 10-13

Fig. 10-14

Figure 10-13. The trabeculae of the pelvis can be followed to the pubic ramus, the ischium, the dome of the acetabulum (primary weight-bearing areas), and the central area of the acetabulum. The trabeculae from the primary weight-bearing area are continuous with the medial trabecular system of the femur, whereas the trabeculae through the central acetabulum are continuous with the lateral trabecular system of the femur. With the addition of the medial accessory system and the lateral accessory system, an area of thin trabeculae (the zone of weakness) is evident.

Figure 10-14. The weight-bearing line from the center of rotation of the femoral head causes a bending force on the shaft of the femur that results in compressive forces medially and tensile forces laterally. (Inset: The weight-bearing through the head of the femur and a contraction of the hip abductors can cause tensile stresses on the femoral neck superiorly and compressive stresses inferiorly.)

shaft of the femur. The medial femoral shaft undergoes compression as the weight-bearing forces pass down the femur from the center of rotation of the femoral head (Fig. 10-14).

The lateral trabecular system of the femur arises from the lateral cortex of the upper femoral shaft and, after crossing the medial system, terminates on the cortical bone on the inferior aspect of the head of the femur (Fig. 10-13). The lateral system is thought to develop in response to forces created during contraction of the abductor muscles and to tensile stresses created by the tendency of the head and neck to bend on the shaft as the body weight arrives on the head of the femur (Fig. 10-14).[8,14] The two accessory trabecular systems are confined primarily to the trochanteric area and neck of the femur (Fig. 10-13). The medial accessory system arises from the medial aspect of the upper femoral shaft, crosses the lateral trabecular system, and fans out into the region of the greater trochanter.[8] The lateral accessory trabecular system runs parallel to the greater trochanter. The areas in which the various trabecular systems within both the pelvis and the femur cross each other at right angles are areas that offer the greatest resistance to stress and strain. There is an area in the femoral neck in which the trabeculae are relatively thin and do not cross each other. This **zone of weakness** has less reinforcement and thus more potential for injury. The zone of weakness of the femoral neck is particularly susceptible to the bending forces across the area and can fracture either when forces are excessive or when the tissues are no longer able to resist normal forces. A more detailed description of the problems of hip fracture will be presented later in the chapter.

FUNCTION OF THE HIP JOINT

Arthrokinematics

The hip joint motions are easiest to visualize as movement of the convex femoral head within the concavity of the acetabulum. The femoral head will glide within the acetabulum in a direction opposite to motion of the distal end of the femur. Flexion and extension occur from neutral position as an almost pure spin of the femoral head around a coronal axis through the head and neck of the femur. The head spins posteriorly in flexion and anteriorly in extension. However, flexion and extension from other positions must include both spinning and gliding of the articular surfaces, depending on the combination of motions. The motions of abduction and adduction and medial-lateral rotation must include both spinning and gliding of one surface on another, but again occur opposite to motion of the distal end of the femur when the femur is the moving segment. Whenever the hip joint is weight-bearing, the femur is relatively fixed and, in fact, motion of the hip joint is produced by movement of the pelvis on the femur. In this more common instance, the concave acetabulum moves in the same direction as the opposite side of the pelvis.

Osteokinematics

Motion of the Femur at the Hip Joint

The range of motion (ROM) available at the hip joint is, once again, most commonly interpreted through movement of the femur and can best be visualized this way. As is true at most joints, ROM is influenced by whether the motion is performed actively or passively and whether passive tension in two-joint muscles is encountered or avoided. The passive ranges of available joint motion are given as found in the summary table presented by Norkin and White.[18] Flexion of the hip is generally considered to have a range of 90° with the knee extended and 120 to 135° when the knee flexed and passive tension in the two-joint hamstrings is released. Hip extension is considered to have a range of 10 to 30°. When hip extension is combined with knee flexion, passive tension in the two-joint rectus femoris may limit the movement. The femur can be abducted 30 to 50° and adducted 10 to 30°. Abduction can be limited by the two-joint gracilis muscle, and adduction limited by the tensor fascia lata muscle and its associated iliotibial band. Medial and lateral rotation of the hip are usually measured with the hip joint in 90° of flexion, giving a range of 45 to 60° of lateral rotation and 30 to 45° of medial rotation. Femoral anteversion is correlated with decreased range of lateral rotation and less strongly with increased range of medial rotation.[7] Normal gait on level ground requires at least the following hip joint ranges: 30° flexion, 10° hyperextension, 5° of both abduction and adduction, and 5° of both medial and lateral rotation.[19,20] Walking on uneven terrain or stairs will increase the need for joint range beyond that required for level ground, as will activities such as sitting in a chair or sitting cross-legged.

Motion of the Pelvis at the Hip Joint

When the proximal segment of a joint moves on the fixed distal segment, the motion across the joint is the same as if the distal segment were the moving part. However, the direction of movement of the lever reverses. For example, elbow

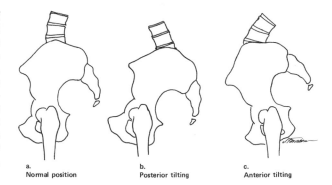

Figure 10-15. Representation of the tilting of the pelvis in the sagittal plane. a. The pelvis is shown in its normal position in erect stance. The lumbar spine is in slight extension. b. Posterior tilting of the pelvis moves the symphysis pubis superiorly and the lumbar spine flexes slightly. The hip joint extends. c. In anterior tilting the symphysis pubis moves inferiorly and the lumbar spine is hyperextended (lordotic). The hip joint is flexed.

a. Normal position   b. Posterior tilting   c. Anterior tilting

flexion can be a rotation of the distal forearm upward or, conversely, a rotation of the proximal humerus downward. At the hip joint, this reversal of motion of the lever is further complicated by the horizontal orientation and nonlever shape of the pelvis and by the new set of terms that refer to pelvic (rather than femoral) motions. The pelvic motions are named because these are the motions that are seen by the observer and, in fact, are key to what occurs at the joints above and below the pelvis. Note that the *arthokinematics and the available ROM at the hip joint remain unchanged, regardless of which segment is moving.*

Anterior and Posterior Pelvic Tilt

Anterior and posterior pelvic tilt are motions in the sagittal plane around a coronal axis. In the normally aligned pelvis, the anterior superior iliac spines of the pelvis lie on a horizontal line with the posterior superior iliac spines and on a vertical line with the symphysis pubis[21] (Fig. 10–15a). Posterior and anterior tilting of the pelvis on the fixed femur produce hip extension and flexion, respectively. Hip joint extension via posterior tilting of the pelvis brings the symphysis pubis up and the posterior aspect of the pelvis closer to the femur, rather than moving the femur posteriorly on the pelvis (Fig. 10–15b). Hip flexion through anterior tilting of the pelvis brings the anterior superior iliac spines anteriorly and inferiorly; the symphysis pubis moves down and closer to the femur rather than moving the femur toward the symphysis pubis (Fig. 10–15c). Anterior and posterior tilting can occur around both hip joints simultaneously or can occur in a single-limb support around one hip joint axis.

Lateral Pelvic Tilt

Lateral pelvic tilt is a frontal plane motion around an AP axis. In the normally aligned pelvis, a line through the iliac crests is horizontal. In lateral tilt of the pelvis, one hip joint serves as the pivot point or axis and the opposite iliac crest elevates

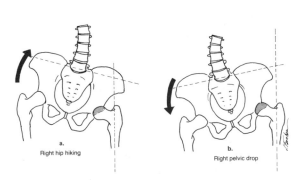

Figure 10-16. Lateral tilting of the pelvis around the left can occur either as hip hiking (elevation of the opposite side of the pelvis) or as pelvic drop (drop of the opposite side of the pelvis). a. Hiking of the pelvis around the left hip joint results in left hip abduction. b. Dropping of the pelvis around the left hip joint results in left hip joint adduction.

a. Right hip hiking   b. Right pelvic drop

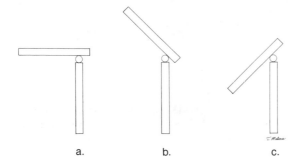

Figure 10–17. a. Neutral pelvis and femur in left stance. b. Hip hiking of the pelvis around the left hip joint results in abduction of the left hip joint. c. Pelvic drop around the left hip joint results in adduction of the left hip joint.

(hip hiking) or drops (pelvic drop) around that pivot point. If one is standing on the left limb and the pelvis laterally tilts up on the right (hip hiking), the left hip is being abducted (Fig. 10–16a). The distance between the femur and the midline of the body (which moves with the pelvis) is increased medially (this is the definition of abduction). If one is again standing on the left leg, dropping the pelvis on the right results in adduction of the left hip joint (Fig. 10–16b). That is, the medial angle between the midline of the body and the left femur is reduced. The relative hip joint motions in lateral pelvic tilt are commonly more difficult to visualize than those in anterior and posterior tilting because the eye tends to follow the iliac crest on the same side as the supporting hip joint. However, it should also be kept in mind that osteokinematic descriptions reference the motion of the *end of the lever farthest from the joint axis.* For example, an osteokinematic description of femoral motion references what is happening to the distal end of the femur, not to the proximally located greater trochanter. In the case of the pelvis, the osteokinematic description of motion always references the opposite side of the pelvis from the hip joint axis. If a woman is standing on her right leg and hikes her pelvis, it should not be necessary to specify that the left side of the pelvis is the one that is rising. Since the right hip joint is the axis, the motion is defined by movement of the left side of the pelvis. Keeping this fact in mind and visualizing the pelvis as a straight lever (through the iliac crests), that pelvic hike is equivalent to hip joint abduction and pelvic drop is equivalent to hip joint adduction is easier to understand and visualize (Fig. 10–17).

Pelvic Rotation

Pelvic rotation occurs in the transverse plane around a vertical axis. Although rotation can occur around a vertical axis through the middle of the pelvis, it most commonly occurs in single-limb support around the axis of the supporting hip joint and will be defined in this way. Forward rotation of the pelvis occurs when the side of the pelvis opposite to the supporting hip joint moves anteriorly (Fig. 10–18a).

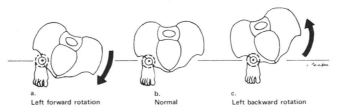

a.
Left forward rotation

b.
Normal

c.
Left backward rotation

Figure 10–18. A superior view of rotation of the pelvis in the transverse plane. a. Forward rotation of the pelvis around the right hip joint results in medial rotation of the right hip joint. b. Neutral position of the pelvis and the right hip joint. c. Backward rotation of the pelvis around the right hip joint results in lateral rotation of the right hip joint.

This motion of the pelvis produces medial rotation of the supporting hip joint. As in lateral pelvic tilt, the reference for the pelvis is always the side farthest from the supporting hip. If the side of limb support is known, it should be redundant to specify *which* side of the pelvis is forwardly rotating. The medial rotation of the hip joint that occurs during forward rotation of the pelvis can best be appreciated by doing the motion yourself. Standing on one leg and rotating the pelvis and trunk forward as much as possible will give a clear "feeling" of the relative medial rotation of the supporting limb. Backward rotation of the pelvis occurs when the side of the pelvis opposite the supporting hip moves posteriorly (Fig. 10–18c). Posterior rotation of the pelvis produces lateral rotation of the supporting hip joint. This again can be best appreciated by doing the motion.

Coordinated Motions of the Femur, Pelvis, and Lumbar Spine

When the pelvis moves on a relatively fixed femur, there are two possible outcomes to consider. Either the head and trunk will follow the motion of the pelvis (moving the head through space), or the head will continue to remain relatively upright and vertical in spite of the pelvic motions. These are open kinematic chain and closed kinematic chain responses, respectively. Each of these two situations produces very different reactions from the joints and segments proximal and distal to the hip joints and pelvis and must be examined separately.

Lumbar-Pelvic Rhythm

Caillet[22] used the term **lumbar-pelvic rhythm** to describe the *open-chain phenomenon* in the hip joint, pelvis, and lumbar spine where coordinated activity of the segments produced a larger ROM than might be available to one segment alone. This is one instance where analogy to the shoulder is functionally relevant. Lumbar-pelvic rhythm is clearly the analog of scapulohumeral rhythm where the combination of motions at several joints serve to increase the range available to the hand. For example, reaching the floor through isolated flexion at the hip joints (anteriorly tilting the pelvis on the femurs) is generally insufficient to reach the ground. If the knees remain extended, the hips can only flex to 90°. The addition of flexion of the

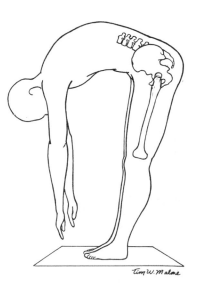

Figure 10–19. Lumbar-pelvic rhythm can increase the range of forward flexion of the body by combining hip flexion, anterior pelvic tilt, and flexion of the lumbar spine. This combination permits the hands to reach the ground.

Figure 10–20. Lumbar-pelvic rhythm increases the range through which the foot can be moved by combining hip abduction, lateral pelvic tilt, and later flexion of the lumbar spine.

lumbar spine will incline the trunk forward about another 45°.[22] The combination of hip flexion and lumbar flexion should be adequate to reach the floor (Fig. 10–19).

Lumbar-pelvic rhythm can increase the ROM available by adding range from the lumbar spine. When a woman is side-lying, for example, an attempt at maximal abduction of the top leg will bring the leg through as much as 90° of motion (Fig. 10–20). This is clearly not all from the hip joint, which can only abduct to 45°, but includes lateral tilting both of the pelvis and of the lumbar spine. Unlike scapulo-humeral rhythm, lumbar-pelvic rhythm in all directions occurs without a specific mandatory sequence of activity and without a fixed proportional contribution of its component parts.

Closed-Chain Responses to Motions of the Pelvis and Hip Joint

In lumbar-pelvic rhythm, the goal is to move the head or the foot through space. During these motions, subtle adjustments or compensations might be required by other joints to assure that the LOG remains within the base of support in spite of the rearrangement of segments. Although the head can be moved through a large ROM while the feet remain fixed, we spend a greater proportion of our time assuring that the head remains upright and over the sacrum. The goal in such an instance is both to keep one or both feet on the ground *and* to maintain the head upright and vertical in spite of motions of the pelvis that might be occurring concomitantly. That is, a closed kinematic chain is formed and motion at one segment within the chain *requires compensatory motion at least at one other segment in the chain.* The relationships between hip joint and pelvic motions and the lumbar spine are now different and generally opposite to those seen in lumbar-pelvic rhythm.

If the pelvis is anteriorly tilted during weight-bearing (hip joint flexion) either purposefully or through tightness of the hip flexor musculature, the head and trunk will be displaced forward. To prevent this, anterior pelvic tilt must be accompanied by *extension* of the lumbar spine to keep the head upright and over the sacrum (Fig. 10–1b). If either the anterior pelvic tilt or the lumbar extension is extreme, further compensation may be required by the thoracic and cervical vertebrae to maintain the head upright. A posterior tilted pelvis in weight bearing (hip joint extension)

Table 10–1. RELATIONSHIP OF PELVIS, HIP JOINT, AND LUMBAR SPINE DURING RIGHT LOWER-EXTREMITY WEIGHT BEARING AND UPRIGHT POSTURE

Pelvic Motion	Accompanying Hip Joint Motion	Compensatory Lumbar Spine Motion
Anterior pelvic tilt	Hip flexion	Lumbar extension
Posterior pelvic tilt	Hip extension	Lumbar flexion
Lateral pelvic tilt (pelvic drop)	Right hip adduction	Right lateral flexion
Lateral pelvic tilt (pelvic hike)	Right hip abduction	Left lateral flexion
Forward rotation	Right hip medial rotation	Rotation to the left
Backward rotation	Right hip lateral rotation	Rotation to the right

will require *flexion* of the lumbar spine to keep the head forward and over the sacrum.

An important example of closed-chain compensatory movements between the pelvis, hip, and lumbar spine can be seen during gait. When we walk, the pelvis will drop slightly (lateral pelvic tilt) around the supporting hip joint (hip joint adduction). If the head and trunk followed the pelvis, the body would lean away from the supporting extremity and the LOG would fall outside the supporting foot. Instead, the lumbar spine laterally flexes *toward* the side of the supporting limb to prevent displacement of the head and trunk.

In any instance in which there is normal or abnormal pelvic motion during weight bearing and the head must remain upright, compensatory motions of the lumbar spine will occur if available. This does not rule out the need for compensation at additional joints as well, but the lumbar spine tends to be the "first line of defense." As we examine the other joints of the lower extremity and move on to posture and gait, other compensatory motions will be encountered and discussed. Table 10–1 presents the compensatory motions of the lumbar spine that accompany given motions of the pelvis and hip joint.

Hip Joint Musculature

There have been numerous studies of the muscles of the hip joint. Most confirm underlying principles of muscle physiology seen at the other joints we have examined so far. That is, hip joint muscles work best in the middle of their contractile range or on a slight stretch (at optimal length-tension); two joint muscles generate greatest force when not required to shorten over both joints simultaneously; and torque generation is best with eccentric contractions, followed by isometric and then concentric contractions. The muscles of the hip joint, however, operate primarily as part of a closed kinematic chain and are somewhat unique in their large areas of attachment, their length and their large cross section. These characteristics, in combination with the large ROM available at the hip joint, result in muscle function that is dependent on the position in the range and availability of motion of the proximal and distal segments. For example, the adductor muscles may be hip flexors in the neutral hip joint, but will be hip extensors when the hip joint is already flexed.[23] Similarly, the lateral rotators of the hip joint may be medial rotators from a position of extreme medial rotation.[8] Such inversions of function are found in a few muscles at the shoulder (the clavicular portion of the pectoralis major, for example), but are fairly common in the hip joint. As a consequence, results of various studies may appear to be contradictory when, in fact, testing conditions explain differing results. Some gender-related differences also have been found that explain differential findings.[24] It is best to examine muscle action at the hip joint in the context of specific functions such as single-limb support, posture, and gait. Given this, the next section will briefly review muscle function, but we will leave more detailed analyses for later in this and other chapters. Although the traditional action of the muscle on the distal lever is described for the most part, it must be emphasized that any of the muscles are as likely to produce joint action by moving the proximal segment instead.

Flexors

The flexors of the hip joint function primarily as mobility muscles that bring the lower extremity forward during ambulation or in various sports (open kinematic chain function). The flexors may function secondarily to resist strong hip extension forces (generally occurring in a closed kinematic chain). There are nine muscles that have action lines crossing the anterior aspect of the hip joint. Of these, the primary muscles of hip flexion are the iliopsoas, rectus femoris, tensor fascia lata, and sartorius. The iliopsoas muscle is considered to be the most important of the primary hip flexors. It consists of two separate muscles, the iliacus muscle and the psoas major muscle, which attach to the femur by a common tendon. The two components of the iliopsoas have many points of origin, including the iliac fossa, lateral portion of the sacrum, the intervertebral disks and bodies of the twelfth thoracic through fourth lumbar vertebrae, and the transverse processes of the first through the fifth lumbar vertebrae. The iliopsoas inserts onto the lesser trochanter of the femur (Fig. 10–21). The attachment of the psoas major to the anterior vertebrae and the iliacus to the iliac fossa make it likely that activity or tension in these muscles may increase lumbar lordosis by pulling the lumbar curvature anteriorly. Basmajian and DeLuca[23] summarize the often contradictory evidence of many investigations by concluding that both segments of the iliopsoas are active in various stages of hip flexion and that any concomitant medial or lateral rotatory function is at best weak and should be ignored.

The rectus femoris muscle is the only portion of the quadriceps muscle that crosses both the hip joint and knee joint. It originates on the anterior inferior iliac spine and inserts by way of a common tendon into the tibial tuberosity. The rectus femoris flexes the hip joint and extends the knee joint. As a two-joint hip flexor, the position of the knee during hip flexion will affect its ability to generate force at the hip. Simultaneous hip flexion and knee extension considerably shorten this muscle and increase the likelihood of active insufficiency. Consequently, the rectus femoris makes its best contribution to hip flexion when the knee is maintained in flexion.

The sartorius muscle is a straplike muscle originating on the anterior superior iliac spine. It crosses the anterior aspect of the femur to insert into the upper portion of the medial aspect of the tibia (Fig. 10–22). The sartorius is considered to be a flexor and abductor of the hip, as well as a flexor and medial rotator of the knee.[25] Although a two-joint muscle, the sartorius does not appear to be affected by the position of the knee given the relatively small proportional change in length with increased knee flexion; it is active when the knee is either flexed or extended.[26]

The tensor fascia lata muscle originates more laterally than the sartorius. Its origin is on the anterolateral lip of the iliac crest. The muscle fibers extend only about one quarter of the way down the lateral aspect of the thigh before inserting into the iliotibial band (Fig. 10–22). The tensor fascia latae is considered to flex, abduct, and medially rotate the femur at the hip,[23] although the tensor's contribu-

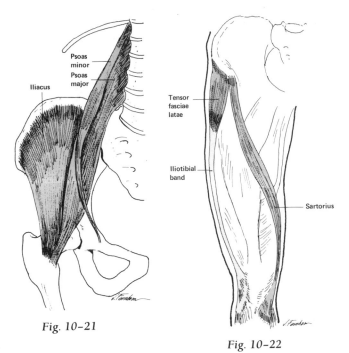

Fig. 10-21

Fig. 10-22

Figure 10-21. Anterior view of the right hip shows the attachments of the iliopsoas muscle.
Figure 10-22. The anterior aspect of the right hip and knee shows the location of the tensor fascia lata and the sartorius. Both are two-joint muscles that act at the hip and the knee.

tion to hip abduction may be dependent on simultaneous hip flexion.[27] The most important contribution of the tensor fascia lata may be in maintaining tension in the iliotibial band. Stretched from the iliac crest to the tibia, the iliotibial band is considered to assist in relieving the femur of some of the tensile stresses imposed on the shaft by weight-bearing forces[11,27] (Fig. 10-14). Since bone more effectively resists compressive than tensile stresses, reduction of tensile stresses is important in maintaining integrity of the bone.[14] Functionally, the tensor fascia lata and iliotibial band are expendable. The iliotibial band may be removed and used for autogenous fascial transplants without any evident change in active or passive hip or knee function.[27]

The secondary flexors are the pectineus, adductor longus, adductor magnus, and the gracilis muscles, which are described in the next section. Each is capable of contributing to hip joint flexion depending on the position of the hip. Kapandji[8] notes that these muscles contribute to flexion only to between 40 to 50° of hip flexion. Once the femur is superior to the point of origin of a muscle, the muscle will become an extensor of the hip joint.[8] We should be aware that lines of pull can change position relative to a joint axis and that function is generally position-dependent. The gracilis is the only two-joint muscle in the adductor group. It is active as a hip flexor when the knee is extended but not when the knee is flexed.[26]

Adductors

The hip adductor muscle group is generally considered to include the pectineus, adductor brevis, adductor longus, adductor magnus, and the gracilis. The contribution of the adductor muscles to hip joint function has been debated for many years. Basmajian and DeLuca[23] believe the variability in study findings supports the

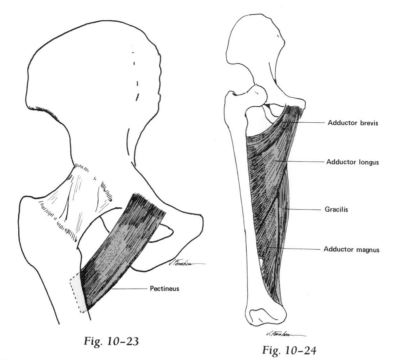

Fig. 10–23

Fig. 10–24

Figure 10–23. Anterior view of the right hip shows the attachments of the pectineus muscle. The pectineus is an adductor and a flexor of the hip.

Figure 10–24. The adductors of the right hip.

theory of Janda and Stara that the adductors function not as prime movers, but by reflex response to gait activities. Activity of this group of muscles cannot be ignored, however, since there are enough muscles that the maximum isometric torque of adduction is greater than that of abduction.[28]

The pectineus muscle is a small muscle located medial to the iliopsoas. It originates on the superior ramus of the pubis and inserts into the femur at a point just below the lesser trochanter (Fig. 10–23). The other adductors are also located anteromedially (Fig. 10–24). The adductors longus, brevis, and magnus muscles arise in a group from the body and inferior ramus of the pubis to insert along the linea aspera. The gracilis muscle originates on the symphysis pubis and pubic arch and inserts on the medial surface of the shaft of the tibia.

Extensors

The one-joint gluteus maximus muscle and the two-joint hamstrings muscle group are the primary hip joint extensors. These muscles may receive assistance from the posterior fibers of the gluteus medius, the superior fibers of the adductor magnus, and from the piriformis muscle. The gluteus maximus is a large, quadrangular muscle that originates from the sacrum, dorsal sacroiliac ligaments, sacrotuberous ligament, and a small portion of the ilium. Its most superior fibers insert into the iliotibial band. The inferior fibers insert into the gluteal tuberosity (Fig. 10–25). The gluteus maximus is a strong hip extensor that appears to be active primarily against a resistance greater than the weight of the limb. Its moment arm (MA) is considerably longer than that of either the hamstrings or the adductor magnus mus-

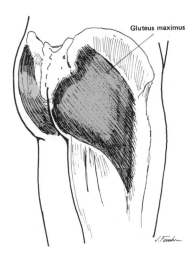

Figure 10–25. The gluteus maximus muscle.

cles and is maximal in the neutral hip joint position.[24] The maximus is also a lateral rotator of the femur, but may reverse this function in extremes of range.[23]

The three two-joint extensors are the long head of the biceps femoris, the semitendinosus, and the semimembranosus muscles. These muscles are known collectively as the **hamstrings**. All of these muscles originate on the ischial tuberosity. The biceps femoris crosses the posterior femur to insert into the head of the fibula and lateral aspect of the lateral tibial condyle. The other two hamstrings insert on the medial aspect of the tibia. All three muscles extend the hip with or without resistance and flex the knee. Although the MA of the combined hamstrings at the hip is less than that of the gluteus maximus at all points in the range, the hamstrings increase their MA as the hip flexes to 35° and decrease it thereafter; the MA of the maximus decreases with any hip flexion beyond neutral.[24] The biceps femoris laterally rotates the knee and the long head laterally rotates the hip; the other two hamstrings medially rotate the knee.[23]

Abductors

Active abduction of the hip is brought about predominantly by the gluteus medius and the gluteus minimus muscles. The superior fibers of the gluteus maximus and the sartorius may assist when the hip is abducted against strong resistance. As already mentioned, the tensor fascia lata may be an effective abductor only during simultaneous hip flexion. The gluteus medius lies deep to the gluteus maximus. It originates on the lateral surface of the wing of the ilium and inserts into the greater trochanter. The gluteus medius has anterior, middle, and posterior parts that function asynchronously during movement at the hip.[29] Analogous to the deltoid muscle of the glenohumeral joint, the anterior fibers of the gluteus medius are active in hip flexion and medial rotation, whereas the posterior fibers function during extension and lateral rotation. All portions of the muscle abduct. The trochanteric bursa of the gluteus medius separates the distal tendon of the muscle from the trochanter over which it must slide. Under some conditions, this bursa may become inflamed and be a source of pain.[13]

The gluteus minimus muscle lies deep to the gluteus medius, arising from the outer surface of the ilium and from the margin of the greater sciatic notch. The muscle fibers converge on an aponeurosis that ends in a tendon on the greater trochanter. The trochanteric bursa of the gluteus minimus allows the tendon to slide over the trochanter. The gluteus minimus and medius muscles function together to either abduct the femur in an open kinematic chain or, more importantly, to stabilize the pelvis (and superimposed HAT) in unilateral stance against the effects of gravity. As will be presented later, the gluteus medius and minimus muscles either prevent or minimize adduction of the pelvis around the stance hip (pelvis drop). Physiologically designed to work at the neutral or slightly adducted hip, the hip abductors generate an isometric torque that is 82 percent greater in neutral position than when the muscle is shortened to 25° of abduction.[28] The concentric maximum torque generated by the hip abductors is also greater in the lengthened position of slight adduction.[30,31]

Lateral Rotators

Six short muscles have lateral rotation as a primary function. These muscles are the obturator internus and externus, the gemellus superior and inferior, quadratus femoris, and the piriformis muscles. Other muscles that have fibers posterior to the axis of motion at the hip (the posterior fibers of the gluteus medius and minimus and superior fibers of the gluteus maximus) may produce lateral rotation combined with the primary action of the muscle. Of the primary lateral rotators, each inserts either on or in the vicinity of the greater trochanter. The obturator internus muscle originates from the inside of the obturator foramen and emerges through the lesser sciatic foramen to insert on the medial surface of the greater trochanter. The obturator internus is closely associated with the two gemelli. The gemellus superior muscle, which originates on the ischial spine, extends along the superior border of the obturator internus. The gemellus inferior muscle lies along the inferior border of the obturator internus muscle. The two gemelli muscles are attached to the greater trochanter by way of the tendon of the obturator internus.

The obturator externus muscle is sometimes considered to be an anteromedial muscle of the thigh because it originates on the external surface of the pelvis on the obturator foramen. However, it crosses the posterior aspect of the hip joint and inserts on the medial side of the greater trochanter in the trochanteric fossa. The quadratus femoris muscle is a small quadrangular muscle that originates on the ischial tuberosity and inserts on the femur between the greater and lesser trochanters. The piriformis muscle is a small muscle located just below the lower posterior edge of the gluteus medius. It originates on the anterior surface of the sacrum and extends through the greater sciatic foramen to insert into the inner aspect of the superior portion of the greater trochanter. The sciatic nerve, the largest nerve in the body, enters the gluteal region just inferior to the piriformis muscle.

The lateral rotator muscle group performs its rotatory function effectively, given the nearly perpendicular orientation to the shaft of the femur (Fig. 10–26).

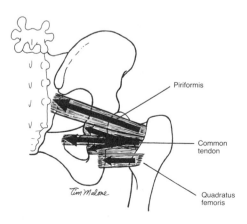

Figure 10–26. The lateral rotators of the hip joint have action lines that lie nearly perpendicular to the femoral shaft (making them excellent rotators) and parallel to the head and neck of the femur (making them excellent compressors). The common tendon is the shared insertion of the gemellus superior, gemellus inferior, and the obturator internus. Not shown in this diagram is the obturator externus.

However, exploration of function of these muscles has been restricted because of the relatively limited access to electromyography (EMG) surface or wire electrodes. Like their rotator cuff counterpart at the glenohumeral joint, these muscles would certainly appear to be effective joint compressors since their combined action line parallels the head and neck of the femur. Hypothetically, the lines of pull of these deep one-joint muscles should make them ideal tonic stabilizers of the joint during most weight-bearing and non-weight-bearing hip joint activities.

Medial Rotators

There are no muscles with the primary function of producing medial rotation of the hip joint. However, muscles with lines of pull anterior to the hip joint axis at some point in the ROM may contribute to this activity. The more consistent medial rotators are the anterior portion of the gluteus medius and the tensor fascia lata muscles. Although controversial, the weight of evidence appears to support the adductor muscles as medial rotators of the joint.[23]

Muscle Function in Stance
Bilateral Stance

In erect bilateral stance, both hips are in neutral or slight hyperextension. In this position the LOG falls just posterior to the axis of the hip joint. The posterior location of the LOG in the sagittal plane creates an extension moment of force (torque) around the hip that tends to posteriorly tilt the pelvis on the femoral heads. The gravitational extension moment is checked by the opposing passive tension in the ligaments and joint capsule, which are taut in full extension. Although intermittent activity of the iliopsoas muscle may occur to relieve the capsule and ligaments of some of the gravitationally imposed stretch, little muscle activity is required to maintain AP equilibrium as long as the LOG remains behind or through the hip. When the LOG is anterior to the hip joints, a flexion moment is created, and activity of the hamstrings and gluteus maximus muscles is required to prevent flexion of the hip through anterior tilting of the pelvis and trunk.

In the frontal plane during bilateral stance, the superincumbent body weight is transmitted through the sacroiliac joints along the pelvic trabecular system to the right and left femoral heads. The weight of the HAT (⅔ of body weight) is distributed so that each femoral head receives one half of the superincumbent weight.[32] In a normally proportioned body, HAT contributes to two thirds of the body weight.[33] If someone weighed 180 lb, there would be 120 lb (⅔ × 180) of compressive force distributed equally (60 lb each) between the two supporting hip joints. As shown in Figure 10–27, the joint axis of each hip lies at an equal distance from the LOG of HAT (DR = DL). Since the body weight on each femoral head is the same (WR = WL), the magnitude of the gravitational torques around each hip must be identical (WR × DR = WL × DL). The two torques, however, occur in opposite direc-

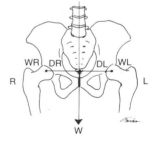

Figure 10-27. An anterior view of the pelvis in normal erect bilateral stance. The weight acting at the right (WR) times the distance from the right hip joint axis to the body's center of gravity (DR) is equal to the weight acting at the left hip (WL) times the distance from the left hip to the body's center of gravity (DL) [WR × DR = WL × DL]).

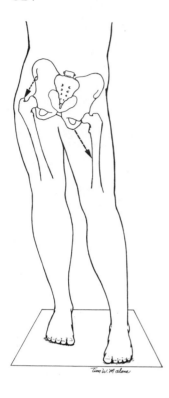

Figure 10-28. The pelvis and weight are shifted to the right. To return to neutral (weight distributed equally on both feet), the right abductor and left adductor can work synergistically to shift the weight to the left.

tions. The weight of the body (W) acting around the right hip tends to drop the pelvis down on the left, while the weight acting around the left hip tends to drop the pelvis down on the right. These two opposing gravitation torques of equal magnitude balance each other and the pelvis is maintained in equilibrium without the need for muscle activity.

When bilateral stance is not symmetrical, muscle activity will be required to either control the motion or to return the hips to symmetrical stance. In Figure 10-28, the pelvis is shifted to the right, resulting in relative adduction of the right hip and abduction of the left hip. To return to neutral position, an active contraction of the right hip abductors and/or left hip adductors is required. In bilateral stance, the contralateral abductors and adductors may function as synergists to control the frontal plane motion of the pelvis. Although the abductors are commonly credited with serving a stability role, the adductors are generally not given such credit. However, *under the condition that both extremities bear at least some of the superimposed body weight*, the adductors may assist the abductors in control of the pelvis against the force of gravity or the ground reaction force. In the absence of adequate hip abductor function, the adductors can contribute to stability in bilateral stance. In unilateral stance, activity of the adductors either in the weight-bearing or non-weight-bearing hip *cannot* contribute to stability of the stance limb. Hip joint stability in unilateral stance is the sole domain of the hip joint abductors.

Unilateral Stance

In Figure 10-29, the left leg has been lifted from the ground and the full superimposed body weight is being supported by the right hip joint. Rather than sharing the compressive force of the superimposed body weight with the left limb, the right hip joint must now carry the full burden. However, in addition to the weight of

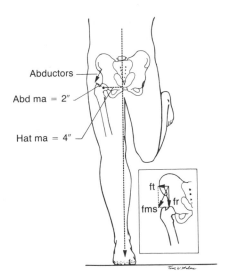

Figure 10–29. In right unilateral stance, the weight of the head, arms, and trunk (HAT) act 4 in from the right hip to produce an adduction torque around the right hip joint. The abductors, acting 2 in from the right hip joint, generate a large force in order to produce an abduction torque sufficient to balance the countertorque of HAT. (Inset: The pull of the abductors (Fms) on the horizontally oriented pelvis will resolve into a translatory component (ft) that will pull the acetabulum into the center of the femoral head and a rotatory component (fr) that will pull the pelvis down on the superior aspect of the femoral head.

HAT (⅔ × 180), there is also the weight of the left limb, which is "hanging" on the left side of the pelvis. Of the one-third portion of the body weight found in the lower extremities,[33] the nonsupporting limb must account for half of that, or one-sixth of the full body weight. The magnitude of body weight compressing the right hip joint is, therefore, five sixths (⅔ + ⅙) of the total body weight, or 150 pounds in an individual weighing 180 lb. By merely lifting the left leg, the force through the right hip joint has gone from 60 lb in bilateral stance to 150 lb in unilateral stance. However, the problem is more complex because the hip joint is not only being compressed by the superimposed gravitational force but gravity is also creating a torque around the hip joint.

The force of gravity (weight of HAT and the left limb) will create an adduction torque around the supporting hip joint. Although the actual distance will vary among individuals, the LOG can be estimated to lie 4 in from the right hip joint axis (MA = 4 in).[22] The magnitude of the gravitational adduction torque at the right hip consequently will be

$$150 \text{ lb} \times 4 \text{ in} = 600 \text{ in-lb}$$

To maintain the single-limb support position, there must be a countertorque (abduction moment) of equivalent magnitude or the pelvis will drop. If the trunk follows the pelvis, the person will fall to the unsupported side. The countertorque must be produced by activity of the hip abductor muscles (gluteus medius and minimus) acting on the pelvis. Assuming that the abductor muscles act through a typical MA of 2 in[22] and knowing the gravitational torque, we can solve for the magnitude of muscle contraction needed to maintain equilibrium:

$$2 \text{ in} \times \text{Fms} = 600 \text{ in-lb}$$

$$\text{Fms} = \frac{600 \text{ in-lb}}{2 \text{ in}} = 300 \text{ lb}$$

To prevent the pelvis from falling to the unsupported side, the abductors must generate a force of 300 lb. Assuming that all the muscular force is transmitted through the acetabulum, the 300-lb muscular force is added to the 150 lb of compression due to body weight passing through the supporting hip. Thus, the total hip joint compression, or joint reaction force, can be estimated at:

$$\begin{array}{r} 150 \text{ lb of body weight compression} \\ + \ 300 \text{ lb of muscular joint compression} \\ \hline 450 \text{ lb of total hip joint compression} \end{array}$$

The location of the joint reaction force can be further defined by knowing the angle of pull of the hip abductors. The action line of the abductors has been estimated to average 22 to 30° from vertical,[32,34] yielding the force components shown in Figure 10–29 (inset). We can estimate that approximately two-thirds of the total hip abductor force (200 lb) acting on the pelvis will bring the pelvis vertically downward on the femoral head while one-third of the force (100 lb) will direct the pelvis laterally into the center of the femoral head. The vertically directed downward force of 200 lb will fall into the same line as the vertical force of the body weight (150 lb), resulting in a net force of 350 lb through the superior aspect of the acetabulum and femoral head. The remaining 100 lb of the total 450 lb of hip joint compression is distributed through the center of the femoral head.

The hypothetical figures used above oversimplify the forces involved in hip joint stresses. Total hip joint compression or joint reaction forces are generally considered to be 2½ to 3 times body weight in static unilateral stance.[20,22,32] Investigators have calculated or measured forces of four and seven times body weight in, respectively, the beginning and end of the stance phase of gait,[35] and seven times the body weight in activities such as stair climbing.[36] Although weight loss can reduce the hip joint reaction force, the larger component of the joint reaction force is generated by the contraction of the hip abductors. The magnitude of hip abductor force required can be affected by individual differences in the angle of pull of the muscles, in the angle of inclination of the femoral head, and in the angle of femoral torsion. Physiologic and biomechanical factors necessitating increased force production by the hip abductor muscles over time may result in joint deterioration as a result of the abnormally large joint compressive forces. Even under conditions of optimal alignment, osteoarthritic changes in the superior acetabulum (the primary weight-bearing surface) are not uncommon.[2,17]

Reduction of Muscle Forces in Unilateral Stance

If the hip joint undergoes osteoarthritic changes leading to pain on weight bearing, the joint reaction force must be reduced to avoid pain. Although a loss of 1 lb of body weight will reduce the joint reaction force by 3 lb or more, greater reductions in compression are generally required than can be realistically achieved through weight loss. The solution must be in a reduction of abductor muscle force requirements that will decrease the amount of muscular compression across the joint. The need to diminish abductor force requirements will also be seen when the abductor muscles are weakened through paralysis or through structural changes in the femur that reduce biomechanical efficiency of the muscles. In fact, paralysis of the hip abductor muscles (gluteus medius and gluteus minimus) is considered the most serious muscular disability in the hip region.[2] Hip abductor muscle absence will inevitably affect gait, whereas paralysis of other hip joint muscles in the presence of intact abductors will permit someone to walk or even run with relatively little disability.

There are several options available when there is a need to decrease hip joint compression. Some compression reduction strategies occur automatically, but at a cost of extra energy expenditure and structural stress. Other strategies require intervention (assistive devices), but minimize the energy cost.

Compensatory Lateral Lean of the Trunk

Gravitational torque at the pelvis is the product of body weight and the distance that the LOG lies from the hip joint (MA). If there is a need to reduce the torque of gravity and body weight cannot be reduced, then the MA of the gravitational force must be reduced or minimized. This can be accomplished by laterally leaning the trunk above the pelvis *toward the side of pain or weakness.* Although leaning toward the side of pain might appear counterintuitive, the compensatory lateral lean of the trunk will swing the LOG closer to the supporting hip joint and reduce its MA. Since all of the superimposed body weight passes through the supporting (painful) hip joint, leaning toward the affected hip does not increase the joint compression due to body weight. However, it does reduce the gravitational torque and, consequently, the need for abductor countertorque. Although it is theoretically possible to laterally lean the trunk enough to bring the LOG *through* the supporting hip (reducing the torque to zero) or to the *opposite side* of the supporting hip (reversing the direction of torque), these are relatively extreme motions that require high energy expenditure and would result in excessive wear and tear on the lumbar spine. More energy-efficient and less structurally stressful compensations can still yield dramatic reductions in the hip abductor force.

If the person weighing 180 lb were to laterally lean enough to bring the LOG within 1 in of the hip joint axis (Fig. 10–30), the torque of gravity would now be

$$\frac{5}{6}(180 \text{ lb}) \times 1 \text{ in} = 150 \text{ in-lb}$$

If only 150 in-lb of adduction torque were produced by the superimposed weight, the abductor force needed would be

$$\text{Fms} = \frac{150 \text{ in-lb}}{2 \text{ in}} = 75 \text{ lb}$$

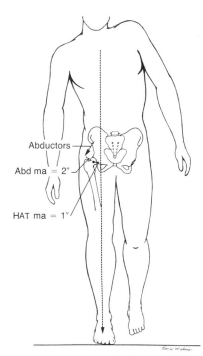

Abductors

Abd ma = 2"

HAT ma = 1"

Figure 10–30. When the trunk is laterally flexed toward the stance limb, the moment arm of HAT is substantially reduced (e.g., 1 in as compared to 4 in with the neutral trunk), whereas that of the abductors remains unchanged (e.g., 2 in). The result is a substantially diminished torque from HAT and, consequently, a substantially decreased need for hip abductor force to generate a countertorque.

If only 75 lb of abductor force were required, the total hip joint compression using the compensatory lateral lean would now be

$$150 \text{ lb body weight compression}$$
$$+ \quad 75 \text{ lb abductor compression}$$
$$\overline{225 \text{ lb total hip joint compression}}$$

The 225-lb joint reaction force is half the 450 lb of compression ordinarily generated in single-limb support. This reduction is enough to relieve some of the pain symptoms experienced by someone with arthritic changes in the joint. The compensatory lean is instinctive and commonly seen in people with hip joint disability. When a lateral trunk lean is seen during gait and is due to hip abductor muscle weakness, it is known as a **gluteus medius gait**. If the same compensation is due to hip joint pain, it is known as an **antalgic gait**.

In some instances, the 75-lb abductor force we calculated as necessary to stabilize the pelvis using a gluteus medius gait is still beyond the work capacity of very weak or completely paralyzed hip abductors. In such cases of extreme abductor muscle weakness, the pelvis will drop to the unsupported side even in the presence of a lateral trunk lean to the supported side. If lateral lean and pelvic drop occur during walking, the gait deviation is commonly referred to as a **Trendelenburg gait**.

Whether a lateral trunk lean is due to muscular weakness or pain, a lateral lean of the trunk during walking still uses more energy than ordinarily required for single-limb support and may result in stress changes within the lumbar spine if used over an extended time period. Using a cane or some other assistive device offers a realistic alternative to the person with hip pain or weakness.

Use of a Cane Ipsilaterally

Pushing downward on a cane held in the hand on the *side of pain or weakness* would reduce the superimposed body weight by the amount of downward thrust. That is, some of the weight of HAT would follow the arm to the cane, rather than arriving on the sacrum and the supporting hip joint. Assuming that our 180-lb subject can push down on the cane with approximately 15 percent of his body weight,[20] he will reduce his 150 lb of gravitational force (HAT and lower limb) acting on the pelvis by about 25 lb. With the gravitational force diminished to 125 lb, the torque of gravity is reduced to 500 in-lb. This assumes that the subject will not laterally lean (the goal of using the cane) and that the MA of gravity will remain at 4 in as previously estimated (125 lb × 4 in = 500 in-lb). With the torque of gravity at 500 in-lb, the required force of the abductors acting through a 2-in MA is reduced to 250 lb (500 in-lb / 2 in). The new hip joint reaction force using a cane *ipsilaterally* would then be

$$125 \text{ lb body weight compression}$$
$$+ \, 250 \text{ lb abductor compression}$$
$$\overline{375 \text{ lb total hip joint compression}}$$

The 375 lb of hip joint compression when using a cane ipsilaterally provides some savings over the 450 lb of compression ordinarily seen in unilateral stance. It is still greater, however, than the 225 lb found with a compensatory lateral trunk lean. Although a cane used ipsilaterally provides some benefits in energy expenditure and structural stress reduction, it is not as effective in reducing hip joint compression as the undesirable lateral lean of the trunk. *Moving the cane to the opposite hand produces substantially different and better results.*

Use of a Cane Contralaterally

When the cane is moved to the side *opposite the painful or weak hip joint*, the reduction in HAT is the same as it is for ipsilateral usage. That is, the superimposed

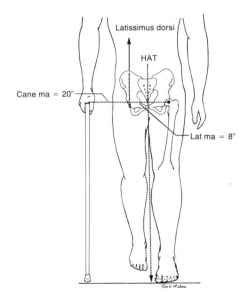

Figure 10–31. When a cane is placed in the hand opposite the painful supporting hip, the weight passing through the left hip is reduced, and activation of the right latissimus dorsi provides a countertorque to that of HAT and diminishes the need for a contraction of the left hip abductors. The moment arm of the cane is estimated to be 20 in whereas the moment arm of the latissimus dorsi is estimated to be 8 in.

body weight is reduced by the 25 lb of downward thrust (approximately 15 percent of body weight) to 125 lb in our 180-lb individual. However, we now have the cane substantially farther from the hip joint than it would be if used on the same side (Fig. 10–31). That is, in addition to relieving some of the superimposed body weight, the cane is now in a position to provide a countertorque to the torque of gravity.[37] We already calculated that a superimposed body weight of 125 lb would yield a gravitational adduction torque of 500 in-lb (noting again that these are hypothetical values that only estimate the actual forces). Estimating an average **MA** of 20 in between the downward thrust on the cane and the supporting hip joint[20,22] (Fig. 10–31), the cane would generate an opposing (abduction) torque of

$$25 \text{ lb} \times 20 \text{ in} = 500 \text{ in-lb}$$

That is, the torque of the cane would be equivalent to the torque produced by gravity, completely eliminating the need for a hip abductor force. The hip joint reaction force without a necessary contraction of the hip abductors would be only that imposed by the body weight itself, or 125 lb.

The classic description of forces using a cane on the side opposite a hip joint disability as presented above can be found in numerous texts and journal articles. However, few, if any, address the question of *how* the downward thrust of the arm on the cane actually acts on the pelvis. The equilibrium of an object such as the pelvis can only be affected by forces actually applied to that object. The explanation of use of the cane lacks cohesiveness if the force cannot be explained in terms of a force applied to the pelvis. Although conjectural, we propose that the force of the downward thrust on the cane arrives on the pelvis through a contraction of the latissimus dorsi.

It is well established that the latissimus dorsi is a depressor of humerus[23] through both its humeral attachment and its more variable scapular attachment,[2] and has been classically defined as the "crutch-walking muscle."[38,39] Since the downward thrust on the cane is accomplished through shoulder depression just as crutch walking is, it is logical to assume that the latissimus dorsi is active. The attachment of the latissimus dorsi to the posterior iliac crest would create an upward pull on the crest on the side of the cane (which is opposite to the affected supporting hip) as shown in Figure 10–31. An upward pull on the side of the pelvis opposite to the supporting hip joint axis creates an abduction torque around the supporting hip joint. The magnitude of latissimus dorsi muscle contraction should be approximately the same as the downward thrust on the cane (25 lb) under the supposition that this muscle initiates the thrust. Although measures of the MA of the pull of the latissimus dorsi on the posterior iliac crest are not readily available, we might estimate that the line of pull is directly above the acetabulum (Fig. 10–31). Since the gravitational force is estimated to lie 4 in from the supporting hip joint, the line of pull of the latissimus should lie about 4 in to the other side of the LOG above the opposite hip joint axis. That is, the MA of the pull of the latissimus can be estimated to be 8 in from the axis of the supporting hip joint. The abduction torque of the latissimus dorsi muscle can then be estimated at

$$25 \text{ lb} \times 8 \text{ in} = 200 \text{ in-lb}$$

If this is a more accurate estimate of the effect of a cane held in the hand opposite to the painful or weak hip, some contraction of the hip abductors would still be needed. Given the estimated 500 in-lb adduction torque created by gravity (removing 25 lb of superimposed weight through the cane), 200 in-lb would be counteracted by the abduction torque of the latissimus dorsi, but 300 in-lb of abduction torque would still need to be generated by the hip abductor muscles if equilibrium is to be maintained. Given the 2-in MA of the abductor muscles, a 150-lb contraction would be needed (300 in-lb/2 in). The total hip joint compression would be

$$
\begin{array}{r}
125 \text{ lb body weight compression} \\
+ \ 150 \text{ lb abductor compression} \\
\hline
275 \text{ lb total hip joint compression}
\end{array}
$$

Note that the hip joint compression estimate does *not* include compression caused by the latissimus dorsi muscle since it does not cross the hip joint. This estimate of joint reaction force is greater than the 125 lb of force estimated by the classic analysis of use of a cane, but more in line with the observation that the hip abductors cannot be completely relieved of their role in unilateral stance by using a cane. A cane is considered to relieve the hip joint of 60 percent of its load in stance,[14] an estimate that is in complete agreement with the forces we calculated for action of the latissimus dorsi on the pelvis.

Adjustment of a Carried Load

When someone with hip joint pain or weakness carries a load, the superimposed weight acting around the supporting hip in unilateral stance will increase by the weight of the load. However, the load also will result in a shift in the LOG toward the painful hip if held on the side of pain, or away from the hip if held in the opposite hand. Any shift of the LOG toward the painful hip should result in a reduction of the weight-bearing torque. Conversely, carrying a load opposite to the painful hip will increase the torque of the superimposed weight and necessitate increased contraction of the hip joint abductors. Neumann and Cook[40] measured EMG activity in the gluteus medius during varying load-carrying conditions. They found that a load of 10 percent of body weight carried ipsilateral to the measured abductors actually reduced the need for hip abductor activity as measured by EMG over the no-load condition, whereas a load of 20 percent of body weight was statistically similar to the activity needed under no-load conditions. That is, a 10 percent load did not add much weight to HAT, but apparently displaced the LOG closer to the hip. Displacing the LOG medially reduced the gravitational torque and, therefore, reduced the need for hip abductors. A 20 percent load similarly displaced the LOG, but apparently added enough weight to HAT that the product of increased load and decreased MA were equivalent to the no-load condition.

When the load was carried in the hand opposite to the measured hip, substantial increases in EMG activity were seen in the abductor muscles as compared to no load. This load condition increased the magnitude of HAT *and* displaced the LOG *away* from the hip joint, increasing the gravitational torque and the need for hip abductor activity. In fact, doubling the contralateral load from 10 to 20 percent of body weight tripled the EMG activity of the abductor muscles. Guidelines for carrying loads for persons with hip joint pain or weakness may help minimize increases in the joint reaction force across the hip (with or without the concomitant use of a cane). Use of a cane contralaterally and carrying a load ipsilaterally can effectively reduce the need for abductor force and minimize the joint reaction force in a weak or painful hip.

HIP JOINT PATHOLOGY

The foundation should be clearly laid at this point for an understanding of dysfunction at the hip joint. The very large active and passive forces crossing the joint make it susceptible to wear and tear of normal components and to failure of weakened components. Small changes in the biomechanics of the femur or the acetabulum can result in increases in passive forces above normal levels or in weakness of the dynamic joint stabilizers. Some of the more common problems and the underlying mechanisms will be discussed in this section.

Arthrosis

The most common painful condition of the hip is due to deterioration of the articular cartilage and to subsequent related changes in articular tissues.[22] Known as **osteoarthritis, degenerative arthritis,** or perhaps most appropriately as **hip joint arthrosis,** prevalence rates are about 10 to 15 percent in those over 55 years of age, with approximately equal distribution among men and women.[41] Although trauma or malalignment such as femoral anteversion[34] may be associated with occurrence, 50 percent of the cases are considered to be idiopathic.[6] That is, half of the cases of hip joint arthrosis have no evident underlying pathology; changes may be due to subtle deviations present from birth, to tissue changes inherent in aging, to the repetitive mechanical stress of loading the body weight on the hip joint over a prolonged period, or to interactions of each of these factors. The factors most closely associated with idiopathic hip joint arthrosis are increased age and increased weight-to-height ratio.[42]

Bullough and associates[4] found in accordance with other studies that cartilaginous deterioration in the hip joint is common at the periphery of the femur and at the perifoveal area, while the most common defects in the acetabular cartilage were at the superior margin (dome). They proposed that the degenerative changes at this area of the acetabulum may well be due to *inadequate* loading since, as described previously, large forces are needed to create full joint congruence. Forces in excess of half the body weight may be needed to compress the head into contact with the superior aspect of the acetabulum. However, forces at the hip generally do not exceed half the body weight other than in unilateral stance and activities requiring unilateral weight bearing. Using a number of other studies as a base, Bullough and associates[4] hypothesized that given the variety of individual activities of daily living, we probably spend no more than 5 to 25 percent of our time in unilateral lower extremity weight-bearing activities. This may be inadequate to maintain flow of nutrients and wastes through the avascular cartilage.

The hypothesis that arthrotic changes in the superior acetabulum may be due to inadequate (rather than excessive) forces at the joint bears some support from the *lack* of conclusive association between increased activity in sports or recreational activities and increased incidence of hip joint osteoarthritis.[43]

Fracture

Although the weight-bearing forces coming through the hip joint may cause deterioration of the articular cartilage, the bony components must also be of sufficient strength to withstand the forces that are acting around and through the hip joint. As noted in the section on the weight-bearing structure of the hip joint, the vertical weight-bearing forces that pass down through the superior margin of the acetabulum in both unilateral and bilateral stance act at some distance from ground reaction force up the shaft of the femur. The result is a bending force across the femoral neck (Fig. 10–14). Normally the trabecular systems are capable of resisting the bending forces, but abnormal increases in the magnitude of the force or weakening of the bone can lead to bony failure. The site of failure is likely to be in areas of thinner trabecular distribution such as the so-called zone of weakness (Fig. 10–13).

Bony failure in the femoral neck is uncommon in the child or young adult even with large applied loads. However, femoral neck fractures occur at the rate of about 98 per 100,000 people in the United States, with the average age of occurrence being in the 70s. Hip joint fractures occur among white adults at a much higher rate than among black adults.[44] There is a predominance of fractures in women, although this is certainly influenced by their greater longevity. In the middle-aged group, women actually suffer fewer hip fractures than do men, although the fractures in this age group are usually attributable to substantial trauma.[44] In 87 percent of cases of hip fracture among the elderly population, the precipitating factor appears to be moderate trauma such as that caused by a fall from standing, from a chair, or from bed. There is consensus that hip fracture is associated with, but not exclusively due to, diminished bone density.[41] Bone density decreases about 2 percent per year after age 50 and trabeculae clearly thin and disappear with aging.[45] Cummings and Nevitt[45] believe the exponential increase in hip fractures with age cannot be accounted for by decreased bone density alone and propose that the slowed gait characteristic of the elderly may play an important part. They contend that the slowing of gait makes it less likely that momentum will carry the body forward in a fall (generally on to an outstretched hand), and more likely that the fall will occur backward on to the hip area weakened by bone loss and no longer padded by the fat and muscle bulk of youth.

Hip fracture will continue to receive considerable attention because of the high health care costs of both conservative and operative treatment. Not only is the condition painful, but malunion of the fracture can lead to joint instability and/or cartilaginous deterioration due to poorly aligned bony segments. Although the femoral head receives blood supply via the ligament of the femoral head in most people, an absent or diminished supply through the head means reliance on anastomoses from the circumflex arteries. This alternative supply may be disrupted by femoral neck fracture, leaving the femoral head susceptible to avascular necrosis and requiring replacement of the head of the femur with an artificial implant. Femoral neck fracture also has an associated mortality rate that may be as high as 20 percent.[44]

Bony Abnormalities of the Femur

When the bony structure of the femur is altered through abnormal angles of torsion or inclination, subsequent changes in the direction and magnitude of the forces acting around the hip can lead to other pathologic conditions such as increased likelihood of joint arthrosis, increased likelihood of femoral neck fracture, or muscular weakness. The normal angles of inclination and torsion appear to represent optimal balance of stresses and muscle alignment. Alterations may actually appear to result in advantages relative to some functions, but are always accompanied by concomitant disadvantages relative to others.

In coxa valga the angle of inclination is greater than the normal adult angle of 125°. The increased angle brings the vertical weight-bearing line closer to the shaft of the femur, diminishing the shear force across the femoral neck. However, the decreased distance between the femoral head and the greater trochanter also decreases the length of the MA of the hip abductor muscles. The decreased muscular MA results in an increased demand for muscular force generation to counterbalance the gravitational moment acting around the supporting hip joint during single-limb support. Either the additional muscular force requirement will increase the total joint reaction force within the hip joint, or the abductor muscles will be unable to meet the increased demand and will be functionally weakened. Although the abductors may be otherwise normal, the reduction in biomechanical effectiveness may produce the compensations typical of primary abductor muscle weakness. Coxa valga also decreases the amount of femoral articular surface in contact with both the superior and central acetabulum, increasing the amount of articular surface exposed superiorly. Consequently, coxa valga decreases the stability of the hip and predisposes the hip to dislocation.[8,9,14]

Coxa vara is considered to give the advantage of improved hip joint stability (if not too extreme an angle reduction). The apparent improvement in congruence occurs because the decreased angle between the neck and shaft of the femur will turn the femoral head deeper into the acetabulum, decreasing the amount of articular surface exposed superiorly. A varus femur, if not due to trauma, also increases the length of the MA of the hip abductor muscles by increasing the distance between the femoral head and the greater trochanter. The increased MA decreases the amount of force that must be generated by the abductor muscles to counterbalance the gravitational moment in single-limb support. The result is a reduction in the joint reaction force. However, coxa vara has the disadvantage of increasing the bending moment along the femoral head and neck. This increase in bending force can actually be seen by the increased density of trabeculae laterally in the femur, due to the increase in tensile stresses.[17] The increased shear force along the femoral neck will increase the predisposition toward femoral neck fracture.[9,14] It may also increase the likelihood in the adolescent child that the femoral head will slide on the cartilaginous epiphysis of the head of the femur. In childhood, the epiphysis is perpendicular to the vertical weight-bearing forces.[2] Consequently, the superimposed weight merely compresses the head into the epiphyseal plate. In adolescence, growth of the bone results in a more oblique orientation of the epiphyseal plate. The epiphyseal obliquity makes the plate more vulnerable to shear forces at a time when the plate is already weakened by the rapid growth that occurs during this period of life.[41] Weight-bearing forces may slide the femoral head inferiorly, resulting in a **slipped capital femoral epiphysis**. As is true for a hip fracture, the altered biomechanics and at-risk blood supply require that normal alignment be restored before secondary degenerative changes can take place.

Variations in the angle of torsion also affect hip biomechanics and function. Retroversion, which results in a relative lateral rotation of the femur, enhances hip stability but may cause out-toeing during walking. Anteversion of the femoral neck predisposes the hip joint to anterior dislocation of the head of the femur and is a common cause of in-toeing among children.[8] Anteversion may also cause the line of the hip abductors to fall more posterior to the joint, reducing the MA for abduction.[34] As is true for coxa valga, the resulting need for additional abductor muscle force may increase joint reaction forces, resulting in predisposition to joint arthrosis, or may functionally weaken the joint, producing energy-consuming and wearing gait deviations.

SUMMARY

The normal hip joint is well designed to withstand the forces that act through and around it, assisted by the trabecular systems, cartilaginous coverings, muscles, and ligaments. Alterations in the direction or magnitude of forces acting around the hip create abnormal concentrations of stress that predispose the joint structures to injury and degenerative changes. The degenerative changes, in turn, can create additional alterations in function that not only affect the hip joint's ability to support the body weight in standing, in locomotor activities, and other activities of daily living but may also result in adaptive changes at more proximal and distal joints. Consequently, the reader must understand both the dysfunction that might occur at the hip *and* the associated dysfunctions that may result in or from dysfunction elsewhere in the lower extremity and spine. The remaining chapters of this text will focus not only on primary dysfunction at a joint complex but on associated dysfunction related to proximal and distal joint problems.

REFERENCES

1. Gowitzke, BA and Milner, M: Scientific Bases of Human Movement, ed. 3. Williams & Wilkins, Baltimore, 1988.
2. Williams, PL and Warwick, R (eds): Gray's Anatomy, ed 37. WB Saunders, Philadelphia, 1985.
3. Brinckmann, P, Frobin, W, and Hierholzer, E: Stress on the articular surface of the hip joint in healthy adults and persons with idiopathic osteoarthrosis of the hip joint. Biomechanics 14:149–156, 1981.
4. Bullough, P, Goodfellow, J, and O'Connor, J: The relationship between degenerative changes and load-bearing in the human hip. J Bone Joint Surg 55B:746–758, 1973.
5. Adna, S, et al: The acetabular sector angle of the adult hip determined by computed tomography. Acta Radiol Diagn 27:443–447, 1986.
6. Brinckmann, P, Hoefert, H, and Jongen, HT: Sex differences in the skeletal geometry of the human pelvis and hip joint. Biomechanics 1:427–430, 1981.
7. Svenningsen, S, et al: Regression of femoral anteversion. Acta Orthop Scand 60:170–173, 1989.
8. Kapandji, IA: The Physiology of the Joints, Vol. 2. Williams & Wilkins, Baltimore, 1970.
9. Singleton, MC and LeVeau, BF: The hip joint: Stability and stress. A review. Phys Ther 55:9, 1975.
10. Rosse, C and Clawson, DK: The Musculoskeletal System in Health and Disease. Harper & Row, Hagerstown, Md., 1980.
11. Radin, EL: Biomechanics of the human hip. Clin Orthop 152:28–34, 1980.
12. Steindler, A: Kinesiology of the Human Body. Charles C Thomas, Springfield, Ill., 1973.
13. D'Ambrosia, RD: Musculoskeletal Disorders: Regional Examination and Differential Diagnosis, ed. 2, JB Lippincott, Philadelphia, 1986.
14. Radin, EL, et al: Practical Biomechanics for the Orthopedic Surgeon. John Wiley & Sons, New York, 1979.
15. Crock, HV: An atlas of the arterial supply of the head and neck of the femur in man. Clin Orthop 152:17–27, 1980.
16. Pauwels, F: Biomechanics of the Normal and Diseased Hip. Berlin, Springer-Verlag, 1976.
17. Bombelli, R, Santore, RF, and Poss, R: Mechanics of the normal and osteoarthritic hip: A new perspective. Clin Orthop 182:69–78, 1984.
18. Norkin, CC and White, DJ: Measurement of Joint Motion: A Guide to Goniometry, FA Davis, Philadelphia, 1985.
19. Pathokinesiology and Physical Therapy Department: Observational Gait Analysis Handbook. Rancho Los Amigos Medical Center, Professional Staff Association, Rancho Los Amigos Hospital, Downey, Calif., 1989.
20. Inman, VT, Ralston, HJ, and Todd, F: Human Walking. Williams & Wilkins, Baltimore, 1981.
21. Kendall, FP and McCreary, EK: Muscles: Testing and Function, ed. 3. Williams & Wilkins, Baltimore, 1983.

22. Cailliet, R: Soft Tissue Pain and Disability, ed 2. FA Davis, Philadelphia, 1988.
23. Basmajian, JV and DeLuca, CJ: Muscles Alive, ed 5. Williams & Wilkins, Baltimore, 1985.
24. Nemeth, G and Ohlsen, H: In vivo moment arm lengths for hip extensor muscles at different angles of hip flexion. J Biomech 18:129–140, 1985.
25. Daniels, L and Worthingham, C: Manual Muscle Testing, ed 5. WB Saunders, Philadelphia, 1986.
26. Wheatly, MD and Jahnke, WD: Electromyographic study of the superficial thigh and hip muscles in normal individuals. Arch Phys Med 32:508, 1951.
27. Kaplan, EB: The iliotibial tract. J Bone Joint Surg 40A:825–832, 1958.
28. Murray, MP and Sepic, SB: Maximum isometric torque of hip abductor and adductor muscles. Phys Ther 48:2, 1968.
29. Soderburg, GL and Dostal, WF: Electromyographic study of three parts of the gluteus medius muscle during functional activities. Phys Ther 58:6, 1978.
30. Jensen, RH, Smidt, GL, and Johnston, RC: A technique for obtaining measurements of force generated by hip muscles. Arch Phys Med 52:207, 1971.
31. Olson, VL, Smidt, GL, and Johnston, RC: The maximum torque generated by the eccentric, isometric and concentric contractions of the hip abductor muscles. Phys Ther 52:2, 1972.
32. Nordin, M and Frankel, VH: Basic Biomechanics of the Skeletal System, ed. 2. Lea & Febiger, Philadelphia, 1989.
33. LeVeau, B: Williams and Lissner's Biomechanics of Human Motion, ed 2. WB Saunders, Philadelphia, 1977.
34. Clark, JM and Haynor, DR: Anatomy of the abductor muscles of the hip as studied by computed tomography. J Bone Joint Surg 69A(7):1021–1031, 1987.
35. Paul, JP and McGrouther, DA: Forces transmitted at the hip and knee joint of normal and disabled persons during a range of activities. Acta Orthop Belg (Suppl) 41:78–88, 1975.
36. Crowinshield, RD, et al: A biomechanical investigation of the human hip. J Biomech 11:75, 1976.
37. Blount, WF: Don't throw away the cane. J Bone Joint Surg 18A:3, 1956.
38. Lehmkuhl, LD and Smith, LK: Brunnstrom's Clinical Kinesiology, ed 4. FA Davis, Philadelphia, 1983.
39. Schenkman, M and DeCartaya, VR: Kinesiology of the shoulder complex. J Orthop Sports Phys Ther 8:438–450, 1987.
40. Neumann, DA and Cook, TM: Effect of load and carrying position on the electromyographic activity of the gluteus medius muscle during walking. Phys Ther 65:305–311, 1985.
41. Kelsey, J: The epidemiology of diseases of the hip: A review of the literature. Int J Epidemiol 6:269–280, 1977.
42. Pogrund, H, Bloom, R, and Mogle, P: Normal width of the adult hip joint: The relationship to age, sex and obesity. Skel Radiol 10:10–12, 1983.
43. Panush, RS and Brown, DG: Exercise and arthritis. Sports Med 4:54–64, 1987.
44. Lewinnek, GE, et al: The significance and a comparative analysis of the epidemiology of hip fractures. Clin Orthop 152:35–43, 1980.
45. Cummings, SR and Nevitt, MC: A hypothesis: The causes of hip fracture. J Gerontol Med Sci 44:M107–M111, 1989.

STUDY QUESTIONS

1. Which trabecular systems in the femoral head and neck are designed to resist compressive forces produced by contractions of the hip abductors? Which trabecular systems are designed to resist the forces created by gravitational torque?

2. How can the compressive forces acting at the hip joint in unilateral stance be reduced?

3. Demonstrate how variations in the angle of inclination affect the MA of the hip abductors by drawing the following: a normal angle of inclination at the hip, the angle in coxa vara, and the angle in coxa valga. Please include the action line and the MA of the hip abductors in the diagram.

4. Describe what would happen to the pelvis in left unilateral stance when the left hip abductors are paralyzed. How is equilibrium maintained in this situation?

5. Describe motion at the right and left hip joints and at the lumbar spine during hiking of the pelvis in right limb stance.

6. Describe the position of greatest hip stability. Explain the reasons for your answer.

7. How do the forces acting at the right hip joint in erect bilateral stance compare with the forces acting at the right hip joint during unilateral stance?

8. Compare the structure and function of the gluteus medius muscle with that of the deltoid.

9. Compare lumbar-pelvic rhythm with scapulohumeral rhythm.

10. What bony abnormality at the hip predisposes the hip to the possibility of anterior dislocation? Why?

11. Which structures at the hip limit the extremes of motion in flexion? extension? lateral rotation? medial rotation? abduction and adduction?

12. How does the position of the knee affect muscle action at the hip?

13. Forward rotation of the pelvis on the right is accompanied by what motion at the stance left hip?

14. If a person has a painful right hip, in which direction should he or she lean his or her trunk to reduce the forces on the right hip during right unilateral support? Explain the reasons for your answer.

15. Why is it preferable to use a cane on the side opposite to a muscle weakness rather than on the same side?

Chapter 11

■ ■ ■

The Knee Complex

■

OBJECTIVES
Following the study of this chapter, the reader should be able to:

Describe
1. The articulating surfaces at the tibiofemoral and patellofemoral joints.
2. The joint capsule.
3. The anatomic and mechanical axes of the knee.
4. Motion of the femoral condyles during flexion and extension in a closed kine-matic chain.
5. Motion of the tibia in flexion and extension in an open kinematic chain.

Draw
1. The Q angle when given an illustration of the lower extremity.
2. Moment arm of the quadriceps at the following degrees of knee flexion: 90, 130, 30, and 10.
3. The action lines of the vastus lateralis and the vastus medialis oblique.

Locate
1. The attachments of all the muscles at the knee.
2. The bursae surrounding the knee.
3. The attachments of the ligaments of the medial and lateral compartments.

Identify
1. Structures that contribute to medial, lateral, and anterior-posterior stability of the knee, including dynamic and static stabilizers.
2. Structures that contribute to rotatory stability at the knee.
3. The normal forces that are acting at the knee.

Compare
1. The knee and the elbow joint on the basis of similarities/dissimilarities in structure and function.
2. The lateral with the medial meniscus on the basis of structure and function.
3. The forces on the patellofemoral joint in full flexion with full extension.
4. The action of the quadriceps in an open kinematic chain with that in a closed kinematic chain.
5. The effectiveness of the hamstring muscles as knee flexors in each of the following hip positions: hyperextension, 10° of flexion, and full flexion (open kinematic chain).
6. The effectiveness of the rectus femoris muscle as a knee extensor at 60° of knee flexion with its effectiveness at 10° of knee flexion.

Explain
1. The function of the menisci.
2. How a tear of the medial collateral ligament may affect joint function.
3. The functions of the suprapatellar, gastrocnemius, infrapatellar, and prepatellar bursae.
4. Why the semiflexed position of the knee is the least painful position.
5. Why the knee may be more susceptible to injury than the hip joint.

The knee complex is similar to the elbow complex in that flexion and extension of the knee produces a functional shortening and lengthening of the extremity. In addition to providing mobility, however, the knee complex plays a major role in supporting the body during dynamic and static activities. In a closed kinematic chain the knee joint works in conjunction with the hip joint and ankle to support the body weight in the static erect posture. Dynamically, the knee complex is responsible for moving and supporting the body in sitting and squatting activities and for supporting and transferring the body weight during locomotor activities. In an open kinematic chain the knee provides mobility for the foot in space. The fact that the knee must fulfill major stability as well as major mobility roles is reflected in its structure and function. The knee is not only one of the largest joints in the body but is also the most complex. Engineers only recently have been able to develop prostheses that are capable of simulating some of the functions of the knee joint.[1]

The knee complex is composed of two distinct articulations located within a single joint capsule: the tibiofemoral joint and the patellofemoral joint. The tibiofemoral joint is the articulation between the distal femur and the proximal tibia. The patellofemoral joint is the articulation between the patella and the femur. Although the patella serves the tibiofemoral mechanism, the characteristics, responses, and problems of the patellofemoral joint are distinct enough from the tibiofemoral joint to warrant separate attention. The superior tibiofibular joint is not considered to be a part of the knee complex since it is not contained within the knee joint capsule and is functionally related to the ankle joint; it will be covered in Chapter 12.

STRUCTURE OF THE TIBIOFEMORAL JOINT

The tibiofemoral, or knee joint, is considered to be a double condyloid joint (defined by the medial and lateral articular surfaces) with 2° of freedom of motion.[2] Flexion and extension occur in the sagittal plane around a coronal axis, while medial and lateral rotation occur in the transverse plane about a vertical axis. Careful examination of the articular surfaces and the relationship of the surfaces to each other will facilitate an understanding of the movements at the knee joint and both the functions and dysfunctions common to the joint.

Femoral Articular Surface

The large medial and lateral condyles on the distal femur form the proximal articular surfaces of the knee joint. The condyles have a large and very obvious curvature anteroposteriorly, but are also each slightly convex in the frontal plane. The AP convexity of the condyles is not a consistent spherical shape but has a smaller radius of curvature posteriorly (Fig. 11–1). The two condyles are separated by the intercondylar notch or fossa through most of their length, but are joined anteriorly by an asymmetric, shallow, saddle-shaped groove called the **patellar groove or surface**; the patellar surface is separated from the tibial articular surface by two slight grooves that run obliquely across the condyles (Fig. 11–2a).[2] The shaft of the femur is not vertical but is angled in such a way that the femoral condyles do not lie immediately below the femoral head, but somewhat medial. Given the obliquity of the shaft of the femur, the lateral condyle lies more directly in line with the shaft than does the medial condyle (Fig. 11–2b). The articular surface of the lateral condyle is also not as long as the articular surface of the medial femoral condyle. When the femur is examined through an inferior view (Fig. 11–2a), the lateral condyle would appear *at first glance* to be longer. However, when the patellofemoral surface is excluded, it can be seen that the lateral tibial surface stops before the medial. The medial femoral condyle is, on average, two thirds of an inch longer than the lateral

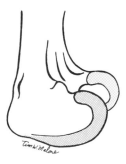

Figure 11–1. The A-P convexity of the condyles is not consistently spherical, being more accentuated posteriorly.

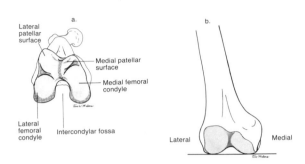

Figure 11–2. a. The patellar surface is separated from the tibial articular surface by two slight grooves that run obliquely across the condyles. The medial femoral condyle is longer than the lateral femoral condyle; the lateral lip of the patellar surface is larger than the medial lip of the patellar surface. b. Given the obliquity of the shaft of the femur, the lateral femoral condyle lies more directly in line with the shaft than does the medial. The medial condyle is more prominent, however, resulting in a horizontal distal femoral surface in spite of the oblique shaft.

condyle.[3] The medial condyle extends further distally than the lateral so that, in spite of the angulation of the shaft of the femur, the distal end of the femur is essentially horizontal (Fig. 11–2b).

Tibial Articular Surface

The articulating surfaces on the tibia that correspond to the femoral articulating surfaces are the two concave, asymmetric medial and lateral tibial condyles or plateaus (Fig. 11–3a). The proximal tibia is enlarged as compared to the shaft and overhangs the shaft posteriorly (the lateral condyle more so than the medial). The articulating surface of the medial tibial condyle is 50 percent larger than that of the lateral condyle (corresponding to the larger medial femoral condyle) and the articular cartilage of the medial tibial condyle is three times thicker.[3] The two tibial condyles are separated by a roughened area and two bony spines called the **intercondylar tubercles.** These tubercles become lodged in the intercondylar notch of the femur during knee extension (Fig. 11–3b).

Tibiofemoral Articulation

When the large articular condyles of the femur are placed on the shallow concavities of the tibial condyle, the incongruence of the knee joint is quite evident. As has been true elsewhere in the body, articular incongruence at the knee is accompanied by an accessory joint structure that enhances congruence and assists

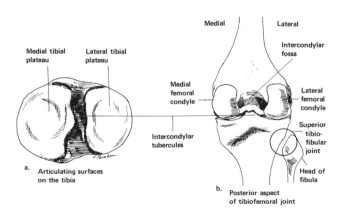

Figure 11–3. a. A superior view of the articulating surfaces on the tibia illustrates differences in size and configuration between the medial and lateral tibial plateaus. b. A view of the posterior aspect of the tibiofemoral joint illustrates how the tibial intercondylar tubercles become lodged in the femoral intercondylar fossa when the joint is fully extended.

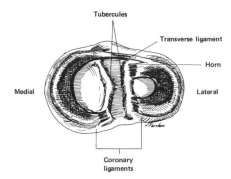

Figure 11-4. Structure of the menisci. A superior view of the menisci illustrates differences in size and configuration between the medial and lateral menisci. The medial meniscus is C-shaped, whereas the lateral meniscus is shaped like a ring, or circle.

in the balance between mobility and stability needed by the joint. Each of the condyles of the knee joint has its own accessory joint structure, together known as the **menisci** of the knee.

Menisci

Two asymmetric fibrocartilaginous joint disks called menisci are located on the tibial condyles (Fig. 11–4). The medial meniscus is a semicircle, while the lateral meniscus is four fifths of a ring.[2] Both menisci are open toward the intercondylar area, thick peripherally and thin centrally, forming concavities into which the respective femoral condyles can sit (Fig. 11–5). The wedge-shaped menisci increase the radius of curvature of the tibial condyles and, therefore, joint congru-

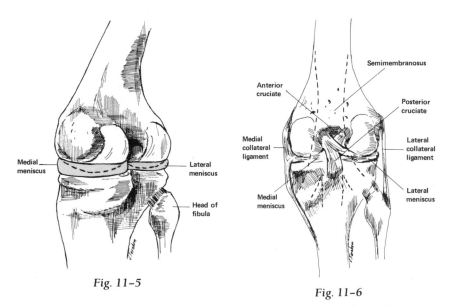

Fig. 11–5

Fig. 11–6

Figure 11-5. A posteromedial view of an extended right tibiofemoral joint shows the menisci tightly interposed between the femur and the tibia. The dotted lines indicate the wedge shape of the menisci and show how the menisci deepen and contour the tibial articulating surface to accommodate the femoral condyles.
Figure 11-6. Meniscal attachments. The medial meniscus is shown with its attachments to the medial collateral ligament, anterior cruciate ligament (ACL), and the outline of the semimembranosus muscle. The lateral meniscus is shown with its attachments to the posterior cruciate ligament (PCL). Its attachments to the popliteus muscle are not shown in the drawing.

Table 11–1. MENISCAL ATTACHMENTS

	Medial Meniscus	Lateral Meniscus
Common attachments	Intercondylar tubercles of the tibia Tibial condyle via coronary ligaments Patella via patellomeniscal or patellotibial ligaments Transverse ligament	
Unique attachments	Medial collateral Semimembranosus muscle	Posterior meniscofemoral ligament Posterior cruciate ligament [Anterior cruciate ligament]° Popliteus muscle

°Variable.

ence. By increasing congruence (articular contact), the menisci also play an important part in distributing weight-bearing forces and in reducing friction between the joint segments. Each meniscus has multiple attachments to surrounding structures, some common to both and some unique to each.

The open ends of the menisci are called **horns** and are attached to their respective tibial intercondylar tubercles. There are a number of common connections of the two menisci to the knee joint capsule. Each is connected around its periphery to the tibial condyle by the coronary ligaments, which are composed of fibers from the knee joint capsule. Both menisci are also attached directly or indirectly to the patella via the so-called **patellomeniscal**[4,5] or **patellotibial ligaments,**[6] which are anterior capsular thickenings. The anterior horns of the menisci are joined to each other by the transverse ligament, which may be connected to the patella via the joint capsule.[4]

The lateral meniscus, in addition to the connections it shares with the medial meniscus, is attached to the posterior cruciate ligament (PCL) (Fig. 11–6) and popliteus muscle via the coronary ligaments and posterior capsule,[7] and to the somewhat variable posterior meniscofemoral ligaments, which attach to the medial femoral condyle.[2,8] Some fibers from the anterior cruciate ligament (ACL) may also join the anterior and posterior horns.[8] The connections of the lateral meniscus are considered to be fairly loose, leaving the lateral menisci a fair amount of mobility on the lateral tibial condyle.

The medial meniscus is attached to the medial collateral ligament (Fig. 11–6) and to the semimembranosus muscle through its capsular connections.[2,9] The medial meniscus is more firmly attached and less movable on the tibial condyle than the lateral meniscus. Its lack of mobility may be one of several reasons why the medial meniscus is torn more frequently than the lateral meniscus. The meniscal attachments are summarized in Table 11–1.

The menisci and meniscoligamentous complex are well established in the 8-week-old embryo.[10] Initially well vascularized, the vascularity of the menisci gradually recedes centrally outward, reaching its adult form at 10 or 11 years of age.[10] The adult meniscus is vascularized only on the periphery by capillaries from the joint capsule and the synovial membrane. The pattern of vascularity may account for the low incidence of meniscal injuries in young children whose menisci have ample blood supply and the ability of the menisci in the adult to regenerate only in the vascularized peripheral region.[2,10]

Tibiofemoral Alignment and Weight-Bearing Forces

The anatomic (longitudinal) axis of the femur, as already noted, is oblique, directed inferiorly and medially from its proximal to its distal end. The anatomic

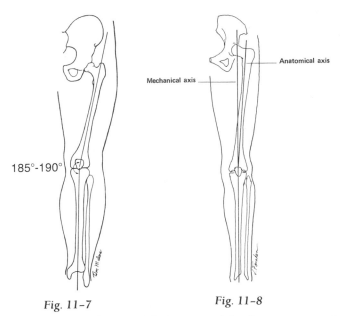

Fig. 11-7

Fig. 11-8

Figure 11-7. The long axis of the femur and the long axis of the tibia intersect to create a physiologic valgus at the knee joint of 185° to 190°.

Figure 11-8. The mechanical axis (weight-bearing line) of the lower extremity passes through the hip, knee, and ankle joints. Because the mechanical axis lies more nearly vertical than the longitudinal (anatomic) axis, weight-bearing forces are about equally distributed between the medial and lateral condyles of the knee joint.

axis of the tibia is directed almost vertically. Consequently, the femoral and tibial longitudinal axes normally form an angle *medially* at the knee joint of 185 to 190° (Fig. 11–7). That is, the femur is angled off vertical 5 to 10°,[4,11] creating a physiologic (normal) valgus angle at the knee. Although this might appear to weight the lateral condyles more than the medial, this is not the case. The mechanical axis of the lower extremity is the weight-bearing line from the center of the head of the femur to the center of the superior surface of the head of the talus[12] (Fig. 11–8). This line normally passes through the center of the knee joint between the inter-condylar tubercles, and averages 3° from the vertical given the width of the hip joints as compared to spacing of the feet.[4] Since the weight-bearing line (ground reaction force) follows the mechanical rather than the anatomic axes, the weight-bearing stresses on the knee joint in *bilateral static stance* are equally distributed between the medial and lateral condyles, without any concomitant horizontal shear forces.[12] This is not necessarily the case in unilateral stance or once *dynamic forces* are introduced to the joint. Deviations in normal force distribution may be caused, among other things, by an increase or decrease in the normal tibiofemoral angle.

If the *medial* tibiofemoral angle is greater than 195° (165° or less measured laterally), an abnormal condition called **genu valgum** ("knock knees") exists (Fig. 11–9). This condition will increase the compressive force on the lateral condyle, while increasing the tensile stresses on the medial structures. If the medial tibiofemoral angle is 180° or less (exceeding 180° as measured laterally), the resulting abnormality is called **genu varum** ("bow legs") (Fig. 11–10). In this condition, the compressive stresses on the medial tibial condyle are increased, whereas the tensile stresses are increased laterally. In either genu valgum or genu varum, constant overloading of, respectively, the lateral or medial articular cartilage may result in damage to the cartilage.

The menisci of the knee are important in distributing and absorbing the large

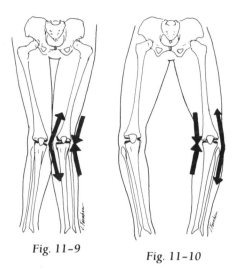

Fig. 11-9

Fig. 11-10

Figure 11-9. An increase in the normal medial valgus angle results in genu valgum or "knock knees." Arrows on the lateral aspect of the left tibiofemoral joint indicate the presence of compression forces, whereas the arrows on the medial aspect indicate the presence of distraction (tensile) forces.

Figure 11-10. A decrease in the normal medial valgus angle results in genu varum or "bow legs." Arrows on the lateral aspect of the left tibiofemoral joint indicate the presence of distraction (tensile) forces, whereas arrows on the medial aspect of the joint indicate the presence of compression forces.

forces crossing the knee joint. Although compressive forces in the dynamic knee joint ordinarily may reach two to three times body weight in normal gait[3] and five to six times body weight in activities such as running and stair-climbing,[13] the menisci assume 40 to 60 percent of the imposed load.[14] If the menisci are removed, the magnitude of the average load per unit area on the articular cartilage nearly doubles on the femur, while it is six or seven times greater on the tibial condyle.[13] Elimination of any angulation between the femur and tibia (a mild genu varum) will increase the compression on the medial meniscus by 25 percent. Five degrees of genu varum (medial tibiofemoral angle of 175°) will increase the forces by 50 percent.[11] The importance of the menisci to function can be highlighted by the body's response to complete removal of a meniscus. In such an instance, regeneration of at least the peripheral portion of the meniscus will occur through intact peripheral vascular structures. However, if the knee is excessively loaded or stressed before tissue regeneration can occur, articular damage may result.[2]

Knee Joint Capsule

Given the incongruence of the knee joint, even with the compensation of the menisci, stability is heavily dependent on the surrounding joint structures. In knee flexion when surrounding passive structures tend to be lax, the incongruence of the joint permits at least some anterior displacement, posterior displacement, and rotation of the tibia beneath the femur.[15,16] The knee joint capsule and its associated ligaments are critical to restricting such motions to maintain joint integrity and normal joint function. Although muscles clearly play a role in stabilization (as we shall examine more closely later in the chapter), it is almost impossible to effectively stabilize the knee with active muscular forces alone in the presence of substantial disruption of passive restraining mechanisms.

The joint capsule that encloses the tibiofemoral and patellofemoral joints is large, complexly attached, and lax with several recesses (Fig. 11-11). Posteriorly,

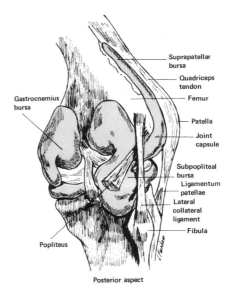

Suprapatellar
bursa

Quadriceps
tendon

Femur

Gastrocnemius
bursa

Patella

Joint
capsule

Subpopliteal
bursa
Ligamentum
patellae
Lateral
collateral
ligament

Fibula

Popliteus

Posterior aspect

Figure 11–11. This view of the posterolateral aspect of the knee complex shows the synovial lining of knee joint capsule and bursae.

the capsule is attached proximally to the posterior margins of the femoral condyles and intercondylar notch and distally to the posterior tibial condyle. The capsule is reinforced posteriorly by a number of muscles and by the oblique popliteal and arcuate ligaments. Medially and laterally, the capsule begins proximally above the femoral condyles to continue distally to the margins of the tibial condyle. The collateral ligaments reinforce the sides of the capsule. Anteriorly, the patella, the tendon of the quadriceps muscles superiorly, and the patellar ligament inferiorly complete the anterior portion of the joint capsule. Anteromedially and anterolaterally, expansions from the vastus medialis and vastus lateralis muscles extend from the patella and patellar ligament to the corresponding collateral ligaments and tibial condyles.[2] The anteromedial and anterolateral portions of the capsule are known as the **extensor retinaculum** or the **medial and lateral patellar retinacula.**[2]

Extensor Retinacula

The capsular and retinacular connections have been variously described and are the subject of some disagreement. There appear to be two layers, the deeper of the two having longitudinally oriented fibers connecting the capsule anteriorly to the menisci and tibia via the coronary ligaments.[5,6] These connections may be called the patellomeniscal[5] or the patellotibial bands.[6,17] The more superficial second layer consists of transversely oriented fibers of which the more proximal blend with fibers of the vastus medialis and lateralis muscles, and the more distal continue to the posterior femoral condyles. The transverse fibers connecting the patella and the femoral condyles are known as the **patellofemoral ligaments.**[6,17,18] The lateral patellofemoral ligament is connected not only to the vastus lateralis muscle but also to the iliotibial band either directly[6,18] or indirectly via an iliopatellar band.[19,20] The iliotibial band and its associated fascia lata are accompanied posteriorly by the tendon of the biceps femoris muscle to provide superficial reinforcement to the capsular and retinacular layers.[8]

Synovial Lining

The intricacy of the fibrous layer of the knee joint capsule is surpassed by its synovial lining, the most extensive and involved in the body.[2] The synovium adheres to the inner wall of the fibrous layer except posteriorly where the synovium invaginates anteriorly following the contour of the femoral intercondylar notch. The invaginated synovium adheres to the anterior aspect and sides of the ACL and the PCL. The infolding of the synovial lining results in the ACLs and PLCs being contained within the fibrous capsule, but not within the synovial sleeve.[2] Embryonically, the synovial lining of the knee joint capsule is actually divided into three separate compartments. There is initially a superior patellofemoral compartment and two separate medial and lateral tibiofemoral compartments. By 12 weeks of gestation, the synovial septa are resorbed to some degree, resulting in a single joint cavity, but retaining the posterior invagination of the synovium that forms some separation of the condyles.[21,22] The superior compartment continues to be recognizable as a superior recess of the capsule known as the **suprapatellar bursa.** The medial and lateral compartments, while not truly separate in the fully formed knee, are still referenced as a way of classifying joint structures that are considered to belong to one compartment or the other.[23,24] Posteriorly, the synovial lining may invaginate laterally between the popliteus muscle and lateral femoral condyle. It may also invaginate medially between the semimembranosus tendon, the medial head of the gastrocnemius muscle, and the medial femoral condyle.

When the synovial septa, which exist embryonically, are not completely resorbed but persist into adulthood, they are known as **synovial** or **patellar plicae.** These vestiges have been observed in 20 to 60 percent of the normal population and are referred to, in order of most frequently to least frequently found, as **infrapatellar plica (or ligamentum mucosum), suprapatellar plica,** and **mediopatellar plica.** The infrapatellar plica has also been described as the **infrapatellar fold.** According to *Gray's,*[2] the infrapatellar pad of fat that separates the synovial lining of the joint from the patellar ligament is itself covered with synovium. The synovial covering of the pad of fat projects into the interior of the joint on either side of the patellar ligament (alar folds) and joins into a single band (the infrapatellar fold or ligamentum mucosum) (Fig. 11–12).

Synovial plica, when they exist, are generally composed of loose, pliant, and elastic loose fibrous connective tissue that easily passes back and forth over the femoral condyles as the knee flexes and extends.[21,22] Occasionally, however, the plica may become irritated, leading to pain and to changes in both the plica itself and in the articular surface of the femoral condyle. So-called plica syndrome generally does not arise from the most common infrapatellar plica, but from one of the other persisting synovial bands.[25]

The knee joint capsule is reinforced by a number of ligaments that play an important part not only in knee joint stability but, as we shall see, in knee joint mobility.

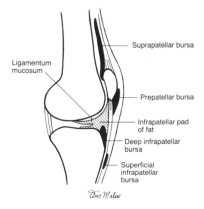

Figure 11–12. The infrapatellar pad of fat which separates the synovial lining of the joint from the patellar ligament is covered with synovium. The synovial covering of the pad of fat projects into the interior of the joint on either side of the patellar ligament and joins into a single band (the infrapatellar fold or ligamentum mucosum).

Knee Joint Ligaments

The roles of the various ligaments of the knee have received extensive attention, reflecting their importance to knee joint stability and the frequency with which function is disrupted through ligamentous injury. Given the lack of bony restraint to virtually any of the knee motions, the ligaments are credited with resisting or controlling:

1. Excessive knee extension
2. Varus and valgus stresses at the knee (attempted adduction or abduction of the tibia, respectively)
3. Anterior or posterior displacement of the tibia beneath the femur
4. Medial or lateral rotation of the tibia beneath the femur
5. Combinations of AP displacements and rotations of the tibia, known as rotatory stabilization.

Although the tibial motions were, for the most part, cited in the aforementioned list, it is also possible that the stresses may occur on the femur while the tibia is fixed (weight-bearing). In such instances, the AP displacements and rotations will reverse. That is, anterior displacement of the tibia is equivalent to posterior displacement of the femur and so forth.

The large body of literature available on ligamentous function of the knee joint can be confusing and appears contradictory. This may be due to some confusion in terms as to whether the tibia or the femur is being referenced, but is more likely due to complex and variable functioning and to dissimilar testing conditions. It is clear that ligamentous function can change depending on the position of the knee joint, on how the stresses are applied, and on what active or passive structures are concomitantly intact. We will credit the ligaments with the ability to provide stabilization in those positions and directions on which there appears to be consensus.

Collateral Ligaments

The medial (tibial) collateral ligament (MCL) attaches to the medial aspect of the medial femoral epicondyle, sloping anteriorly to insert into the medial aspect of the proximal tibia (Fig. 11–13). The posterior medial fibers of the ligament blend with fibers of the joint capsule and some fibers extend medially to attach to the

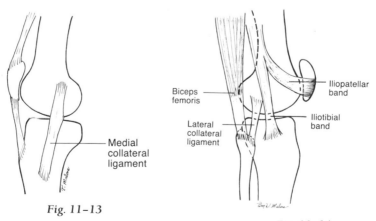

Fig. 11–13

Fig. 11–14

Figure 11–13. The medial collateral ligament (MCL) runs from the medial condyle of the femoral down and anteriorly to the right.

Figure 11–14. Lateral knee joint structures.

medial meniscus. The lateral (fibular) collateral ligament (LCL) is a strong cordlike structure attaching the lateral femoral epicondyle and attaching posteriorly to the head of the fibula (Fig. 11–14). Unlike the MCL, the LCL has no attachment either to the meniscus or to the joint capsule. Both collateral ligaments are taut in full extension and, therefore, help resist hyperextension of the knee joint.

Medial Collateral Ligament

The MCL resists valgus stresses across the knee joint, being especially effective in the extended knee when the ligament is taut. However, it may play a more critical role in resisting valgus stresses in the slightly flexed knee when other structures make a lesser contribution. Grood and associates[26] found that the MCL carried 57 percent of the valgus stress when the knee was at 5° of flexion, but 78 percent of the load when the knee was flexed to 25°. The MCL is also aligned in such a way as to check lateral rotation of the tibia.[8] Nielsen and associates[27,28] found that the MCL made a major contribution throughout the knee joint range of motion (ROM) to checking lateral rotation of the tibia combined with either anterior or posterior tibial displacement. The MCL is also a backup restraint to pure anterior displacement of the tibia when the primary restraint of the ACL is absent.[8]

Lateral Collateral Ligament

The LCL resists varus stresses (attempted adduction of the tibia) across the knee. Given its alignment, it also appears to limit lateral rotation of the tibia, making its most substantial contribution at about 35° of flexion, in conjunction with the posterolateral capsule.[27,28] The LCL also resists combined lateral rotation with posterior displacement of the tibia in conjunction with the tendon of the popliteus muscle.[8]

Iliotibial Band

The iliotibial band (ITB) or iliotibial tract is formed proximally from the fascia investing the tensor fascia lata, the gluteus maximus, and the gluteus medius muscles. It continues distally to attach to the linea aspera of the femur via the lateral intermuscular septum and inserts into the lateral tubercle of the tibia, reinforcing the anterolateral aspect of the knee joint (Fig. 11–14). Although there are muscular connections to the iliotibial band, Kaplan[29] considered the ITB to be essentially a passive structure at the knee joint since contraction of either the tensor fascia lata or the gluteus maximus muscles did not produce any longitudinal excursion of the distal ITB. The ITB appears to be consistently taut regardless of position of the hip joint or knee joint, although it falls anterior to the knee joint axis in extension and posterior to the axis in flexion.[7,30] The direction of the ITB is comparable to that of the MCL,[7] and its strength comparable to that of the ACL.[20] The ITB's fibrous connections to the biceps femoris and vastus lateralis muscles through the lateral intermuscular septum form a sling behind the lateral femoral condyle, assisting the ACL in preventing posterior displacement of the femur when the tibia is fixed and the knee joint is near extension.[20] The ITB sends fibers from its anterior margin to attach to the patella, forming an iliopatellar band[20,29] that may be implicated in abnormal lateral forces on the patella.[30]

Cruciate Ligaments

The ACL and PCL are centrally located within the articular capsule but lie outside the synovial cavity. These ligaments are named according to their tibial attachments. The ACL arises from the anterior aspect of the tibia, while the PCL arises from the posterior aspect of the tibia. Both ligaments have main posterolateral and smaller anteromedial bands that behave differently in different movements.[2]

Anterior Cruciate Ligament

The ACL attaches to the anterior tibia, passes under the transverse ligament,[8] and extends superiorly and posteriorly to attach to the posterior part of the inner aspect of the lateral femoral condyle (Fig. 11–15a and b). The numerous fascicles of the ACL may be grouped into an anteromedial band (AMB) and a posterolateral band (PLB), with the names taken again from the points of tibial origin. Although some portion of the ACL is tight throughout the knee joint range,[31] the AMB is considered to be moderately lax in extension while the PLB is taut. In flexion, the AMB is taut (maximally tensed at 70° of flexion[30]) and the PLB is lax.[8,32,33]

The ACL is generally considered to be the primary restraint to anterior displacement of the tibia on the femoral condyles. There would appear to be essentially no anterior translation of the tibia possible in full extension when many of the supporting passive structures of the knee are taut (including the PLB of the ACL). However, a cadaver study that employed serial sectioning of ligaments and application of measured loads concluded that the ACL carried 87 percent of the load when an anterior translational force was applied to the extended knee.[34] Forces producing anterior translation of the tibia will result in maximal excursion of the tibia at about 30° of flexion[15] when neither of the ACL bands are particularly tensed. The pattern of reciprocal tension between the AMB and the PLB is such that the PLB checks and, therefore, tends to be injured with excessive knee hyperextension, while the AMB would tend to be injured with trauma to the flexed knee.[32] The ACL would also appear to make at least a minor contribution to restraining both varus and valgus stresses across the knee joint.[26,27]

Both cruciate ligaments appear to play a role in producing and controlling rotation of the tibia. The ACL appears to twist around the PCL in medial rotation of the tibia, thus *checking* excessive medial rotation.[32] However, other investigators have also found that stress on the ACL produced by an anterior translational force on the tibia will *create* a concomitant medial rotation of the tibia.[8,15,16] When the ACL was sectioned experimentally in cadavers, anterior displacement increased and the amount of medial rotation decreased with the application of an anterior translational force.[16] When Lipke and associates[35] loaded an ACL-deficient cadaver limb under conditions simulating weight bearing, they found both excessive anterior dis-

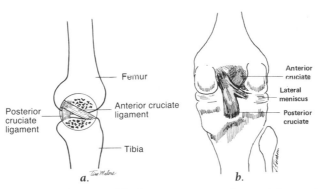

Figure 11–15. a. A schematic drawing of the knee joint ignores the condyles to which each ligament attaches, but shows the longer and more oblique ACL as it crosses the shorter, thicker, and more vertically oriented PCL. b. A posterior view of the knee joint shows the femoral condyles to which the ACL and PCL each attach.

placement of the tibia and excessive medial rotation. Unlike most other situations where motions of the free tibia (femur fixed) will be mirrored by reverse motions of the free femur (tibia fixed), tension in the ACL does not appear to produce rotation of the femur.[15] Regardless of the rotatory effect of the ACL on the tibia, injury to the ACL appears to occur most commonly when the knee is flexed and the tibia rotated *in either direction.* In flexion and medial rotation, the ACL is tensed as it winds around the PCL. In flexion and lateral rotation, the ACL is tensed as it is stretched over the lateral femoral condyle.[36] When attempting to determine whether there has been a tear of the ACL, the presence of *both* anteromedial and anterolateral instability is the most diagnostic. Of the knees presenting with instability in both directions, Terry and Hughston[37] found 100 percent to have confirmed torn or nonfunctional ACLs.

Posterior Cruciate Ligament

The PCL runs superiorly and somewhat anteriorly from its posterior tibial origin to attach to the inner aspect of the medial femoral condyle. It is shorter and less oblique than the ACL (Fig. 11–15b). Like the ACL, its fascicles can be divided into an AMB and a PLB named by the tibial origin. The AMB is lax in extension, while the PLB is taut. At 80 to 90° of flexion, the AMB is maximally taut and the PLB is relaxed.[33]

There is consensus that the PCL is the primary restraint to posterior displacement of the tibia beneath the femur, with little or no displacement possible in full extension. The PCL was found to carry 93 percent of the load in the extended knee when a posterior translational force was applied to the tibia.[34] In the flexed knee, maximal displacement of the tibia with a posterior translational force occurs at 75 to 90° of flexion, although sectioning of the PCL increased posterior translation at all angles of flexion.[16] The PCL also has some role in restraining varus and valgus stresses at the knee.[26,27]

As is true for the ACL, the PCL appears to play a role in both restraining and producing rotation of the tibia. Posterior translatory forces on the tibia are consistently accompanied by concomitant lateral rotation of the tibia,[16] with little or no rotation produced at the femur.[15] Tension in the PCL with knee extension may be instrumental in creating the lateral rotation of the tibia that is critical to locking of the knee for stabilization.

Posterior Capsular Ligaments

The posteromedial aspect of the capsule is reinforced by the tendinous expansion of the semimembranosus muscle, which is known as the **oblique popliteal ligament** (Fig. 11–16). This ligament passes from a point posterior to the medial tibial

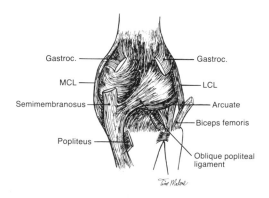

Figure 11–16. A view of the posterior capsule of the knee joint shows the reinforcing oblique popliteal ligament and the arcuate ligament. Also seen are the collateral ligaments (MCL and LCL) and some of the reinforcing posterior musculature (semimembranosus, biceps femoris, medial and lateral heads of the gastrocnemius, and the upper and lower sections of the popliteus).

condyle and attaches to the central part of the posterior aspect of the joint capsule. The posterolateral aspect of the capsule is reinforced by the arcuate popliteal (arcuate) ligament (Fig. 11–16). The arcuate ligament arises from the posterior aspect of the head of the fibula and passes over the tendon of the popliteus muscle to attach to the intercondylar area of the tibia and to the lateral epicondyle of the femur. Both the oblique popliteal and the arcuate ligaments are taut in full extension and assist in checking hyperextension of the knee. The arcuate and oblique popliteal ligaments were found to play an important role in checking varus and valgus stresses, respectively, in the extended knee.[27] Other investigators have similarly found contributions by these structures to checking varus/valgus stresses[26] and providing secondary restraint to other tibial motions.[27,28,35]

Knee Joint Bursae

The extensive ligamentous apparatus of the knee joint and the large excursion of the bony segments set up substantial frictional forces between muscular, ligamentous, and bony structures. However, numerous bursae prevent or limit such degenerative forces. Three bursae have already been mentioned in discussion of the knee joint capsule. These are the suprapatellar bursa, the subpopliteal bursa, and the gastrocnemius bursa. These bursae are not usually separate entities, but are either invaginations of the synovium within the joint capsule (Fig. 11–11) or communicate with the capsule through small openings.[38] The suprapatellar bursa lies between the quadriceps tendon and the anterior femur; the subpopliteal bursa lies between the tendon of the popliteus muscle and the lateral femoral condyle; and the gastrocnemius bursa lies between the tendon of the medial head of the gastrocnemius muscle and the medial femoral condyle. The gastrocnemius bursa may also continue beneath the tendon of the semimembranosus muscle to protect it from the medial femoral condyle.

The lubricating synovial fluid contained in the knee joint capsule moves from recess to recess during flexion and extension of the knee, lubricating the articular surfaces. In extension, the posterior capsule and ligaments are taut and the gastrocnemius and subpopliteal bursae are compressed. This shifts the synovial fluid anteriorly[38] (Fig. 11–17a). In flexion, the suprapatellar bursa is compressed anteriorly by tension in the anterior structures and the fluid is forced posteriorly (Fig. 11–17b). When the joint is in the semiflexed position, the synovial fluid is under the least amount of tension (Fig. 11–17c). When there is an excess of fluid in the joint cavity due to injury or disease, the semiflexed knee position helps to relieve tension in the capsule and therefore helps to reduce pain.

Several other bursae are associated with the knee but do not communicate with the synovial capsule (Fig. 11–18). The prepatellar bursa, located between the skin and the anterior surface of the patella, allows free movement of the skin over the

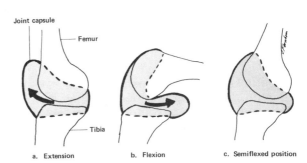

Figure 11–17. a. The synovial fluid is forced anteriorly during extension. b. In flexion, the synovial fluid is forced posteriorly. c. In the semiflexed position, the capsule is under the least amount of tension.

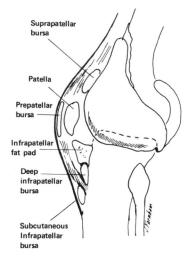

Figure 11-18. The prepatellar bursa, deep infrapatellar bursa, and superficial infrapatellar bursa are separate from the knee joint cavity.

patella during flexion and extension. The subcutaneous infrapatellar bursa lies between the patellar ligament and the overlying skin. The subcutaneous infrapatellar bursa and the prepatellar bursa may become inflamed as a result of direct trauma to the front of the knee or through activities like kneeling. The deep infrapatellar bursa, which is located between the patellar ligament and the tibial tuberosity, is separated from the synovial cavity of the joint by the infrapatellar pad of fat. The deep infrapatellar bursa helps to reduce friction between the patellar ligament and the tibial tuberosity.

There are also several small bursae that are associated with the ligaments of the knee joint. There is commonly a bursa between the LCL and the tendon of the biceps femoris muscle and between the LCL and the popliteus muscle. Also, there is a bursa deep to the MCL protecting it from the tibial condyle, and one superficial to the MCL protecting it from the tendons of the semitendinosus and gracilis muscles that cross the MCL.

KNEE JOINT FUNCTION

Knee Joint Motion

The primary motions of the knee joint are flexion/extension and, to a lesser extent, medial-lateral rotation. These motions occur about changing but definable axes and serve the weight-bearing functions of the lower extremity. The knee joint can also undergo tibial or femoral displacement anteriorly and posteriorly, and some abduction and adduction through varus and valgus forces. However, these movements are generally not considered part of the *function* of the joint but are, rather, part of the cost of the tremendous compromise between mobility and stability. The small amounts of AP displacement and varus/valgus forces that can occur in the normal flexed knee are the result of joint incongruence and variations in ligamentous elasticity. The magnitude of such motions varies among individuals and from side to side in the same individual.[8,15,16] Excessive amounts of such motions are abnormal and generally indicate ligamentous incompetence. We will focus on normal knee joint motions, including both osteokinematics (degrees of freedom) and arthokinematics (intra-articular movements within the joint).

Figure 11–19. The tibia moves from a position slightly lateral to the femur in extension to a position slightly medial to the femur in flexion.

Flexion/Extension

The axis for flexion and extension at the tibiofemoral joint passes horizontally through the femoral condyles at an angle to the mechanical and anatomic axes.[4] The obliquity of the axis (lower on the medial side of the joint) is similar to that found at the elbow, causing the tibia to move from a position slightly lateral to the femur in full extension to a position medial to the femur in full flexion (Fig. 11–19). However, unlike the elbow, the axis of motion for flexion and extension at the knee is not relatively fixed, but moves to a considerable extent through the ROM. The instant axis of rotation (IAR) for each point in the knee joint ROM can be found in a series of roentgenograms and the path of these sequential centers plotted. The pathway of the IAR of the tibiofemoral joint for flexion and extension forms a semicircle, moving posteriorly and superiorly on the femoral condyles with increasing flexion (Fig. 11–20).[3]

Since many of the muscles associated with the knee are two-joint muscles that cross both the hip and the knee, hip joint position can influence knee ROM. Passive range of knee flexion is generally considered to be 130 to 140°.[39] Knee flexion may

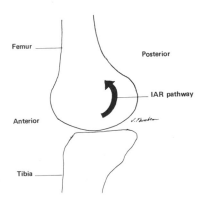

Figure 11–20. Schematic drawing of the knee joint. The arrow represents the path of the instantaneous axis of rotation (IAR) for the joint as it moves from extension into flexion.

Figure 11–21. If the ankle were to be fused in dorsiflexion, knee flexion and extension could not occur without lifting some or all of the foot from the ground.

be limited to 120° or less when the hip joint is simultaneously hyperextended and the stretched rectus femoris muscle becomes passively insufficient.[2] Knee flexion may also reach as much as 160° in activities like squatting when the hip and knee are flexing at the same time and the body weight is superimposed on the joint.[2,4] Normal gait on level ground requires approximately 60° of knee flexion.[40] This requirement increases to about 80° for stair climbing,[41] and to 90° or more for sitting down into a chair and arising from it. Activities beyond simple mobility tasks require 115° of knee flexion or more.[8] Knee joint extension (or hyperextension) of 5 to 10° is considered within normal limits.[2] Excessive knee hyperextension is termed **genu recurvatum.**

When the lower extremity is weight bearing and the knee is part of a closed kinematic chain, range limitations at the ankle joint may cause restriction in knee joint flexion or extension. For example, a limitation in ankle dorsiflexion (due to tight plantarflexors) may prevent the knee from being flexed; a limitation in plantarflexion (due to tight dorsiflexors) may restrict the ability of the knee to fully extend. If the ankle were fixed in the position shown in Figure 11–21, the knee would be unable to either flex or extend without lifting all or some part of the foot from the ground.

Rotation

The axis of motion for rotation at the tibiofemoral joint is a longitudinal axis that runs through or close to the medial tibial intercondylar tubercle.[4,42] During lateral rotation of the knee joint, the medial tibial condyle and intercondylar tubercle act as a pivot point. When knee lateral rotation is produced with the tibia free (open kinematic chain), the medial tibial condyle moves only slightly anteriorly on the femoral condyle, while the lateral tibial condyle moves a large distance posteriorly on the lateral femoral condyle. In medial rotation, the direction of motion of the lateral tibial condyle reverses while the medial intercondylar tubercle and medial tibial condyle continue to act as a pivot. In weight bearing (closed kinematic chain), the **lateral femoral condyle** moves posteriorly on the **lateral tibial condyle** in the lateral rotation of the *femur* and anteriorly on the **lateral tibial** condyle in the medial rotation of the *femur*. The pivot point remains at the medial condyles, however, with lateral motion exceeding medial motion.

The range of knee joint rotation is dependent on the position of the knee. When the knee is in full extension, it is in the close-packed (locked) position and the ligaments are taut; no rotation is possible. The tibial tubercles are lodged in the intercondylar notch and the menisci are tightly interposed between the articulating surfaces. However, when the knee is flexed to 90°, the ligaments are lax. The tibial tubercles are no longer in the intercondylar notch and the menisci are free to move. At 90° of knee flexion, approximately 60 to 70° of either active or passive rotation is considered to be possible.[2] The range for lateral rotation (0 to 40°) is slightly greater than the range for medial rotation (0 to 30°[3,4,9]). With maximum rotation available at 90° of knee flexion, rotation diminishes as the knee approaches both full extension and full flexion.

Arthrokinematics

Flexion/Extension

The large articular surface of the femur and the relatively small tibial condyle create a potential problem as the femur begins to flex on the tibia. If the femoral condyles were permitted to roll posteriorly on the tibial condyle, the femur would run out of tibial condyle before much flexion had occurred. This would result in a limitation of flexion, or the femur would roll off the tibia (Fig. 11–22). For the femoral condyles to continue to roll with increased flexion of the femur, the condyles must simultaneously glide anteriorly on the tibial condyle to prevent them from rolling posteriorly off the tibial condyle (Fig. 11–23a). The first part of flexion of the femur from full extension (0 to 25°) is primarily rolling of the femoral condyles on the tibia,[43] bringing the contact of the femoral condyles posteriorly on the tibial condyle. As flexion continues, the rolling is accompanied by a simultaneous anterior glide just sufficient to create a nearly pure spin of the femur. That is, the magnitude of posterior displacement that would occur with the rolling of the condyles is offset by the magnitude of anterior glide, resulting in little linear displacement of the femoral condyles after 25° of flexion.

The anterior glide of the femoral condyles results in part from the tension encountered in the ACL as the femur rolls posteriorly on the tibial condyle. The glide may be further facilitated by the menisci whose wedge shape forces the femoral condyle to roll "uphill" as the knee flexes. As shown in Figure 11–24, the oblique contact force of the wedged meniscus (meniscus-on-femur [MF]) creates

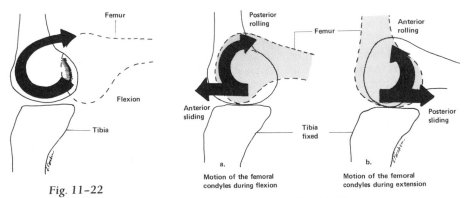

Fig. 11–22

Fig. 11–23

Figure 11-22. Schematic illustration of pure rolling of the femoral condyles on a fixed tibia shows the femur rolling off of the tibia.

Figure 11-23. a. A schematic representation of rolling and sliding of the femoral condyles on a fixed tibia. The femoral condyles roll posteriorly, while simultaneously sliding anteriorly. b. Motion of the femoral condyles during extension. The femoral condyles roll anteriorly, while simultaneously sliding posteriorly.

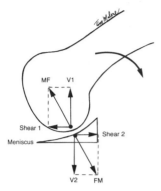

Figure 11–24. Schematically represented, the oblique contact of the femur with the wedge-shaped meniscus results in the forces of **meniscus-on-femur** (MF) and **femur-on-meniscus** (FM). These can be resolved into vertical and shear components. Shear$_1$ assists the femur in its forward glide during flexion while shear$_2$ assists in the posterior migration of the menisci that occurs with knee flexion.

an anterior shear (Shear$_1$) on the femur. Similarly, the oblique reaction force of femur-on-meniscus (FM) also has a shear component (Shear$_2$) that forces the menisci posteriorly on the tibial condyle. The result is that the menisci accompany the femoral condyles as the condyles move posteriorly on the tibial condyle, maintaining the increased congruence the menisci provide in the fully extended knee. The menisci cannot move in their entirety since they are attached at their horns to the intercondylar tubercles of the tibial condyle. Rather, the posterior migration is a posterior distortion, with the anterior aspect of the menisci remaining relatively fixed. Given the closer attachment of the two horns of the lateral meniscus to each other, the lateral meniscus will distort slightly more than the medial.

Extension of the knee from flexion occurs initially as a rolling of the femoral condyles on the tibial condyle, displacing the femoral condyles anteriorly back to neutral position. After the initial forward rolling, the femoral condyles glide posteriorly just enough to continue extension of the femur as an almost pure spin (roll plus posterior glide) of the femoral condyles on the tibial condyles (Fig. 11–23b). Tension in the PCL and the shape of the menisci facilitate the intra-articular movements of the femoral condyles during knee extension. The condyles are once again accompanied in displacement by distortion of the wedge-shaped menisci. As extension begins from full flexion, the posterior margins of the menisci return to their neutral position. As extension continues, the anterior margins of the menisci move anteriorly with the femoral condyles.

The motion (or distortion) of the menisci with flexion and extension are an important component of the motions. Given the need of the menisci to reduce friction and absorb forces of the large femoral condyles on the small tibial condyle, the menisci must remain beneath the femoral condyles to continue their function. Failure of the menisci to distort in the proper direction can also result in limitation of joint motion. If the femur literally rolls up the wedge-shaped menisci in flexion (without either the anterior glide of the femur or the posterior distortion of the menisci), the increasing thickness of the menisci and the threat of rolling off the posterior margin will cause flexion to be limited. Similarly, failure of the menisci to distort anteriorly with the femoral condyles in extension will cause the thick anterior margins to become wedged between the femur and tibia as the segments are drawn together in the final stages of extension. The interposition of the menisci will prevent extension from being completed.

Locking and Unlocking

Although the incongruence of the femoral condyles and tibial condyle results in a rolling and gliding of the condylar surfaces on each other, the asymmetry in the size of the medial and lateral condyles also causes complex intra-articular motions. Using weight-bearing closed-chain motion as an example, extension of the femur on the relatively fixed tibia results in additional motions to those described in the previous section. As the femur extends to about 30° of flexion, the shorter lateral femoral condyle completes its rolling-gliding motion. As extension continues, the longer medial condyle continues to roll and to glide posteriorly although the lateral condyle has halted. This continued motion of the medial femoral condyle results in medial rotation of the femur on tibia, pivoting about the fixed lateral condyle. The medial rotatory motion of the femur is most evident in the final 5° of extension.[2] Increasing tension in the knee joint ligaments as the knee approaches full extension may also contribute to the rotation within the joint.

Since the medial rotation of the femur that accompanies the final stages of knee extension is not voluntary or produced by muscular forces, it is referred to as **automatic** or **terminal rotation** of the knee joint. This rotation within the joint that accompanies the end of extension also brings the knee joint into the close-packed or locked position. The tibial tubercles are lodged in the intercondylar notch, the menisci are tightly interposed between the tibial and femoral condyles, and the ligaments are taut. Consequently, automatic rotation is also known as the **locking mechanism or screw home mechanism** of the knee. To initiate flexion, the knee must first be *unlocked.* That is, the medially rotated femur cannot flex in the sagittal plane, but must laterally rotate before flexion can proceed. A flexion force will *automatically* result in lateral rotation of the femur since the longer medial side will move before the shorter lateral side of the joint. If there is an external restraint to unlocking or derotation of the femur, the joint, ligaments, and menisci can be damaged as the femur is forced into flexion oblique to the sagittal plane in which its structures are oriented.

Automatic rotation or locking of the knee occurs in both open-chain and closed-chain knee joint function. In an open kinematic chain, the freely moving tibia *laterally rotates on the relatively fixed femur* during the last 30° of extension. Unlocking, consequently, is brought about by *medial rotation of the tibia on the femur* before flexion can proceed.

Axial Rotation

Automatic rotation (locking) of the knee joint can be differentiated from what we will call the **axial rotation** of the knee that occurs in knee flexion, although both are clearly transverse plane motions around a vertical axis. Axial rotation of the knee has the greatest potential range at 90° of knee flexion; it is *available,* but does not have to be used. The vertical axis for this motion is at the medial intercondylar tubercle, rather than at the lateral tibial condyle as found in automatic rotation. Axial rotation is due to joint incongruence and ligamentous laxity, while in contrast, automatic rotation is *obligatory* and is produced by asymmetry of surfaces and by ligamentous tension.

Whether rotation is axial or automatic, the menisci of the knee joint maintain their relationship to the femoral condyles just as they did in flexion and extension. That is, in rotation of the knee, the menisci will distort in the direction of movement of the corresponding femoral condyle. In medial rotation of the *femur*, the posterior **portion** of the medial meniscus will distort posteriorly on the tibial condyle to remain beneath the medial femoral condyle, while the anterior portion of the lateral meniscus will distort anteriorly to remain beneath the lateral femoral condyle. In this way, the menisci continue to reduce friction and distribute the forces the femoral condyle create on the tibial condyle *without restricting motion* of the femur as more solid or firmly attached structures would do.

The motions of the knee joint, exclusive of automatic rotation, are produced to a great extent by the muscles that cross the joint. We will complete our examination of the tibiofemoral joint by first examining the individual contribution of the muscles, emphasizing their mobility role in producing knee joint motion. We will then re-examine both the passive knee joint structures and the muscles in their combined role as stabilizers of this very complicated joint.

Muscles

Flexors

There are seven muscles that flex the knee. The knee flexors are the semimembranosus, semitendinosus, biceps femoris, sartorius, gracilis, popliteus, and gastrocnemius muscles. All of the knee flexors, except for the short head of the biceps femoris and the popliteus, are two-joint muscles. As two-joint muscles, their ability to produce effective force can be influenced by the relative position of the two joints over which they pass. Four of the flexors (the popliteus, gracilis, semimembranosus, and semitendinosus muscles) are considered to medially rotate the tibia on the fixed femur, while the biceps femoris is considered to be a lateral rotator of the tibia.

The semitendinosus, semimembranosus, and the biceps femoris muscles are known collectively as the **hamstrings**. These muscles all originate on the ischial tuberosity of the pelvis. The semimembranosus and the semitendinosus insert on the posteromedial and anteromedial aspects of the tibia, respectively. The semimembranosus muscle has fibers that attach to the medial meniscus. This attachment assists in knee flexion by facilitating posterior motion of the medial meniscus during active knee flexion. The significance of this contribution will be discussed with the popliteus muscle. The semitendinosus muscle has a fibrous septum that separates it into distinct proximal and distal compartments.[44] This may give it some specificity of action at the hip joint and at the knee joint.

The biceps femoris muscle has two heads, both of which insert on the lateral condyle of the tibia and the head of the fibula (Fig. 11–25). As mentioned, the biceps femoris tendon may be attached to the iliotibial band and retinacular fibers of the lateral joint capsule, a set of attachments that implies that the biceps femoris has a stabilizing role at the posterolateral aspect of the joint. The short head of the biceps femoris does not cross the hip joint and, therefore, has a unique action at the knee joint.

Most of the hamstrings, crossing both the hip (as extensors) and the knee (as flexors), work most effectively at the knee joint if they are lengthened over a flexed hip. Electromyographic (EMG) recordings of the biceps femoris show that there is

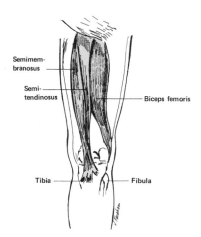

Figure 11–25. The hamstring muscles are shown in a posterior view of the knee.

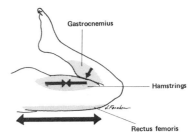

Figure 11–26. During active knee flexion in the prone position, the rectus femoris is stretched over the hip and the knee and becomes passively insufficient. The hamstrings are actively shortened over the hip and the knee and are likely to develop active insufficiency. In addition, the bulk of the contracting muscles (hamstrings and gastrocnemius) also limits the range of active knee flexion in the prone position.

a decrease in EMG activity as the muscle is lengthened, and an increase in activity as the muscle is shortened. Since the muscle generates greater tension when it is elongated, fewer motor units are required to produce the same amount of torque. Shortening of the muscle causes it to approach active insufficiency and more motor units are required to produce the same torque. With active knee flexion with the body in the prone position, the hamstrings muscles are forced to attempt to shorten over both the hip (which will be extended) and over the knee. The hamstrings will weaken as knee flexion proceeds since the muscle group is approaching active insufficiency and must overcome the increasing tension in the rectus femoris, which is approaching passive insufficiency (Fig. 11–26).

The gastrocnemius muscle arises from the posterior aspects of the medial and lateral condyles of the femur by two heads. It inserts into the calcaneus by way of the calcaneal tendon. Except for the plantaris muscle (which is commonly absent), the gastrocnemius is the only muscle at the knee that crosses the ankle and the knee (Fig. 11–27). Although the gastrocnemius generates a large plantarflexor torque at the ankle, it makes a relatively small contribution to knee flexion. In fact, when someone goes up on their toes (plantarflexes their ankles) and then only slightly flexes the knee, the gastrocnemius will relax (leaving the maintenance of the position to the soleus muscle). Apparently the gastrocnemius becomes actively insufficient quite easily. Rather than working to produce knee flexion, the gastrocnemius appears to be effective in preventing knee joint hyperextension. Paralysis of the plantarflexors is classically accompanied by a snapping back of the knee into hyperextension in the final stages of single-limb support during walking. From observing this abnormal response, we can conclude that the gastrocnemius must

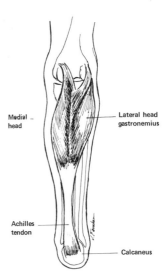

Figure 11–27. The posterior aspect of the knee complex showing the gastrocnemius muscle. The gastrocnemius helps to provide support for the posterior aspect of the knee.

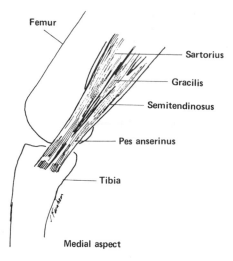

Femur

Sartorius

Gracilis

Semitendinosus

Pes anserinus

Tibia

Medial aspect

Figure 11-28. The sartorius, gracilis, and semitendinosus insert as a conjoined tendon known as the pes anserinus on the anteromedial aspect of the tibia.

contribute substantially to resisting the very large extension torque at the knee joint at this point in the gait cycle. The gastrocnemius appears to be less a mobility muscle at the knee joint than a dynamic stabilizer.

The sartorius muscle arises anteriorly from the anterior superior spine of the ilium and crosses the femur to insert into the anteromedial surface of the tibial shaft posterior to the tibial tuberosity. Although a potential flexor and medial rotator of the tibia, activity in the sartorius is more common with hip motion than with knee motion. It appears to be relatively impervious to active insufficiency, since it is equally active at the hip with the knee flexed and with the knee extended.[46] This may be accounted for somewhat by the fact that it consists of a large group of three or four fibers in series joined by fibrous septa,[44] rather than being a group of single fibers, continuous from proximal attachment to distal. Variations in distal attachment of the sartorius muscle are not uncommon. When attached just anterior to its more usual location, it may fall anterior to the knee joint axis, serving as a mild knee joint extensor rather than as a knee flexor.

The gracilis muscle arises from the inferior half of the symphysis pubis arch and

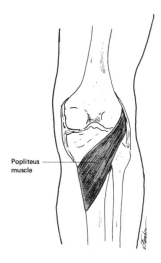

Popliteus muscle

Figure 11-29. A posterior view of the knee joint showing the popliteus muscle (more superficial structures have been removed). The direction of pull will unlock the fully extended knee by medially rotating the tibia (open chain) or laterally rotating the femur (closed chain).

Figure 11-30. A schematic representation of the semimembranosus muscle
and its attachment to the medial meniscus is shown. The arrow in the insert
represents the direction of pull of the
muscle on the medial meniscus during
flexion.

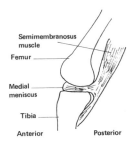
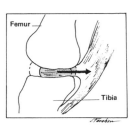

inserts on the medial tibia by way of a common tendon with the sartorius and the semitendinosus muscles. It is not only a hip joint flexor and adductor, but it can also flex the knee joint and produce slight medial rotation of the tibia. The gracilis apparently becomes actively insufficient readily, however, ceasing activity if the hip and knee are permitted to flex simultaneously.[46]

The gracilis, semitendinosus, and sartorius muscles attach to the tibia by a common tendon on the anteromedial aspect of the tibia (Fig. 11–28). The common tendon is called the **pes anserinus** because of its shape (*pes anserinus* means "goose's foot"). The three muscles of the pes anserinus appear to function effectively as a group to stabilize the medial aspect of the knee joint.

The only other one-joint knee flexor besides the short head of the biceps femoris is the relatively small popliteus muscle. This muscle originates on the posterior aspect of the lateral femoral condyle and attaches on the medial aspect of the tibia. The fibers of the muscle run medially across the posterior aspect of the knee joint (Fig. 11–29). The popliteus muscle is a medial rotator of the tibia on the femur in an open kinematic chain (or a lateral rotator of the femur on the tibia in a closed kinematic chain). The active popliteus muscle is considered to play an important role in initiating unlocking of the knee since it reverses the direction of automatic rotation that occurred in the final stages of knee extension. It is important to note, however, that unlocking of the knee joint will still occur effectively if the knee flexion takes place passively.

The popliteus muscle is commonly attached to the lateral meniscus as the semimembranosus muscle is to the medial meniscus. Since both the semimembranosus and the popliteus are knee flexors, activity in these muscles will not only generate a flexion torque but will actively contribute to the posterior movement of the two menisci on the tibial condyles that should occur during knee flexion as the femur begins its rolling motion. The menisci's ability to distort during motion ensures that the slippery surface is present throughout the femoral ROM. The medial meniscus is drawn posteriorly by tension in the semimembranosus muscle (Fig. 11–30). The lateral meniscus is drawn posteriorly by tension in the popliteus expansion. Although the menisci will move posteriorly on the tibial condyle even during passive flexion, the assistance of the semimembranosus and popliteus muscles reinforces the movement and minimizes the chance that the menisci will become entrapped and limit knee flexion.

Extensors

The four extensors of the knee are known collectively as the **quadriceps femoris** muscle. The only portion of the quadriceps that crosses two joints is the rectus femoris, which originates on the inferior spine of the ilium. The vastus intermedius, vastus lateralis, and vastus medialis muscles originate on the femur and merge into a common tendon, the quadriceps tendon (Fig. 11–31). The fibers of the quadriceps tendon continue distally as the patellar ligament. The patellar lig-

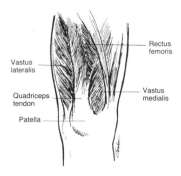

Rectus
femoris

Vastus
lateralis

Quadriceps
tendon

Vastus
medialis

Patella

Figure 11–31. Three of the four segments of the quadriceps femoris can be seen in this anterior view of the knee joint as the muscles insert into the common quadriceps tendon.

ament runs from the apex of the patella, across the anterior surface of the patella, into the proximal portion of the tibial tubercle.[2] The vastus medialis and vastus lateralis also insert directly into the medial and lateral aspects of the patella by way of the retinacular fibers of the joint capsule.

Together, the muscles of the quadriceps femoris extend the knee. Lieb[47] found the resultant pull of the muscle fibers in relation to the long axis of the femur to be 7 to 10° medially and 3 to 5° anteriorly. The pull of the vastus lateralis alone was found to be 12 to 15° lateral to the long axis of the femur, with distal fibers more angulated yet. The pull of the vastus intermedius was parallel to the shaft of the femur, making it the purist knee extensor of the group. The angulation of the pull of the vastus medialis depended on which segment of the muscle was assessed. The upper fibers were angulated 15 to 18° medially to the femoral shaft, while the distal fibers were angulated as much as 50 to 55° medially. The drastically different orientation of lower fibers of the vastus medialis muscle has resulted in reference to the upper fibers as the **vastus medialis longus (VML)** and the lower fibers as the **vastus medialis oblique (VMO)**.

Mechanically, the efficiency of the quadriceps muscle is affected by the patella; the patella lengthens the MA of the quadriceps by increasing the distance of the quadriceps tendon and patellar ligament from the axis of the knee joint. The patella, as an anatomic pulley, deflects the action line of the quadriceps femoris away from the joint, increasing the angle of pull and the ability of the muscle to generate a flexion torque. Interposing the patella between the quadriceps tendon and the femoral condyles also reduces friction between the tendon and condyles.[17] The femoral condyles encounter not the quadriceps tendon, but the hyaline-cartilage-covered posterior surface of the patella. The patella is tied to the tibial tuberosity by the patellar ligament, making the patella almost like an anterior wall to the tibia. As the femur flexes on the tibia, the patella remains essentially still while relatively sliding down the moving femoral condyles. The position of the patella relative to the joint axis varies as the instantaneous axis shifts and as the contour of the femoral condyles changes. The patella's effect on the MA of the quadriceps muscle, therefore, will vary through the knee joint ROM. Regardless of joint position, however, substantial decreases in the strength (torque) of the quadriceps of up to 49 percent have been found following removal of the patella[48] since the MA of the quadriceps is substantially reduced at most points in the ROM (Fig. 11–32).

Although increasing the magnitude of torque produced by the quadriceps muscle appears to serve function well, it does have a cost. Increasing the angle of pull of the quadriceps on the tibia and, thus, the size of the rotatory component of the pull of the quadriceps, also increases the magnitude of shear of that same component. That is, the rotatory component not only creates rotation of the tibia around an axis but also creates a translatory force that attempts to shear the tibia anteriorly beneath the femur. As shown in Chapter 1, Figure 1–61, the anterior

Figure 11–32. The lower portion of the illustration shows a normal knee in extension with the patella positioned anterior to the tibiofemoral joint. The arrow indicates the action line of the quadriceps and the dotted line indicates the moment arm (MA) of the quadriceps. The top drawing shows a knee with the patella removed. The arrow indicates the changed action line of the quadriceps, and the dotted line indicates the decreased moment arm that results from removal of the patella.

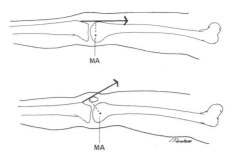

shear created by the quadriceps must be offset by a pull similar to that provided by the ACL. Increases and decreases in the angle of pull of the quadriceps are accompanied by concomitant increases and decreases in stress in the ACL.

The contribution of the patella to improving torque production by the quadriceps will vary with the joint ROM. In full knee flexion, the patella slides into the intercondylar notch of the femur, effectively eliminating the patella as a pulley. The rounded contour of the femoral condyles at this point, however, is already deflecting the muscle action line. Also, in full flexion the instantaneous axis of rotation has moved well posteriorly into the femoral condyles, further away from the action line of the quadriceps muscle. The movement of the IAR, therefore, already results in an increased MA for extension without the mediation of the patella.[11]

During knee extension, the MA of the quadriceps muscle lengthens as the patella leaves the intercondylar notch and must travel over the rounded femoral condyles. At about 60° of knee flexion, the patella is pushed by the rounded femoral condyles as far from the IAR as it will get. With continued extension, the MA once again begins to diminish. Evidence of the change in the MA is seen in the findings that the mean maximum isometric torque of the knee extensors is greater at 60° of knee flexion than at either 30 or 45°.[49] This is influenced, however, not only by the MA, but by the changing length-tension of the muscle and by the type of contraction. When peak torque is assessed during active isokinetic knee extension, the greatest amount of torque generated by the knee extensor muscles occurs at 45° of knee flexion.[49]

The decrease in the length of the MA of the quadriceps muscle and the decrease in the length-tension of the muscle during the last 15° of knee extension place the quadriceps at a mechanical and physiologic disadvantage. A 60 percent increase in quadriceps force over that needed in the rest of the ROM is required to complete the last 15° of knee joint extension.[46] Consequently, although the effect of the patella on improving the MA of the quadriceps femoris is diminished in the final stages of knee extension, the small improvement in MA provided by the patella may be most important here. Given the reduced ability of the muscle to generate active tension, the relative size of the MA is critical to torque production. Removal of the patella will have almost no effect on the strength of the quadriceps muscle as the muscle initiates extension from full flexion since the patella has little effect at this point in the range. Loss of the patella in the 60 to 30° range of flexion will produce a noticeable weakness in extension, but the roundness of the femoral condyles will still deflect the line of pull of the muscle and the length-tension is still favorable. Loss of the patella has its most profound effect in the last stages of joint extension when the decrease in MA may reduce torque to the point where knee extension cannot be completed by an active quadriceps muscle contraction alone.

The loss of the patella is most evident when the tibia is the moving segment and the quadriceps muscle must work against the resistance of gravity. In weight bearing, however, the diminished quadriceps extensor function can be supplemented by other forces affecting the closed chain. In fact, patterns of quadriceps activity may appear quite different in the weight-bearing position. In weight bearing, the quadriceps controls knee flexion (rather than creating extension) by acting eccentrically during activities. The quadriceps then works concentrically in extension to return the body to the erect posture. When the erect posture has been attained, activity of the extensors ceases. No knee extensor muscle activity is necessary to maintain knee extension in normal erect stance because the LOG is located anterior to the axis of flexion and extension at the knee joint. The resulting gravitational torque created around the knee joint is sufficient to maintain the knee in extension. The posterior joint capsule, ligaments, and musculotendinous structures are able to maintain equilibrium at the knee by counterbalancing the gravitational torque and preventing hyperextension. If the LOG passes posterior to the knee joint axis, the gravitational torque will tend to cause knee flexion and activity of the knee extensor muscles is necessary to counterbalance the gravitational torque and maintain the knee joint in equilibrium. The extensor muscles, which have the responsibility of supporting the body weight and resisting the force of gravity, are about two times stronger than the flexor muscles.

In a closed kinematic chain, movement of the knee is accompanied by movement at the hip and ankle. Thus, knee flexion usually occurs in weight bearing in conjunction with hip flexion and ankle dorsiflexion. Activity of the soleus muscle, which is continuously active in erect stance, exerts a posterior pull on the tibia as it acts in reverse action and contributes to knee joint stability in the erect posture. Secondary support may also be given to the knee in extension by the flexor muscles as they develop passive tension as a result of being stretched over the posterolateral and posteromedial aspects of the knee.

Further consideration of the effect of the quadriceps femoris muscles will be given in discussion of the patellofemoral joint. Although the patella primarily serves the quadriceps mechanism, the quadriceps mechanism can have a substantial effect on the ability of the patella to fulfill its function in an effective, pain-free way.

Stabilization

The supporting structures of the knee may be classified on the basis of *function, structure,* or *location.* Classification systems based on function use a static/dynamic differentiation, while those based on structure use a capsular/extracapsular method. Systems based on location use a compartmental approach, referring to the embryonic medial and lateral joint compartments. According to the functional classification system, the **static stabilizers** include the passive structures such as the joint capsule and the ligaments. Included as static stabilizers are components of the joint capsule and associated structures such as the coronary ligaments and the meniscopatellar and patellofemoral ligaments. Ligaments that are static stabilizers include the MCL and LCL, the ACL and PCL, the oblique popliteal and arcuate, and the transverse ligament. Since the ITB is considered to be a passive force at the knee in spite of its muscular connections, we will consider it as part of the static stabilizers. The **dynamic stabilizers** of the knee include the following muscles and aponeuroses: the quadriceps femoris and extensor retinaculum, pes anserinus (semitendinosus, sartorius, and gracilis muscles), popliteus, biceps femoris, and the semimembranosus.[23]

According to the classification system based on location, the supporting structures of the knee joint that are located on the anteromedial, medial, and posteromedial aspects of the knee are **medial compartment structures.** Structures located in the same respective areas on the lateral aspect are **lateral compartment structures.**[23,24] The medial compartment structures include the following: the medial patellar retinaculum, MCL, oblique popliteal ligament, and the PCL. The medial compartment structures also include the medial head of the gastrocnemius, the pes anserinus, and the semimembranosus muscles. The **lateral compartment structures** include the following static and dynamic stabilizers: ITB; the biceps femoris and popliteus muscles; LCL; the meniscofemoral, arcuate, and ACL; and the lateral patellar retinaculum.

Regardless of the classification system used, attempting to credit structures with contributing primarily to one type of stabilization is extremely difficult and generally requires oversimplification of effect. The many studies and literature reviews already referenced in this chapter make it clear that the contribution of both muscles and ligaments are dependent on joint position (not only of the knee joint, but of the surrounding joints), magnitude and direction of force, availability of reinforcing structures, and nature of the testing conditions. *Almost all knee joint structures can contribute to stability in all directions under specific normal or abnormal conditions.* The variations among individuals (and between knees in the same individuals) also contribute to the variation in findings. The following summary should be considered, therefore, a reiteration of some (but not all) contributors to stability of the knee joint.

Some Contributors to Anterior-Posterior Stabilization

Anterior-posterior stability of the knee is provided by static and dynamic stabilizers and lateral and medial compartment structures. The contribution of the ligaments to AP stability was discussed previously. Some stabilizers, however, are particularly critical and bear reiteration of function. The extensor retinaculum, which is composed of fibers from the quadriceps femoris, fuses with fibers of the joint capsule to provide dynamic support for the anteromedial and anterolateral aspects of the knee. The medial and lateral heads of the gastrocnemius reinforce the medial and lateral aspects of the posterior capsule. The popliteus is considered to be a particularly important posterolateral stabilizer, complementing the function of the PCL.[46] The ACL and the hamstrings work in a complementary manner to resist forces that are attempting to displace the tibia anteriorly or shear the femur posteriorly.[36] Such forces are exemplified by the pull of the quadriceps and by the effect of the ground reaction force on the tibia when the heel hits the ground.[50] Kaplan[7] placed particular emphasis on the semimembranosus, contending that the knee could not be stable in flexion unless this structure and its multiple connections remained intact.

The role of the patella itself cannot be ignored when examining anterior-posterior stability of the knee. The patella prevents the femur from sliding forward off the tibia, actually serving as part of the tibia connected by an elastic tendon. This combination of patella and tibia cradles the femur.[51]

Some Contributors to Medial-Lateral Stabilization

Medial-lateral stability at the knee is provided for by static and dynamic soft tissue structures and by the tibial tubercles and menisci when the knee is in full extension. The knee, like the elbow, is reinforced on its medial and lateral aspects by collateral ligaments. The collaterals clearly play a critical role in resisting varus-valgus stresses, especially in the more extended knee. Both cruciates contribute, although the magnitude and balance of the contribution varies with many factors. As knee flexion increases, the dynamic stability provided by the musculature such as the muscles of the pes anserinus on the medial aspect of the knee become increasingly important. Laterally, the iliotibial tract, LCL, popliteus tendon, and biceps tendon form a quadruple complex that contributes to stability.[7] The posterolateral capsule is particularly important in varus stability in extension, whereas the popliteus is a major stabilizer in 0 to 90° of flexion.[27]

The menisci are particularly important to medial-lateral stability since the knee remains stable in full extension regardless of sectioning of ligamentous structures.[27] Removing both menisci would appear to have its greatest effect in stabilization during varus and valgus stresses.[52]

Some Contributors to Rotational Stabilization

The complex nature of rotational stabilization of the knee makes it particularly difficult to isolate certain structures as major contributors. It would appear, however, that the role of the passive mechanisms predominate over the dynamic mechanisms. The cruciates are most often credited with rotational stability of the joint, especially in the extended knee.[16] However, rotational instability may occur even in the presence of intact cruciates.[28] Credit is also given to the medial collateral, lateral collateral, posteromedial capsule, posterolateral capsule, and the popliteus tendon by investigators exploring rotational stability under varied conditions.[27,28,35]

THE PATELLOFEMORAL JOINT

The role of the patella has been well covered in discussion of the quadriceps femoris. It is primarily an anatomic pulley and a mechanism to reduce friction between the quadriceps tendon and the femoral condyles. The ability of the patella to perform its functions without restricting knee motion is dependent on its ability to slide on the femoral condyles while remaining seated between them. In full knee extension, the patella sits on the anterior surface on the distal femur. With knee flexion, the patella slides distally on the femoral condyles, seating itself between the femoral condyles. In full flexion, the patella sinks into the intercondylar notch. Extension reverses the sliding of the patella and brings it back to the patella surface of the femur. As the patella travels (or "tracks") down the femur, it undergoes some rotation about its vertical axis (patellar tilt). This motion helps the patella accommodate to some of the asymmetry of the femoral condyles. The patella tilts medially an average of 11° as the knee flexes from 25 to 130°.[53] The patella also must rotate about an anterior-posterior axis (rotation of the patella) to remain seated in the intercondylar notch as the femur undergoes automatic or axial rotation. Since the inferior aspect of the patella is tied to the tibial tuberosity, the inferior patella continues to point to the tibial tuberosity while moving with the femur. That is, when the femur medially rotates on the tibia, the upper portion of the patella will follow the femur medially while the lower portion will remain laterally with the tibia.[54] This will be referred to as **lateral rotation of the patella. Medial rotation of the patella** occurs with lateral rotation of the femur. The patella laterally rotates 6 to 7° as the knee flexes from 25 to 130°, with most of the rotation having occurred by 60° of knee flexion.[53] Failure of the patella to slide, tilt, or rotate appropriately can lead to restriction in knee joint ROM, to instability of the patellofemoral joint, or to pain caused by erosion of the patellofemoral surfaces.

We must closely examine the oddly shaped patella, the uneven surface on which it sits, and the tremendous forces to which the patella and patellofemoral surfaces are subject in order to understand the many potential problems encountered by the patella in performing what would appear to be a relatively simple function. A comprehension of the structures and forces that influence patellofemoral function leads readily to an understanding of the common clinical problems found at the patellofemoral joint.

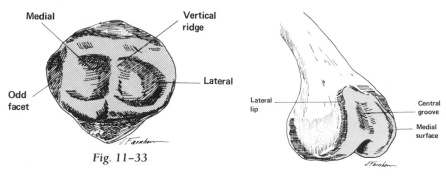

Fig. 11–33

Fig. 11–34

Figure 11–33. Articulating surfaces on the patella.

Figure 11–34. Articulating surfaces on the femur. Note the well-developed lateral lip on the lateral aspect of the articulating surface.

Patellofemoral Articular Surfaces

The triangularly shaped patella is distinguished by being the largest sesamoid bone in the body. Together with the femoral surface on which it sits, the patellofemoral joint is also the least congruent joint in the body.[55] The posterior surface of the patella is covered by articular cartilage and divided by a vertical ridge. The ridge may be situated approximately in the center of the patella, dividing the articular surface into approximately equally sized medial and lateral patellar facets. Occasionally, the ridge may be situated slightly toward the medial border of the patella, making the medial facet smaller than the lateral.[56] Regardless of size, the medial and lateral facets are flat to slightly convex side to side and top to bottom. At least 30 percent of the patellae also have a second vertical ridge toward the medial border, separating the medial facet from an extreme medial edge known as the **odd facet** of the patella[55] (Fig. 11–33).

The patellar articulating surface of the femur is the intercondylar groove or femoral sulcus on the anterior aspect of the distal femur. The groove or sulcus corresponds to the vertical ridge on the patella, dividing the femoral surface into lateral and medial portions. The femoral surfaces are concave side to side, but convex top to bottom.[2] The lateral facet is slightly more convex than the medial surface and has a more highly developed lip than the medial surface (Fig. 11–34 and Fig. 11–2a). The angle formed by the medial and lateral facets (angle of the femoral sulcus) has been found to average 138°, but varies widely among individuals (116 to 151°).[57,58]

Patellofemoral Joint Congruence

In the fully extended or neutral knee, the patella lies on the femoral sulcus. The vertical position of the patella in the femoral sulcus is related to the length of the patellar tendon. Ordinarily, the ratio of the length of the tendon (LT) to the length of the patella (LP) is approximately 1:1. This ratio (LT/LP) is referred to as the **index of Insall and Salviti**.[59,60] It is specifically calculated by measuring LT from the apex of the patella to the small notch just proximal to the tibial tuberosity on lateral radiograph and the greatest diagonal LP.[59] The ratio may be affected by gender, with females having slightly larger ratios (or slightly longer patellar tendons).[57,61,62] The limit of normal for the ligament to patella ratio is 1.3,[59] or the ligament should not exceed the patella in length by more than 20 percent.[63] An excessively long tendon produces an abnormally high position of the patella on the femoral sulcus known as **patella alta**. The ramifications of this will be seen later in the chapter.

The patella in the neutral or extended knee has little or no contact with the femoral sulcus beneath it. There is *at most* only a narrow band of contact between the inferior pole of the patella and the femoral sulcus.[64] The first consistent contact of the patella is made at 10 to 20° of flexion on inferior margin of the patella across both medial and lateral facets. With increasing flexion, the area of contact increases and shifts from distal to proximal, spreading from the ridge separating the medial and odd facet to the lateral facet. By 90° of knee flexion, all portions of the patella have experienced some (although inconsistent) contact, with the exception of the odd facet. As flexion continues past 90°, the medial facet enters the intercondylar notch, and the odd facet achieves contact for the first time.[65] At 135° of flexion, contact is on the lateral and odd facets,[53] with the medial facet completely out of contact.[66]

Overall, the medial patellar facet normally receives the most consistent contact with the femoral surfaces,[56,67] while the odd facet receives the least. Since any imbalance in compression and release of pressure on cartilage can lead to degenerative changes, it is not surprising that the most common cartilaginous changes on the patella are, in fact, found on the medial and odd facets. The changes, however, appear to be part of the normal aging process, are not necessarily progressive, and are commonly asymptomatic.[13,55,68] Although the lack of sufficient compression on the odd facet may be obvious (it is in contact only in full knee flexion), the excessive compression on the medial facet must be understood in the context of the patellofemoral joint reaction forces.

Patellofemoral Joint Reaction Forces

The patella is pulled on simultaneously by the quadriceps tendon superiorly and by the patella tendon inferiorly. When the pulls of these two structures are vertical or in line with each other, the patella may be suspended between them, making little or no contact with the femur. This is essentially the case when the knee joint is in full extension. Cadaver studies have confirmed that there is little or no contact between the patella and the femur in this position.[67] Even a strong contraction of the quadriceps in full extension will produce little or no patellofemoral compression. This is the rationale for use of straight leg-raising exercises as a way of improving quadriceps muscle strength without creating or exacerbating patellofemoral problems.

As knee flexion proceeds from full extension, the pull of the quadriceps tendon (F_Q) and the pull of the patellar ligament (F_{pl}) become increasingly oblique, compressing the patella into the femur (Fig. 11–35). The increasing compression caused by the quadriceps mechanism with increased joint flexion occurs whether the muscle is active or passive. If the quadriceps muscle is inactive, the elastic tension alone will increase with increased knee joint flexion. If the quadriceps muscle is active, both the active tension and passive elastic tension will contribute increasingly to compression as the knee flexion angle increases. The compression, of course, creates a joint reaction force across the patellofemoral joint. The total joint reaction force is influenced by both the magnitude of active and passive pull of the quadriceps and by the angle of knee flexion.

The patellofemoral joint reaction force found in gait when the foot first contacts the ground and the knee flexes slightly to 10 to 15° is 50 percent of body weight.[65] The increased knee flexion and quadriceps muscle activity seen with stair climbing or with running hills may increase the patellofemoral joint reaction force to 3.3 times body weight at 60°.[17,65] The joint reaction force may reach 7.8 times

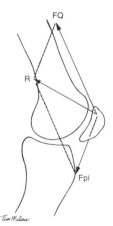

Figure 11–35. The combined pull of the quadriceps (F_Q) and the patellar ligaments (Fp_l) can be composed into a single resultant vector (R) that will clearly compress the patella into the femur. The magnitude of R will increase with an increase in magnitude of (F_Q) and (Fp_l) and with increased knee flexion.

the body weight at 130° of knee joint flexion in such activities as deep knee bends when knee flexion is extreme and a strong quadriceps contraction is required.[65] Although reaction forces at other lower-extremity joints may reach these same magnitudes, they do so over much more congruent joints. That is, the compressive forces are distributed over larger areas. At the normal patellofemoral joint, the medial facet bears the brunt of the compressive force. Several mechanisms help minimize or dissipate the patellofemoral joint compression on the patella in general and on the medial facet specifically.

Since there is essentially no compressive force on the patella in full extension, no compensatory mechanisms are necessary. As knee joint flexion proceeds, the area of patellar contact gradually increases, spreading out the increased compressive force. From 30 to 70° of flexion, contact occurs at the thick cartilage of medial facet near the central ridge. In fact, the cartilage of the medial facet is the thickest hyaline cartilage in the human body.[67] The thick cartilage can better withstand the substantial compressive forces. Within this same range of motion, the patella has its greatest effect as a pulley, maximizing the MA of the quadriceps. With a large MA, less quadriceps muscle force (and less patellofemoral joint compression) is needed to produce the same torque. As flexion proceeds, the MA diminishes, necessitating an increase in force production by the quadriceps. Between 70 and 90°, however, the patella is no longer the only structure contacting the femoral condyles. At this point in the flexion range, the quadriceps tendon contacts the femoral condyles, dissipating some of the patellofemoral compression.[65,66]

The adaptive mechanisms of the patella in dealing with the high joint reaction forces appear to be fairly successful. Although some cartilaginous deterioration is common at both the odd and the medial facet, it bears reiteration that these changes rarely cause problems.

Medial-Lateral Patellofemoral Joint Stability

In the extended knee, the patella perches precariously on the femoral sulcus. Until the patella begins sliding down the femoral condyles with knee joint flexion and is drawn into the intercondylar notch (at about 20° of flexion), medial-lateral stability rests solely on active and passive tension in the structures around the patella. Ficat[69] identified that the patellofemoral joint is under the permanent control of two restraining mechanisms that cross each other at right angles: a transverse group of stabilizers and a longitudinal group of stabilizers. The position of the patella and its mobility will be determined by the relative tension in these two stabilizing systems.

The transverse stabilizers of the patella have been variously described. The medial and lateral extensor (patellar) retinacula join the vastus medialis and lateralis muscles, respectively, directly to the patella.[2] Several investigators have described medial and lateral patellofemoral ligaments that may be part of or blend with the retinacular fibers.[6,17,18] There may also be an iliopatellar band attaching the patella directly to the iliotibial tract.[19,20]

The longitudinal stabilizers of the patella are the patellar tendon inferiorly and the quadriceps tendon superiorly. The patellotibial ligaments are thickenings of the capsule anteriorly, which extend from the inferior border of the patella distally to the anterior coronary ligaments and anterior margins of the tibia on each side of the patellar tendon.[6,17] As has been demonstrated, the longitudinal structures can stabilize the patella through patellofemoral compression. The compression is essentially absent in the extended knee, leaving the patella relatively unstable in this knee joint position. When extension is exaggerated, as in genu recurvatum, the pull of the quadriceps muscle and patellar ligament may actually distract the patella from the femoral sulcus, further aggravating the instability of the patella. There are also other contributing longitudinal structures and other effects of the longitudinal structures to be examined. Both the transverse and the longitudinal structures will influence the medial-lateral positioning of the patella within the femoral sulcus and the so-called **patellar tracking** or path of the patella as it slides down the femoral condyles within the intercondylar notch.

Medial-Lateral Positioning of the Patella

All the passive and dynamic transverse and longitudinal stabilizing mechanisms of the patella can influence the medial-lateral position of the patella. The passive mobility of the patella and its medial-lateral positioning are largely governed by the passive and dynamic pulls of the structures surrounding it. When the knee is fully extended and the musculature relaxed, some investigators have concluded that the patella should be able to be passively displaced medially or laterally no more than one half the width of the patella[70] and that the excursion should be symmetrical.[71] Others, however, have not found either relationship to hold true[72] or have found considerable variation among individuals.[18] Imbalance in passive tension or changes in the line of pull of the dynamic structures will substantially influence the patella. This is predominantly true when the knee joint is in extension and the patella sits on the relatively shallow femoral sulcus. Abnormal forces, however, may influence the excursion of the patella even in its more secure location within the intercondylar notch during knee flexion.

Medial-Lateral Forces on the Patella

During active extension or when the quadriceps muscle is passively stretched, the patella is pulled by the quadriceps. The force on the patella is determined by the resultant pull of the four segments of the quadriceps (F_Q) and by the pull of the patellar ligament (F_{pl}). Since the action lines of F_Q and F_{pl} do not coincide, the patella tends to be pulled slightly laterally by the two forces (Fig. 11–36). Anything that might increase the obliquity of the resultant pull of the quadriceps or the obliquity of the patellar ligament in the frontal plane may increase the lateral force on the patella. An increase in this lateral component may increase the compression on the lateral patellar facet as it pushes harder into the lateral lip of the femoral sulcus (in knee extension) or the lateral aspect of the intercondylar notch (in knee joint flexion). A large lateral force on the patella may actually cause it to sublux or dis-

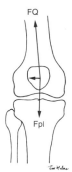

Figure 11-36. The pull of the quadriceps (F_Q) and the pull of the patellar ligament (Fp_l) lie at a slight angle to each other, producing a slight lateral force on the patella.

locate off the lateral lip of the femoral sulcus when the knee is extended. However, even a very large lateral force on the patella would be unlikely to result in dislocation once the patella is in the intercondylar notch.

The pull of the vastus lateralis muscle is normally 12 to 15° lateral to the long axis of the femur with even greater obliquity in its lower fibers.[47] The pull of the vastus medialis longus muscle is approximately 15 to 18° medial to the femoral shaft, with the vastus medialis oblique (VMO) pulling 50 to 55° medially.[47] Since these two muscles pull not only on the common quadriceps tendon but also exert a pull on the patella through their retinacular connections, complementary function is critical. That is, relative weakness of the vastus medialis muscle (especially the VMO fibers) may substantially increase the resultant lateral forces on the patella.

The obliquity of the pull of the quadriceps muscle and patellar tendon can increase with problems other than imbalance between the vastus lateralis and medialis. Genu valgum increases the obliquity of the femur and, concomitantly, the obliquity of the pull of the quadriceps. Femoral anteversion (internal femoral torsion) in the older child or adult generally results in the femoral condyles being turned in (medially rotated) relative to the tibia. This creates an increased obliquity in the patellar tendon that may also be seen with lateral tibial torsion. Each of these conditions can predispose the patella to excessive pressure laterally or to subluxation or dislocation.

The net effect of the pull of the quadriceps and the patellar ligament is commonly assessed clinically using the **Q (quadriceps) angle** of the knee. The Q angle is the angle formed between a line connecting the anterior superior iliac spine (ASIS) to the midpoint of the patella and a line connecting the tibial tubercle and the midpoint of the patella (Fig. 11-37). An angle of 15° is considered to be normal, with females having a slightly greater Q angle than males.[17,57] The increased angle among females is due to the wider pelvis, the increased femoral anteversion, and the relative knee valgus. The Q angle is usually measured with the knee in extension, since excessive lateral forces may be more of a problem here, and because as the knee is flexed, the Q angle will reduce as the tibia rotates medially relative to femur.[73]

A Q angle of 20° or more is considered to be abnormal, creating excessive lateral forces on the patella that may predispose the patella to pathologic changes.[68,74] Although an excessively large Q angle is usually an indicator of some structural misalignment, an apparently normal Q angle is *not* necessarily consistent with the absence of problems. The line between the ASIS of the pelvis and the midpatella is only an estimate of the line of pull of the quadriceps. If substantial imbalance exists between the vastus medialis and lateralis muscles, the Q angle may underestimate the lateral force on the patella since the actual pull of the quadriceps muscle is no longer on the estimated line. Similarly, a patella that is already subluxed or dislo-

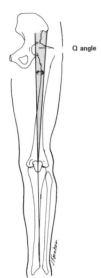

Figure 11–37. The Q angle is the angle between a line connecting the anterior superior iliac spine to the midpoint of the patella and the extension of line connecting the tibial tubercle and the midpoint of the patella. The size of the Q angle may be proportional to the lateral force imposed on the patella by the pull of the quadriceps.

cated may be seen with an inaccurately small Q angle and be misinterpreted as normal.[75]

Forces other than the alignment and balance of the components of the quadriceps muscle may influence patellar positioning. Excessive tension in or adaptive shortening of the lateral retinaculum or stretch of the medial retinaculum have been implicated in patellar dysfunction. Excessive tension in the lateral retinaculum (and/or weakness in the VMO) may cause the patella to tilt laterally, thus increasing compression laterally and reducing it medially. There is some indication that the patellar connections to the ITB may exert an excessive lateral pull on the patella when the ITB band is tight.[76] It is unknown whether changes in the passive structures are primary or are secondary to changes in the dynamic stabilizers.

Any misalignment of the medial-lateral stabilizers of the patella may lead to excessive pressure on the lateral patellar facet. Even large lateral forces can be prevented from subluxing or dislocating the patella as long as the lateral lip of the femoral sulcus is of sufficient height. When it is underdeveloped, however, even relatively small lateral forces may create patellar subluxation or full dislocation. This may be a causative factor in the higher incidence of patellofemoral problems seen in adolescence, when the patellofemoral joint is still developing.[23] The height of the lateral lip of the femoral sulcus may also be a factor in patella alta. In this condition, the lateral lip is not necessarily underdeveloped (although it may be), but the high position of the patella places the patella proximal to the high lateral wall. The upper aspect of the femoral sulcus is less developed and, therefore, makes it easier to sublux the patella.

The result of changes in patellofemoral alignment or imbalance of forces across the joint will be discussed in the following section on injury and disease.

EFFECTS OF INJURY AND DISEASE

The joints of the knee complex, like other joints in the body, are subject to developmental defects, injury, and disease processes. However, there are a number of factors that make the knee joint unique. The knee, unlike the shoulder, elbow, and wrist, must support the body weight and at the same time provide mobility. Although the hip and ankle joints similarly support the body weight, the knee is a more complex structure than either the hip or ankle. The knee joint also joins two of the longest levers in the body and is located in a more exposed position than either the hip or the ankle joints.

Knee Joint Injury

The interest and participation in physical fitness and sports activities currently in vogue among all age groups and both sexes is subjecting the knee complex to an increased risk of injury. Sports such as jogging, skating, skiing, football, and tennis may cause either direct or indirect injury. Injuries to the knee complex may involve the menisci, the ligaments, the bones, or the musculotendinous structures. Meniscal injuries, especially of the medial meniscus, are common and usually occur as a result of sudden rotation of the femur on the fixed tibia when the knee is in flexion.[77] Axial rotation in the flexed knee occurs with the medial meniscus as the pivot point. The more rigidly attached medial meniscus may tear under the sudden load.

Ligamentous injuries may occur as a result of a force that causes the joint to exceed its normal ROM. A blow to the lateral aspect of the knee joint or the tibia may cause a valgus stress that results in a tearing of the ligaments restraining valgus motion. Likewise, forced hyperextension of the knee may cause tearing of the posterior ligaments. Although excessive forces may cause ligamentous tears, lower-level forces may similarly cause disruption in ligaments weakened by aging, disease, immobilization, steroids, or vascular insufficiency. Each of these may affect the collagen or ground substance of the ligaments. Cyclic loading (whether short term and intense or over a prolonged period) may also affect viscoelasticity and stiffness.[36] A weakened ligament may take 10 months or more to return to normal stiffness once the underlying problem has been resolved.[32]

The bony and cartilaginous structures may be injured either by the application of a direct force, such as bumping the patella or falling on the knees, or by indirect forces that are exerted by abnormal ligamentous and muscular forces. Knee joint instability, as frequently seen in the knee following ACL injury, can lead to progressive changes in the articular cartilage, in the menisci, and in the other ligaments attempting to restrain the increased joint mobility.[50,78]

The numerous bursae and tendons at the knee are also subject to injury. The cause of injuries to these structures may be either a direct blow or prolonged compressive or tensile stresses. Bursitis most commonly occurs in the prepatellar bursa and the superficial infrapatellar bursa (known as **housemaid's knee**), but may also occur in high-friction areas such as at the bursa beneath the pes anserinus. The localized tenderness generally associated with bursitis may also be found if the fat pad between the patellar ligament and anterior synovial membrane becomes inflamed.[9]

Another potential source of pain and dysfunction in the knee joint is the presence of a patellar plica. Classic symptoms include pain with prolonged sitting, with stair climbing, and during resisted extension exercises. More than half the patients will also complain of a snapping sensation.[25] In flexion, the plica is drawn tightly over the medial femoral condyle and pressed under the patella. The resultant tension in the band may cause patellar misalignment (leading to pain) or the plica itself may become inflamed.[22,25] If the inflamed plica becomes fibrotic, it may create a secondary synovitis around the femoral condyle and deterioration of the condylar cartilage may occur.[21,22,25]

Patellofemoral Joint Injury

We have presented and discussed the mechanics of a number of problems that may predispose the knee to patellofemoral dysfunction. Any one problem in isolation, or various combinations of problems (which may include primary and secondary changes), may lead to excessive pressure on the lateral facets of the patella, to lateral subluxation, or to lateral dislocation. Whether excessive pressure, subluxation, and dislocation are separate clinical entities or part of a continuum of patellofemoral dysfunction is arguable.[75,79,80] Each, however, is commonly associated with knee pain, poor tolerance of sustained passive knee flexion (as in sitting for long periods), "giving way" of the knee and exacerbation of symptoms by repeated use of the quadriceps on a flexed knee. These symptoms are similar to the complex of symptoms found with patella plica, which may occur as a related disorder. Differentiation in symptomatology may occur, however, when patellar subluxation or dislocation is present. Tenderness of the medial retinaculum and medial border of the patella develop with repeated subluxation or dislocation. The medial retinaculum is stretched as the patella deviates toward or slips over the lateral lip of the femoral sulcus or condyle. The return of the patella into the intercondylar notch may impact the medial patella (occasionally causing osteochondral fracture).[23,69,70]

Until recently, cartilaginous changes seen on the lateral patellar facet were considered to be diagnostic of patellofemoral dysfunction, and the term **chondromalacia patella** (softening of the cartilage) assigned. With the knowledge that sim-

Table 11–2. INJURY TO THE MEDIAL COLLATERAL LIGAMENT OF THE KNEE

Normal Ligament	Effects of Injury
Normal function	Lack of normal function
1. Medial stability. Provides resistance to tensile stress at the medial aspect of tibiofemoral joint.	1. Decrease in medial stability.
2. Rotatory stability. Provides resistance to rotation at the knee.	2. Decrease in rotatory stability.
Normal stresses	Abnormal stresses
1. Normal tibiofemoral valgus. Tensile stress on medial aspect.	1. Possible increase in physiologic valgus, increased tensile stress on medial aspect.
Anatomic relationships	Disturbed anatomic relationships
1. Attachments to the joint. Capsule, medial meniscus, tibia, and femur.	1. Possible tear of medial meniscus and capsular ligaments or avulsion of bony attachment on tibia or femur.
Functional relationships	Disturbed functional relationships
1. Works in conjunction with cruciates to provide anteromedial and posteromedial rotatory stability. Works in conjuction with the medial compartment structures.	2. Increased stress on cruciates and medial compartment structures, because these structures must provide additional support.
2. Helps to prevent excessive compression on lateral joint surfaces by restraining widening of medial joint space.	2. Increased compression forces on lateral aspect of joint surfaces.

ilar cartilaginous changes can be found in asymptomatic knees and that the medial patellar facet frequently shows greater change without symptoms or progressive cartilage deterioration, more general diagnoses have been used, including patellofemoral arthralgia or patellofemoral pain syndrome.[17] Although in fact cartilage changes on the medial patellar facet are more common, changes found on the lateral facet will more commonly progress to osteoarthritis.[13,68] The mechanism of pain is presumed to be the disruption of cartilage, the by-products of which irritate the synovium, which when inflamed, may cause stretching of sensitive surrounding structure.[55,64] Pain may also arise from the innervated subchondral bone, which is subjected to increased load as the cartilage deteriorates.[17] Such cartilage deterioration is generally found when there is substantial misalignment or instability. Minor structural, stability, or overuse problems typically found among adolescents and young adults may cause patellofemoral pain. In such instances, however, the pain frequently resolves spontaneously over time and will *not* necessarily progress to later osteoarthritic problems.[17]

A Model for Dysfunction

Given the range of possible problems that can occur in the knee joint, an exhaustive discussion is beyond the scope of this text. However, a thorough knowledge of normal structure and function can be used to predict or understand the immediate impact on the joint of a specific injury and the secondary effects on intact structures. An example of such an analysis is presented in Table 11–2, using the example of rupture of the MCL. The four aspects of normal structure and function to be considered are:

- The normal function that the structure is designed to serve
- The stresses that are present during normal situations
- Anatomic relationship of the structure to adjacent structures
- Functional relationship of the structure to other structures

Any injury or disease process can be considered by using the model and normal structure and function as the basis for analysis. The model in Table 11–2 can also be applied to such injuries as a torn meniscus or torn cruciate ligament.

REFERENCES

1. Sonstegard, DA, Matthews, LS, and Kaufer, H: The surgical replacement of the human knee joint. Sci Am, 1978.
2. Williams, PL and Warwick R (eds): Gray's Anatomy, ed 37. WB Saunders, Philadelphia, 1985.
3. Nordin, M and Frankel, VH: Basic Biomechanics of the Skeletal System, ed 2. Lea & Febiger, Philadelphia, 1989.
4. Kapandji, IA: The Physiology of the Joints, Vol. 2, ed 2. Williams & Wilkins, Baltimore, 1970.
5. Seebacher, JR, et al: Structure of the posterolateral aspect of the knee. J Bone Joint Surg 64A:536–540, 1982.
6. Larson, RL, et al: Patellar compression syndrome: Surgical treatment by lateral retinacular release. Clin Orthop 134:158–167, 1970.
7. Kaplan, EB: Some aspects of functional anatomy of the human knee joint. Clin Orthop 23:18–29, 1962.
8. Nicholas, JA and Hershman, EB: The Lower Extremity and Spine in Sports Medicine. CV Mosby, St Louis, 1986.
9. Cailliet, R: Soft Tissue Pain and Disability, ed 2. FA Davis, Philadelphia, 1988.
10. Clark, CR and Ogden, FA: Development of menisci of human knee joint. J Bone Joint Surg 65A:538–546, 1983.
11. Reilly, DT: Dynamic loading of normal joints. Rheum Dis Clin North Am 14(3):497–502, 1988.
12. Johnson, F, Leitl, S, and Waugh, W: The distribution of the load across the knee: A comparison of static and dynamic measurements. J Bone Joint Surg 62B, 1980.
13. Radin, EL, DeLamotte, F, and Maquet, P: Role of the menisci in distribution of stress in the knee. Clin Orthop 185:290–293, 1984.

14. Seedhom, BB: Loadbearing function of the menisci. Physiotherapy 62:7, 1978.
15. Torzilli, PA, Greenberg, RL, and Insall, J: An in vivo biomechanical evaluation of anterior-posterior motion of the knee. J Bone Joint Surg 63A:960–968, 1981.
16. Fukubayashi, T, et al: An in vitro biomechanical evaluation of anterior-posterior motion of the knee. J Bone Joint Surg 64A:258–264, 1982.
17. Cox, JS: Patellofemoral problems in runners. Clin Sports Med 4:699–715, 1985.
18. Paulos, L, et al: Patellar malalignment, a treatment rationale. Phys Ther 60:1624–1632, 1980.
19. Blauth, M and Tillman, B: Stressing on the human femoro-patellar joint. Anat Embryol 168:117–123, 1983.
20. Terry, GC, Hughston, JC, and Norwood, LA: Anatomy of the iliopatellar band and iliotibial tract. Am J Sports Med 14:39–45, 1986.
21. Bogdan, RF: Plicae syndrome of the knee. J Am Pod Soc 75:377–381, 1985.
22. Blackburn, TA, Eilnad, WG, and Bandy, WD: An introduction to the plicae. J Orthop Sports Phys Ther 3:171–177, 1982.
23. Hughston, JC, et al: Classification of knee ligament instabilities, Part I: The medial compartment and cruciate ligaments. J Bone Joint Surg 58A:2, 1976.
24. Hughston, JC, et al: Classification of knee ligament instabilities, Part II: The lateral compartment. J Bone Joint Surg 58A:2, 1976.
25. Hardaker, WT, Whipple, TL, and Bassett, FH: Diagnosis and treatment of the plicae syndrome of the knee. J Bone Joint Surg 62A:221–225, 1980.
26. Grood, ES, et al: Ligamentous and capsular restraints preventing straight medial and lateral laxity in intact human cadaver knees. J Bone Joint Surg 63A:1257–1269, 1981.
27. Nielsen, S, et al: Rotatory instability of cadaver knees after transection of collateral ligaments and capsule. Arch Orthop Trauma Surg 103:165–169, 1984.
28. Nielsen, S: Kinesiology of the knee joint: An experimental investigation of ligamentous and capsular restraints preventing knee instability. Dan Med Bull 34:297–309, 1987.
29. Kaplan, EB: The iliotibial tract. J Bone Joint Surg 40A:817–832, 1958.
30. Jeffreys, TE: Recurrent dislocation of the patella due to abnormal attachment of ilio-tibial tract. J Bone Joint Surg 45B:740–743, 1963.
31. Arnoczky, SP: Anatomy of the anterior cruciate ligament. Clin Orthop 172:19–25, 1983.
32. Cabaud, HE: Biomechanics of the anterior cruciate ligament. Clin Orthop 172:26–30, 1983.
33. France, EP, et al: Simultaneous quantitation of knee ligament forces. J Biomech 16:553–564, 1983.
34. Piziali, RL, et al: The function of primary ligaments of the knee in anterior-posterior and medial-lateral motions. J Biomech 13:777–784, 1980.
35. Lipke, JM, et al: Role of incompetence of the anterior cruciate and lateral ligs in anterolateral and anteromedial stability. J Bone Joint Surg 63A:954–959, 1981.
36. Feagin, JA and Lambert, KL: Mechanism of injury and pathology of anterior cruciate ligament injuries. Orthop Clin North Am 16:41–45, 1985.
37. Terry, GC and Hughston, JC: Associated joint pathology in the anterior cruciate ligament-deficient knee with emphasis on a classification system and injuries to the meniscocapsular ligament–musculotendinous unit complex. Orthop Clin North Am 16:29–38, 1985.
38. Rauschning, W: Anatomy and function of the communication between knee joint and popliteal bursae. Ann Rheum Dis 39:354–358, 1980.
39. Norkin, CC and White, DW: Measurement of Joint Motion: A Guide to Goniometry. FA Davis, Philadelphia, 1985.
40. Pathokinesiology and Physical Therapy Department: Observational Gait Analysis Handbook. Rancho Los Amigos Medical Center, Professional Staff Association, Rancho Los Amigos Hospital, Downey, California, 1989.
41. Inman, VT, Ralston, HJ, and Todd, F: Human Walking. Williams & Wilkins, Baltimore, 1981.
42. Lehmkuhl, LD and Smith, LK: Brunnstrom's Clinical Kinesiology, ed 4. FA Davis, Philadelphia, 1983.
43. Wismans, J, Veldpaus, F, and Janssen, J: A three-dimensional mathematical model of the knee joint. J Biomech 13:677, 1980.
44. Wickiewics, TL, et al: Muscle architecture of the human lower limb. Clin Orthop 179:275–283, 1983.
45. Dunnen, JD, Yack, J, and LeVeau, BF: Relationship between muscle length, muscle activity and torque of the hamstring muscles. Phys Ther 61:2, 1981.
46. Basmajian, JV and DeLuca, CJ: Muscles Alive, ed 5. Williams & Wilkins, Baltimore, 1985.
47. Lieb, FJ and Perry, J: Quadriceps function: An anatomical and mechanical study using amputated limbs. J Bone Joint Surg 50A:1535–1548, 1968.
48. Sutton, FS, et al: The effect of patellectomy on knee function. J Bone Joint Surg 58A:, 1976.
49. Murray, MP, et al: Strength of isometric and isokinetic contractions: Knee muscles of men aged 20–86. Phys Ther 60:, 1980.
50. Tamea, CD and Henning, CE: Pathomechanics of the pivot shift maneuver. Am J Sports Med 9:31–37, 1981.
51. McLeod, WD and Hunter, S: Biomechanical analysis of the knee. Phys Ther 60:1561–1564, 1980.

52. Markolf, KL, et al: Role of joint load in knee stability. J Bone Joint Surg 63A:570–585, 1981.
53. Fujikawa, K, Seedhom, BB, and Wright, V: Biomech of patello-femoral joint. Part I: a study of contact and congruity of the patello-femoral compartment and movement of the patella. Engr Med, 12:3–11, 1983.
54. Sikorski, JM, Peters, J, and Watt, I: The importance of femoral rotation in chondromalacia patellae as shown by serial radiography. J Bone Joint Surg 61B:, 1970.
55. Radin, EL: A rational approach to the treatment of patellofemoral pain. Clin Orthop 144:107–109, 1979.
56. Wiberg, G: Roentgenographic and anatomic studies on the femoropatellar joint. Acta Orthop Scand 12:319–409, 1941.
57. Aglietti, P, Insall, JN, and Cerulli, G: Patellar pain and incongruence. Clin Orthop 176:217–223, 1983.
58. Merchant, AC, et al: Roentgenographic analysis of patellofemoral congruence. J Bone Joint Surg 56A:1391–1396, 1974.
59. Jacobsen, K and Bertheussen, K: The vertical location of the patella. Acta Orthop Scand 45:436–445, 1974.
60. Insall, J, and Salvati, E: Patella position in the normal knee joint. Radiology 101:101–104, 1971.
61. Norman, O, Eglund, N, and Runow, A: Vertical position of the patella. Acta Orthop Scand 54:908–913, 1983.
62. Runow, A: The dislocating patella: Etiology and prognosis in relation to generalized joint laxity and anatomy of the patellar articulation. Acta Orthop Scand (Suppl)201:1–53, 1983.
63. Insall, J: Proximal realignment in the treatment of patellofemoral pain. In Pickett, JC and Radin, EL (eds): Chondromalacia of the Patellae. Williams & Wilkins, Baltimore, 1983.
64. Goodfellow, J, Hungerford, DS, and Woods, C: Patello-femoral joint mechanics and pathology: Chondromalacia patellae. J Bone Joint Surg 58B:291–299, 1976.
65. Hungerford, DS and Barry, M: Biomechanics of the patellofemoral joint. Clin Orthop 144:9–15, 1979.
66. Hungerford, DS: Patellar subluxation and excessive lateral pressure as a cause of fibrillation. In Pickett, JC and Radin, EL (eds): Chondromalacia of the Patellae. Williams & Wilkins, Baltimore, 1983.
67. Henche, R, Junze, HU, and Morscher, E: The areas of contact pressure in the patello-femoral joint. Int Orthop (SICOT) 4:279–281, 1981.
68. Insall, J, Falvo, KA, and Wise, DW: Chondromalacia patellae: A prospective study. J Bone Joint Surg 58A:1–8, 1976.
69. Ficat, P: Lateral fascia release and lateral hyperpressure syndrome. In Pickett, JC and Radin, EL (eds): Chondromalacia of the Patellae. Williams & Wilkins, Baltimore, 1983.
70. Carson, WG Jr, et al: Patellofemoral disorders: Physical and radiographic examination—Part 1. Clin Orthop 135:174, 1984.
71. Harwin, SF and Stern, RE: Subcutaneous lateral retinacular release for chondromalacia patellae: A preliminary report. Clin Orthop 156:207–210, 1981.
72. Levangie, PK, et al: Assessment of medial and lateral patellar mobility in subjects with and without knee pain. Poster presentation, American Physical Therapy Association, Annual Conference, Nashville, Tennessee, June 1989.
73. Hvid, I: Stability of the human patellofemoral joint. Engr Med 12:55–59, 1983.
74. Levine, J: Chondromalacia patellae. Phys Sports Med 7, 1979.
75. Kettelkamp, DB: Current concepts review: Management of patellar malalignment. J Bone Joint Surg 63A:1344–1348, 1981.
76. Ober, FR, Brown, JB, and Brown, B: Recurrent dislocation of the patella. Am J Surg 43:497–500, 1939.
77. Derscheid, FL and Malone, RT: Knee disorders. Phys Ther 60:12, 1980.
78. Noyes, FR, McGinness, FH, and Grood, ES: Variable functional disability of the anterior cruciate ligament-deficient knee. Orthop Clin North Am 16:47–67, 1985.
79. Abernethy, PJ, et al: Is chondromalacia patel a separate clinical entity? J Bone Joint Surg 60B:205–210, 1978.
80. Ficat, P: Lateral fascia release and lateral hyperpressure syndrome. In Pickett, JC and Radin, EL (eds): Chondromalacia of the Patellae. Williams & Wilkins, Baltimore, 1983.

STUDY QUESTIONS

1. Describe the congruency of the tibiofemoral joint. What factors add to or detract from stability?

2. Describe the menisci of the knee, including shape, attachments, and function.

3. Describe the intra-articular movement of the femur on the tibia, as the femur

moves from full extension into flexion. Where is the axis generally located and how does it change during the described motion?

4. Describe the locking mechanism of the knee, including the structure(s) responsible.

5. What happens to the menisci during motions of the knee? How do their attachments contribute to the movement?

6. Identify the bursae of the knee joint. Which of these are generally separate from and which are part of the capsule?

7. Which knee joint ligaments contribute to AP stability of the knee joint and how does each make its contribution? To which compartment (medial or lateral) do each of the structures belong?

8. Which ligaments contribute to medial-lateral stability of the knee joint and how does each make its contribution? To which compartment do the structures belong?

9. What are the dynamic stabilizers of the knee, and in what plane do they contribute to stability?

10. What is the axis given for rotation of the knee joint? How does this differ between automatic and axial rotation? What implications does the location have for stress on the joint?

11. What is the normal physiologic angulation of the knee joint? What terms are used to describe pathologic increases or decreases in this angulation?

12. When is axial rotation of the knee greatest? Which muscles produce active medial rotation? Lateral rotation?

13. What is a patella plica and what implications does it have for knee joint dysfunction?

14. Describe the patellofemoral articulation, including the number and shape of the surfaces.

15. What function(s) does the patella serve at the knee joint?

16. How does the patella move in relation to the femur in normal motions? How would function be affected if the patella could not slide on the femur?

17. Describe the contact of the patella with the femur at rest in full extension. Describe the contact as knee flexion proceeds.

18. Is the patella equally effective as an anatomic pulley at all points in the knee ROM? At which point(s) is it most effective? Least effective?

19. Which facet(s) of the patella are subject to the earliest degenerative changes under *normal* conditions? Why?

20. Which facet of the patella is most likely to undergo excessive degenerative changes when there is misalignment? Describe the misalignment and the condition(s) that may predispose these changes.

21. What is the Q angle of the knee joint? How is it measured, and what implications does it have for patellofemoral problems?

22. What changes will the condition of genu recurvatum produce at the patellofemoral joint?

23. Why is ascending stairs commonly cited as producing knee pain? Relate this to patellofemoral joint compression.

24. Which aspect of the knee joint is most prone to injury? Why is this true, and what structures are most commonly involved?

Chapter 12

■ ■ ■

The Ankle-Foot Complex

OBJECTIVES

Following the study of this chapter, the reader should be able to:

Define

1. The terminology unique to the ankle-foot complex, including supination/pronation, inversion/eversion, dorsiflexion/plantarflexion, flexion/extension, and adduction/abduction.

Describe

1. The compound articulations of the ankle, subtalar, talocalcaneonavicular, transverse tarsal, and tarsometatarsal joints.
2. The role of the tibiofibular joints and supporting ligaments.
3. The degrees of freedom and range of motion available at the joints of the ankle and foot.
4. The significant ligaments that support the ankle, subtalar, and transverse tarsal joints.
5. The triplanar nature of ankle joint motion.
6. The articular movements that occur in the weight-bearing subtalar joint during supination/pronation.
7. The relationship between tibial rotation and subtalar/talocalcaneonavicular supination/pronation.
8. The relationship between hindfoot supination/pronation and mobility/stability of the transverse tarsal joint.
9. The function of the tarsometatarsal joints, including when motion at these joints is called on.
10. Supination/pronation twist of the forefoot at the tarsometatarsal joints.
11. Distribution of weight within the foot.
12. The structure and function of the plantar arches, including the primary supporting structures.
13. When muscles supplement arch support, including those muscles that specifically contribute to arch support.
14. The effects of toe extension on the plantar arches.
15. The general function of the extrinsic muscles of the ankle-foot.
16. The general function of the intrinsic muscles of the foot.

INTRODUCTION

The ankle/foot complex is structurally analogous to the wrist/hand complex of the upper extremity. Although the hand is undeniably more critical to uniquely human functions, the ankle and foot receive more attention from both medical and lay communities. The interdependence of the ankle and foot with the more proximal joints of the lower extremities and the great weight-bearing stresses to which these joints are subjected have resulted in a greater frequency and diversity of difficulties in the joints of the ankle and foot than in its more vital upper extremity counterpart. The prevalence of ankle/foot problems, in fact, has resulted in the formation of a branch of medicine, podiatry, that focuses on the correction and prevention of problems in structures distal to the knee. Specialists in orthopedics and sports medicine similarly find that ankle/foot problems or problems attributable to ankle or foot dysfunction make up a substantial percentage of their practice.

The frequency of ankle/foot problems can be traced readily to the foot's complex structure, to the need to sustain large weight-bearing stresses, and to the multiple and somewhat conflicting functions that the foot must perform. The ankle/foot complex must meet the stability demands of: (1) providing a stable base of support

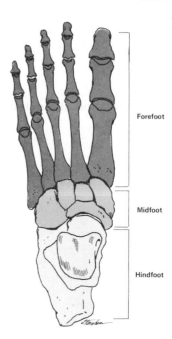

Forefoot

Midfoot

Hindfoot

Figure 12–1. Functional segments of the foot.

for the body in a variety of weight-bearing postures without undue muscular activity and energy expenditure and (2) acting as a rigid lever for effective push-off during gait. The stability requirements can be contrasted to the mobility demands of: (1) dampening of rotations imposed by the more proximal joints of the lower limbs, (2) being flexible enough to absorb the shock of the superimposed body weight as the foot hits the ground, and (3) permitting the foot to conform to the changing and varied terrain on which the foot is placed.[1] The ankle/foot complex meets its diverse requirements through its 28 bones that form 25 component joints. These joints include the proximal and distal tibiofibular joints, the talocrural or ankle joint, the talocalcaneal or subtalar joint, the talonavicular and the calcaneocuboid joints, the five tarsometatarsal joints, five metatarsophalangeal joints, and nine interphalangeal joints.

To facilitate description and understanding of the ankle/foot complex, the bones of the foot are traditionally divided into three functional segments. These are the hindfoot (posterior segment), composed of the talus and calcaneus; the midfoot (middle segment), composed of the navicular, cuboid, and three cuneiforms; and the forefoot (anterior segment), composed of the metatarsals and the phalanges (Fig. 12–1).[2] These terms are commonly used when describing ankle or foot dysfunction or deformity and are similarly useful in understanding normal ankle and foot function.

ANKLE JOINT

The term **ankle** specifically refers to the **talocrural joint**; that is, the articulation between the talus and the distal tibia (tibiotalar surface) and the talus and fibula (talofibular surface) (Fig. 12–2). The ankle is a synovial hinge joint with a joint capsule and associated ligaments. It is generally considered to have a single oblique axis with 1° of freedom: dorsiflexion/plantarflexion.

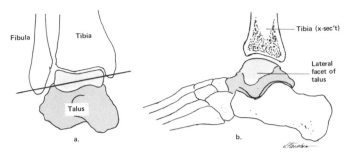

Figure 12–2. The ankle joint. a. Anterior view showing the mortise astride the body of the talus and the superior-inferior inclination of the ankle axis. b. Lateral view of the talus and cross section of the tibia showing the large lateral articulation on the talus for the fibular malleolus.

Ankle Joint Structure
Proximal Articular Surface

The proximal segment of the ankle is composed of the concave surface of the distal tibia and of the tibial and fibular malleoli. These three facets form an almost continuous concave joint surface that extends more distally on the fibular (lateral) side (Fig. 12–2a) and on the posterior margin of the tibia (Fig. 12–2b). The structure of the distal tibia and the malleoli resembles and is referred to as a **mortise**. A common example of a mortise is the gripping part of a wrench. The wrench can either be fixed (fitting only one size bolt) or it can be adjustable (permitting use of the wrench on a variety of bolt sizes). The adjustable mortise is more complex than a fixed mortise since it combines mobility and stability functions. The mortise of the ankle is adjustable, relying on the **proximal** and **distal tibiofibular joints** to both permit and control the changes in the mortise.

The proximal and distal tibiofibular joints (Fig. 12–3) are anatomically distinct from the ankle joint, but function almost exclusively to serve the ankle. Unlike their upper extremity counterparts—the proximal and distal radioulnar joints—the tib-

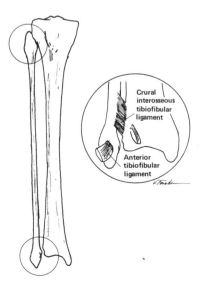

Figure 12–3. Proximal and distal tibiofibular joints. (Inset: Ligaments of the distal tibiofibular joint.)

iofibular joints do not add any degrees of freedom to the more distal segments of the extremity. However, while fusion of the radioulnar joints would have little effect on wrist function, fusion of the tibiofibular joints might impair normal ankle function.

Proximal Tibiofibular Joint

The proximal tibiofibular joint is a plane synovial joint formed by the articulation of the head of the fibula with the posterolateral aspect of the tibia. Although the facets are fairly flat and vary in configuration among individuals, a slight convexity to the tibial facet and slight concavity to the fibular facet seem to predominate.[3] The inclination of the facets may vary from nearly vertical to nearly horizontal in orientation.[3,4] Each proximal tibiofibular joint is surrounded by a joint capsule that is reinforced by anterior and posterior ligaments and in all but rare instances is anatomically separate from the knee joint.[3] Motion at the proximal tibiofibular joint is variable but consistently small: it has been described as superior and inferior sliding of the fibula and as fibular rotation.[4,5] The relevance of motion at the proximal tibiofibular joint will be seen when the ankle articulation is examined more closely.

Distal Tibiofibular Joint

The distal tibiofibular joint is a syndesmosis, or fibrous union, between the concave facet of the tibia and the convex facet of the fibula. The tibia and fibula do not actually come into contact with each other at this point but are separated by fibroadipose tissue. Although there is no joint capsule, there are several associated ligaments.

The ligaments of the distal tibiofibular joint are primarily responsible for restricting motion at both the proximal and the distal tibiofibular joints and for maintaining a stable mortise. The strongest and most important of the ligaments found at the distal tibiofibular joint is the crural tibiofibular interosseous ligament (Fig. 12–3, inset).[6] Its oblique fibers run for a short distance between the tibia and fibula, maintaining proximity of the bones. The crural interosseous ligament also serves as a fulcrum for fibular motion; as a consequence, small movements of the fibular malleolus result in magnified movements at the proximal tibiofibular joint. The other ligamentous structures that support the distal tibiofibular joint are the anterior and posterior tibiofibular ligaments and the interosseous membrane. The interosseous membrane directly supports both tibiofibular articulations. Considering each of the tibiofibular ligaments, the distal tibiofibular joint is an extremely strong articulation. Stresses that tend to move the talus excessively in the mortise may tear the ankle collaterals first. Continued force may fracture the fibula proximal to the distal tibiofibular ligaments before the tibiofibular ligaments will tear.[7]

The function of the talocrural or ankle joint is dependent on the tibiofibular mortise. The tibia and fibula would be unable to grasp and hold on to the distal joint segment if the bones were permitted to separate or if one side of the mortise were missing. The analogous mortise of a wrench could not perform its function of grasping a bolt if the two pincer segments moved apart every time a force was applied to the wrench. The fibula, in fact, has little weight-bearing function, bearing no more than 10 percent of the weight that comes through the femur.[8,9] The fibula is present almost exclusively to serve as one of the pincers in the mortise.

Distal Articular Surface

The distal articular surface of the ankle joint is formed by the body of the talus. The body of the talus has three articular surfaces: a large lateral facet, a smaller medial facet, and a trochlear or superior facet. The large convex trochlear surface

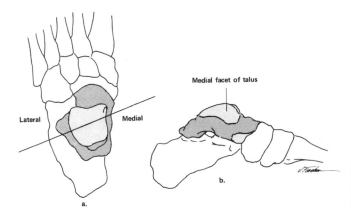

Figure 12–4. a. Superior view of the talus shows its wedge-shaped trochlea and the anterior-posterior inclination of the ankle axis. b. Medial view of the talus shows the small medial facet for articulation with the tibial malleolus.

has a central groove that runs at a slight angle to the head and neck of the talus. The body of the talus is also wider anteriorly than posteriorly, giving it a wedge shape (Fig. 12–4a). The degree of wedging may vary among individuals, with no wedging at all in some and a 25 percent decrease in width anteriorly to posteriorly in others.[10] The articular cartilage covering the trochlea is continuous with the cartilage covering the more extensive lateral (fibular) facet (Fig. 12–2b) and the smaller medial (tibial) facet (Fig. 12–4b).

The ankle joint is the most congruent joint in the human body.[6] The structural integrity is maintained throughout the range of motion (ROM) of the joint by a number of important ligaments.

Ligaments

The capsule of the ankle joint is fairly thin and especially weak anteriorly and posteriorly. Therefore, the stability of the ankle is dependent on an intact ligamentous structure. Several ligaments important to ankle joint stability have already been presented. These are the crural tibiofibular interosseous ligament, the anterior and posterior tibiofibular ligaments, and the tibiofibular interosseous membrane. The tibiofibular ligaments maintain the grasp of the tibiofibular mortise on the body of the talus. Two other major ligaments maintain the contact of the ankle joint surfaces and control medial-lateral joint stability. These are the medial collateral ligament (MCL) and the lateral collateral ligament (LCL).

The MCL is most commonly called the **deltoid ligament.** As its name implies, it is a fan-shaped ligament. It has both superficial and deep fibers that arise from the borders of the tibial malleolus and insert in a continuous line on the navicular anteriorly and on the talus and calcaneus distally and posteriorly (Fig. 12–5). The deltoid ligament is extremely strong. Forces that would open the medial side of the ankle may actually fracture off (avulse) the tibial malleolus before the deltoid ligament tears. This ligament not only helps to control medial distraction stresses on the joint but also helps check motion at the extremes of joint range.

The LCL is composed of three separate bands that are commonly referred to as separate ligaments. These are the **anterior** and **posterior talofibular** ligaments and the **calcaneofibular** ligament (Fig. 12–6). Generally, the LCL is weaker and more prone to injury than is the MCL. The LCL helps to control varus stresses resulting in lateral distraction of the joint and to check extremes of joint ROM. Many experiments have been conducted on serial sectioning of the lateral ligaments, with some differing results. There is consensus, however, on several factors:

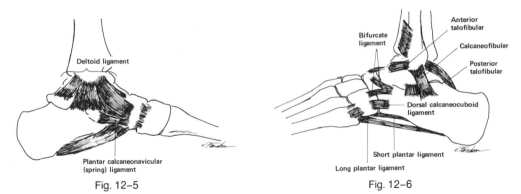

Fig. 12–5   Fig. 12–6

Figure 12–5. Medial ligaments of the posterior ankle-foot complex.

Figure 12–6. Lateral ligaments of the posterior ankle-foot complex.

1. The ability of the LCL to check either dorsiflexion/plantarflexion, talar tilt (AP axis), or talar rotation (vertical axis) is dependent on the position of the ankle joint.[11–13]
2. The anterior talofibular ligament is the weakest and most commonly torn of the LCLs, followed by tears in the calcaneofibular ligament. Rupture of the anterior talofibular ligament invariably results in anterolateral rotatory instability of the ankle.[13–15]
3. The posterior talofibular ligament is the strongest of the collaterals and is rarely torn in isolation.[13,14]

Ankle Joint Function

The ankle joint is classically considered to move around a single axis, contributing 1° of freedom to the foot when the foot is free to move, or to the tibia and fibula when the foot is weight bearing. However, many investigators have concluded from both in vivo and in vitro investigations that the ankle joint is capable of some rotation of the talus within the mortise in both the transverse plane and the frontal plane.[13,16–18] Such motions result in a moving or instantaneous axis of rotation for the ankle joint. Rasmussen and Tovborg-Jensen[13] measured transverse plane talar motion (ligaments intact) to be as much as 7° of medial rotation and 10° of lateral rotation, while talar tilt (AP axis) averaged 5°. Lundberg and associates[17–19] found similar motions in their studies of the weight-bearing foot in vivo, although the magnitudes were smaller and depended on the position of the talus in the mortise.[17–19] There is consensus among investigators that the motions of dorsiflexion and plantarflexion predominate at the talocrural or ankle joint and that these motions do not occur purely in the sagittal plane. Dorsiflexion/plantarflexion of the ankle, like some other motions in the foot, is a single motion that crosses three planes. The triplanar motion is attributed to the obliquity of the ankle joint axis and the shape of the body of the talus.

Axis

In neutral position of the ankle joint, the axis passes approximately through the fibular malleolus and the body of the talus, and through or just below the tibial malleolus.[20] The fibular malleolus, however, extends more distally than does the tibial malleolus and lies more posteriorly. This more posterior position of the fibular mal-

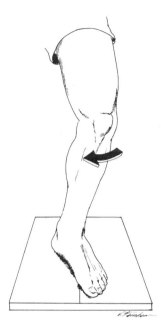

Figure 12-7. Dorsiflexion of the leg on the fixed foot results in inclination and rotation of the tibia medially.

leolus is due to the normal torsion or twist that exists in the distal tibia relative to its proximal plateau. The torsion in the tibia is similar to the torsion found in the shaft of the femur, although reversed in direction. The distal tibia is twisted laterally compared with its proximal portion, accounting for the toe-out position of the foot in normal standing. Given the relationship of the malleoli, the axis of the ankle is considered to be rotated laterally 20 to 30° in the transverse plane (see Fig. 12–4a) and inclined 10° down on the lateral side (see Fig. 12–2a).[4,21] The inclination of the ankle joint axis results in motion across two planes: dorsiflexion of the foot brings the foot up and slightly laterally (increased toe-out), whereas plantarflexion brings the foot down and medially (decreased toe-out). The same relative pattern of motion exists whether the foot is the moving segment or the tibia moves on the fixed foot (Fig. 12–7). The third plane of ankle joint dorsiflexion/plantarflexion can be seen when one closely examines movement of the tibiofibular mortise over the talus.

Arthrokinematics

It has long been noted that the trochlea of the talus is wider anteriorly than posteriorly in most individuals. Although it would appear reasonable to assume that the mortise must adjust to the variable size of the talus, investigations have shown that there is little change in the intermalleolar space with ankle joint motion.[22] Rather than perceiving the talus as a wedge-shaped surface, it can be thought of as a segment of a cone lying on its side with its base directed laterally.[23] The cone should be visualized as "truncated" or cut off on either end (Fig. 12–8). This representation takes into account that both the surface of the lateral facet and its radius of curvature are larger than those of the medial facet. Using the concept of a truncated cone, it can be seen that the fibula moving on the lateral facet of the talus must have a greater displacement anteroposteriorly than the tibial malleolus moving on the smaller medial facet of the talus. The greater excursion of the fibula results in the imposition of medial rotation on the leg as the leg passes over the foot in weight-bearing dorsiflexion and lateral rotation of the superimposed leg in plantarflexion. When the foot is free to move, the talus appears to abduct (rotate laterally on a

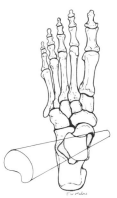

Figure 12-8. The trochlea, smaller medial facet, and larger lateral facet of the talus can be pictured as part of a conical surface, with the ends of the cone cut off (the larger end of the cone facing laterally).

Truncated cone

vertical axis) in dorsiflexion and adduct in plantarflexion. Barnett and Napier[4] also noted that the larger radius of curvature of the lateral talar facet produces an actual rotation of the fibula around its long axis during ankle joint motion that must be reflected at the tibiofibular joints. The range of rotation of the fibula is greater as the orientation of the proximal tibiofibular facets become more vertical,[3] and may account for some of the individual differences in ankle joint range. Others have attributed to the fibula a superior gliding motion as the wider portion of the talus enters the mortise.[5] Given that the fibula must glide anteriorly on the lateral facet of the talus in dorsiflexion and posteriorly in plantarflexion, the ankle joint axis cannot be fixed as it would be in a true hinge joint but must change from dorsiflexion to plantarflexion.[4,20,22]

The mechanism of the mortise traveling on a truncated cone accounts for the stability of the normal ankle joint throughout ankle joint ROM and without any changes in the intermalleolar space.[1,6] The increased frequency of ankle injuries in the position of ankle plantarflexion should not be attributed to instability of the talus in the mortise but to other factors that will be explored later. It can also be seen now that fusion of one or both of the tibiofibular joints would limit ankle joint ROM since the ability of the mortise to glide over the cone would be impaired.

With the large and congruent articular surfaces of the ankle, the ankle is able to withstand compression forces during gait as much as 450 percent of body weight with little incidence of primary (nontraumatic) degenerative arthritis.[24] Greenwald and Matejczyk[25] demonstrated changes in contact across the joint surfaces with ankle motion and hypothesized that some incongruence in the ankle is necessary for normal load distribution, cartilage nutrition, and lubrication of the ankle joint.

Range of Motion

The normal ankle joint range is generally given as 20° of dorsiflexion from neutral, and a more variable 30 to 50° of plantarflexion from neutral.[5,13] The ankle is in neutral when the foot is at a right angle to the tibia.[5] Talocrural joint measurements made in living subjects in the weight-bearing position find dorsiflexion about the same as found in cadaver studies, but plantarflexion to be a more limited 23 to 28°.[17,22] Lundberg and associates[17] found the range to vary among individuals, with anywhere from 60 to 90 percent of the total plantarflexion of the foot attributable to the ankle. As much as 40 percent of the plantarflexion of the foot occurred in more distal joints in some individuals. Hornsby and associates[26] found active non–weight-bearing dorsiflexion to be limited to an average of just over 4°.

Normal checks to ankle dorsiflexion and plantarflexion are primarily muscular.

Passive or active tension in the triceps surae (gastrocnemius and soleus muscles) is the main check to dorsiflexion, with dorsiflexion more limited with knee extension than with knee flexion.[26] Tension in the tibialis anterior, extensor hallucis longus, and extensor digitorum longus muscles is the primary check to plantarflexion. Although the ligaments of the ankle assist in checking dorsiflexion and plantarflexion,[12,13] a more important function appears to be in minimizing side-to-side movement or rotation of the mortise on the talus. The ligaments are assisted in that function by the muscles that pass on either side of the ankle. The tibialis posterior, flexor hallucis longus, and flexor digitorum longus muscles help protect the medial aspect of the ankle, while the peroneus longus and peroneus brevis muscles protect the lateral aspect. Bony checks to any of the potential ankle motions are rarely encountered unless there is extreme hypermobility (as may be found among gymnasts or dancers) or a failure of one or more of the other restraint systems. A more complete analysis of the function of the muscles crossing the ankle will be presented later, since all muscles of the ankle cross at least two and generally three or more joints of the ankle and foot.

THE SUBTALAR JOINT

The talocalcaneal, or subtalar, joint is a composite joint formed by three separate plane articulations between the talus superiorly and the calcaneus inferiorly. Together, the three surfaces provide a triplanar movement around a single joint axis. Function at the weight-bearing subtalar joint is critical for dampening the rotational forces imposed by the body weight while maintaining contact of the foot with the supporting surface.

Subtalar Joint Structure

The posterior talocalcaneal articulation is the largest of the three articulations found between the talus and calcaneus. The posterior articulation is formed by a concave facet on the undersurface of the body of the talus and a convex facet on the body of the calcaneus. The smaller anterior and middle talocalcaneal articulations are formed by two convex facets on the inferior body and neck of the talus and two concave facets on the calcaneus. The anterior and middle articulations, therefore, have an intra-articular configuration that is the reverse of that found at the posterior facet. Between the posterior articulation and the anterior and middle articulations, there is a bony tunnel formed by concave grooves, or sulci, in the inferior talus and superior calcaneus. This funnel-shaped tunnel, known as the tarsal canal, runs obliquely across the foot. Its large end lies just anterior to the fibular malleolus; its small end lies posteriorly below the tibial malleolus and above a bony outcropping on the calcaneus called the sustentaculum tali. There are ligaments running the length of the tarsal canal that completely divide the posterior articulation and the anterior and middle articulation into two separate noncommunicating joint cavities.[27] The posterior articulation has its own capsule, while the anterior and middle articulations share a capsule with the talonavicular joint.

Ligaments

The subtalar joint is a stable joint that rarely dislocates. Its ligamentous support (Fig. 12–9) is ample and strong. The interosseous talocalcaneal ligament lies within the tarsal canal and has both anterior and posterior bands.[5] It is extremely strong, being predominantly composed of collagen with little elastin; this ligament must be severed to fully open the subtalar joint.[27] Occasional reference is also made to a

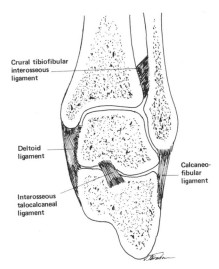

Crural tibiofibular interosseous ligament

Deltoid ligament

Interosseous talocalcaneal ligament

Calcaneo-fibular ligament

Figure 12-9. Ligaments of the subtalar joint (cross section, posterior view).

smaller ligament lying lateral to the interosseous talocalcaneal ligament, but still within the tarsal canal. This is the **ligamentum cervicis.**[2,27,28] The MCL and LCL of the ankle also cross and, presumably, lend some support to the subtalar joint. Although sectioning of the collateral ligaments does not appear to result in substantial decrease in subtalar joint stability,[29] the amount of subtalar motion has been found to vary somewhat with ankle joint position and differential tension in the collateral ligaments.[11] Additional reinforcement at the subtalar joint is provided by the **posterior and lateral talocalcaneal ligaments.**[5]

Subtalar Joint Function
Arthrokinematics

Although the subtalar joint is composed of three articulations, the alternating convex-concave facets limit the potential mobility of the joint. When the talus moves on the posterior facet of the calcaneus, the articular surface of the talus should slide in the same direction as the bone moves (concave surface moving on a stable convex surface). However, at the middle and anterior joints, the talar surfaces should glide in a direction opposite to movement of the bone (convex surface moving on a stable concave surface). Motion of the talus, therefore, is a complex twisting (or screwlike motion) that can continue only until the posterior and the anterior and middle facets can no longer accommodate simultaneous and opposite motions. The result is a triplanar motion of the talus around a single oblique joint axis. The subtalar joint is, therefore, a uniaxial joint with 1° of freedom: supination/pronation.

Supination and pronation are composite motions that can best be defined by describing what is occurring in each of the three planes. It should be understood, however, that although each of the composite motions can be broken down into its components, these component motions *cannot and do not occur independently.* The components occur simultaneously as the talus twists across its three articular surfaces and cannot be separated other than for purposes of discussion.

Non–Weight-bearing Joint Movements

Supination of the non–weight-bearing subtalar joint contains the component motions of adduction (vertical axis); inversion (longitudinal axis through the foot);

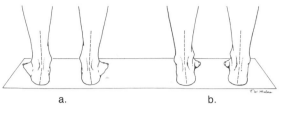

a.   b.

Calcaneovalgus   Calcaneovarus

Figure 12–10. a. Subtalar pronation is accompanied by eversion of the calcaneus. This can be seen in a posterior view as an increase in the medial angle formed by a line through the leg and through the tuber of the calcaneus (calcaneovalgus). b. Subtalar supination is accompanied by inversion of the calcaneus. This can be seen in a posterior view as a decrease in the medial angle formed by a line through the leg and through the tuber of the calcaneus (calcaneovarus).

and plantarflexion (coronal axis) of the calcaneus (and the more distal segments of the foot). Pronation of the non–weight-bearing subtalar joint contains the component motions of abduction (vertical axis), eversion (longitudinal axis), and dorsiflexion (coronal axis) of the calcaneus (and the more distal segments of the foot). Viewing the calcaneus from behind when the subtalar joint is pronated will show eversion of the calcaneus or a valgus angulation of the calcaneus with respect to the tibia. The component of calcaneal eversion is seen as an increase in the medial angulation between the long axis of the tibia and an axis through the tuberosity of the calcaneus (Fig. 12–10a). The movement of eversion is, therefore, equivalent to valgus of the calcaneus and the terms may be used interchangeably. When the posterior aspect of the calcaneus is viewed in subtalar supination, there is a varus angulation between the calcaneus and the tibia (Fig. 12–10b). Because of the nature of the subtalar joint surfaces, the component motions of subtalar supination and pronation always occur together and cannot be combined differently.°

The Subtalar Axis

The axis for subtalar supination/pronation has been the subject of many investigations. Studies have found different average inclinations for the axis, but more importantly these studies have shown that large differences exist among apparently normal individuals. Manter[30] found the axis inclined 42° upward and anteriorly from the transverse plane (with a range of 29 to 47°), and inclined medially 16° from the sagittal plane (with a range of 8 to 24°) (Fig. 12–11). Because the subtalar axis is about half way between longitudinal and vertical, the component motions of eversion/inversion (longitudinal axis) and abduction/adduction (vertical axis) are about equal in magnitude. The subtalar axis is inclined only slightly toward the frontal plane and therefore has only a small component of dorsiflexion/plantarflexion. It is important to note that the contribution of each of the component move-

---

°In the first edition of this book, inversion/eversion were used as the triplanar motions, while supination/ pronation were used as one of the three component movements. This was done to be consistent with the use of supination/pronation in the upper extremity where the terms refer to a uniaxial and uniplanar set of motions. Since there is not an upper-extremity analog to the uniaxial triplanar motion of the subtalar joint, it would seem appropriate to use the term inversion/eversion, as we did in the first edition, for the unique set of component movements found only in the foot.

Members of the disciplines most likely to use this book are finding supination/pronation in more common usage as the triplanar subtalar motions, with inversion/eversion used as the frontal plane component movements. We bow to this more common usage in the second edition. Regardless, it should be noted that inversion is invariably linked with supination and eversion is invariably linked with pronation. Mandatory coupling of the triplanar motion with its components may make the terminology controversy moot to those who understand foot function, but failure to define terms can lead to confusion among less sophisticated readers. Terms used in research and published literature should carefully be defined to impart the most and clearest information.

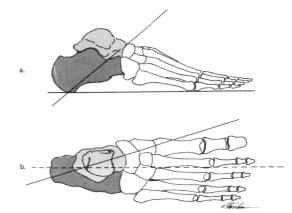

Figure 12–11. Axis of the subtalar joint. a. Inclined up anteriorly, approximately 42°. b. Inclined in medially, approximately 16°.

ments to supination or pronation will vary with individual differences in inclination of the subtalar axis. If the subtalar axis is inclined upwardly only 30° (rather than the average of 42°), the axis will be less vertical and more longitudinal. In such a foot, supination and pronation would consist of nearly twice as much inversion and eversion as compared to adduction and abduction.

Subtalar Neutral

Regardless of the inclination of the subtalar axis in any individual, every subtalar joint has a so-called neutral position from which its range can be assessed and from which its relationship to more proximal and distal joints can be evaluated. The so-called **subtalar neutral position** is generally considered to be the point from which the calcaneus will invert twice as many degrees as it will evert.[31] That is, subtalar neutral is encountered when the subtalar joint is fully supinated and then carried two thirds of the way through to maximum pronation. Baily, Perillo, and Forman[32] used radiologic evidence to demonstrate that the neutral position of the subtalar joint was quite varied among individuals. It was not always found two thirds of the way from maximum supination, although the average neutral subtalar position for their subjects was close to this value.

Weight-Bearing Subtalar Joint Motion

When the foot is non–weight-bearing, the motion across the subtalar joint is accomplished by the distal calcaneal segment. When the calcaneus is on the ground and weight-bearing, it is *not* free to accomplish each of the component motions. The calcaneus can still evert and invert (valgus and varus motions, respectively), although this will result in some side-to-side motion of the foot on the ground. However, the calcaneus cannot dorsiflex/plantarflex or abduct/adduct while weight bearing since the superimposed body weight effectively prevents these movements. Since subtalar motion cannot consist of inversion/eversion in isolation, the other two components of the subtalar motion are accomplished by the proximal talar segment rather than by the distal calcaneal segment.

The motion accomplished at a joint remains unchanged whether the distal segment of the joint moves or whether the proximal segment moves. When switching segments, however, the direction of movement of the new segment reverses to accomplish the same joint motion. In weight-bearing subtalar motion, the direction of movement described for the calcaneus in supination/pronation is reversed when the motion is accomplished by the talus. That is, subtalar supination that consisted in non–weight bearing of adduction and plantarflexion of the calcaneus is, in weight bearing, accomplished by abduction and dorsiflexion of the talus. Weight-bearing subtalar supination, therefore, is accomplished by the component movements of inversion (varus) of the calcaneus and dorsiflexion and abduction of the talus. Weight-bearing subtalar pronation is accomplished by the component movements of eversion (valgus) of the calcaneus and plantarflexion and adduction of the talus. The component motions of abduction and adduction of the talus occur around a vertical axis and are sometimes referred to as lateral rotation and medial rotation of the talus, respectively.

Subtalar Joint Motion and Rotation of the Leg

During weight-bearing subtalar supination/pronation, the components of dorsiflexion and plantarflexion of the talus may be absorbed by the ankle joint as the trochlea of the talus slides posteriorly in the tibiofibular mortise in dorsiflexion and anteriorly in plantarflexion. The tibia remains unaffected by the talar motion. The component motions of abduction and adduction of the talus that must occur as part of weight-bearing supination and pronation, however, cannot be absorbed by the ankle joint. When the talus must abduct as part of weight-bearing supination, the body of the talus cannot laterally rotate within the mortise. Consequently, as the talus is forced to rotate by the other component subtalar movements, the talus carries the mortise laterally with it. That is, abduction of the talus laterally rotates the superimposed tibia and fibula. Conversely, weight-bearing subtalar pronation causes adduction of the talus that carries the tibia and fibula into medial rotation.

Through the component movements of abduction and adduction of the talus, weight-bearing subtalar joint motion directly influences the segments and joints superior to it. A weight-bearing subtalar joint maintained in pronation creates a medial rotatory force on the leg that may influence the knee joint in a number of ways. For example, the medial rotatory force may carry the tibial tuberosity medially, resulting in an increased obliquity in the patellar tendon and an increased Q angle. Rotation of the leg may also influence the subtalar joint. When a lateral rotatory force is imposed on the leg (as in rotating to the right around a planted right foot), the lateral motion of the tibia carries the mortise and its mated trochlea of the talus laterally as well. The talus cannot be rotated laterally (abducted) without also causing the other component movements that are part of subtalar supination. When the talus is abducted by the rotating leg, the talus also will dorsiflex in the mortise and the calcaneus will move into inversion. A medial rotatory force imposed on the leg will necessarily result in subtalar pronation as the talus is medially rotated (adducted) by the rotating tibiofibular mortise. The interdependence of the leg and talus were mechanically represented by Inman and Mann[33] using the concept of the subtalar joint as a mitered hinge. Figure 12–12 presents a good visualization of this concept.

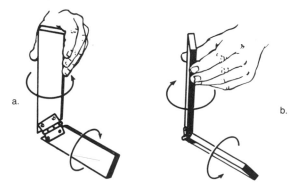

a.   b.

Figure 12–12. The subtalar joint can be visualized as a mitered hinge between the leg and the foot. a. Medial rotation of the leg proximally imposes pronation on the distally located foot. b. Lateral rotation of the leg proximally imposes supination on the distally located foot. (From Mann, RA: Biomechanics of running. In Mann, RA [ed]: Surgery of the Foot, ed 5. CV Mosby, St Louis, p 19, 1986, with permission.)

Table 12–1. SUMMARY OF SUBTALAR COMPONENT MOTIONS

COMPONENT MOVEMENTS OF SUBTALAR SUPINATION/PRONATION		
	Non–Weight-Bearing	Weight-Bearing
Supination	Calcaneal inversion (or varus)	Calcaneal inversion (or varus)
	Calcaneal adduction	Talar abduction (or lateral rotation)
	Calcaneal plantarflexion	Talar dorsiflexion
		Tibiofibular lateral rotation
Pronation	Calcaneal eversion (or valgus)	Calcaneal eversion (or valgus)
	Calcaneal abduction	Talar adduction (or medial rotation)
	Calcaneal dorsiflexion	Talar plantarflexion
		Tibiofibular medial rotation

One of the primary functions of the subtalar joint is seen in the interdependence of the subtalar joint and the leg. When the foot is weight bearing, the subtalar joint motion absorbs the imposed lower-extremity rotations that would otherwise either spin the foot on the ground or disrupt the ankle joint by rotating the mortise around a fixed talus. When the subtalar joint is non–weight-bearing (in an open kinematic chain), the motions of the subtalar joint and the leg are independent and do not influence each other. Table 12–1 summarizes non–weight-bearing and weight-bearing subtalar component movements.

Range of Motion

The range of subtalar supination and pronation are difficult to assess because of the triplanar nature and because they vary with the inclination of the subtalar axis. The calcaneal component of subtalar motion, however, is relatively easy to measure in both weight-bearing and non–weight-bearing movements as the angle of varus or valgus of the calcaneus with respect to the tibia. The available range of the component of eversion (valgus) of the calcaneus has been measured as 10° and inversion (varus) as 20° for a total range of 30°.[31]

Tibial rotation during the weight-bearing portion of gait has been measured as about 10°.[34] This 10° range would also represent the amount of abduction/adduction of the talus and is equivalent to the range of calcaneal eversion/inversion measured during gait.[31] The equivalence of the subtalar ranges of eversion/inversion and abduction/adduction is in agreement with the 4:4:1 contribution of eversion/inversion, abduction/adduction, and dorsiflexion/plantarflexion projected by Sgarlato[35] for an average subtalar joint axis. That is, for every 4° of calcaneal eversion, one should see 4° of adduction of the talus (4° of medial rotation of the tibia) and 1° of plantarflexion of the talus.

Although the range of triplanar subtalar pronation and supination can only be estimated by assessing measurable component movements, the restraints to subtalar motion are generally accepted. The ligamentum cervicis is credited with limiting supination and the interosseous talocalcaneal ligament with limiting pronation.[27,28] Supination is considered to be the position in which ligamentous tension draws the talocalcaneal joint surfaces together, resulting in locking of the articular surfaces. Supination is considered to be the close-packed position for the subtalar joint.

Pronation of the subtalar joint is a position of mobility. The adduction and plantarflexion of the talus that occurs in weight-bearing pronation causes a splaying (spreading) of the adjacent tarsal bones. Subtalar pronation is limited by the ligaments that maintain talocalcaneal joint integrity, as well as by those that are part of the talonavicular joint.

Close examination of the subtalar joint can leave little doubt as to the significance of this joint in total foot function. It has, in fact, been called the determinative joint of the foot, influencing joint motion both proximal and distal to it.[6] Discussion of the subtalar joint, however, is not complete until one examines the talonavicular joint that lies immediately anterior to it.

TALOCALCANEONAVICULAR JOINT

The talonavicular joint is classically considered to be part of the compound joint known as the **transverse tarsal joint** (Fig. 12–13). In weight bearing, however, the navicular, which is the distal segment of the talonavicular joint, has little mobility as compared to the talus.[36] Talonavicular joint motion, therefore, is movement of the talus on the relatively fixed navicular. The talar motion is exclusively that motion just described as belonging to the subtalar joint. Since the talus is moving simultaneously on the calcaneus and the navicular, the concept of the talocalcaneonavicular (TCN) joint has evolved.[5,28]

TCN Joint Structure

The TCN joint name ties together the talonavicular joint and the subtalar (talocalcaneal) joint, which are both anatomically and functionally related. The talonavicular articulation is formed proximally by the large convex head of the talus and distally by the concave posterior navicular. The concept of the TCN joint permits the large head of the talus to be received not only by the concavity of the navicular, but by a "socket" enlarged and deepened by the plantar calcaneonavicular (spring) ligament inferiorly, by the deltoid ligament medially, and by the bifurcate ligaments laterally (Fig. 12–14). The spring ligament is a triangular sheet of ligamen-

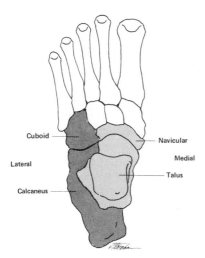

Figure 12–13. Talonavicular joint and calcaneocuboid joint form a compound joint known as the transverse tarsal joint line that transects the foot.

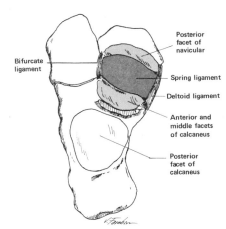

Figure 12-14. Superior view looking into the talocalcaneonavicular joint. The talus has been removed, showing the enlarged concavity formed by the navicular, the deltoid ligament, and the medial band of the bifurcate ligament.

tous connective tissue arising from the sustentaculum tali of the calcaneus and inserting on the inferior navicular. Its central portion supports the talar head and is covered with fibrocartilage. The spring ligament is continuous medially with a portion of the deltoid ligament of the ankle. It also joins laterally with the medial band of the bifurcate ligament (also known as the lateral calcaneonavicular ligament). Through these ligamentous connections, the navicular is tied to the calcaneus, with the two forming a relatively fixed unit on which the superimposed talus moves. A functional unit called the TCN joint is formed.

The head of the talus and its large socket are enclosed by the same capsule that houses the anterior and middle facets of the subtalar joint. The capsule, therefore, anatomically joins the subtalar and talonavicular joints into a TCN joint. Ligamentous support for the TCN joint is provided by those ligaments that reinforce the subtalar joint (ankle joint collaterals, interosseous talocalcaneal); those that help form that concavity for the head of the talus (spring, bifurcate); and by the dorsal talonavicular ligament. Additional support is also received from the ligaments that reinforce the adjacent calcaneocuboid joint.

TCN Joint Function

The talus acts as a ball bearing placed between the tibiofibular mortise superiorly, the calcaneus inferiorly, and the navicular anteriorly. Motion of the weight-bearing talus at one articulation *must* cause motion at each of its inferior articulations. The TCN joint, therefore, like its subtalar component, is a triplanar joint with 1° of freedom: supination/pronation. The axis for TCN supination and pronation is inclined 40° upward and anteriorly and 30° medially and anteriorly (Fig. 12–15).[37] This is, as one would expect, essentially the same as the axis for the subtalar joint alone. The axis of the talonavicular joint, however, is inclined medially more than the subtalar joint axis, which gives the TCN joint more dorsiflexion/plantarflexion than attributed to the subtalar joint. TCN joint motions may have a slightly greater range than those of the subtalar joint alone, but the movements of the talus and calcaneus and the weight-bearing relationship to the tibia are virtually identical to those presented in discussion of the subtalar joint.

Functionally, the subtalar joint and talonavicular joint exist as components of the more complete TCN joint. *The TCN joint is the key to foot function.* The bones and joints distal to the TCN essentially form a single elastic unit that, for the most

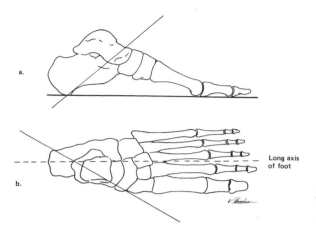

Long axis
of foot

Figure 12–15. Talocalcaneonavicular joint axis. a. Inclined up anteriorly, approximately 40°. b. Inclined in anteriorly, approximately 30°.

part, moves in response to and as compensation for movement of the talus and the calcaneus.[34,37,38]

TRANSVERSE TARSAL JOINT

The transverse tarsal (midtarsal) joint is a compound joint formed by the talonavicular and calcaneocuboid joints (see Fig. 12–13). Since the talonavicular joint is classically considered to be part of the transverse tarsal joint, it belongs to two joint complexes: the TCN joint and the transverse tarsal joint. The other component of the transverse tarsal joint is the calcaneocuboid joint. The two joints together present an S-shaped joint line that transects the foot horizontally, dividing the hindfoot from the midfoot and forefoot. Like the navicular, the cuboid is essentially immobile in the weight-bearing foot. Transverse tarsal joint motion, therefore, is considered to be motion of the talus and of the calcaneus on the relatively fixed naviculocuboid unit.[36,39] Motion at the compound transverse tarsal joint, however, is more complex than the simple joint line might suggest and occurs predominantly in response to action at the TCN.

Transverse Tarsal Joint Structure

The structure of the talonavicular joint has already been presented as part of the TCN joint. The calcaneocuboid joint is formed proximally by the anterior calcaneus, and distally by the posterior cuboid (see Fig. 12–13). The articular surfaces of both the calcaneus and the cuboid are complex, being reciprocally concave/convex across both dimensions. The reciprocal shape makes available motion at the calcaneocuboid joint more restricted than that of the ball-and-socket–shaped talonavicular joint, since the calcaneus (usually the moving segment in weight bearing) must meet the conflicting arthrokinematic demands of the saddle-shaped surfaces. The calcaneocuboid articulation has its own capsule that is reinforced by several major ligaments. These are the lateral band of the bifurcate ligament (also known as the calcaneocuboid ligament); the dorsal calcaneocuboid ligament; and the plantar calcaneocuboid (short plantar) and the long plantar ligaments (see Fig. 12–6). The long plantar ligament is the most important of these ligaments, because it extends inferiorly between the calcaneus and the cuboid and then continues on distally to the bases of the second, third, and fourth metatarsals. It makes a significant contribution both to transverse tarsal joint stability and to related support of the longitudinal arch of the foot. Important support for the transverse tarsal joint is also provided by the extrinsic muscles of the foot, which pass medial and lateral to the joint, as well as by the intrinsic muscles inferiorly.

Transverse Tarsal Joint Function

Axes

The transverse tarsal joint has been analyzed in a number of different ways, usually resulting in function around two independent axes. Manter[30] and Hicks[21] both proposed longitudinal and oblique axes around which the relatively fixed naviculocuboid unit moves. The longitudinal (or AP) axis is nearly horizontal, being only slightly inclined upward and medially anteriorly (Fig. 12–16). Motion around this axis is triplanar, producing supination/pronation as seen at the subtalar/TCN joints. Unlike the axis of the subtalar/TCN joint, the longitudinal axis of the transverse tarsal joint approaches a true AP axis, so the inversion/eversion components of the movement predominate. The oblique (transverse) axis of the transverse tarsal joint nearly parallels the axis of the TCN joint, providing triplanar supination/pronation with predominating dorsiflexion/plantarflexion and abduction/adduction components. Motion around this axis is more restricted than motion around the longitudinal axis because the axis is more inclined across all three planes. The two axes together provide a total range of supination/pronation that is one third to one half of the range available at the TCN joint.[21] Root[31] has proposed that the longitudinal axis around which most of the movement can occur provides less than 10° of inversion/eversion.[31] The two joints of the transverse tarsal joint can function somewhat independently, although motion at one is generally accompanied by at least some motion of the other.

The TCN joint and the transverse tarsal joint are mechanically linked by the shared talonavicular joint. Any subtalar, and therefore, TCN, motion must include motion at the talonavicular joint. Since talonavicular motion is interdependent with calcaneocuboid motion, subtalar/TCN motion will involve the entire transverse tarsal joint. As the TCN inverts, its linkage to the transverse tarsal joint carries the calcaneocuboid with it. When the TCN is fully supinated and locked (bony surfaces drawn together), the transverse tarsal joint is also carried into full supination and its bony surfaces drawn together into a locked position. When the TCN is pronated and loose-packed, the transverse tarsal joint is also mobile and loose-packed.

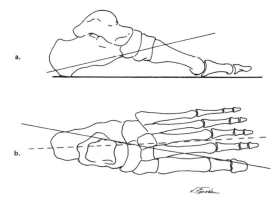

Figure 12–16. Longitudinal axis of the transverse tarsal joint.

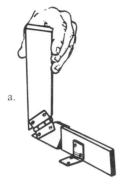

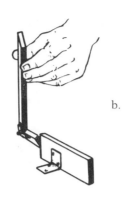

a.   b.

Figure 12–17. The transverse tarsal joint permits the forefoot to remain evenly on the ground, absorbing the pronation occurring at the subtalar joint through medial rotation of the leg (a) or the supination occurring at the subtalar joint through lateral rotation of the leg (b). (From Mann, RA: Biomechanics of running. In Mann, RA [ed]: Surgery of the Foot, ed 5. CV Mosby, St Louis, p 15, 1986, with permission.)

Action

The transverse joint is the transitional link between the hindfoot and the forefoot serving (1) to add to the supination/pronation range of the TCN joint and (2) to compensate the forefoot for hindfoot position. Compensation in this context refers to the ability of the forefoot to remain flat on the ground while the hindfoot is in varus or valgus. The first of the transverse tarsal joint functions can occur either in the weight-bearing or in the non–weight-bearing foot and is self-explanatory. The second function requires closer analysis.

In the weight-bearing position, medial rotation of the tibia imposes pronation on the TCN joint. If the pronation force continued distally through the foot, the lateral border of the foot would tend to lift from the ground, diminishing the stability of the base of support and resulting in unequal weight bearing. This undesirable effect of TCN pronation may be avoided when the transverse tarsal joint can absorb the rotation. That is, the talus and calcaneus move on the essentially fixed naviculocuboid unit, resulting in a relative supination of the forefoot distal to the transverse tarsal joint. The transverse tarsal joint maintains normal weight-bearing forces on the forefoot, while allowing the hindfoot to absorb the rotation of the lower limb (Fig. 12–17).

As long as the hindfoot is pronated (and the TCN joint is mobile), the transverse tarsal joint is free to accommodate to the demands of the supporting surface. In static bilateral stance on level ground, both the TCN and the transverse tarsal joints pronate. As a person moves into single-limb support and begins to walk, the TCN will continue to pronate while the transverse tarsal joint will supinate approximately an equal amount to maintain proper weight bearing in the forefoot. A rock under the medial forefoot may require even greater supination of the transverse tarsal joint to maintain appropriate contact of the forefoot with the ground, if additional supination range is available. If the range is not available at the transverse tarsal joint, the rock may force the hindfoot into supination as well. With other surface demands such as standing sideways on a steep hill, the uphill foot must pronate substantially to maintain contact; pronation may be required at both the TCN and the transverse tarsal joints. *When the hindfoot (TCN) is pronated, the joints of both the hindfoot and midfoot are mobile and free to make necessary compensatory changes required to maintain contact of the foot with the ground . . . within the limits of the joint ROM.*

Supination of the hindfoot (TCN) restricts the ability of the transverse tarsal joint to compensate or counterrotate the forefoot. The transverse tarsal joint is carried into increasing supination with increasing TCN supination. Consequently, transverse tarsal joint mobility is increasingly restricted as the TCN supinates. In full TCN supination, such as when the tibia is laterally rotated on the weight-bearing foot, supination locks not only the TCN but the transverse tarsal joint as well. Although the tarsometatarsal joints are capable of minor compensatory changes, the forefoot may follow into supination depending on the magnitude of the imposed force.

In normal gait, hindfoot supination occurs at a time when the foot is required to serve as a rigid lever (during the second half of the stance phase). The locking of the subtalar/TCN and the transverse tarsal joints facilitate transfer of weight through the tarsometatarsal joints to the forefoot by converting the foot to a rigid lever. Locking of the hindfoot and midfoot may be undesirable, however, if the hindfoot has been supinated by uneven terrain. The entire medial border of the foot may lift, and, unless the muscles on the lateral side of the foot are active, a supination sprain of the lateral ligaments will result. *When the locked TCN and transverse tarsal joints are unable to absorb the rotation superimposed by the weight-bearing limb or by uneven ground, the forces must be dissipated at the ankle* and may result in injury to the ankle joint structures.

TARSOMETATARSAL JOINTS

Tarsometatarsal Joint Structure

The tarsometatarsal (TMT) joints are plane synovial joints formed by the distal tarsal row posteriorly and by the bases of the metatarsals anteriorly (Fig. 12–18). The first TMT joint is the articulation between the base of the first metatarsal and the medial cuneiform. It has its own articular capsule. The second TMT joint is the articulation of the base of the second metatarsal with a mortise formed by the middle cuneiform and the sides of the medial and lateral cuneiforms. This joint is set back more posteriorly than the other TMT joints; it is stronger and its motion more restricted. The third TMT joint, formed by the third metatarsal and the lateral

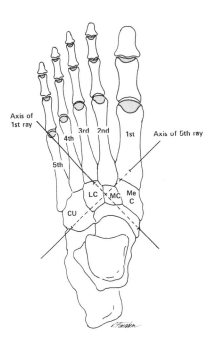

Figure 12–18. Tarsometatarsal, metatarsophalangeal, and interphalangeal joints of the foot (CU = cuboid; MeC = medial cuneiform; MC = middle cuneiform; LC = lateral cuneiform).

cuneiform, shares a capsule with the second TMT joint. The fourth and fifth TMT joints are formed by the bases of the fourth and fifth metatarsals with the cuboid. These two joints also share a joint capsule. There are also small plane articulations between the bases of the metatarsals to permit motion of one metatarsal on the next. Each tarsometatarsal joint is reinforced by numerous dorsal, plantar, and interosseous ligaments.

Tarsometatarsal Joint Function

TMT joint function is primarily a continuation of the function of the transverse tarsal joint. That is, these joints attempt to regulate position of the metatarsals and phalanges relative to the weight-bearing surface. As long as transverse tarsal joint motion is adequate to compensate for the hindfoot position, TMT joint motion is not required. However, when the hindfoot position is extreme and the transverse tarsal joint is inadequate to provide full compensation, the TMT joints may rotate to provide further adjustment of forefoot position.[31]

Axes

Each TMT joint is considered to have a unique, although not fully independent axis of motion. Hicks[21] examined the axes for the five rays. A ray is a functional unit formed by a metatarsal and its associated cuneiform (first through third rays). The cuneiform is included as part of the movement unit since the small amount of motion contributed by a cuniometatarsal joint has not been accounted for in some other way. The cuneonavicular motion, therefore, becomes part of the available TMT motion. Where there is not an associated cuneiform, the ray is formed by the metatarsal alone (fourth and fifth rays).

The axes for the first and fifth rays are shown in Figure 12–18. Each is oblique and, therefore, triplanar. ROM of the first ray is the largest of the metatarsals. The axis of the first ray is inclined so that dorsiflexion (extension) of the first ray is accompanied by inversion and adduction, while plantarflexion (flexion) is accompanied by eversion and abduction. The abduction/adduction components normally are minimal. Movements of the fifth ray around its axis are more restricted and occur with the opposite arrangement of components: dorsiflexion is accompanied by eversion and abduction and plantarflexion by inversion and adduction. The TMT joint motions are not considered to be pronation/supination movements since the components are not as consistent as the composite supination/pronation motions of the TCN and transverse tarsal joints.

The axis for the third ray nearly coincides with a coronal axis; the predominant motion, therefore, is dorsiflexion/plantarflexion. The axes for the second and fourth rays were not determined by Hicks[21] but were considered to be intermediate between the adjacent axes for the first and fifth rays, respectively. That is, the second ray moves around an axis that is inclined toward, but is not as oblique as, the first axis. The fourth ray moves around an axis that is similar to, but not as steep as, the fifth axis. The second ray is considered to be the least mobile of the five. Table 12–2 provides a summary of the motions of the five rays during TMT dorsiflexion and plantarflexion.

Actions

The motions of the TMT joints are somewhat interdependent, as are the motions of the CMC joints in the hand. The TMT joints also contribute to hollowing and flattening of the foot, just as the CMC joints did for the hand.

An active dorsiflexion force across the non–weight-bearing TMTs will simul-

Table 12–2. SUMMARY OF MOTIONS OF THE RAYS OF THE FOOT

	Dorsiflexion	Plantarflexion
First ray	Inversion	Eversion
	Slight adduction	Slight abduction
Second ray	Slight inversion	Slight eversion
Third ray	—	—
Fourth ray	Slight eversion	Slight inversion
Fifth ray	Eversion	Inversion
	Slight abduction	Slight adduction

taneously extend the five metatarsals while also creating inversion at the first two rays and eversion at the last two rays. The opposite motions accompanying TMT dorsiflexion flatten the contour of the plantar surface of the foot. An active plantarflexion force across the non–weight-bearing TMTs will flex the TMTs while everting the first two rays and inverting the last two. This action increases the "cupping" of the plantar surface of the foot. However, the relevance of TMT joint motions is not found in the free foot (open kinematic chain) but in the weight-bearing foot (closed kinematic chain).

Supination Twist

When the hindfoot pronates to any substantial degree, the transverse tarsal joint will generally supinate to counter-rotate the forefoot. If the range of transverse tarsal supination is not sufficient to meet the demands of the pronating force, the medial forefoot will press into the ground, and the lateral side will tend to lift. The first and second ray will be pushed into dorsiflexion while the fourth and fifth rays will plantarflex in an attempt to maintain contact with the ground. The component rotation that accompanies dorsiflexion of the first and second rays and plantarflexion of the fourth and fifth rays is inversion. Consequently, the entire forefoot undergoes an inversion rotation around a hypothetical axis at the second ray. This rotation is referred to as **supination twist of the TMT** joints.[21]

As an example of supination twist of the forefoot, Figure 12–19 shows the response of the segments of the foot to a strong pronation torque across the TCN. The transverse tarsal joint can only supinate to a limited degree in response to the hindfoot motion and requires a supination twist of the TMT joints to fully adjust the forefoot. The configuration of the forefoot in a supination twist is not constant, but can vary according to the weight-bearing needs of the foot and the terrain.

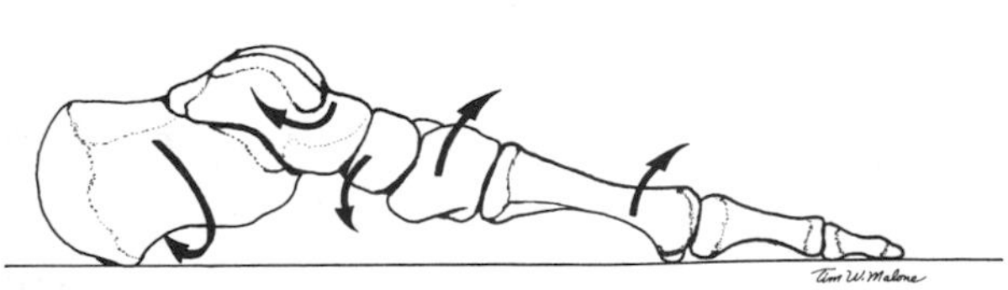

Figure 12-19. Extreme pronation of the foot is accompanied by adduction of the head of the talus, eversion of the calcaneus, and pronation at the transverse tarsal joint (not mandatory). If the forefoot is to remain on the ground, the tarsometatarsal joints must undergo a counteracting supination twist.

Figure 12–20. Extreme supination of the foot is accompanied by abduction of the head of the talus, inversion of the calcaneus, and mandatory supination of the transverse tarsal joint. If the forefoot is to remain on the ground, the tarsometatarsal joints must undergo a counteracting pronation twist.

Pronation Twist

When the hindfoot and the transverse tarsal joints are both locked in supination, the adjustment of forefoot position is left entirely to the TMT joints. With hindfoot supination, the forefoot tends to lift on its medial side and press into the ground on its lateral side. The first and second ray will plantarflex to maintain contact with the ground while the fourth and fifth rays are forced into dorsiflexion. Since eversion accompanies plantarflexion of the first and second ray and dorsiflexion of the fourth and fifth ray, the forefoot as a whole undergoes a **pronation twist.**[21]

Pronation twist, like supination twist, can vary in configuration. Although the pronation twist may provide adequate counterrotation for moderate hindfoot supination, it may be inadequate to maintain forefoot stability in extreme supination. Figure 12–20 shows the rotations imposed on the weight-bearing foot by lateral rotation initiated in the lower extremity or by supination initiated at the TCN joint.

- ■ Pronation and supination twist of the TMT joints occur only when the transverse tarsal joint function is inadequate; that is, when the transverse tarsal joint is unable to counterrotate or its range is insufficient to fully compensate for hindfoot position.

METATARSOPHALANGEAL JOINTS

The five metatarsophalangeal (MTP) joints (see Fig. 12–18) are condyloid synovial joints with 2° of freedom: extension (dorsiflexion)/flexion (plantarflexion) and abduction/adduction. Although both degrees of freedom might be useful to the MTP joints in the rare instances the foot participates in grasplike activities, flexion and extension are the predominate functional movements at these joints. In the weight-bearing foot, toe extension permits the body to pass over the foot while the toes dynamically balance the superimposed body weight as they press into the supporting surface through activity of the toe flexors.

MTP Joint Structure

The MTP joints are formed proximally by the heads of the metatarsals and distally by the bases of the proximal phalanges. The metatarsals may vary in length. In 56 percent of individuals, the second metatarsal is the longest of the metatarsals, with the first metatarsal being next, followed in order by the third through fifth metatarsals. This pattern of metatarsal length is referred to as an **index minus** foot. In 28 percent of the individuals, the first metatarsal is equivalent in length to the second metatarsal, a pattern identified as an **index plus minus** foot. The second metatarsal is shorter than the first in 16 percent of individuals (referred to as an **index plus** foot).[40] The pattern of metatarsal length in an individual may predispose him or her to a particular set of problems with the MTP joints and the toes.

The MTP joints of the foot are structurally analogous to the MCP joints of the hands, although there are some exceptions to the analogy. Unlike what we see at the MCP joints, the MTP extension range exceeds the flexion range. The heads of the metatarsals each bear weight in stance. Consequently, the articular cartilage must stop short of the weight-bearing surface on the plantar aspect of the metatarsal head. This restricts the available range of MTP flexion. Since there is no opposition available at the first TMT joint, the first toe (hallux) moves in the same plane as the other four digits. The first MTP joint, like the first MCP joint, has two sesamoid bones located at the joint. In neutral MTP position, the sesamoid bones lie in two grooves separated by the intersesamoid ridge. The ligaments associated with the sesamoid bones form a triangular mass that stabilizes the sesamoids within their grooves.[41] The sesamoid bones serve as anatomic pulleys for the flexor hallucis brevis muscle and protect the tendon of the flexor hallucis longus muscle from weight-bearing trauma as it passes through a tunnel formed by the sesamoid bones and the intersesamoidal ligament that connects their plantar surfaces.[41] Unlike those of the hand, the sesamoid bones of the foot share in weight bearing with the relatively large quadrilaterally shaped head of the first metatarsal.[42] In toe extension greater than 10°, the sesamoids no longer lie in their grooves and may become unstable. Chronic lateral instability may lead to MTP deformity.[42]

The analogous fibrocartilage palmar plates in the hand are referred to at the MTP joints as **plantar pads** or **ligaments** and are interconnected between the four lesser toes by the deep transverse metatarsal ligament as the palmar plates are in the hand. The sesamoids and thick plantar capsule of the first MTP joint replace the plantar pads found at the other toes.[41]

MTP Joint Function

Although the MTP joints have 2° of freedom, flexion/extension is clearly more important than abduction/adduction. The first MTP joint was found in one study to have 82° of extension and 17° of flexion.[43] The range will vary somewhat depending on the relative lengths of the metatarsals and whether the motions occur in weight-bearing or non–weight-bearing activities. The ROM of the first MTP has also been found to be influenced by the degree of dorsiflexion or plantarflexion of the TMT joints and to be more restricted with increasing age.[43] The MTP joints serve primarily to allow the foot to "hinge" at the toes so that the heel may rise off the ground, while still maintaining the small but dynamic base of support afforded by the toes and the toe musculature. This function is enhanced by two structural aspects of the MTP joints: the metatarsal break and the effect of MTP extension on the plantar aponeurosis.

Extension

The Metatarsal Break

The **metatarsal break** refers to the single oblique axis for MTP flexion/extension that lies through the second to fifth metatarsal heads. The inclination of the axis is produced by the diminishing lengths of the metatarsals from the second through the fifth toes and varies among individuals. The metatarsal "break" may range from 54 to 73° compared to the long axis of the foot.[1] It is called the break because it is where the foot hinges as the heel rises in weight bearing.

For the weight-bearing heel to rise, there must be an active contraction of ankle plantarflexor musculature. Most of these muscles, as shall be discussed later, contribute to supination of the hindfoot and directly, or indirectly through the TCN, to transverse tarsal supination. The musculature cannot lift the body weight

Figure 12–21. The metatarsal break distributes weight across the metatarsal heads as the heel lifts.

unless the joints of the hindfoot and midfoot are fully supinated and locked; that is, the heel will rise when the foot has become a rigid lever from the calcaneus through the metatarsals. The rigid lever will rotate around the metatarsal break (the **MTP** axis) (Fig. 12–21). During this period of MTP extension, the metatarsal heads glide in a plantar direction on the phalanges, which are stabilized by the supporting surface. The toes become the base of support and the line of gravity of the body must lie within this base. The obliquity of the combined MTP joint axis serves the particular purpose of more evenly distributing the body weight across the toes than would occur if the axis were truly coronal. If the body weight passed through the foot along the longitudinal axis of the foot and the foot lifted around a coronal MTP axis, an excessive amount of weight would be placed on the first metatarsal head and on the long second metatarsal. These two toes would also require a disproportionately large extension range. The obliquity of the metatarsal break shifts the weight laterally, minimizing the large load on the first two digits.

The Plantar Aponeurosis

The plantar aponeurosis is a dense fascia that runs nearly the entire length of the foot. It begins posteriorly on the calcaneus and continues anteriorly to attach by digitations to the proximal phalanx of each toe via the deep transverse metatarsal ligament and the plantar pads just deep to the ligament. When the toes are extended at the MTP joints (regardless of whether the motion is active or passive, weight-bearing or non–weight-bearing), the aponeurosis is pulled increasingly tight as the proximal phalanges glide dorsally in relation to the metatarsals. The large metatarsal heads end up acting as pulleys around which the plantar aponeurosis is tightened. The tension in the plantar aponeurosis can contribute to supination of the foot as the heel is drawn toward the toes by its action (Fig. 12–22).

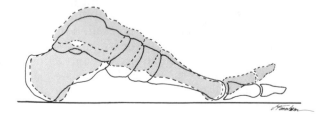

Figure 12–22. Elevation of the arch with toe extension occurs through the windlass effect of the metatarsophalangeal joints on the plantar aponeurosis.

When the joints of the hindfoot and midfoot supinate and lock through a strong active plantarflexion force, continued force will cause the heel to lift and the toes to extend at the metatarsal break. The plantar aponeurosis will tighten as the MTPs extend, supporting the locked hindfoot and midfoot structures through which the body weight must pass to reach the toes. The tightened aponeurosis will also resist excessive toe extension by creating a passive flexor force across the MTP joint. The passive flexor force will assist the active toe musculature in pressing the toes into the ground to support the body weight on its now diminished base of support. The mechanism of the plantar aponeurosis is considered to be most effective at the first MTP and progressively less effective from the second to fifth MTPs.[44] The effect of MTP extension on the plantar aponeurosis will be further discussed when the arches of the foot are presented.

Flexion, Abduction, and Adduction

MTP flexion from neutral can occur to a limited degree but has relatively little use in the weight-bearing foot other than when the supporting terrain drops away distal to the metatarsal heads. Most MTP flexion occurs as a return to neutral from extension. Similarly, MTP abduction and adduction do not serve an obvious function. The joint motions apparently remain to absorb some of the force that would be imposed on the toes by the metatarsals as they move in a pronation or supination twist. The first toe is normally adducted on the first metatarsal about 15°.[42] An increase in this normal valgus angulation of the first MTP joint is referred to as **hallux valgus**. Hallux valgus may be associated with an index minus foot or with one of several other conditions. Exaggerated MTP abducted or adducted positions of the toes may also be seen when there are abduction or adduction deformities in bones or joints further back in the foot. The resultant MTP deviations may be an attempt to compensate toe position to prevent excessive weight on any one toe, while also maintaining the metatarsal break in an appropriate position for gait.

INTERPHALANGEAL JOINTS

The interphalangeal (IP) joints of the toes are synovial hinge joints with 1° of freedom: flexion/extension. There are five proximal IP joints and four distal IP joints. Each phalanx is virtually identical in structure to its counterpart in the hand, although substantially shorter in length. The toes function to smooth the weight shift to the opposite foot in gait and help maintain stability by pressing against the ground both in static posture when necessary, and in gait. The relative lengths of the toes may vary. The most common pattern is to find the first toe longer than the others (69 percent of individuals). The second toe may be longer than the first in 22 percent of the people, with 9 percent having first and second toes of equal lengths.[40] Viladot[40] has proposed that each configuration predisposes the foot to different problems, although the best configuration for modern footwear might be the index minus foot (second metatarsal shorter than the first) and a longer second toe.

PLANTAR ARCHES

The bony and ligamentous configuration of the talocalcaneonavicular joint, the transverse tarsal joint, and the tarsometatarsal joints combine to produce a structural vault within the foot. The toes are not part of the vault but, as shall be seen, may indirectly affect the shape of the vault. Although we have examined the function of the joints of the foot individually and discussed the effect of each joint on

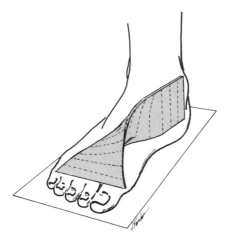

Figure 12–23. Twisted osteoligamentous plate of the foot, resulting in longitudinal and transverse arches.[18]

contiguous joints, combined function can also be investigated by looking at the archlike structure of the foot. The design of the arch, or arches, can best be understood by visualizing the foot as a single twisted osteoligamentous plate (Fig. 12–23). The anterior margin of the plate (formed by the metatarsal heads) is horizontal and in full contact with the ground. The posterior margin of the plate (the posterior calcaneus) is vertical. The resulting twist in the plate between its horizontal and vertical margins imposes both longitudinal and transverse arcs.[1,45] Loading the plate (weight bearing) will tend to untwist it, flattening the arches slightly. As the plate is unloaded (body weight removed), the resilient arches return to their original shape. The twisted plate, through the contributing joints, is a supporting mechanism designed to facilitate absorption and distribution of the superimposed body weight through the foot under changing weight-bearing conditions and changing terrain. The actual mechanism of twisting and untwisting occurs through motion at the talocalcaneonavicular, transverse tarsal, and tarsometatarsal joints that link the bones of the osteoligamentous plate. The adult configuration of the arch is not present at birth, but evolves with the progression of weight bearing. Gould and associates[46] found a flattened arch in all children examined between 11 and 14 months of age. By 5 years of age, the majority, but not all, of the children had developed an adultlike arch.

Structure

Although the concept of a single-twisted osteoligamentous plate gives the best representation of the vault of the adult foot, the vault more traditionally is considered to be composed of two different arches, the longitudinal and the transverse arches of the foot. The longitudinal arch has been described as an arch based posteriorly at the calcaneus and anteriorly at the metatarsal heads. The arch is continuous both medially and laterally through the foot, but because the arch is higher medially, the medial side is usually the side of reference (Fig. 12–24).

The transverse arch, like the longitudinal, is a continuous structure. It is easiest to visualize at the level of the anterior tarsals and at the bases of the metatarsals. At the anterior tarsals (Fig. 12–25a), the middle cuneiform forms the keystone of the arch. The arch continues distally to the metatarsals with slightly less curvature (Fig. 12–25b). The second metatarsal, recessed into its mortise, is at the apex of this arc. At the level of the metatarsal heads, the transverse arch is completely reduced with all metatarsal heads parallel to the weight-bearing surface.

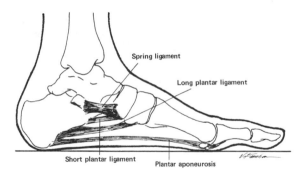

Figure 12–24. Medial longitudinal arch with its associated ligamentous support. (The plantar ligaments are projected through from the lateral side of the foot.)

The shape and arrangement of the bones are partially responsible for stability of the plantar arches. However, without additional support, the intertarsal and tarsometatarsal joints would permit collapse of the arch; the plate would untwist. This is prevented by the supporting ligaments of the joints and, in some instances, by the muscles spanning these joints. The primary support ligaments of the plantar arches are, in order of importance: the spring (plantar calcaneonavicular) ligament; the long plantar ligament; the plantar aponeurosis; and the short plantar (plantar calcaneocuboid) ligament.[47] These ligaments not only serve the stabilizing function of supporting the arches but also provide some mobility by permitting some untwisting of the resilient plate to absorb shock and to conform to the uneven supporting surface.

The concept of the twisted osteoligamentous plate requires that all of the ligaments that support the plate in one location must contribute to support throughout the plate. The plate cannot untwist at one joint without also untwisting at others. Supporting structures, however, can be thought of as longitudinally or transversely oriented as they provide support. The spring ligament provides important longitudinal support of the arch. Located on the medial side of the plate, it spans and supports the talocalcaneonavicular joint (particularly the head of the talus) and it checks joint motion that contributes to flattening of the arch. Some elasticity is credited to the spring ligament, but it would appear to be weight-bearing joint motion rather than stretch of elastic ligaments that accounts for springiness of the arch. The actual elongation of the loaded spring ligament can be compared to the elongation that a soft steel wire would undergo under similar load.[37]

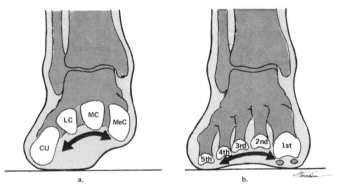

Figure 12–25. The transverse arch. a. At the level of the anterior tarsals. b. At the level of the metatarsals. (CU = cuboid; MeC = medial cuneiform; MC = middle cuneiform; LC = lateral cuneiform.)

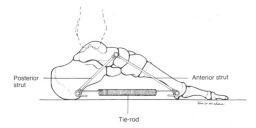

Figure 12-26. The foot can be considered to function as a truss and tie-rod, with the calcaneus and talus serving as the posterior strut, the remainder of the tarsals and the metatarsals serving as the anterior strut, and the plantar aponeurosis serving as a tensed tie-rod. Weighting of the foot will compress the struts and create additional tension in the tie-rod.

The long and the short plantar ligaments provide longitudinal support on the lateral side of the plate. Each supports the calcaneocuboid joint and, therefore, checks motion at the transverse tarsal joint that would result in depression of the arch. The long and short plantar ligaments are less critical than the spring ligament, because the weight-bearing compression through the calcaneocuboid joint is only half that encountered by the talonavicular joint.[48]

The plantar aponeurosis is particularly important to the arches of the foot since the aponeurosis serves the function of a tie-rod on a truss.[37] The truss and the tie-rod form a triangle; the two struts of the truss form the sides of the triangle, while the tie-rod is the bottom. The talus and calcaneus form the posterior strut, while the anterior strut is formed by the remaining tarsal and the metatarsals (Fig. 12-26). As does a tie-rod, the plantar aponeurosis holds together the anterior and posterior struts when the body weight is loaded on the triangle. The struts in weight bearing are subjected to compression forces, while the tie-rod is subjected to tension forces. Increasing the load on the truss, or actually causing flattening of the triangle, will increase tension in the tie-rod.

The tension in the plantar aponeurosis in the loaded foot is evident if active or passive MTP extension is attempted while the triangle is flattened. The range of MTP extension will be limited. Conversely, raising the height of the triangle independent of the truss can unload the tie-rod. For example, when the tibia is subjected to a lateral rotatory force, the hindfoot will supinate and the plantar aponeurosis will be unloaded. With the reduction in tension in the plantar aponeurosis, the range of available toe hyperextension will increase. Finally, increasing tension in the tie-rod independent of loading the foot will draw the two struts of the truss together, shortening and raising the triangle. This phenomenon can occur when the MTP joints are extended. Whether the toes are extending with the distal lever free or the toes are being extended as the heel rises in weight bearing, the aponeurosis is pulled tighter and the arch can be raised simply through an increase in its passive tension. Through the mechanism of the plantar aponeurosis, the MTP joints (not actually part of the osteoligamentous plate) act interdependently with the joints of the hindfoot and may contribute to supination of the foot through the effect of MTP extension on the plantar aponeurosis.

Muscles contribute little support to the osteoligamentous plate in the normal *static* foot.[45] In gait, however, both the longitudinally and transversely oriented muscles become active and contribute to support of the twisted plate.

Function

Although the archlike structure of the foot is similar to the structure of the palmar arches of the hand, the purpose served by each of these systems is quite different. The arches of the hand are predominantly structured to facilitate grasp and manipulation, but must also assist the hand in some weight-bearing functions. In contrast, the foot in most individuals is rarely called on to perform any grasp activities. The plantar arches are adapted exclusively to serve the weight-bearing functions of the foot. The following *stability* functions could be performed by a foot with a fixed arch structure: (1) distribution of weight through the foot for proper weight bearing; and (2) conversion of the foot to a rigid lever. However, the following *mobility* functions can only be performed by a nonrigid structure: (1) dampening of the shock of weight bearing; (2) adaptation to changes in the supporting surface; and (3) dampening of superimposed rotations.

Weight Distribution

Since the foot is *not* a fixed arch, the distribution of body weight through the foot depends on the shape of the arch and the location of the line of gravity at a given moment. Consistently, however, distribution of superimposed body weight begins with the talus, since the body of the talus receives all the weight that passes down through the leg. In bilateral stance each talus receives 50 percent of the body weight; in unilateral stance, the weight-bearing talus receives 100 percent of the superimposed body weight. In static unilateral or bilateral stance, 50 percent of the weight received by the talus passes through the posterior subtalar articulation to the calcaneus, and 50 percent passes anteriorly through the TCN and calcaneocuboid joints to the forefoot. The pattern of weight distribution can be seen easily by looking at the trabeculae in the bones of the foot (Fig. 12–27). Because of the more medial location of the talar head, twice as much weight passes through the talonavicular joint as through the calcaneocuboid. The weight bearing at the anterior margin of the osteoligamentous plate follows a similar pattern. In static standing, the distribution of weight at the metatarsal heads occurs in a 2:1:1:1:1 proportion from the first ray medially to the fifth ray laterally.[49] Loads on the first and second ray increase substantially in the later stages of the stance of gait when the body weight shifts medially. At this point in the gait cycle, the axis of MTP extension is no longer across the five joints, but is a transverse axis across the first and second MTPs[10] (the relative length of the second metatarsal may substantially influence the distribution of weight between the first and second rays).

The large amount of weight imposed on the calcaneus in both static standing and in gait is partially dissipated by the heel pad, which lies on the plantar surface of the calcaneus. The heel pad is composed of fat cells that are located in chambers formed by fibrous septa attached to the calcaneus above and the skin below. The role of the heel pad is even more critical in gait when the loads on the calcaneus at contact of the heel will vary from 85 to 100 percent of body weight. Running may

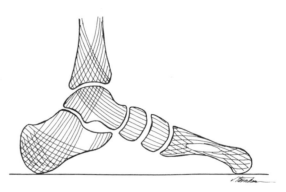

Figure 12–27. Trabeculae of the bones on the medial aspect of the foot.

increase this force to 250 percent of body weight. The effectiveness of the cushioning action of the heel pad decreases with age and with concomitant loss of collagen, elastic tissue, and water. The change is evident in most people past 40 years of age.[10,50]

Mobility

The mobility component of weight distribution through the foot (shock absorption) and adaptation to the terrain can be seen in the response of the linkages and supporting structures of the arches to loading of the foot. In non–weight-bearing position, the subtalar/TCN joint is normally slightly inverted and the transverse tarsal joint is neutral. With weight bearing, the osteoligamentous plate is loaded. The response to loading of the plate is eversion (valgus) of the calcaneus and adduction and plantarflexion of the head of the talus. The talar motion causes slight depression (anterior/inferior movement) of the navicular. Further depression of the talar head is checked by tension in the spring ligament. If no more than the body weight is introduced (such as in static bilateral stance), the TCN joint pronates to a neutral position and the transverse tarsal joint pronates fully.[31] The net effect is to slightly flatten the longitudinal arch and to absorb some of the shock of the superimposed weight through the untwisting of the plate and the elasticity of supporting structures like the spring ligament and the plantar ligaments and aponeurosis. If the ground is not level and forefoot pronation is required, pronation twist of the forefoot at the TMTs can be called on before the hindfoot is forced to pronate further (maximally untwisting the plate). If there is a supination demand on the forefoot, both the transverse tarsal and TMT joints are available to respond before the hindfoot must begin to reverse its pronation position. Given the resiliency of the supporting structures of the arches, unloading the foot will restore the original arch and joint alignment once again.

The ability of the foot to absorb rotations of the leg is considered a mobility function. This is clear when twist of the tibia is a medial rotatory force. Not only does the TCN pronate in response, but the foot retains its ability to contour itself to other possible demands of the terrain. However, when the tibia is torsioned laterally, the absorption response of the TCN simultaneously locks the foot and severely limits the foot's ability to respond to changes in the terrain. The TCN joint will supinate as will the transverse tarsal joint. The TMT joint will undergo a pronation twist to maintain the metatarsal heads on the ground and the osteoligamentous plate will increase its twist. Although the tibial rotation has been absorbed, the plate cannot be twisted any further and is not mobile. The TMTs are free to respond *only* to a change in terrain requiring a supination twist (which would unwind the plate slightly). Any other terrain change would tip the entire twisted plate laterally or require unlocking of the hindfoot and reversal of the lateral rotatory force on the tibia.

MUSCLES OF THE ANKLE AND FOOT

There are no muscles in the ankle or foot that cross and act on only one joint. All act on at least two joints or joint complexes. Muscle activity will produce joint actions that are dependent on the angle of pull of the muscle in relation to the joint axis. In the foot, a joint may be intersected by two axes, and the instantaneous axis of rotation may vary considerably between extremes of joint range. Muscle action is further complicated by the interdependent nature of the ankle-foot joints. Therefore, although a brief review of muscle function is presented, muscle activity is best examined in the context of actual function in posture and in gait.

Extrinsic Musculature

Ankle Plantarflexors

The gastrocnemius muscle arises from two heads of origin on the condyles of the femur and inserts via the achilles (calcaneal) tendon into the most posterior aspect of the calcaneus. The soleus muscle is deep to the gastrocnemius, originating on the tibia and fibula and inserting with the gastrocnemius into the posterior calcaneus. The two heads of the gastrocnemius and the soleus muscles together are known as the **triceps surae** and are the main plantarflexors of the ankle. The achilles tendon occupies a position on the calcaneus that is far from the ankle axis and provides a large moment arm (MA) for plantarflexion. The achilles tendon also crosses the TCN joint and, therefore, acts on it. The resultant combined action line of the gastrocnemius and soleus muscles passes medial to the axes of the subtalar/TCN joints, so that *activity of the gastrocnemius and soleus muscle each produce strong hindfoot supination.* Activity of the gastrocnemius and soleus on the weight-bearing foot will lock the foot into a rigid lever both through direct supination of the TCN and through indirect supination of the transverse tarsal joint. Continued plantarflexion force will raise the heel and cause elevation of the arch. The elevation of the arch by the triceps surae is easily observable in most people when they actively plantarflex the weight-bearing foot (Fig. 12–28). Elevation of the arch occurs through the twisting of the osteoligamentous plate as the hindfoot is inverted and through tightening of the plantar aponeurosis with MTP joint extension.

The other ankle plantarflexion muscles are the plantaris, the tibialis posterior, the flexor hallucis longus, the flexor digitorum longus, and the peroneus longus and brevis muscles. Although each of these muscles passes posterior to the ankle axis, the MA for plantarflexion for these muscles is so small that they provide only 5 percent of the total plantarflexor force at the ankle.[2] The plantaris muscle is so weak that its function can essentially be disregarded. The tibialis posterior muscle is predominantly a supinator of the foot. It has a large MA for both TCN and transverse tarsal joint supination, although its relatively small cross-sectional area enables it to produce only half the supination torque of the soleus muscle.[50] The tibialis posterior plays a significant role in controlling and reversing the pronation of the foot that occurs during gait.

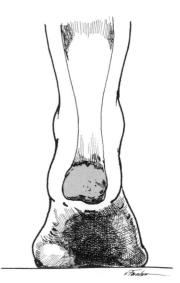

Figure 12–28. Activity of the triceps surae muscles on the fixed foot will cause ankle plantarflexion, talocalcaneonavicular inversion (varus of the calcaneus), and elevation of the longitudinal arch.

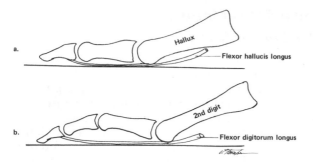

Figure 12–29. a. Action of the flexor hallucis longus causes the distal phalanx of the hallux to press against the ground. b. Activity of the flexor digitorum longus causes the four lesser toes to grip the ground.

The flexor hallucis longus and the flexor digitorum longus muscles both span the medial longitudinal arch and help support the arch during gait. These muscles attach to the distal phalanges of each digit and, through their actions, cause the toes to flex. Flexion of the IP joint of the hallux by the flexor hallucis longus muscles produces a press of the toe against the ground (Fig. 12–29a). Flexion of the distal and proximal IP joints of the four lesser toes by the flexor digitorum longus causes a gripping action that may produce a clawing (MTP extension with IP flexion) similar to what occurs in the fingers when the proximal phalanx is not stabilized by intrinsic musculature (Fig. 12–29b). As is true in the hand, activity of the interossei muscles can stabilize the MTP joint and prevent MTP hyperextension.

The peroneus longus and brevis muscles are the primary pronators of the foot. Although the MA for subtalar pronation is small, the peroneus longus will plantarflex and pronate the first ray to which it attaches. This action facilitates transfer of weight from the lateral to the medial side of the foot and stabilizes the first ray as the ground reaction force attempts to dorsiflex it. The peroneus longus muscle is also credited with support of the transverse and lateral longitudinal arches. This appears to be contradictory to its function as a pronator of the foot, but the tendon of the peroneus longus spans the tarsals transversely and passes under the cuboid, thus helping to limit its depression (Fig. 12–30).

Ankle Dorsiflexors

The dorsiflexor muscles of the ankle are the tibialis anterior, the extensor hallucis longus, the extensor digitorum longus, and the peroneus tertius muscles. The tibialis anterior and the extensor hallucis longus muscles are both strong dorsiflexors of the ankle and may also weakly supinate the TCN joint. The tibialis anterior

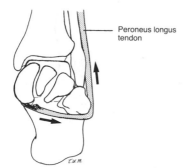

Figure 12–30. The tendon of the peroneus longus passes transversely beneath the foot. An active contraction of the muscle can support the transverse arch of the foot and therefore the twisted osteoligamentous plate.

and extensor hallucis longus are the only supinators active when the heel first contacts the ground in gait, a period when there is a strong pronation force on the hindfoot. The tibialis anterior, however, can exert only slightly more than half the supination force of the tibialis posterior.[50] The extensor hallucis longus is a weaker supinator than is the tibialis anterior.[5] The supination action of the tibialis anterior muscle may reverse if the line of pull falls to the lateral side of the subtalar joint axis, as it may in excessive foot pronation. The extensor hallucis longus also extends the MTP joints of the hallux. When the extensor hallucis longus attempts to dorsiflex the foot without the assistance of the tibialis anterior, the first toe will tend to claw simultaneously.

The extensor digitorum longus and the peroneus tertius muscles are relatively weak dorsiflexors of the ankle and pronators of the foot. The extensor digitorum longus also extends the MTP joints of the lesser toes. Its structure and function at the MTP and IP joints are identical to that of the extensor digitorum of the hand. Taken together, the musculature that produces supination of the foot is stronger than that producing pronation. This is why the relaxed and unweighted foot returns to a position of slight subtalar inversion and a neutral transverse tarsal joint.

Intrinsic Musculature

The function of the intrinsic muscles of the foot can best be understood and appreciated by comparing each foot muscle with its corresponding hand muscle. Although most people are not able to utilize the muscles of the foot with the facility of those in the hand, the potential for similar function is limited only by the unopposable hallux and the length of the digits. The intrinsic muscles of the foot most often are relegated to their roles as stabilizers of the toes and as *important dynamic supporters of the transverse and longitudinal arches during gait.* The intrinsics of the hallux attach either directly or indirectly to the sesamoids and contribute to the stabilization of these weight-bearing bones.[41] The extensor mechanism of the toes is essentially the same as that of the fingers. The extensor digitorum longus and brevis are MTP extensors. Activity in the lumbricals and the dorsal and plantar interossei musculature maintains or produces IP extension. Perhaps more importantly, the toe intrinsics (as MTP flexors) can eccentrically assist in control of the toe extension produced indirectly by the ankle plantarflexors.[51] Table 12–3 summarizes the functions of the intrinsic muscles of the foot.

DEVIATIONS FROM NORMAL STRUCTURE AND FUNCTION

The complex interdependency of the joints of the ankle and foot make it almost impossible to have dysfunction or abnormality in only one joint or structure. Once present, deviations from normal will affect both proximal and distal joints. The large number of congenital and acquired problems cannot each be described, but two examples of the "domino effect" of deformity will be given.

Flatfoot (Pes Planus)

The key to the foot in both function and dysfunction are the joints of the hindfoot. A pronated or flat foot (pes valgus) is marked by excessive unwinding of the

Table 12–3. INTRINSIC MUSCLES OF THE FOOT

Muscle	Function	Analog in Hand
Extensor digitorum brevis	Extends the MTP joints	None
Abductor hallucis	Abducts and flexes MTP of hallux	Abductor pollicis brevis
Flexor digitorum brevis	Flexes PIP of four lesser toes	Flexor digitorum superficialis°
Abductor digiti minimi	Abducts and flexes small toe	Abductor digiti minimi
Quadratus plantae	Adjusts oblique pull of flexor digitorum longus into line with long axes of digits	None
Lumbricals	Flex MTPs, extend IPs of four lesser toes	Lumbricals
Flexor hallucis brevis	Flexes MTP of hallux	Flexor pollicis brevis
Adductor hallucis	Oblique head: aducts and flexes MTP of hallux	Adductor pollicis
	Transverse head: adducts metatarsal heads transversely	
Flexor digiti minimi	Flexes MTP of small toe	Flexor digiti minimi
Plantar interossei	Adduct MTPs of 3rd–5th toes, flex MTPs, extend IPs of four lesser toes	Volar interossei
Dorsal interossei	Abduct MTPs of 2nd toe (either way), abduct MTPs 3rd and 4th toes, flex MTPs, extend IPs of four lesser toes	Dorsal interossei

°The flexor digitorum superficialis is an extrinsic muscle, whereas the flexor digitorum brevis is an intrinsic foot muscle.

osteoligamentous plate with weight bearing. The pronated position of the TCN/subtalar and the transverse tarsal joints create (or result from) a medial rotatory stress on the leg. The medial rotatory stress or position of excessive medial rotation of the leg may result in several possible problems around the knee joint, including excessive angulation of the patellar tendon and excessive pressure on the lateral patellar facet. The lowering of the arch that accompanies TCN and transverse tarsal joint pronation may result in a functional leg length inequality if the problem is asymmetrical. Pronation of the subtalar joint can lower the ankle joint axis and result in a slight reduction in overall limb length.[52] Lowering the arch also tenses the plantar ligaments and the plantar aponeurosis (the tie-rods of the triangle). Prolonged stress on these structures can result in a cycle of microtears, pain, and inflammation. With hindfoot pronation, the forefoot must adjust by supinating at the TMT joints. This may result in a dorsiflexion of the first ray that prevents it from providing some of its normal weight-bearing support, subsequently overloading the more laterally located rays.[40] The supinated position of the first ray may also create additional valgus stress at the first MTP, resulting in a hallux valgus. Hallux valgus, in turn, changes the line of pull of the flexor muscles of the first toe and may affect the power of push-off in the final stages of stance.

The most common form of flatfoot is termed a **flexible flatfoot** and is marked by an arch that reappears when the foot is non–weight-bearing. In such a foot, however, the non–weight-bearing forefoot may continue to appear relatively supinated. Treatment is focused around prevention of excessive pronation when the foot is loaded by controlling valgus of the calcaneus. If this can be done, the cycle of tension in the passive structures can be interrupted.

Supinated Foot (Pes Cavus)

A less common but potentially more serious problem exists when the weight-bearing foot is excessively supinated (a cavus foot). In a cavus foot, the TCN/subtalar and transverse tarsal joints may be locked into supination, prohibiting these joints from participating in shock absorption or in adapting to uneven terrain. The hindfoot supination that results in added twist in the osteoligamentous plate may cause or be caused by a lateral rotatory stress on the leg. The lateral rotatory stress may, in turn, affect knee joint structures. The inability to absorb additional limb rotations across the hindfoot puts a strain on the ankle joint structures, especially the lateral collateral ligaments. The plantar aponeurosis remains slack and may adaptively shorten over time. The TMTs must undergo a pronation twist to maintain appropriate weight bearing of the forefoot. This may result in chronic plantarflexion of the first ray. There is not an effective conservative intervention for a cavus foot as there is for a flexible flatfoot. The only exception would be in an instance where there is a correctable rotatory deficit in one of the superimposed leg segments and where secondary changes in the bones or soft tissue structures of the foot have not occurred.

Both function and dysfunction of the numerous structures and joints making up the foot and ankle are complex. It is frequently difficult to determine whether problems are primary or are secondary to problems proximal or distal to the site of pain. Regardless, increased participation in sports will inevitably result in an increased number of people seeking medical attention for foot and ankle problems. Offsetting this trend somewhat is the improving research and technology associated with footwear. It is unlikely, however, that even the most sophisticated footwear will be able to take into consideration simultaneously the many mobility and stability demands of the joints of the lower extremity, including the large number of individual differences.

REFERENCES

1. Morris, JM: Biomechanics of the foot and ankle. Clin Orthop 122:10, 1977.
2. Cailliet, R: Foot and Ankle Pain, ed 2. FA Davis, Philadelphia, 1983.
3. Eichenblat, M and Nathan, H: The proximal tibio fibular joint. Int Orthop (SICOT) 7:31–39, 1983.
4. Barnett, CH and Napier, JR: The axis of rotation at the ankle joint in man: Its influence upon the form of the talus and mobility of the fibula. J Anat 86:1, 1952.
5. Kapandji, IA: The Physiology of the Joints, Vol II—The Lower Limb. Churchill-Livingstone, New York, 1970.
6. Rocce, D: The leg, ankle and foot. In The Musculoskeletal System in Health and Disease. Harper & Row, Hagerstown, Maryland, 1980.
7. Rasmussen, O, Tovborg-Jensen, I, and Boe, S: Distal tibiofibular ligaments. Acta Orthop Scand 53:681–686, 1982.
8. Takebe, K, et al: Role of the fibula in weight-bearing. Clin Orthop 184:2899–2892, 1984.
9. Segal, D, et al: The role of the lateral malleolus as a stabilizing factor of the ankle joint: A preliminary report. Foot Ankle 2:25–29, 1981.
10. Nuber, GW: Biomechanics of the foot and ankle during gait. Clin Sports Med 7:1–12, 1988.

11. Kjaersgaard-Anderson, P, Wethelund, J, and Nielsen, S: Lateral talocalcaneal instability following section of the calcaneofibular ligament: A kinesiologic study. Foot Ankle 7:355–361, 1987.
12. Johnson, EE and Markolf, KL: The contribution of the anterior talofibular ligament to ankle laxity. J Bone Joint Surg 44A:81–88, 1983.
13. Rasmussen, O and Tovborg-Jensen, I: Mobility of the ankle joint. Acta Orthop Scand 53:155–160, 1982.
14. Rasmussen, O, Tovberg-Jensen, I, and Hedeboe, J: An analysis of the function of the posterior talofibular ligament. Int Orthop (SICOT) 7:41–48, 1983.
15. Rasmussen, O and Kromann-Andersen, C: Experimental ankle injuries. Acta Orthop Scand 54:356–362, 1983.
16. Siegler, S, Chen, J, and Schneck, CD: The three dimensional kinematics and flexibility characteristics of the human ankle and subtalar joints—part I: Kinematics. Biomech Eng 110:364–373, 1988.
17. Lundberg A, et al: Kinematics of the ankle/foot complex: Plantarflexion and dorsiflexion. Foot Ankle 9:194–200, 1989.
18. Lundberg, A, et al: Kinematics of the ankle/foot complex—part 3: Influence of leg rotation. Foot Ankle 9:304–309, 1989.
19. Lundberg, A, et al: Kinematics of the ankle/foot complex—part 2: Pronation and supination. Foot Ankle 9:245–253, 1989.
20. Lundberg, A, et al: The axis of rotation of the ankle joint. J Bone Joint Surg 71B(1): 1989.
21. Hicks, JH: Mechanics of the foot I: The joints. J Anat 87:345, 1953.
22. Sammarco, W, Burstein, AH, and Frankel, VH: Biomechanics of the ankle: A kinematic study. Orthop Clin North Am 4:76, 1973.
23. Inman, VT and Mann, RA: Biomechanics of the foot and ankle. In Mann, RA (ed): Duvries Surgery of the Foot, ed 4. CV Mosby, St. Louis, 1978.
24. Stauffer, RN, Chao, EYS, and Brewster, RC: Force and motion analysis of normal, diseased and prosthetic ankle joints. Clin Orthop 127:189, 1977.
25. Greenwald, AS and Matejczyk, MB: Pathomechanics of the human ankle joint. Bull Hosp Joint Dis 38:105, 1977.
26. Hornsby, TM, et al: Effect of inherent muscle length on isometric plantar flexion torque in healthy women. Phys Ther 67:1191–1196, 1987.
27. Viladot, AV, et al: The subtalar joint: Embryology and morphology. Foot Ankle 5:54–65, 1984.
28. Gray's Anatomy, ed 37. Churchill Livingstone, Edinburgh, 1989.
29. Cass FR, Morrey, BF, and Chao, EY: Three-dimensional kinematics of ankle instability following serial sectioning of lateral collateral ligaments. Foot Ankle 5:142–149, 1984.
30. Manter, JT: Movements of the subtalar and transverse tarsal joints. Anat Rec 80:397, 1941.
31. Root, ML, Orien, WP, and Weed, JH: Normal and Abnormal Function of the Foot: Clinical Biomechanics, Vol II. Clinical Biomechanics Corp, Los Angeles, 1977.
32. Bailey, DS, Perillo, JT, and Forman, M: Subtalar joint neutral. J Am Podiatr Assoc 74:59–64, 1984.
33. Inman, VT: Joints of the Ankle. Williams & Wilkins, Baltimore, 1976.
34. Close, JR, Inman, VT, Poor, PM, and Todd, FN: The function of the subtalar joint. Clin Orthop 50:59, 1967.
35. Sgarlato, TE: The angle of gait. Phys Ther 55:645, 1965.
36. Elftman, H: The transverse tarsal joint and its control. Clin Orthop 16:41–45, 1960.
37. Lapidus, PW: Kinesiology and mechanical anatomy of the transverse tarsal joints. Clin Orthop 30:20, 1963.
38. Inkster, RG: Inversion and eversion of the foot and the transverse tarsal joints. J Anat 72:612, 1938.
39. Lewis, OJ: The joints of the evolving foot. Part II: The intrinsic joints. J Anat 130:833–857, 1980.
40. Viladot, A: Metatarsalgia due to biomechanical alterations of the forefoot. Orthop Clin North Am 4(1):165–178, 1973.
41. McCarthy, DJ and Grode, SE: The anatomical relationships of the first metatarsophalangeal joint: A cryomicrotomy study. J Am Podiatr Assoc 70:493–504, 1980.
42. Yoshioka, Y, et al: Geometry of the first metatarsophalangeal joint. J Orthop Res 6:878–885, 1988.
43. Buell, T, Green, DR, and Risser, J: Measurement of the first metatarsophalangeal joint range of motion. J Am Podiatr Med Assoc 78:439–448, 1988.
44. Mann, RA: Foot problems in adults: AAOS Instructional Course Lectures. 31:167–180, 1982.
45. MacConaill, MA and Basmajian, JV: Muscles and Movements: A Basis for Human Kinesiology. Williams & Wilkins, Baltimore, 1969.
46. Gould, N, et al: Development of the child's arch. Foot Ankle 9:241–245, 1989.
47. Moore, KL: Clinically Oriented Anatomy, ed 2. Williams & Wilkins, Baltimore, 1985.
48. Sarrafian, SK: Functional characteristics of the foot and plantar aponeurosis under tibiotalar loading. Foot Ankle 8:4–18, 1987.
49. Manter, JT: Distribution of compression forces in joint of the human foot. Anat Rec 96:313, 1946.
50. Perry, J: Anatomy and biomechanics of the hindfoot. Clin Orthop 177:7–15, 1983.

51. Kalin, PJ and Hirsch, BE: The origins and function of the interosseous muscles of the foot. J Anat 152:83–91, 1987.
52. Sanner, WH, et al: A study of ankle joint height changes with subtalar joint motion. J Am Pod Soc 71:156–161, 1981.

STUDY QUESTIONS

1. Identify the proximal and distal articular surfaces that comprise the ankle (talocrural) joint. What is the joint classification?
2. Describe the superior and inferior tibiofibular joints, including classification and their composite function.
3. Identify the ligaments that support the tibiofibular joints, including which is strongest.
4. Describe the ligaments that support the ankle joint, including the names of components, when relevant.
5. Why is ankle joint motion considered triplanar?
6. Why does the fibula move during dorsiflexion/plantarflexion?
7. What are the primary checks to ankle joint motion?
8. Which muscles crossing the ankle are single-joint muscles?
9. Describe the three articular surfaces of the subtalar joint, including the capsular arrangement.
10. Which ligaments support the subtalar joint?
11. Describe the axis for subtalar motion. What movements take place around that axis and how are these motions defined?
12. When the foot is weight-bearing, the calcaneus (the distal segment) of the subtalar joint is not free to move in all directions. Describe the movements that take place during weight-bearing subtalar inversion and eversion.
13. What is the close-packed position for the subtalar joint? Which motion of the tibia will lock the subtalar joint?
14. Describe the relationship between the subtalar and the TCN joint with regard to articular surfaces, axes, and available motion.
15. Describe the articulations of the transverse tarsal joint.
16. What is the general function of the transverse tarsal joint in relation to the subtalar joint?
17. What are the TMT rays? Describe the axis for each ray and the movements that occur around each axis.
18. What is the function of the TMT joints in relation to the TCN and the transverse tarsal joints?
19. How does pronation twist of the TMT joints relate to inversion of the subtalar joint?
20. What ligaments contribute to support of the osteoligamentous arch of the foot on either the medial or lateral side?
21. What is the weight distribution through the various joints from the ankle through the metatarsal heads in unilateral stance?
22. How does extension of the MTPs contribute to stability of the foot?
23. In terms of structure, compare the MTPs of the foot to the MCPs of the fingers.

24. What is the metatarsal break? What function does it serve and when does this function occur?

25. What is the role of the triceps surae muscle group at each joint it crosses?

26. What is the non–weight-bearing posture of the subtalar (TCN) and transverse tarsal joints?

27. What other muscles besides the triceps surae exert a plantarflexion influence at the ankle? What is the primary function of each of these muscles?

28. Which muscles may contribute to support of the arch(es) of the foot?

29. What is the function of the quadratus plantae? What is the analog in the hand?

30. Drawing an analogy between the foot and the hand, describe the function of each of the intrinsic and extrinsic foot muscles.

31. If a person has a pes planus, describe two possible causes for this condition.

32. Identify at least three possible *effects* of pes planus.

Chapter 13

■ ■ ■

Posture

■

OBJECTIVES

Following the study of this chapter, the reader should be able to:

Describe

1. The position of the hip, knee, and ankle joints in optimal erect posture.
2. The position of the body's gravity line in optimal erect posture, using appropriate points of reference.
3. The "sway envelope."
4. The sequence of events in the development of the erect posture.
5. The basic elements of postural control.
6. The body's center of pressure and its relation to the ground reaction force.
7. The gravitational moments acting around the vertebral column, pelvis, hip, knee, and the ankle in optimal erect posture.
8. The effects of moments on body segments in optimal erect posture.

419

9. Muscle and ligamentous structures that counterbalance moments in optimal erect posture.
10. The following postural deviations: pes planus, hallux valgus, pes cavus, forward head, idiopathic scoliosis, kyphosis, and lordosis.
11. The effects of the above postural deviations on body structures, that is, ligaments, joints, and muscles.
12. Effects of age on posture.

Determine
1. How changes in location of the body's gravity line will affect gravitational moments acting around specified joint axes.
2. How changes in the alignment of body segments will affect either the magnitude or the direction of gravitational moments.
3. How changes in alignment will affect supporting structures such as ligaments, joint capsules, muscles, various joint structures, and articular surfaces.

INTRODUCTION

The basic elements of individual joint structures and associated muscles have been explored in the preceding chapters. The principles of biomechanics and knowledge of muscle physiology have been applied to various body segments for the purpose of gaining an understanding of joint and muscle function. In this chapter, the focus is on discovering how the various body structures are integrated into a system of levers that permits effective and efficient functioning of the body as a whole. Knowledge of individual joint and muscle structure and function is used as the basis for determining how each structure contributes to stability of the body in posture.

Static and Dynamic Posture

Posture can be either static or dynamic. In static postures the body and its segments are aligned and maintained in certain positions. Examples of static postures include standing, lying, or sitting. Dynamic posture refers to postures in which the body and/or its segments are moving, that is, walking, running, jumping, throwing, and lifting. An understanding of static posture forms the basis for understanding dynamic posture, and therefore static posture will be emphasized in this chapter. The dynamic postures of walking and running will be discussed in Chapter 14.

The study of any particular posture includes kinetic and kinematic analyses of all segments in the kinematic chain. Humans have the ability to arrange and to rearrange body segments to form a large variety of postures such as bilateral and single-leg erect standing, sitting, lying down, and kneeling. Many living creatures also are capable of assuming different postures, but maintenance of erect bipedal stance is unique to humans. The erect posture allows persons to use their upper extremities for the performance of large and small motor tasks. When crutches, canes, or other assistive devices must be used to maintain the erect posture, an important human attribute is lost.

Erect bipedal stance gives us freedom of the upper extremities, but in comparison with the quadrupedal posture, erect stance has certain disadvantages. Erect bipedal stance increases the work of the heart, places increased stress on the vertebral column, pelvis, and lower extremities, and reduces stability. In the quadruped posture the body weight is distributed between the upper and lower extremities. In human stance the body weight is distributed only between the two lower extremities. The human species' base of support, defined by an area bounded pos-

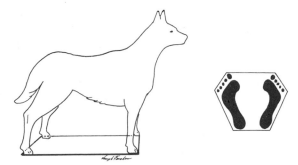

Figure 13-1. Bases of support. The quadripedal stance and the erect bipedal stance.

teriorly by the tips of the heels and anteriorly by a line joining the tips of the toes, is considerably smaller than the quadrupedal base[1] (Fig. 13–1). The human's center of gravity (COG), located within the body approximately at the level of the second sacral segment, is relatively distant from the base of support. Despite the instability caused by a small base of support and a high COG, maintaining a static erect posture requires very little energy expenditure in the form of muscle contraction. The bones, joints, and ligaments are able to provide the major torques needed to counteract gravity, and frequent changes in body position assist in promoting circulatory return.

Postural Control

The ability to attain and maintain the erect posture is one aspect of the human motor control system. Although the focus of this text is not on the motor control aspects of human function, a brief discussion of some features of postural control should help to improve the reader's understanding of posture. **Postural control,** which can be either static or dynamic, refers to a person's ability to maintain stability of the body and body segments in response to forces that threaten to disturb the body's structural equilibrium. Static postural control involves maintenance of a particular posture against gravity, while dynamic postural control involves maintenance of stability during movements of the body or body segments, and/or changes in the supporting surface. The pattern and timing of muscle activities that are used by individuals in response to changes that affect the static equilibrium are currently being investigated.[2–12]

Maintenance and/or control of posture depends on the integrity of the central nervous system (CNS), the visual system, the vestibular system, and the musculoskeletal system and inputs from receptors located in and around the joints and in tendons and ligaments. The CNS must be capable of receiving and processing information from all of the systems and must be able to interpret information from the receptors regarding the position of the body in space. The CNS must be able to respond to all of this input with appropriate output to maintain the equilibrium of the body. The joints in the musculoskeletal system must have a range of motion (ROM) that is adequate for specific tasks, and the muscles must be able to respond with appropriate speeds and forces. When inputs are altered or absent, such as occurs either in the absence of the normal gravitational force in weightless conditions during spaceflight, or when someone has decreased sensation in the lower extremities, the control system must respond to incomplete or distorted data and thus the person's posture may be altered.

Example: The astronauts aboard the US Space Shuttle *Discovery* in June 1985 demonstrated erect postures, both in space and immediately on their return to earth, that were very different from their normal preflight postures. When the astronauts' feet were fastened to the floor in space, and immediately on their return to earth, the astronauts assumed a position in which the neck, hip, and knee were flexed significantly more than in preflight posture. The observed postural changes have been attributed to altered inputs from tactile, articular, and proprioceptive cues.[13]

A more common example of altered inputs occurs when one attempts to attain and maintain the erect standing posture when one's foot has "fallen asleep." Attempts at standing may result in a fall partly because both input regarding the position of the foot and ankle and information from contact of the "asleep" foot with the supporting surface are missing. Another instance in which inputs may be disturbed is following injury. A disturbance in the kinesthetic sense about the ankle and foot following ankle sprains has been implicated as a cause of poor balance or loss of stability.[14]

In addition to altered inputs, a person's ability to maintain the erect posture may be affected by the inability of the muscles to respond appropriately to signals from the CNS. For example, in the elderly, the muscle's response time to the signal may be decreased in comparison to a younger individual.

Investigators of postural control have suggested that for any particular task many different combinations of muscles may be activated to complete the task. A normally functioning CNS selects the appropriate combination of muscles to complete the task based on an analysis of sensory inputs. Variations in an individual's past experience and customary patterns of muscle activity will also affect the response.

Monitoring of muscle activity patterns through electromyography (EMG), response times, and determinations of muscle peak torque and power outputs are used to study postural responses during horizontal perturbations of upright postural stability. One method of producing perturbations experimentally is by placing subjects on a movable platform. The platform can be moved forward or backward or from side to side. Some platforms can be tipped and the velocity can be varied. The postural responses have been found to be task-specific and to vary with (1) size of the supporting surface; (2) direction of motion of the supporting surface (anterior-posterior or medial-lateral); (3) location of the perturbing force (movements of the supporting surface or push or pull on a part of the body); (4) the magnitude of the applied force; (5) initial posture at the time of the perturbation; and (6) velocity of perturbation.

As a result of these studies, the following three patterns of muscle activity have been identified as occurring in response to perturbations of standing postures in the sagittal plane: (1) ankle, (2) hip, and (3) stepping patterns.[2,12]

The ankle pattern consists of discrete bursts of muscle activity on either the anterior or posterior aspects of the body that occur in a distal-to-proximal pattern in response to forward and backward movements of the supporting platform, respectively. Forward motion of the platform moves the feet anterior to the body's COG (Fig. 13–2a). In response to the disturbance of postural stability, bursts of muscle activity occur in the ankle dorsiflexors, hip flexors, abdominal muscles, and neck flexors. The tibialis anterior contributes to the restoration of stability by pulling the tibia anteriorly (reverse muscle action) and hence the body forward so that the line of gravity (LOG) remains within the base of support (Fig. 13–2b). Backward movement of the platform moves the feet posterior to the body's COG (Fig. 13–3a). Bursts of activity in the plantarflexors, hip extensors, trunk extensors, and neck extensors are used to restore the body's COG over the base of support (Fig. 13–3b). The hip pattern of muscle activity consists of discrete bursts of muscle activity on the side of the body opposite to the ankle pattern in a proximal-to-distal pattern of activation.[12]

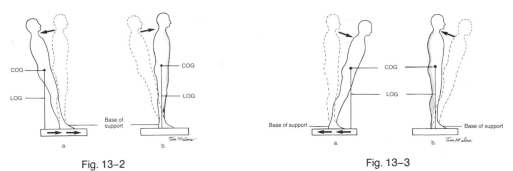

Fig. 13–2   Fig. 13–3

Figure 13-2. Perturbation of erect stance equilibrium by forward horizontal platform movement. a. Anterior (forward) movement of the platform causes posterior (backward) movement of the body and, as a consequence, displacement of the body's center of gravity (COG) posterior to the base of support. b. Employment of the ankle strategy (activation of dorsiflexors, hip flexors, abdominals, and neck flexors) brings the body's COG back over the base of support and reestablishes stability.

Figure 13-3. Pertubation of erect stance equilibrium by backward horizontal platform movement. a. Posterior movement of the platform causes anterior movement of the body and, as a consequence, displacement of the body's COG anterior to the base of support. b. Employment of the ankle strategy (activation of the plantarflexors, hip extensors, back and neck extensors) brings the body's COG over the base of support and reestablishes stability.

The stepping pattern is used to prevent a loss of equilibrium when the limits of stability have been exceeded and ankle and hip patterns of muscle activity are insufficient to bring and maintain the LOG over the base of support. Stepping forward or backward in response to movements of the platform moves the body's base of support so that it is under the body's COG (Fig. 13–4).[2,12]

EXTERNAL AND INTERNAL FORCES

The muscle activity patterns described in the preceding section are employed to counteract forces that affect the equilibrium of the body in the erect standing posture. The following section will examine the effects of both external and internal forces on the body and body segments on a joint-by-joint basis in order to under-

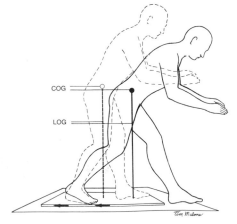

Figure 13-4. Perturbation of erect stance equilibrium by backward platform movement. The person in this illustration is employing a stepping strategy to keep from falling forward in response to backward movement of the platform. Stepping forward brings the body's COG over a new base of support.

stand how static and dynamic postures are maintained. The external forces that will be considered are inertia, gravity, and ground reaction forces (GRFs). The internal forces are produced by muscle activity and passive tension in ligaments, tendons, joint capsules, and other soft-tissue structures. According to the definition of equilibrium presented in Chapter 1, the external and internal forces must be balanced and the sum of all the forces acting on the body and its segments must be equal to zero for the body to be in equilibrium.

Inertial Forces

Inertial forces are ignored in static postures because little or no acceleration is occurring. However, in the erect standing posture the body undergoes a constant swaying motion called **postural sway or sway envelope**.[2] The extent of the sway envelope for a normal individual standing with about 4 in between the feet can be as large as 12° in the sagittal plane and 16° in the frontal plane.[2] The inertial forces that may result from this swaying motion are generally ignored and not considered in the analysis of forces for static postures.[15] Inertial forces must be considered in postural analysis of all dynamic postures such as walking, running, and jogging.

Gravitational forces act downward from the body's COG (center of mass). In the static erect standing posture the vertical projection of the body's COG falls within the base of support, which is the space defined by the two feet (Fig. 13–1). The gravity line must fall within the borders of the feet to maintain equilibrium in the static erect posture.[2] In dynamic postures such as walking and running, the gravity line falls outside of the borders of the feet during a large portion of the activity.[5]

Ground Reaction Forces

GRFs, which are the forces that act on the body as a result of interaction with the ground, are composed of three forces: a vertical force and two forces directed horizontally. One of the two horizontal forces is in a medial-lateral direction, whereas the other horizontal force is in an anterior-posterior direction along the ground. The resultant of these forces, the ground reaction force vector (GRFV), is equal in magnitude but opposite in direction to the gravitational force in the erect static standing posture.[7] The GRF resultant vector indicates the magnitude and direction of loading applied to the foot.[16] The point of application of the GRFV is at the body's center of pressure (COP), which is located in the foot in unilateral stance and between the feet in bilateral standing postures. If one were doing a handstand, the COP would be located between the hands. The COP, like the COG, is the theoretical point where the force is considered to act, although the body surface that is in contact with the ground may have forces acting over a large portion of its surface area.[17] The path of the COP that defines the extent of the sway envelope can be determined by plotting the COP at regular intervals when a person is standing on a force plate system (Fig. 13–5a and b).[5,18]

The GRFV and the gravity line form a common action line in the static erect posture.[15] In Figure 1–25, the GRFV (scale on man) and the gravity line representing the man on the scale form a common action line. In many dynamic postures, the intersection of the gravity line with the supporting surface may not coincide with the point of application of the GRFV. The horizontal distance from the point on the supporting surface where the LOG intersects the ground and the COP (where the GRFV acts) indicates the magnitude of the moment that must be opposed to maintain a posture and keep the person from falling.

The technology required to obtain GRFs, COP, and muscle activity is expensive and not available to the average evaluator of human function. Therefore, in the

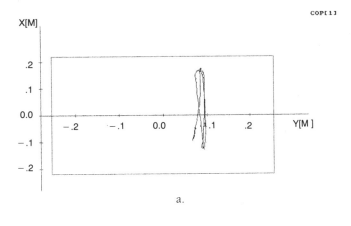

a.

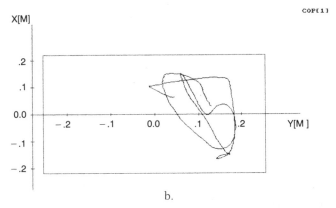

b.

Figure 13-5. Path of the center of pressure (COP) in erect stance. a. A COP tracing plotted for a person standing on a force plate. The rectangle represents the outline of the force plate. The tracing shows a normal rhythmic anterior-posterior "sway envelope" during approximately 30 seconds of stance. b. A COP tracing showing relatively uncontrolled postural sway. (COP tracings courtesy of Leonard Elbaum, Director of Research at the Physical Therapy Laboratory, Florida International University, Miami, Fla. Data were collected with an AMTI Force Platform, Newton, Mass. Analysis and display software were provided by Ariel Life Systems, Inc, La Jolla, Calif.)

following sections a simplified method of analyzing posture will be presented using diagrams and the combined action of the LOG and the GRF as a reference.

Combined Action Line

In an ideal erect posture, body segments are aligned so that the torques and stresses are minimized throughout the kinematic chain. The combined action line formed by the GRFV and gravity line serves as a reference for the analysis of the effects of these forces on body segments (Fig. 13–6). The combined action line will be referred to as the LOG in the remainder of this chapter. The location of the LOG shifts continually (as does the COP) because of the postural sway. As a result of the continuous motion of the LOG, the moments acting around the joints are continually changing. Receptors in and around the joints and on the soles of the feet detect these changes and relay this information to the CNS. The CNS analyzes the inputs and makes an approriate response to maintain postural stability.

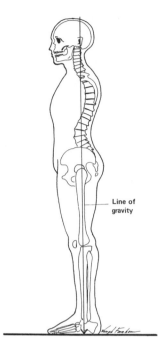

Figure 13–6. The location of the combined action line formed by the ground reaction force vector (GRFV) and the line of gravity (LOG) in optimal erect posture.

Sagittal Plane

The effect of forces on body segments in the sagittal plane is determined by the location of the gravity action line relative to the axis of motion of body segments. When the LOG passes directly through a joint axis, no gravitational torque is created around that joint. However, if the LOG passes at a distance from the axis, a gravitational torque is created. This torque will cause motion of the superimposed body segments around that joint axis unless the gravitational torque is opposed by a counterbalancing torque. The magnitude of the gravitational moment of force increases as the distance between the LOG and the joint axis increases. The direction of the gravitational moment of force depends on the location of the gravity line relative to a particular joint axis. If the gravity line is located *anterior* to the joint axis, the torque will tend to cause anterior motion of the proximal segment of the

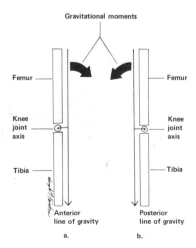

Figure 13–7. A schematic representation of the gravitational moments in the lower extremities of a person facing to the viewer's right. a. The gravitation moment tends to move the proximal segment (femur) anteriorly when the LOG passes anterior to the joint axis. b. The gravitational moment tends to move the proximal segment (femur) posteriorly when the LOG is located posterior to the joint axis.

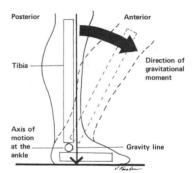

Figure 13–8. The anterior location of the LOG relative to the ankle joint axis creates a dorsiflexion moment. The arrow indicates the direction of the dorsiflexion moment. The dotted line indicates the direction in which the tibia would move if the dorsiflexion moment is unopposed.

body supported by that joint (Fig. 13–7a). If the gravity line falls *posterior* to the joint axis, the torque will tend to cause motion of the proximal segment in a posterior direction (Fig. 13–7b). In a postural analysis, gravitational moments producing sagittal plane motion of the proximal joint segment are referred to as either flexion or extension moments. The structure of the joint under consideration determines the type (flexion or extension) of moment.

    Example 1: If the LOG passes anterior to the center of rotation of the ankle joint, the gravitational torque will tend to rotate the tibia (proximal segment) in an anterior direction (Fig. 13–8). Anterior motion of the tibia on the fixed foot will result in dorsiflexion of the ankle. Therefore, the moment of force is called a **dorsiflexion moment**.

    Example 2: If the LOG passes anterior to the axis of rotation of the knee joint, the gravitational torque will tend to rotate the femur (proximal segment) in an anterior direction (Fig. 13–9). An anterior movement of the femur will cause extension of the knee. In this instance the moment of force is called an **extension moment**.

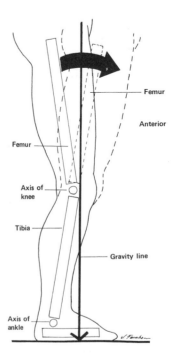

Figure 13–9. The anterior location of the LOG relative to the knee joint axis creates an extension moment. The arrow indicates the direction of the extension moment. The dotted line indicates the direction in which the femur would move if the gravitational moment were unopposed.

OPTIMAL POSTURE

Since the force of gravity is constantly acting on the body, an ideal posture is one in which the body segments are aligned vertically and the LOG passes through all joint axes. Normal body structure makes such an ideal posture almost impossible to achieve, but it is possible to attain a posture that is close to the ideal. In normal optimal standing posture, the LOG falls close to, but not through, most joint axes.[19] Therefore, in the normal optimal standing posture the gravitational forces may be balanced by countertorques generated by passive ligamentous tension and minimal muscle activity. The body segments in the normal optimal posture are in or near vertical alignment; the compression forces are distributed optimally over the weight-bearing surfaces of the joints; and no excessive tension is exerted on the ligaments and muscles. Slight deviations from the optimal posture are to be expected in a normal population because of the many variations found in body structure.

ANALYSIS OF POSTURE

Skilled observational analysis of posture involves identification of the location of body segments relative to the LOG. A plumbline, or line with a weight on one end, is used to represent the LOG. Analysis must include all segments and joints in the kinematic chain, since they are interdependent. Evaluators of posture should be able to determine if a body segment or joint deviates from the normal optimal postural alignment by using their observational skills. More sophisticated analyses may be performed using radiography, photography, and electromyography. However, a skilled observational analysis yields large amounts of information without the use of any instrumentation but a plumbline.

Lateral View—Optimal Alignment Segmental Analysis

Ankle and Knee

In the optimal erect posture the ankle joint is in the neutral position, or midway between dorsiflexion and plantarflexion. The knee joint is in full extension. The LOG falls slightly anterior to the lateral malleolus and just anterior to midline of the knee, posterior to the patella. The anterior position of the gravitational line relative to the ankle joint axis creates a dorsiflexion moment. In the neutral ankle position there are no ligamentous checks capable of counterbalancing the dorsiflexion moment; therefore, muscle activity of the plantarflexors is necessary to prevent forward motion of the tibia. The soleus muscle acting in reverse action exerts a posterior pull on the tibia and in this way is able to oppose the dorsiflexion moment (Fig. 13–10).

Electromyographic studies have demonstrated that soleus[20,21] and gastrocnemius[21] activity is fairly continuous in normal subjects during erect standing. This activity suggests that these muscles are exerting a light and constant torque about the ankles to oppose the dorsiflexion moment that exists at the ankle. The continuous character of the muscle activity suggests that the muscles are not acting in response to postural sway. Other ankle joint muscles that have shown inconsistent activity in EMG recordings during standing are the tibialis anterior, peroneals, and tibialis posterior.[22] Since the tibialis anterior, tibialis posterior, and peroneals have primary actions other than plantarflexion at the ankle joint, it is probable that these muscles are helping to provide transverse stability in the foot during postural sway rather than acting to oppose the gravitational moment around the ankle joint.

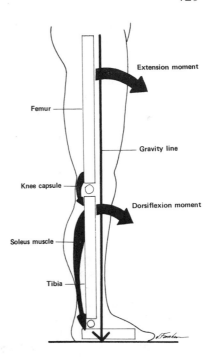

Figure 13-10. The extension moment acting around the knee joint is balanced by an opposing moment created by passive tension in the posterior joint capsule. The dorsiflexion moment at the ankle is counterbalanced by activity of the soleus muscle.

The anterior location of the gravitational line relative to the knee joint axis creates an extension moment. Passive tension in the posterior joint capsule and associated ligaments is sufficient to balance the gravitational moment and prevent hyperextension. Little or no muscle activity is required to maintain the knee in extension in the optimal erect posture. However, a small amount of activity has been identified in the hamstrings. Activity of the soleus muscle may augment the extension moment created around the knee through its posterior pull on the tibia.

Hip and Pelvis

The hip is in neutral position and the pelvis should be level with no anterior or posterior tilt (Fig. 13-11a). In the optimal position, lines connecting the symphysis pubis and the anterior superior iliac spines are vertical; and the lines connecting the anterior superior iliac and posterior superior iliac spines are horizontal.[23] In this

Figure 13-11. The location of the LOG relative to the axis of the hip joint. a. The LOG passes through the greater trochanter and posterior to the axis of the hip joint. b. The posterior location of the gravity line creates an extension moment at the hip, which tends to rotate the pelvis posteriorly on the femoral heads. The arrows indicate the direction of the gravitational moment.

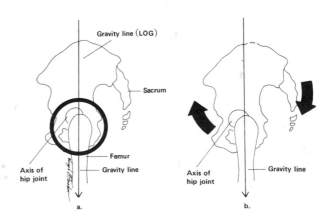

optimal position, the LOG passes slightly posterior to the axis of the hip joint, through the greater trochanter.[24] The posterior location of the gravitational line relative to the hip joint axis creates an extension moment at the hip that tends to rotate the pelvis (proximal segment) posteriorly on the femoral heads[25] (Fig. 13–11b). EMG studies have reported activity of the iliopsoas muscle during standing,[26] and it is possible that the iliopsoas is acting to create a balancing flexion moment at the hip. If the gravitational extension moment at the hip were allowed to act without muscular balance, as in a relaxed standing posture, hip hyperextension ultimately would be checked by passive tension in the iliofemoral, pubofemoral, and ischiofemoral ligaments. The relaxed standing posture does not require any muscle activity at the hip, but causes an increase in the tension stresses on the anterior hip ligaments and increases the magnitude of the gravitational torque at other joints in the body.

Sacroiliac and Lumbosacral Joints

The optimal position at the lumbosacral joint is determined by the amount of angulation that is present. The lumbosacral angle is the angle formed by two lines: one drawn parallel to the ground and the other drawn in line with the superior plateau of the first sacral vertebra (Fig. 13–12a). The optimal lumbosacral angle is about 30°.[27] Anterior tilting of the sacrum increases the lumbosacral angle and results in an increase in the shearing stress at the lumbosacral joint and an increase in the anterior lumbar convexity.

When the lumbosacral angle is in the optimal position, the LOG passes slightly anterior to the sacroiliac joint. The gravitational moment that is created at the sacroiliac joint tends to cause the superior portion of the sacrum to rotate anteriorly and inferiorly. Since the sacral vertebrae form a rigidly linked segment, the moment acting on the superior portion tends to force the inferior portion in a posterior direction (Fig. 13–12b). Tension in the sacrospinous and sacrotuberous ligaments counterbalances the gravitational torque and prevents the inferior portion of the sacrum from moving posteriorly. The superior portion of the sacrum is kept from being thrust anteriorly by the sacroiliac ligament.[28] The LOG passes through the body of the fifth lumbar vertebra and close to the axis of rotation of the lumbosacral joint. Gravity therefore creates a very slight extension moment that is opposed by the anterior longitudinal ligament.

Vertebral Column

The curves of the vertebral column should represent a normal configuration such as described in Chapter 4. When the vertebral curves are in optimal alignment, the reference line will pass through the midline of the trunk (Fig. 13–13). The location of the gravity line to the vertebrae above the fifth lumbar level is controversial. Cailliet[29] reports that the line transects the vertebral bodies at the level of the first and twelfth thoracic vertebrae and at the odontoid process of the second cervical vertebra.

Using Cailliet's frame of reference, the LOG will pass posterior to the axes of rotation of the cervical and lumbar vertebrae, anterior to the thoracic vertebrae, and through the body of the fifth lumbar vertebra. In this situation, the gravitational moments would tend to increase the natural curves in the lumbar, thoracic, and cervical regions. The maximal gravitational torque occurs at the apex of each curve or at C-5, T-8, and L-3, since the apical vertebrae would be farthest from the LOG. Kendall,[30] however, believes that the LOG passes through the bodies of the lumbar and cervical vertebrae and anterior to the thoracic vertebrae in the optimal posture.[30] In this instance, the stress on the supporting structures would be greatest in

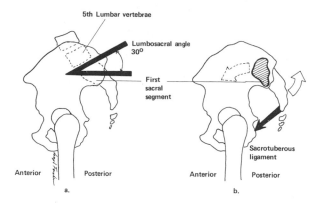

Figure 13-12. a. The lumbosacral angle in optimal erect posture is about 30°. b. The gravitational moment tends to rotate the superior portion of the sacrum anteriorly. Consequently, the inferior portion tends to thrust posteriorly. Passive tension in the sacrotuberous ligament prevents posterior motion of the inferior sacral segment.

the thoracic area, where the LOG falls at a distance from the vertebrae. Stress in the lumbar and cervical regions is comparatively less since the LOG falls close to or through the joint axes of these regions.

EMG studies have shown that the longissimus dorsi, rotatores, erector spinae, and neck extensor muscles exhibit intermittent electrical activity during normal standing.[31] Although not confirming either Cailliet's or Kendall's hypotheses, this evidence suggests that ligamentous structures alone are unable to provide enough force to oppose any gravitational moments acting around the joint axes of the vertebral column. In the lumbar region, where minimal muscle activity appears to occur, tension in the anterior longitudinal ligament apparently is sufficient to balance the gravitational extension moment.

Head

The LOG relative to the head passes through the external auditory meatus posterior to the coronal suture and through the odontoid process. The LOG falls anterior to the transverse axis of rotation for flexion and extension of the head and cre-

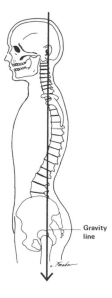

Figure 13-13. Location of the LOG in relation to the trunk.

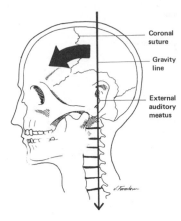

Figure 13-14. The anterior location of the LOG relative to the transverse axis for flexion and extension of the head creates a flexion moment.

ates a **flexion moment** (Fig. 13–14). The gravitational moment, which tends to tilt the head forward, is counteracted by tension in the ligamentum nuchae and tectorial membrane, and by activity of the neck extensors.[32] When a postural analysis is being performed, the gravity line should pass through the lobe of the ear. A summary of normal alignment in the sagittal plane is presented in Table 13–1.

Summary

The swaying motion that occurs in the normal erect posture will change the position of the LOG relative to individual joint axes. The COP also will move during swaying. For example, during forward sway of the body the LOG may move from the optimal posterior location relative to the hip joint axis to a position anterior to the hip joint axis (if the amount of swaying is sufficient). The COP will move anteriorly toward the toes. The flexion moment created by the change in position of the LOG may be counteracted by a brief burst of activity of the hip extensors, which will move the LOG and COP posteriorly. On the other hand, a burst of activity in the soleus muscles rather than the hip extensors might be used to bring the entire body and thus the LOG back into a position posterior to the hip joint axis.

A sudden backward movement of the supporting surface causes a similar, but much larger and more forceful movement of the LOG as the body is thrust forward. Flexion moments are created at the neck and head; cervical, thoracic, and lumbar spine; hip; and ankle. To counteract these moments, the neck, back, hip extensors, and ankle plantarflexors may have to contract. The CNS will decide which muscle or pattern of muscles will be used to counteract the inertial and flexion moments, to bring the LOG over the COP and to re-establish static erect equilibrium. If the movement is of sufficient magnitude, the stepping pattern may be selected to prevent falling (loss of equilibrium).

Lateral View—Deviations from Optimal Alignment

Reduction of energy expenditure and stress on the supporting structures is the primary goal of any posture. Any change in position or malalignment of one body segment will cause changes to occur in other segments. Changes in normal alignment increase stress or increase force per unit area on body structures. If stresses are maintained over long periods of time, the body structures may be altered. Mus-

Table 13–1. NORMAL ALIGNMENT IN THE SAGITTAL PLANE

Joints	Line of Gravity	Gravitational Moment	OPPOSING FORCES	
			Passive Opposing Forces	Active Opposing Forces
Atlanto-occipital	Anterior Anterior to transverse axis for flexion and extension	Flexion	1. Ligamentum nuchae 2. Tectorial membrane	Posterior neck muscles
Cervical	Posterior	Extension	1. Anterior longitudinal ligament	
Thoracic	Anterior	Flexion	1. Posterior longitudinal ligament 2. Ligamentum flavum 3. Supraspinous ligament	Extensors
Lumbar	Posterior	Extension	1. Anterior longitudinal ligament	
Sacroiliac joint	Anterior	Flexion type motion	1. Sacrotuberous ligament 2. Sacrospinous ligament 3. Sacroiliac ligament	
Hip joint	Posterior	Extension	1. Iliofemoral ligament	Iliopsoas
Knee joint	Anterior	Extension	1. Posterior joint capsule	
Ankle joint	Anterior	Dorsiflexion		Soleus

cles may lose sarcomeres if held in a shortened position and thus accentuate and perpetuate the abnormal posture and also will prevent full ROM from taking place. Muscles also may add sarcomeres if maintained in a lengthened position and thus the muscle's length-tension relationships will be altered. Shortening of the ligaments will limit normal ROM, while stretching of ligamentous structures will reduce the ligament's ability to provide sufficient tension to protect the joints. Prolonged weight-bearing stresses on the joint surfaces cause deformation of cartilage and interfere with the nutrition of the cartilage. As a result, the joint surfaces may become susceptible to early degenerative changes. The following examples illustrate how deviation from normal alignment of one or two body segments causes changes in other segments and increases the amount of energy required to maintain erect standing posture. Postural problems may originate in any part of the body and cause problems all along the kinematic chain. Therefore, it is important that the evaluator determine the causes of a problem.

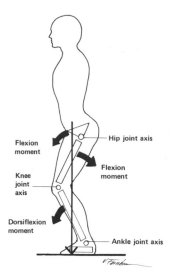

Flexion moment

Hip joint axis

Flexion moment

Knee joint axis

Dorsiflexion moment

Ankle joint axis

Figure 13–15. Gravitation moments in a flexed knee posture. Flexion moments are present, acting around the hip and knee joints, while a dorsiflexion moment is acting around the ankle.

Flexed Knee Posture

In this posture the LOG falls posterior to the knee joint axes. The posterior location of the gravity line creates a flexion moment at the knees that must be balanced by activity of the quadriceps muscles to maintain the erect position. The quadriceps force required to maintain equilibrium at the knee in erect stance increases from zero, with the knee extended, to 22 percent of maximum with the knee in 15° of flexion. A rapid rise in the amount of quadriceps force is required between 15 and 30° of knee flexion. When the knee is in 30° of flexion, the necessary quadriceps force reaches 51 percent of maximum.[33] The increase in muscle activity needed to maintain a flexed-knee posture subjects the tibiofemoral and patellofemoral joints to greater than normal compressive stress.

Other consequences of a knee-flexed erect standing posture are related to the ankle and hip. Since knee flexion is accompanied by hip flexion and ankle dorsiflexion, the location of the gravity line also will be altered in relation to these joint axes. At the hip, the gravity line will fall anterior to the hip joint axes. Activity of the hip extensors may be necessary to balance the flexion moment acting around the hip, and increased soleus activity may be required to counteract the increased magnitude of the dorsiflexion moment at the ankle (Fig. 13–15). The additional muscle activity subjects the hip and ankle joints to greater-than-normal compression stress.

Excessive Anterior Pelvic Tilt

In this posture as the pelvis tilts forward the lumbar vertebrae are forced anteriorly, thereby increasing the lumbar anterior convexity (lordotic curve). The LOG therefore is at a greater distance from the joint axes than is optimal and the extension moment is increased. The posterior convexity of the thoracic curve increases and becomes kyphotic in order to balance the lordotic lumbar curve. The anterior convexity of the cervical curve increases to bring the head over the sacrum (Fig. 13–16). Table 13–2 illustrates the changes that may result from an excessive anterior tilt.

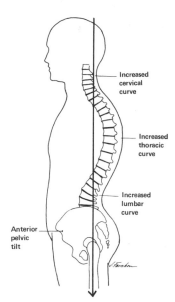

Figure 13-16. An excessive anterior pelvic tilt results in an increase in the lumbar anterior convexity. To compensate for the increased lumbar convexity, there is an increase in the posterior convexity of the thoracic region and an increase in the anterior convexity of the cervical curve.

The lumbar disks are subject to anterior tension and posterior compression in normal erect standing and erect unsupported sitting postures. A greater diffusion of nutrients into the anterior compared to the posterior portion of the disk occurs in the normal erect posture. Increases in the anterior convexity of the lumbar curve during either erect standing or unsupported erect sitting postures increases both the compressive forces on the posterior annuli and the tensile forces on the anterior annuli in the lumbar region. Although the increase in forces may benefit the nutrition of the anterior portion of the disk, the nutrition of the posterior portion of the intervertebral disks may be adversely affected and excessive compressive forces may be applied to the zygapophyseal joints.[34,35]

A flexed posture of the lumbar spine reverses the imbalance caused by the erect posture and enhances nutrient flow to the posterior portion of the disk and reduces compressive force on the zygapophyseal joints. The disadvantages of the flexed posture are that compressive forces are present on the anterior portion of the lumbar disks and the hydrostatic pressure in the nucleus pulposus is increased over pressures in the erect posture at low loads. However, the hydrostatic pressure in the disk can be higher in lordotic unsupported sitting postures than in flexed unsupported sitting postures. Apparently, the adoption of a postural strategy that includes alternating between flexed and extended sitting and standing postures at regular intervals may be necessary to ensure adequate nutrition of the lumbar disks and to avoid excessive stress on other lumbar structures.

Anterior-Posterior View—Optimal Alignment

In an anterior view the LOG bisects the body into two symmetrical halves (Fig. 13–17). The joint axes of the hip, knee, and ankle are equidistant from the LOG, and the gravitational line transects the central portion of the vertebral bodies. When postural alignment is optimal, little or no muscle activity is required to maintain medial-lateral stability. The gravitational torques acting on one side of the body are opposed by equal torques acting on the other side of the body.

TABLE 13-2. EFFECTS ON BODY STRUCTURES

Deviation	Compression	Distraction	Stretching	Shortening
Excessive anterior tilt of pelvis	Posterior vertebral bodies Interdiskal pressure L-5–S-1 increased	Lumbosacral angle increased Shearing forces at L-5–S-1 Likelihood of forward slippage of L-5 on S-1 increased	Abdominals	Iliopsoas
Excessive lumbar lordosis	Posterior vertebral bodies and facet joints Interdiskal pressures increased Intervertebral foramina narrowed	Anterior annulus fibers	Anterior longitudinal ligament	Posterior longitudinal ligament Interspinous ligaments Ligamentum flavum Lumbar extensors
Excessive dorsal kyphosis	Anterior vertebral bodies Intradiskal pressures increased	Facet joint capsules and posterior annulus fibers	Dorsal back extensors Posterior ligaments Scapular muscles	Anterior longitudinal ligament Upper abdominals Anterior shoulder girdle musculature
Excessive cervical lordosis	Posterior vertebral bodies and facet joints Interdiskal pressure increased Intervertebral foramina narrowed	Anterior annulus fibers	Anterior longitudinal ligament	Posterior ligaments Neck extensors

Anterior-Posterior View—Deviations from Optimal Alignment

Any asymmetry of body segments caused either by movement of a body segment or by a unilateral postural deviation will disturb optimal muscular and ligamentous balance. Symmetric postural deviations, such as bilateral genu valgum, that disturb the optimal vertical alignment of body segments, cause an abnormal distribution of weight-bearing or compressive forces on one side of a joint and increased tensile forces on the other side. The increased gravitational torques that may occur require increased muscular activity and cause ligamentous stress.

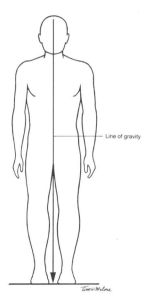

Figure 13–17. In an anterior view of the human body, the LOG, in optimal posture, divides the body into two symmetric parts.

Knee

In genu valgum the mechanical axes of the lower extremities are displaced laterally. The anatomic axes of the femur and tibia are deviated away from optimal vertical alignment. The gravitational moments, which tend to produce motion of the proximal femur laterally and motion of the proximal tibia medially, are greater than normal. As a result of the increased torque acting around the knee, the medial knee joint structures are subjected to tensile or distraction stress, and the lateral portion of the femurs are subjected to compressive stress (Fig. 13–18). These abnormal stresses may cause atrophic changes in the medial meniscus and hypertrophic changes in the lateral meniscus.

The gravitational torque acting on the foot in genu valgum tends to produce pronation of the foot with an accompanying stress on the medial longitudinal arch and abnormal weight bearing on the posterior medial aspect of the calcaneus.

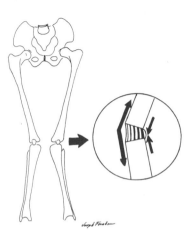

Figure 13–18. In genu valgum (''knock knees'') the medial aspect of the knee complex is subjected to tensile stress and the lateral aspect is subjected to compressive stress.

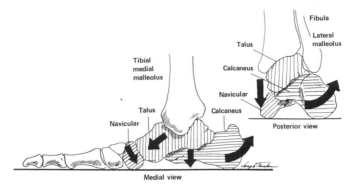

Figure 13-19. In pes planus ("flatfoot") there is displacement of the talus anteriorly, medially, and inferiorly; depression and pronation of the calcaneus; and depression of the navicular.

Foot and Toes

An evaluation of standing posture from the anterior-posterior aspect should include a careful evaluation of the feet. Normally the plumbline should lie equidistant from the malleoli, and the malleoli should appear to be of equal size and directly opposite from one another. When one malleolus appears more prominent or lower than the other, it is possible that a common foot problem known as **pes planus,** or **flatfoot,** may be present. Flatfoot, which is characterized by a reduced or absent arch, may be either rigid or flexible. A rigid flatfoot is a structural deformity that may be hereditary. In the rigid flatfoot the medial longitudinal arch is absent in non–weight-bearing, toe standing, and normal weight-bearing situations. In the flexible flatfoot, the arch is reduced during normal weight-bearing situations but reappears during toe standing or non–weight-bearing situations.

In either the rigid or flexible type of pes planus, the talar head is displaced anteriorly, medially, and inferiorly. The displacement of the talus causes depression of the navicular and stretching of the plantar calcaneonavicular (spring) ligament and the tibialis posterior muscle (Fig. 13–19). The degree of flatfoot may be estimated by noting the location of the navicular relative to the head of the first metatarsal. Normally the navicular should be intersected by the Feiss line (Fig. 13–20). If the navicular is depressed, it will lie below the Feiss line and may even rest on the floor in a severe degree of flatfoot. The pronated flatfoot results in a relatively overmobile foot that may require muscular contraction during standing. It also may result in increased weight bearing on the second through fourth metatarsal heads with subsequent callus formation. Pes planus interferes with push-off during walking because the foot is unable to assume the supinated position and become a rigid

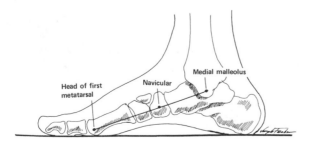

Figure 13-20. In the normal foot the medial malleolus, tuberosity of the navicular, and the head of the first metatarsal lie in a straight line called the *Feiss line.*

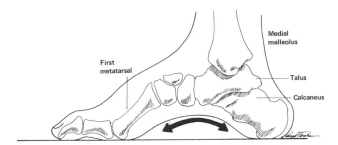

Figure 13-21. Pes cavus.

lever for push-off in gait. Pronation in a closed kinematic chain also causes medial rotation of the tibia and may affect knee joint function.

The medial longitudinal arch of the foot instead of being low may be unusually high. A high arch is called **pes cavus** (Fig. 12–21). This condition also may be rigid or flexible. Pes cavus is a more stable position of the foot than pes planus, but the weight borne on the lateral borders of the foot may stretch the lateral ligaments and the peroneus longus muscle. In walking the foot is unable to adapt to the supporting surface.

Three pathologic conditions of the toes may also be observed. They are **hallux valgus, claw toe,** and **hammer toe.** Hallux valgus is a deformity in which there is a lateral deviation of the great toe at the metatarsophalangeal joint (Fig. 13–22). An excess growth of bone may form on the medial aspect of the first metatarsal head or the joint may actually become dislocated. As a result of abnormal stress, bunions are formed at the metatarsophalangeal joint. They are a source of pain and may require surgical intervention.

Claw toe is a deformity of the toes characterized by hyperextension of the metatarsophalangeal joint combined with flexion of the distal and proximal interphalangeal joints.[36] This condition is often associated with pes cavus (Fig. 13–23a). The causes of this condition are unknown, but it has been attributed to the restrictive effect of shoes, muscular imbalance, ineffectiveness of the intrinsic flexors, or age-related deficiencies in the plantar structures.[36]

Hammer toe is a deformity characterized by hyperextension of the metatarsophalangeal joint and the distal interphalangeal joint and flexion of the proximal interphalangeal joint (Fig. 13–23b).[36] Callosities often develop under the heads of the metatarsophalangeal joints because of the excess weight bearing on these structures. Callosities also may be found on the superior surface of the interphalangeal

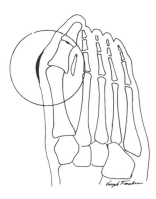

Figure 13-22. Hallux valgus.

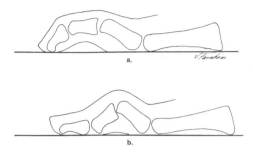

Figure 13–23. Claw toes and hammer toes. a. The drawing of claw toes shows hyperextension at the metatarsophalangeal joint and flexion of the interphalangeal joints. The abnormal distribution of weight may result in callous formation either under the heads of the metatarsals or under the end of the distal phalanx. Abnormal pressure between the superior surfaces of the flexed interphalangeal joint and the lining of the shoe also may result in callous formation. b. Hammer toes are characterized by hypertension of the metatarsophalangeal and distal interphalangeal joints and flexion of the proximal interphalangeal joints. Callous formation caused by the pressure of the shoe develops on the superior surface of the proximal interphalangeal joints.

joints as a result of pressure from shoes or on the tips of the toes because of abnormal weight bearing.

Vertebral Column

Scoliosis

Another segment of the body that requires special consideration when evaluating posture from the anterior or posterior view is the vertebral column. Normally, when viewed from the posterior aspect, the vertebral column is vertically aligned and bisected by the LOG. The gravity line falls through the midline of the occiput, through the spinous processes of all vertebrae, and directly through the gluteal cleft. In an optimal posture the vertebral structures, ligaments, and muscles are able to maintain the column in vertical alignment. If one or more of the supporting structures fails to provide adequate support, the column will bend to the side. The lateral bending will be accompanied by rotation of the vertebrae because lateral flexion without rotation does not occur below the level of the second cervical vertebrae.

Consistent lateral deviations of a series of vertebrae from the LOG in one or more regions of the spine may indicate the presence of a lateral spinal curvature called **scoliosis** (Fig. 13–24). There are different types of scoliosis, but the adolescent idiopathic type makes up 80 percent of all scolioses.[37] Idiopathic scoliotic curves are defined as structural curves[38] (Fig. 13–25). These curves involve changes in the vertebral bodies, transverse and spinous processes, intervertebral disks, ligaments, and muscles. Nonstructural curves are called **functional curves** in that they can be reversed if the cause of the curve is corrected. These curves are the result of correctable imbalances such as a leg length discrepancy or a muscle spasm. The curves in scoliosis are named according to the direction of the convexity and location of the curve. If the curve is convex to the left in the cervical area, then it is designated as a left curve. If more than one region of the vertebral column is involved, the superior segment is named first. The curve shown in Figure 13–24 is called a right thoracic, left lumbar curve.

The effects of unequal torques on the structures of the body are dramatically illustrated in adolescent idiopathic scoliosis. The following example depicts a hypothetical series of events. The first step in the process is unknown, because researchers have been unable to identify the supporting structure involved in the initial failure or the cause of the failure. Investigators have postulated that adolescent idiopathic scoliosis may result from a dysfunction in the vestibular system,[39] a disturbance in control of the muscle spindle,[40] a primary defect in the collagen of the

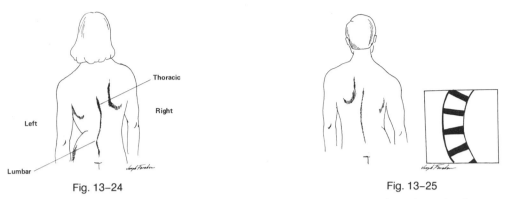

Fig. 13–24

Fig. 13–25

Figure 13–24. A lateral curvature of the vertebral column that is convex to the right in the thoracic region and convex to the left in the lumbar region.

Figure 13–25. A lateral curvature of the vertebral column that is convex to the left in the thoracic region. Wedging of the vertebrae may be seen in the portion of the curve shown in the inset. This wedging illustrates one of the structural changes observed in idiopathic scoliosis.

annulus fibrosus,[41] and subcortical brain stem abnormalities.[42] Lidstrom[18] found differences in postural sway in 100 children aged 10 to 14 years: 35 of the children were siblings of scoliotic patients and 65 were control subjects.[18]

Possible sequence of events for adolescent idiopathic scoliosis:

1. Failure in the supporting and/or control systems
2. Lateral flexion moment
3. Deviation of the vertebrae with rotation
4. Compression of the vertebral body on the side of the concavity of the curve
5. Inhibition of growth of vertebral body on the side of the concavity of the curve
6. Wedging of the vertebra
7. Head out of line with sacrum
8. Compensatory curve
9. Adaptive shortening of trunk musculature on the concavity
10. Stretching of muscles, ligaments, and joint capsules on the convexity.

The structural changes listed above may progress to produce a severe deformity unless intervention occurs at the appropriate time. Deformities can interfere with breathing and other internal organs as well as being cosmetically unacceptable. It has been estimated that about 10 percent of the adolescent population in the United States has some degree of scoliosis. About one quarter of this population has a curvature that will need intervention in the form of observation, bracing, or surgery. If a curvature is recognized early in its development, then measures may be instituted either to correct the curve or to prevent its increase.[43] According to the second phase of the Utah study in which a visual assessment (scoliosis screening) was performed of 3000 college-aged women (19 to 21 years of age) in 34 states and five foreign countries, 12 percent of this population had a lateral deviation of the spine.[44] This curvature had not been detected previously. Therefore, the initiation of screening programs in the schools appears to be of critical importance. As reported in 1986 only 20 states required screening programs.[38]

The vertebral deviations in scoliosis cause asymmetric changes in body structures. Several of these changes may be detected through simple observation of body contours. Typical screening programs usually are designed for identification

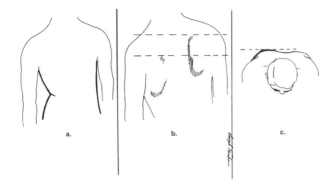

Figure 13-26. Typical changes in body contours used in scoliosis screening programs. a. Uneven waist angles or difference in arm to body space. b. Unequal shoulder height or unequal scapula level. c. A rib hump during forward trunk flexion.

of the following features: unequal waist angles (Fig. 13–26a); unequal shoulder levels or unequal scapulae (Fig. 13–26b); rib hump; and obvious lateral spinal curvature (Fig. 13–26c).

EFFECTS OF AGE, PREGNANCY, OCCUPATION, AND RECREATION ON POSTURE

Age

Infants and Children

Postural control in infants develops progressively during the first year of life from control of the head, to control of the body in a sitting posture, to control of the body in a standing posture. Stability in a posture or the ability to fix and hold a posture in relation to gravity must be accomplished before the child is able to move within a posture. The child learns to maintain a certain posture usually through cocontraction of antagonist and agonist muscles around a joint and then is able to move in and out of the posture (sit to stand and stand to sit). Once stability is established, the child procedes to controlled mobility and skill. Controlled mobility refers to the ability to move within the posture, that is, weight shifting in the standing posture. Skill refers to performance of activities like walking, running, and hopping, which are dynamic postural activities.[45]

According to Woollacott,[4] by the time a child reaches 7 to 10 years of age postural responses to platform perturbations are comparable to adults in patterns of muscle activity and timing of responses. Woollacott found considerably less variability in postural responses to platform movements in 7 to 10-year-olds than was present in the younger age ranges. Also, as might be predicted, responses of the younger children included greater coactivation of agonists and antagonists and slower response times for muscle activation than adults or older children.[6]

Postural alignment in children should be similar to adult alignment by the time the child reaches 10 or 11 years of age.[46] However, poor posture in a 7- or 8-year-old can be recognized because it is similar to adults, that is, kyphosis, excessive anterior pelvic tilt and accompanying lordosis, forward head, and/or hyperextended knees.[30,46]

Elderly

Postural responses of older adults (ages 61 to 78 years) to platform perturbations show differences in timing and amplitude and include greater coactivation of

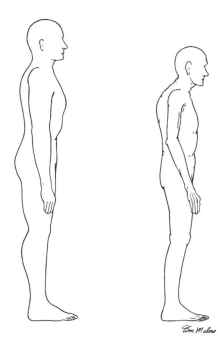

Figure 13-27. Changes in posture as a result of aging.

antagonist and agonist muscles when compared to younger subjects (ages 19 to 38 years). Iverson and associates,[47] who tested noninstitutionalized men 60 to 90 years of age on two types of balance tests that involved one-legged stance, found that balance time and torque production decreased significantly with age.[47] In some of the tests, the authors found that torque production was a significant predictor of balance time, that is, the greater the torque production, the longer the balance time. These authors also found that men who exercised five to six times per week had greater torque production than men who exercised less frequently. This finding suggests that high levels of fitness and activity may have beneficial effects on the aging person's ability to perform one-legged balancing activities.[47]

Postural alignment in the elderly often shows a more flexed posture than in the young adult (Fig. 13–27). The typical elderly posture is probably due to a number of factors that may be attributed either to the aging process or to a sedentary life-style, or to a combination of aging and sedentary life-style. Changes in the vertebral column such as loss of fluid from the vertebral disks and degeneration of the disks and vertebral bodies causes an increase in the normal thoracic curve that results in shortening of the trunk flexor muscles and lengthening of trunk extensors. The head is often forward, which causes an increase in flexion in the cervical region and increase in extension at the atlanto-occipital joint. The ROM at knees, hips, ankles, and trunk may be restricted due to muscle shortening and disuse atrophy. Furthermore, as postural response times in the elderly appear to be longer than in the young, the elderly may elect to stand with a wide base of support to have a margin of safety.

Pregnancy

Postural changes in pregnancy include an increase in the lordotic curves in the cervical and lumbar areas of the vertebral column, protraction of the shoulder girdle, and hyperextension of the knees. These changes are caused partly by the combination of a weight gain and an increase in weight distribution in the breasts and abdomen and partly to the effects of softening of the ligamentous and connective tissue that accompanies pregnancy. These changes predispose the pregnant women to strains in supporting structures.[48]

Occupation

The static and dynamic postures assumed at work and during recreational activities may have adverse effects on joint structure and function. Injuries to joints and their supporting structures can lead to impaired function, decreased productivity, absences from work, and sometimes loss of work and permanent disability. Recognition by the medical profession that postures assumed at work and during recreational activities produce unique injuries requiring specialized knowledge for management has led to the establishment of medical specialties in the following areas: industrial or occupational medicine, performing arts medicine or simply "arts medicine," and sports medicine.

Each particular occupational and recreational activity has unique postures and resulting injuries. Bricklayers, surgeons, carpenters, and cashiers assume and perform tasks in standing postures for a majority of the working day. Others, such as secretaries, typists, accountants, computer programmers, and receptionists assume sitting postures for a large proportion of the day. Performing artists often assume asymmetrical postures while playing a musical instrument, dancing, or acting. Running, jogging, and walking are other postures that have very specific associated injuries.

The high incidence of back problems among the population has led to a large amount of research related to lifting and stooping postures and to the development of various treatment programs.[49-55] Different sitting postures and their effects on interdiskal pressures in the lumbar spine have been analyzed.[52] Wheelchair postures and the effects of different degrees of anterior-posterior and lateral pelvic tilt on the vertebral column and trunk muscle activity in sitting and standing postures in selected work activities also have been investigated.[56] A large portion of the research suggests that many back problems are preventable because they result from mechanical stresses produced by prolonged static postures in the forward stooping or sitting positions and the repeated lifting of heavy loads.

Many of the injuries sustained during both occupational and recreational activities fall into the category of "overuse injuries." This type of injury is caused by repetitive stress that exceeds the physiologic limits of the tissues. Muscles, ligaments, and tendons are especially vulnerable to the effects of repetitive tensile forces, whereas bones and cartilage are susceptible to injury from the application of excessive compressive forces. A random sample of professional musicians in New York found that violin, piano, cello, and bass players were frequently affected by back and neck problems.[57] In a larger study involving 485 musicians, the authors found that 64 percent had painful overuse syndromes. The majority of problems were associated with the musculotendinous unit, and others involved bones, joints, bursae, and muscle. String players experienced shoulder and neck problems due to the maintenance of abnormal head and neck positions, whereas flute players had shoulder problems associated with maintaining an externally rotated shoulder position that has to be assumed for prolonged periods during performances and practices.[58]

Each occupational and recreational activity requires a detailed biomechanical analysis of the specific postures involved to determine how abnormal and excessive stresses can be relieved. Sometimes the analysis involves not only a person's posture but also features of the work site such as chair or table height, weight of objects to be lifted or carried, and weight and shape of a musical instrument or tool. Treatment may involve a combination of modifications of the environment, adaptations of the instrument or tools, and modifications of posture. Knowledge of biomechanics and normal human structure and function forms the basis for determining the potentially harmful effects of asymmetrical postures on joint structure and function in static and dynamic postures.

REFERENCES

1. Schenck, JM and Cordova, FK: Introductory Biomechanics, ed 2. FA Davis, Philadelphia, 1980.
2. Nashner, LM: Sensory, neuromuscular and biomechanical contributions to human balance. Proc APTA Forum, APTA, Alexandria, Va,1990.
3. Keshner, EA: Reflex, voluntary and mechanical processes in postural stabilization. Proc APTA Forum, APTA, Alexandria, Va,1990.
4. Woollacott, M: Postural control mechanisms in the young and old. Proc APTA Forum, APTA, Alexandria, Va,1990.
5. Patla, AE, et al: Identification of age-related changes in the balance-control system. Proc APTA Forum, APTA, Alexandria, Va, 1990.
6. Studenski, S, et al: The role of instability in falls among older persons. Proc APTA Forum, APTA, Alexandria, Va, 1990.
7. Whipple, R and Wolfson, LI: Abnormalities of balance, gait and sensorimotor function in the elderly population. Proc APTA Forum, APTA, Alexandria, Va, 1990.
8. Moore, SP, et al: Human automatic postural responses:Responses to horizontal perturbations of stance in multiple directions. Ex Brain Res, 73:648–658,1988.
9. Nashner, LM and McCollum, G: The organization of human postural movements:A formal basis and experimental synthesis. Behav Brain Sci 8:135–150, 1985.
10. Hansen, PD, Woollacott, MH, and Debu, B: Postural responses to changing task conditions. Ex Brain Res 73:627–636,1988.
11. Johansson, R, Magnusson, M, and Akeson, M: Identification of human postural dynamics. IEEE Trans Biomed Eng 35:858–869,1988.
12. Horak, FB: Measurement of movement patterns to study postural coordination. Proc 10th Annu Eugene Michels Res. Forum, APTA Section on Research, New Orleans, La, 1990.
13. Clement, G and Lestienne, F: Adaptive modifications of postural attitude in conditions of weightlessness. Exp Brain Res 72:381–389,1988.
14. Garn, SN and Newton, RA: Kinesthetic awareness in subjects with multiple ankle sprains. Phys Ther 68:1667–1671, 1988.
15. Schenkman, M: Interrelationship of neurological and mechanical factors in balance control. Proc APTA Forum Balance APTA, Alexandria, Virginia, 1990.
16. Rogers, M: Dynamic biomechanics of normal foot and ankle during walking and running. Phys Ther 68:1822–1830,1988.
17. Rogers, M and Cavanaugh, PR: Glossary of biomechanical terms, concepts and units. Phys Ther 64:82–98,1984.
18. Lidstrom, J, et al: Postural control in siblings to scoliosis patients and scoliosis patients. Spine 13:1070–1074,1988.
19. Lehmkul, D and Smith, LK: Brunnstrom's Clinical Kinesiology, ed 4. FA Davis, Philadelphia, 1983.
20. Carlsoo, S: The static muscle load in different work positions: An electromyographic study. Ergonomics 4:193,1961.
21. Soames, RW and Atha, J: The role of antigravity muscles during quiet standing in man. Appl Phys 47:159–167,1981.
22. Gray, ER: The role of the leg muscles in variations of the arches in normal and flat feet. Phys Ther 49:1084–1088,1969.
23. Williams and Worthingham, C: Therapeutic Exercise for Body Alignment and Function, ed 2. WB Saunders, Philadelphia,1977.
24. Cailliet, R: Soft Tissue Pain and Disability, ed.2. FA Davis, Philadelphia, 1988.
25. Don Tigny, RL: Dysfunction of the sacroiliac joint and its treatment. J Ortho Sp Phys Ther 1:1,1979.
26. Basmajian, JV: Muscles Alive, ed 4. Williams & Wilkins, Baltimore,1978.
27. Cailliet, R: Low Back Pain Syndrome, ed 3. FA Davis, Philadelphia, 1981.
28. Grieve, GP: The sacroiliac joint. Physiotherapy 62:12, 1976.
29. Cailliet, R: Neck and Arm Pain. FA Davis, Philadelphia, 1981.
30. Kendall, F and McCreary, EK: Muscles: Testing and Function, ed 3. Williams & Wilkins, Baltimore, 1983.
31. Morris, JM, Benner, G, and Lucas, DB: An electromyographic study of intrinsic muscles of the back in man. J Anat 96:509,1962.
32. Kapandji, IA: The Physiology of the Joints, Vol 3. Churchill-Livingstone, Edinburgh, 1974.

33. Perry, J, Antonelli, MS, and Ford, W: Analysis of knee joint forces during flexed-knee stance. J Bone Joint Surg 57A:7, 1975.

34. Adams, MA and Hutton, WC: The effect of posture on the diffusion into lumbar intervertebral discs. J Anat 147:121–134, 1986.

35. Adams, MA and Hutton, WC: The effect of posture on the lumbar spine. J Bone Joint Surg 47B:625–629,1985.

36. Myerson, MS and Shereff, MJ: The pathological anatomy of claw and hammer toes. J Bone Joint Surg 71A:45–49,1989.

37. Cailliet, R: Scoliosis. FA Davis, Philadelphia, 1975.

38. National Scoliosis Foundation: States that Require Postural Screening for Scoliosis. Belmont, Mass., July, 1986.

39. Jensen, GM and Wilson KB: Horizontal postrotatory nystagamus response in female subjects with adolescent idiopathic scoliosis. Phys Ther 59:10,1979.

40. Yekutiel, M, Robin, GC, and Yarom, R: Proprioceptive function in children with adolescent scoliosis. Spine 6:560–566, 1981.

41. Bushell, GR, et al: The collagen of the intervertebral disc in adolescent idiopathic scoliosis. J Bone Joint Surg 61B:4,1979.

42. Dretakis, EK, et al: Electroencephalographic study of schoolchildren with adolescent idiopathic scoliosis. Spine 13:143–145,1988.

43. Winter, RB and Moe, JH: A plea for routine school examination of children for spinal deformity. Minn Med 57:419,1974.

44. Francis, RS: Scoliosis screening of 3,000 college-aged women: The Utah study. Phase 2. Phys Ther 68:1513–1516,1988.

45. O'Sullivan, SB: Motor control assessment. In O'Sullivan, S and Schmitz, T (eds): Physical Rehabilitation: Assessment and Treatment. FA Davis, Philadelphia,1988.

46. Connolly, B: Postural applications in the child and adult. In Kraus, S (ed): Clinics in Physical Therapy, Vol 11. Churchill-Livingstone, New York,1988.

47. Iverson, B, et al: Balance performance, force production and activity levels in noninstitutionalized men 60–90 years of age. Phys Ther 70:348–355,1990.

48. Gleeson, PB and Pauls, JA: Obstetrical physical therapy:Review of literature. Phys Ther 68:1699–1702,1988.

49. Platts, RGS: Spinal mechanics. Physiotherapy 63:7,1977.

50. Hall, H: The Canadian back education unit. Physiotherapy 66:4,1980.

51. Edgar, M: Pathologies associated with lifting. Physiotherapy 65:245, 1979.

52. Anderson, GBJ, Ortengren, R, and Schultz, A: Analysis and measurement of the loads on the lumbar spine during work at a table. J Biomech 17:513,1979.

53. Matmiller, AW: The California back school. Physiotherapy 66:4,1980.

54. Kennedy, B: An Australian programme for management of back problems. Physiotherapy 66:4,1980.

55. Forsell, MZ: The Swedish back school. Physiotherapy 66:4,1980.

56. Borello-France, DF, Burdett, RG, and Gee, ZL: Modification of sitting posture of patients with hemiplegia using seat boards and backboards. Phys Ther 68:67–71,1988.

57. Brody, JE: For artists and musicians creativity can mean illness and injury. New York Times, Oct 17,1989.

58. Lockwood, AH: Medical problems of performing artists. N Engl J Med 320:221–227,1989.

STUDY QUESTIONS

1. What is a "sway envelope"?

2. Is quadriceps muscle activity necessary to maintain knee extension in static erect stance? Explain your answer.

3. Is activity of the abdominal muscles necessary to keep the pelvis level in static standing posture? Explain your answer.

4. What is the function of the sacrotuberous ligament in the erect standing posture?

5. In which areas of the vertebral column would you expect to find the most stress in the erect standing posture?

6. In the erect standing posture identify the type of stresses that would be affecting the following structures: apophyseal joints in the lumbar region, apophyseal joint capsules in the thoracic region, annulus fibrosus L-5–S-1, anterior longitudinal ligament in the thoracic region, and the sacroiliac joints.

7. What effect would tight hamstrings have on the alignment of the following structures during erect stance: pelvis, lumbosacral angle, hip joint, knee joint, and the lumbar region of the vertebral column?

8. How would you describe a typical idiopathic lateral curvature of the vertebral column?

9. Describe the moments that would be acting at all body segments as a result of an unexpected forward movment of a supporting surface. Describe the muscle activity that would be necessary to bring the body's LOG over the COP.

10. Identify the changes in body segments that are commonly used in scoliosis screening programs.

11. Compare hammer toes with claw toes.

12. How do postural responses to perturbations of the erect standing posture in the elderly compare with responses of children who are 1 to 6 years of age?

13. Compare a flexed lumbar spine posture with an extended posture in terms of the nutrition of the disks and stresses on ligaments and joint structures.

14. What is the relationship between the GRFV, LOG, and COG in the erect static posture?

15. Draw the Feiss line on a foot.

Chapter 14

■ ■ ■

Gait

■

OBJECTIVES
Following the study of this chapter, the reader should be able to:

Define
1. The stance, swing, and double support phases of walking gait.
2. The subdivisions of the stance and swing phases of walking gait.
3. The time and distance parameters of walking gait.

Describe
1. Joint motion at the hip, knee, and ankle for one extremity during a walking and running gait cycle.

2. The location of the ground reaction force vector in relation to the hip, knee, and ankle joints during the stance phase of walking gait.
3. The moments of force acting at the hip, knee, and ankle joints during the stance phase of walking gait.

Explain
1. Muscle activity at the hip, knee, and ankle throughout the walking gait cycle, including why and when a particular muscle is active and the type of contraction required.
2. The role of each of the determinants of gait.
3. The muscle activity that occurs in the upper extremities and trunk during walking gait.

Compare
1. Motion of the upper extremities and trunk with motion of the pelvis and lower extremities during walking gait.
2. The traditional gait terminology with the new terminology.
3. Normal walking gait with a gait in which there is a weakness of the hip extensor and hip abductor muscles.
4. Normal gait with a gait in which there is unequal leg length.
5. Normal walking gait with stair gait in relation to range of motion and muscle activity.
6. Normal walking gait with running gait.

GENERAL FEATURES

Preceding chapters in this text introduced the reader to the basic elements of human structure and function. In Chapter 13 the reader learned how individual joints and muscles function cooperatively to maintain the body in the erect posture. Knowledge of body structure and function in the static posture forms the basis for study of the human structure in a dynamic situation. In human locomotion (ambulation, gait) the reader is given the opportunity to discover how individual joints and muscles function in an integrated fashion to produce motion of the body as a whole. Knowledge of the kinematics and kinetics of normal ambulation provides the reader with a foundation for analyzing, identifying, and correcting abnormalities in gait.

KINEMATICS

One of the most distinctive features of human gait is the fact that it is individualistic. Each person has his or her own characteristic gait pattern. Novelists and dramatists often employ gait patterns to help portray the characters in their writings. Gait patterns may reflect a person's occupation, body structure, health status, and personality, as well as many other physical and psychological attributes. For example, the rolling gait that is used to describe a sailor's movement reflects the wide base of support needed to maintain balance at sea. The term "waddling" used to describe an obese person's gait reflects his or her unique body structure. Staggering gaits usually are associated with either drunkenness or weakness, while bouncy gaits are associated with vitality and strength. Swaggering gaits imply an aggressive personality, while mincing gaits may imply timidity. Investigations of gait indicate that potential assault victims display unique patterns of movement during walking. These particular patterns presumably communicate vulnerability and, as a result, predispose the person to attack.[1]

The descriptive terms used by nonscientific writers to describe gait are successful in conveying images of a unique gait but do not provide the information necessary for evaluators of human function. Scientists from many different disciplines have contributed to our present knowledge of gait and have provided us with detailed methods for analyzing gait.

Human locomotion, or gait, may be described as a translatory progression of the body as a whole, produced by coordinated, rotatory movements of body segments.[2] Normal gait is rhythmic and characterized by alternating propulsive and retropulsive motions of the lower extremities. The alternating movements of the lower extremities essentially support and carry along the head, arms, and trunk (HAT).[3] HAT constitutes about 75 percent of total body weight. The head and arms combined constitute about 25 percent of total body weight, while the trunk accounts for the remaining 50 percent.[4]

In standing posture, HAT is supported by both lower extremities; in gait, HAT not only must be balanced over one extremity but also must be transferred from one extremity to the other. The weight of HAT (75 percent of total body weight) plus the weight of the swinging lower extremity (about 10 percent of total body weight) must be supported by one extremity during single-limb support. Although single-limb support alternates with periods in which both lower extremities are in contact with the ground, gait places many more demands on the lower extremities than does the static erect posture. Before individuals can walk they must be able to balance HAT in the erect standing posture, transfer HAT from one lower extremity to another, and lift one lower extremity off the ground and place it in front of the other extremity in an alternating pattern. These activities require coordination, balance, intact kinesthetic and proprioceptive senses, and integrity of the joints and muscles.

To simplify the understanding of gait, several authors have defined gait in terms of certain tasks that must be accomplished.[5,6] According to the professional staff at Rancho Los Amigos Medical Center in California, the following three main tasks are involved in walking: (1) weight acceptance, (2) single-limb support, and (3) swing limb advancement.[5] In other words, a person must be able to accept and support the weight of the body on one lower extremity and to swing one extremity forward in order to progress.

Winter[6] also has proposed three main tasks for walking gait that include (1) maintenance of support of HAT against gravity, (2) maintenance of upright posture and balance, and (3) control of the foot trajectory to achieve safe ground clearance and a gentle heel contact. The tasks described by these authors are similar except that Winter has included gentle heel contact and foot clearance as essential tasks.

Gait is an extremely complex activity to analyze. Therefore, gait has been divided into a number of segments that make it possible to identify the events that are taking place. It is also worth noting that gait analysis is merely a special case of movement analysis as a whole. Knowledge of the terminology used to describe the events is necessary to be able to understand and analyze gait. Generally, gait is described by using the activities of one lower extremity (referred to as the reference extremity) from the beginning to the end of one gait cycle.

Phases of the Gait Cycle

The **gait cycle** includes the activities that occur from the point of initial contact of one lower extremity to the point at which the same extremity contacts the ground again (Fig. 14–1). During one gait cycle each extremity passes through two phases, a single **stance phase** and a single **swing phase.**

The stance phase begins at the instant that one extremity contacts the ground (heel strike) and continues only as long as some portion of the foot is in contact with

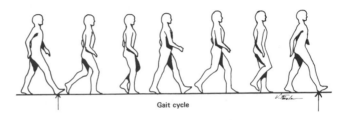

Figure 14–1. A gait cycle consists of the events that take place between initial contact of the reference extremity (right) and the successive contact of the same extremity.

Gait cycle

the ground (toe off) (Fig. 14–2). During the stance phase of gait some portion of the foot is in contact with the supporting surface at all times. The stance phase makes up approximately 60 percent of the gait cycle during normal walking.[7]

The swing phase begins as soon as the toe of one extremity leaves the ground and ceases just prior to heel strike or contact of the same extremity (Fig. 14–3). When the reference extremity is in the swing phase, it does not contact the ground at any time. Swing phase makes up 40 percent of the gait cycle.[8]

A period of double-limb support occurs in walking, when the lower extremity of one side of the body is beginning its stance phase and the lower extremity on the opposite side is ending its stance phase. Therefore, there are *two* periods of double support in a single gait cycle (Fig. 14–4). During double support *both* lower extremities are in contact with the ground at the same time. At a normal walking speed, stance of one leg and stance of the opposite leg overlap for about 22 percent of the gait cycle (Fig. 14–5).

Subdivisions

Traditionally, the stance and swing phases of gait have been divided into the following subunits: stance, which consists of heel strike, foot flat, midstance, heel off and toe off, and swing, which consists of acceleration, midswing, and deceleration. Another set of terms has been introduced to describe the subunits of gait by the Gait Laboratory at Rancho Los Amigos (RLA) Medical Center.[5] The RLA terminology may eventually replace the older descriptive terms. The reader should become familiar with both the traditional and newer terms because both methods of describing gait are being used at the present time.[8,9] In the following section definitions will be given for the traditional terminology first and for the RLA terminology second. It should be noted, however, that the two terminologies are not exactly equivalent but are placed in the following order to make comparison easier. The traditional terminology refers to points in time while the RLA terminology refers to lengths of time.

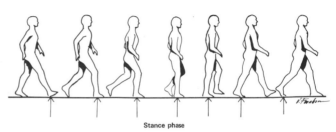

Figure 14–2. The stance phase of a gait cycle is defined as the period in which some portion of the foot of the reference extremity is in contact with the supporting surface. The period extends from the point of initial

Stance phase

foot contact of the reference extremity (right lower extremity in the diagram) to the point at which only the toes of the same (right) extremity are in contact with this supporting surface.

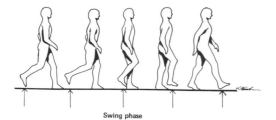

Figure 14-3. The swing phase of gait is defined as the period in which the foot of the reference extremity is not in contact with the supporting surface. The swing phase extends from the instant that the toe of the reference extremity (right lower extremity) leaves the ground to just prior to initial contact of the reference extremity.

Stance Phase

1. *Heel strike* refers to the instant at which the heel of the leading extremity strikes the ground (Fig. 14–6).
   RLA: Initial contact refers to the instant the foot of the leading extremity strikes the ground.[5] In normal gait the heel is the point of contact. In abnormal gait it is possible for either the whole foot or the toes rather than the heel to make initial contact with the ground.
2. *Foot flat* occurs immediately after heel strike and is the point at which the foot fully contacts the ground (Fig. 14–7).
   RLA: Loading response occurs immediately following initial contact and continues until the contralateral extremity lifts off the ground at the end of the double-support phase.[5]
3. *Midstance* is the point at which the body weight is directly over the supporting lower extremity (Fig. 14–8).
   RLA: Midstance begins when the contralateral extremity lifts off the ground and continues to a position in which the body has progressed over and ahead of the supporting extremity (Fig. 14–9).

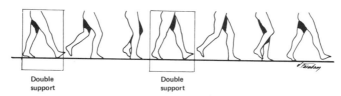

Figure 14-4. Double support is defined as the period in which some portion of the feet of both extremities are in contact with the supporting surface at the same time. Two periods of double support occur within a single gait cycle. One period occurs early in the stance phase of the reference extremity and the other occurs late in the stance phase of the reference extremity.

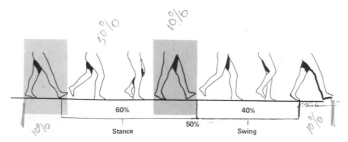

Figure 14-5. The two periods of double support overlap the stance phases of the two lower extremities for about 22 percent of the gait cycle at a normal walking speed. The stance phase constitutes 60 percent of the gait cycle and the swing phase constitutes 40 percent of the cycle at normal walking speeds. Increases or decreases in walking speeds will alter the percentages of time spent in each phase.

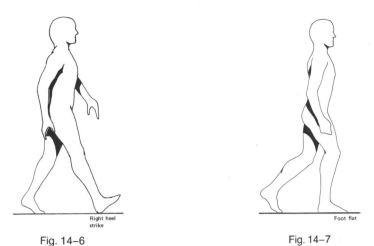

Fig. 14–6   Fig. 14–7

Figure 14–6. Heel strike refers to the instant at which the heel of the reference extremity contacts the supporting surface. Right heel strike in the diagram constitutes the beginning of the stance phase of gait for the right lower extremity. Heel strike is analogous to initial contact.

Figure 14–7. Foot flat occurs immediately after heel strike and is defined as the point at which the foot is flat on the ground. The period of loading response (RLA) extends from initial contact until the contralateral extremity leaves the ground at the end of the double support period.

Figure 14–8. Midstance is the point at which the body weight passes directly over the supporting lower extremity. The dotted line in the diagram outlines the point of midstance for the right lower extremity. Midstance encompasses the period from the end of foot flat to the beginning of heel off.

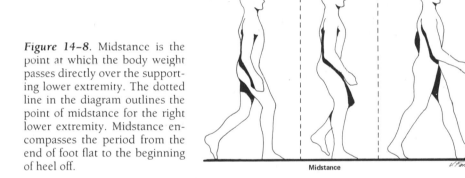

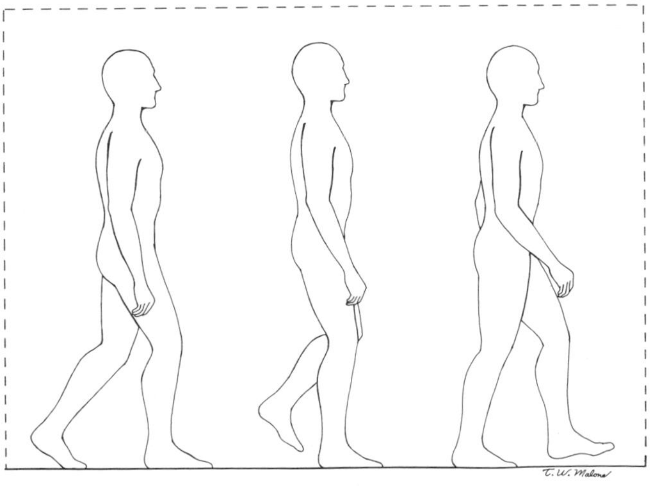

Figure 14–9. Midstance (RLA) begins when the contralateral extremity lifts off the ground and continues to a position in which the body has progressed over and ahead of the supporting extremity.

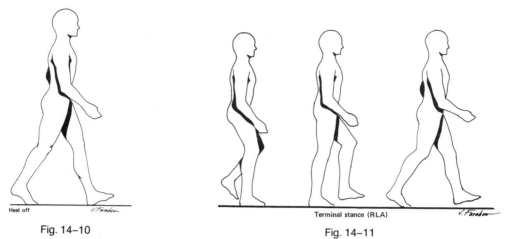

Heel off

Fig. 14–10

Terminal stance (RLA)

Fig. 14–11

Figure 14–10. Heel off is the point at which the heel of the reference extremity (right extremity in the diagram) leaves the supporting surface.

Figure 14–11. The period of terminal stance (RLA) includes the interval of the stance phase from the end of midstance (RLA) to just after the heel of the reference extremity (right extremity in the diagram) leaves the ground. Therefore heel off is included in terminal stance.

4. *Heel off* is the point at which the heel of the reference extremity leaves the ground (Fig. 14–10).
    RLA: Terminal stance is the period from the end of midstance to a point just prior to initial contact of the contralateral extremity or following heel off of the reference extremity (Fig. 14–11).[9]
5. *Toe off* is the point at which only the toe of the ipsilateral extremity is in contact with the ground (Fig. 14–12).

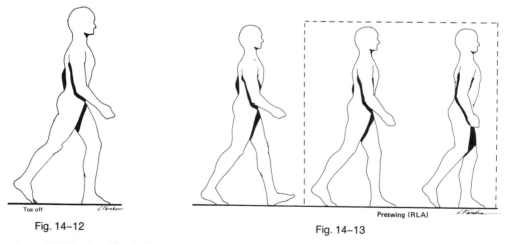

Toe off

Fig. 14–12

Preswing (RLA)

Fig. 14–13

Figure 14–12. Toe off is defined as the point in which only the toe of the reference extremity (right extremity) is touching the ground. The period from heel off to toe off often is referred to as the push off period of the stance phase.

Figure 14–13. Preswing (RLA) includes the interval of stance from the end of terminal stance (just after the right heel leaves the ground and the left heel contacts the ground) until the toe leaves the ground. Preswing is not exactly comparable to push off because preswing does not include heel off. Heel off occurs during terminal stance when using RLA terminology.

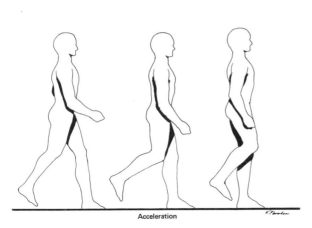

Acceleration

Fig. 14–14

Midswing

Fig. 14–15

Figure 14-14. Acceleration begins once the toe of the reference (right extremity in the diagram) has left the supporting surface. Initial swing (RLA) begins at approximately the same point as acceleration and continues until maximum flexion of the knee of the reference extremity.

Figure 14-15. Midswing occurs when the reference extremity passes directly below the body. In the diagram, the right lower extremity has just passed directly beneath the trunk. The midswing period extends from the end of acceleration to the beginning of deceleration.

RLA: Preswing encompasses the period from just following heel off to toe off (Fig. 14–13).

Swing Phase

1. *Acceleration* begins once the toe of the reference (ipsilateral) extremity leaves the ground and continues until midswing or the point at which the swinging extremity is directly under the body (Fig. 14–14).
   RLA: Initial swing begins at the same point as acceleration and continues until maximum knee flexion of the reference (ipsilateral) extremity occurs.
2. *Midswing* occurs when the ipsilateral extremity passes directly beneath the body (Fig. 14–15).
   RLA: Midswing encompasses the period immediately following maximum knee flexion and continues until the tibia is in a vertical position (Fig. 14–16).

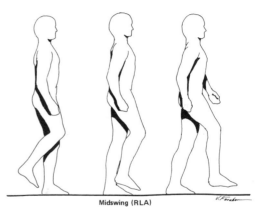

Midswing (RLA)

Figure 14-16. Midswing (RLA) is the period from maximal knee flexion to a point at which the tibia attains a vertical position.

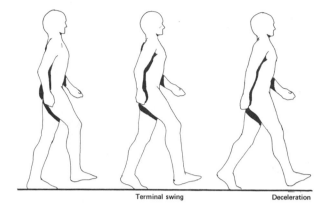

Terminal swing   Deceleration

Figure 14–17. Deceleration is the point at which the knee is extending in preparation for heel strike. Terminal swing (RLA) includes the period in which the tibia moves from its vertical position to the point at which the knee is in full extension just prior to initial contact.

3. *Deceleration* occurs after midswing when the tibia passes beyond the perpendicular and the knee is extending in preparation for heel strike.

RLA: Terminal swing includes the period from the point at which the tibia is in the vertical position to a point just prior to initial contact (Fig. 14–17).

A comparison of the traditional and RLA terminology reveals some differences between the two. However, these differences appear to be based primarily on the fact that the RLA terminology better defines a particular interval than the traditional. Sometimes the traditional terminology does not adequately define the end and the beginning of an interval; for example, where foot flat ends and midstance begins. In the newer RLA terminology each segment has a fairly well-defined beginning and observable endpoint and reference is made to an interval of gait rather than to a point. The following summary compares the traditional and new terminologies. A detailed description of the joint and muscle activity during each subdivision of gait will be present later in the chapter.

COMPARISON OF GAIT TERMINOLOGY

Traditional	RLA
Heel strike	Initial contact
Heel strike to foot flat	Loading response
Foot flat to midstance	Midstance
Midstance to heel off	Terminal stance
Toe off	Preswing
Toe off to acceleration	Initial swing
Acceleration to midswing	Midswing
Midswing to deceleration	Terminal swing

Distance and Time Variables

Time and distance are two of the basic parameters of motion, and measurements of these variables provide a basic description of gait. **Temporal variables** include stance time, single-limb and double-support time, swing time, stride and step time, cadence, and speed. The **distance variables** include stride length, step

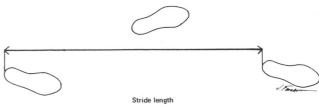

Figure 14–18. Stride length is determined by measuring the distance from the point of heel strike of the reference extremity to the next heel strike of the same extremity. Two successive heel strikes of a right lower extremity are shown in the drawing, and a left step.

length, width of walking base, and degree of toe out. These variables provide essential quantitative information about a person's gait and should be included in any gait description. Each variable may be affected by such factors as age, sex, height, size and shape of bony components, distribution of mass in body segments, joint mobility, muscle strength, type of clothing and footgear, habit, and psychologic status. However, a discussion of all the factors affecting gait is beyond the scope of this text.

Stance time is the amount of time that elapses during the stance phase of one extremity in a gait cycle.

Single support time is the amount of time that elapses during the period when only one extremity is on the supporting surface in a gait cycle.

Double support time is the amount of time that a person spends with both feet on the ground during one gait cycle. The percentage of time spent in double support may be increased in the elderly and in those with balance disorders. The percentage of time spent in double support decreases as the speed of walking increases.

Stride length is the linear distance between two successive events that are accomplished by the *same* lower extremity during gait.[10] Generally, stride length is determined by measuring the linear distance from the point of heel strike of one lower extremity to the next heel strike of the same extremity (Fig. 14–18). The length of one stride is traveled during one gait cycle and includes all of the events of one gait cycle. Stride length also may be measured by using other events, such as two successive toe offs of the same extremity, but in normal gait two successive heel strikes are used as the reference points. A stride includes two steps, a right step and a left step. However, stride length is not always twice the length of a single step because right and left steps may be unequal. Stride length varies greatly among individuals, since it is affected by leg length, height, age, sex, and other variables. Stride length can be normalized by dividing stride length by leg length or by total body height. Stride length usually decreases in the elderly[11–13] and increases as the speed of gait increases.[14] *The length of one stride is traveled during one gait cycle.*

Stride duration refers to the amount of time it takes to accomplish one stride. Stride duration and gait cycle duration are synonymous.

Step length is the linear distance between two successive points of contact of *opposite* extremities. It is usually measured from heel strike of one extremity to heel strike of the opposite extremity (Fig. 14–19). A comparison of right and left step

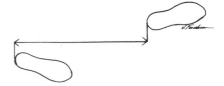

Figure 14–19. Step length is the linear distance between two successive points of contact of opposite extremities. Two successive points of contact of a right and left lower extremity are shown in the drawing.

lengths will provide an indication of gait symmetry. The more equal the step lengths, the more symmetric is the gait.

Step duration refers to the amount of time spent during a single step. Measurement usually is taken in seconds per step. When there is weakness of an extremity or pain, step duration may be decreased on the affected side and increased on the unaffected (stronger or less painful) side.

Cadence is the *number* of steps taken by a person per unit of time. Cadence may be measured as the number of steps per second or per minute.

$$\text{Cadence} = \text{number of steps/time.}$$

A shorter step length will result in an increased cadence at any given velocity.[14] As one walks with increased cadence, the duration of the double support period decreases. When the cadence of walking approaches 180 steps per minute, the period of double support disappears and running commences. A step frequency or cadence of about 110 steps per minute can be considered as "typical" for adult men, while a typical cadence for women is about 116 steps per minute.[4]

Walking velocity is the rate of linear forward motion of the body, which can be measured in centimeters per second, meters per minute, or in miles per hour.

$$\text{Walking velocity} = \text{distance walked/time.}$$

Women tend to walk with shorter and faster steps than men at the same velocity.[14] Increases in velocity are brought about by both increases in cadence and step length.[15]

Acceleration is the rate of change of velocity with respect to time.

Speed is measured in centimeters per minute. Speed of gait usually is referred to as slow, free, and fast. **Free speed** of gait refers to a person's normal walking speed, while **slow** and **fast** speeds of gait refer to speeds slower or faster than the person's normal comfortable walking speed. A fast speed of gait is generally accompanied by increased cadence and stride length, and decreased angle of toe out. However, there is a certain amount of variability in the way an individual elects to increase walking speed. Some individuals increase stride length and decrease cadence to achieve a fast walking speed. Other individuals decrease stride length and increase cadence to increase the speed of gait.

The **width of the base of support** is found by measuring the linear distance between the midpoint of the heel of one foot and the same point on the other foot (Fig. 14–20). The width of the walking base has been found to increase when there is an increased demand for side-to-side stability such as occurs in the elderly and in small children. In children the COG is higher than in the adult and a wide base of support is necessary for stability. In the normal population the mean width of the base of support falls within a range of 1 to 5 in.[9]

Degree of toe out represents the angle of foot placement and may be found by measuring the angle formed by each foot's line of progression and a line intersecting the center of the heel and the second toe. The angle for men normally is about 7° from the line of progression of each foot at free speed walking (Fig. 14–21).[11] The degree of toe out decreases as the speed of walking increases in normal men.[11]

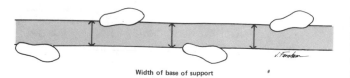

Width of base of support

Figure 14–20. Width of the base of support. The midpoint of the heel is used as a point of reference for measuring the width of the base of support.

Figure 14-21. Degree of toe out. The angle
formed by each foot's line of progression and a
line intersecting the center of the heel and the
second toe represents the angle of foot placement. The normal angle is 7° for males at free
speed walking.

Degree of toe out

Joint Motion

Another way in which gait may be described is through measuring the trajectories of the lower extremities and the joint angles. Sophisticated equipment, such
as stroboscopic flash photography, electrogoniometers, and computers have been
used to provide the information about joint angles and limb trajectories in normal
and abnormal gait.[16,17] Diagrams that plot the joint angles of the lower limbs against
each other, called **angle-angle diagrams,** are used to analyze and display data
obtained from photography.[18] The angle-angle diagrams can provide an objective
measure of a patient's progress, since various gaits produce characteristic angle-
angle diagram displays.[18,19]

Less sophisticated and less objective methods are used in observational gait
analysis, whereby an observer makes a judgment as to whether or not a particular
joint angle or motion adheres to a norm. For example, an observer may estimate
that the walking subject has more knee flexion at the end of midstance than the 5°
that represent the norm. The observer then must determine why the subject has
increased knee flexion.

One disadvantage of the observational method of analysis is that it requires a
great deal of training and practice to be able to identify the particular segment of
gait in which a particular joint angle deviates from a norm while a person is walking.
Observational gait analysis is used in the clinical setting because it has the advantage of being less cumbersome to walking subjects and less expensive than other
more sophisticated methods of analysis. In addition, well-trained observers can
derive a great deal of information from simple observation.

The gait laboratory at the RLA medical center has developed observational
gait analysis recording forms that help the observer to focus his or her attention
systematically on different portions of the body during gait.[5] In our experience
these forms have been useful for educational purposes. The joint ranges of motion
(ROM) that are presented in Table 14-1 have been adapted from the RLA gait analysis forms.[5] The degrees of motion presented in the table are the values achieved
at the end of each phase. However, neither the population that was used to determine the ranges of joint motion nor the speed of gait used are defined by RLA.[5]
Therefore, the reader should expect to find that the degrees of angulation will vary
among investigators because the populations used to establish norms may differ in
such characteristics as age and sex. Sometimes the norms will differ because of differences in terminology. For example, the *New York University Manual,*[8] which uses
the traditional terminology, describes the knee as being in 15° of flexion at midstance (body directly over the supporting extremity), while the RLA group
describes the knee as being in 5° of flexion at midstance (period from contralateral
toe off to point where body has progressed ahead of the supporting extremity).[5]

The approximate total ROM needed for normal gait may be determined by
looking at the value of the joint angle at each joint throughout the gait cycle. For
example, in Table 14-1 it can be seen that the knee joint is in extension (0°) at initial
contact in the stance phase and in 60° of knee flexion in the swing phase. Therefore,
one may conclude that a person needs full extension and up to 60° of knee flexion

Table 14-1. RANGE OF MOTION (RLA)

Joints	Initial Contact	End of Loading Response	End of Midstance	End of Terminal Stance	End of Preswing
			Stance Phase		
Hip	30° of flexion	25° of flexion	0°	10–20° of hyperextension	0°
Knee	0°	15° of flexion	5° of flexion	0°	35–40° of flexion
Ankle	0° (neutral position)	15° of plantarflexion	5–10° of dorsiflexion	0° of dorsiflexion	20° of plantarflexion
Toes (MTP joints)	0°	0°	0°	30° of hyperextension	50–60° of hyperextension

	End of Initial Swing	End of Midswing	End of Terminal Swing
		Swing Phase	
Hip	20° of flexion	30° of flexion	30° of flexion
Knee	60° of flexion	30° of flexion	0°
Ankle	10° of plantarflexion	0°	0°

Total Range of Motion

	Stance Phase	Swing Phase
Hip	0–30° of flexion 0–10 or 20° of hyperextension	20–30° of flexion
Knee	0–40° of flexion	0–60° of flexion
Ankle	0–10° of dorsiflexion 0–20° of plantarflexion	0–10° of plantarflexion

for normal knee motion to occur during gait. If an individual's knee motion were limited to 10 or 15° of knee flexion, one would expect his or her gait pattern to show considerable deviations from the norm.

Kinematic data on the norms of joint angles at each segment of gait also enable one to describe the changes in joint motion that occur during each phase. For example, a study of Table 14–1 shows that the knee is in extension at initial contact and in 15° of flexion at the end of loading response. Therefore, the knee must be flexing during the loading response period of gait. In the period of midstance the knee is extending because it moves from 15° of flexion at the end of loading response to 5° of flexion at the end of midstance. Again, by referring to Table 14–1, it is possible to determine that the knee continues to extend through terminal stance and flexes again in preswing. The hip begins the stance phase in flexion and extends throughout the phase until it reaches 10° of extension (hyperextension) by the end of terminal stance. Flexion of the hip begins and proceeds through preswing.

Similar descriptions of joint motion may be derived for the ankle in stance and for the hip, knee, and ankle during the swing phase by studying the average values given for joint angles at each gait segment at each period of gait in Table 14–1.

Determinants of Gait

Although the sagittal plane view of the lower extremity during gait has been described, there is still another group of components of gait usually considered in a kinematic analysis. These components are called **determinants of gait**.[4] The determinants represent adjustments made by the body that help to keep movement of the body's center of gravity (COG) to a minimum. There are six determinants: lateral pelvic tilt in the frontal plane, knee flexion, knee interactions, ankle interactions, pelvic rotation in the transverse plane, and physiologic valgus of the knee. The determinants are credited with minimizing the up-and-down and side-to-side movements of the COG and producing a smooth sinusoidal curve of the COG (Fig. 14–22). The order of presentation of the determinants that follows is based on their function and is not necessarily related to the order in which they appear in gait. The first four determinants help to keep the vertical rise of the body's COG to a minimum. The fifth determinant prevents a drop in the body's COG, while the sixth determinant reduces the side-to-side movement of the COG.

Lateral Pelvic Tilt (Pelvic Drop in the Frontal Plane)

In single-limb support (unilateral weight bearing) the combined weight of HAT and the swinging leg must be balanced over one lower extremity. During this period the COG reaches its highest point in the sinusoidal curve. Lateral tilting of

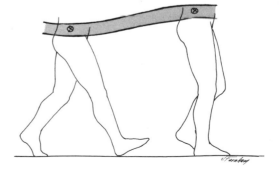

Figure 14–22. The vertical displacement of the body's center of gravity (COG) produces a smooth sinusoidal curve in normal walking. The lowest point in the curve is during the period of double support. The highest point in the curve coincides with midstance when the trunk is directly over the stance extremity. The drawing shows the lowest and highest points in the curve.

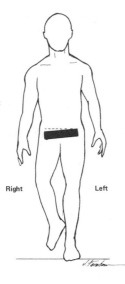

Right   Left

Figure 14–23. Lateral pelvic tilt in the frontal plane keeps the peak of the sinusoidal curve lower than it would have been if the pelvis did not drop, because it produces adduction at the stance hip. Lateral pelvic tilt (drop) to the right is controlled by the left hip abductors.

the pelvis (pelvic drop) on the side of the unsupported extremity (swing leg) keeps the peak of the rise lower than if the pelvis did not drop, because the drop produces adduction of the stance hip (Fig. 14–23). The tilting of the pelvis is controlled by the hip abductor muscles of the stance extremity. For example, pelvic drop on the side of the right swing extremity is controlled by isometric and eccentric contractions of the left hip abductor muscles.

Knee Flexion

Knee flexion at midstance when the COG is at its highest point represents another adjustment that helps to keep the COG from rising as much as it would have to if the body had to pass over a completely extended knee.

Knee, Ankle, and Foot Interactions

Movements at the knee occur in conjunction with movements at the ankle and foot and are responsible for smoothing the pathway of the body's COG so that it forms a sinusoidal curve. Combined knee, ankle, and foot movements prevent abrupt changes in the vertical displacement of the body's COG from a downward to an upward direction. The change from a downward motion of the COG at heel strike to an upward motion at foot flat (loading response) is accomplished by knee flexion, ankle plantarflexion, and foot pronation. These combined motions serve to relatively shorten the extremity and thus prevent an abrupt rise in the body's COG after heel strike. If these motions did not occur in conjunction with each other, the COG would rise abruptly after heel strike as the tibia rides over the talus.

Another instance in which knee, ankle, and foot interactions play an important role is when the body's COG falls after midstance. A combination of ankle plantarflexion, foot supination, and knee extension at heel off slow the descent of the body's COG by a relative lengthening of the stance extremity.

Forward and Backward Rotation of the Pelvis

Forward and backward rotations of the pelvis in the transverse plane accompany forward and backward movements of the lower extremities during gait (Fig.

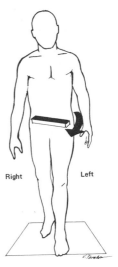

Figure 14–24. Pelvic rotation in the transverse plane. The drawing shows left forward rotation of the pelvis on the side of the swinging extremity. The right hip joint serves as the axis for motion. The bar, which represents the pelvis, shows the apparent backward rotation that is occurring simultaneously on the right side of the pelvis. The pelvic rotation relatively lengthens the extremities and therefore minimizes the drop of the body's COG that occurs at double support.

14–24). Forward rotation occurs on the side of the swinging extremity with the hip joint of the weight-bearing extremity serving as the axis for pelvic rotation. The pelvis begins to move forward at preswing and continues as the swinging extremity moves forward during initial swing. At the point of maximal elevation of the body's COG in midstance, the forward pelvic rotation has brought the pelvis to a neutral position with respect to rotation. Forward rotation of the pelvis continues beyond neutral on the swing side through terminal swing to initial contact.

The total amount of rotation of the pelvis is small and averages about 4° on the swing and stance sides for a total of 8°. The result of pelvic rotation is an apparent lengthening of the lower extremities. The swinging lower extremity is lengthened in terminal swing by the forwardly rotating pelvis, and the weight-bearing extremity is lengthened in preswing by the posterior position of the pelvis. Therefore both the stance and swing extremities are lengthened as the COG descends to its lowest

Figure 14–25. Physiologic valgus at the knee. The normal physiologic valgus at the knee reduces the width of the base of support from what it would be if the lower extremity were aligned vertically. The darkened left lower extremity in the drawing is supposed to represent a hypothetical vertical alignment of the tibia and femur. It is evident from the drawing that the base of support is considerably wider when the leg is aligned vertically than when the leg is normally aligned.

level in the period of double support. The relative lengthening helps to prevent an excessive drop of the COG and maintains the COG at a higher level than would be possible if no pelvic rotation occurred. Pelvic rotation functions to minimize the depression of the COG, while the first two determinants function to minimize the elevation of the COG.

Physiologic Valgus at the Knee

The physiologic valgus at the knee reduces the width of the base of support from what it would be if the femoral and tibial shafts formed a vertical line from the greater tuberosity of the femur (Fig. 14–25). Therefore, because the base of support is relatively narrow, little lateral motion of the body is necessary to shift the COG from one lower extremity to another over the base of support.

ENERGY REQUIREMENTS

The kinematic aspects of human gait have been covered in the preceding sections. In the following sections the reader will be introduced to some of the forces involved in producing gait. Force must be used to produce accelerations and decelerations of the body and its segments. Muscles use metabolic energy to perform mechanical work by converting metabolic energy into mechanical energy. The overall metabolic cost incurred during locomotion may be measured by assessing the body's oxygen consumption per unit of distance traveled.

The main objective of locomotion is to move the body through space with the least expenditure of energy. If a long distance is traveled, but only a small amount of oxygen is consumed, the metabolic cost of that particular gait is low. Oxygen consumption for a person walking at 4 to 5 km/h averages 100 mL per kilogram of body weight per minute. The highest efficiency is attained when the least amount of energy is required to travel a unit of distance. If the speed of walking increases, the energy cost per unit of distance walked increases.[20]

Mechanical Energy

The mechanical energy cost of gait involves assessments of mechanical energy exchanges between various segments of the body. The two types of mechanical energy are kinetic energy and potential energy. **Kinetic energy** has translational and rotational components. **Translational energy** refers to energy related to the linear velocity of a segment in space. **Rotational energy** is due to the rotational velocity of a segment in space. **Potential energy** is the quantity of mass raised, multiplied by the height to which it is raised. In other words, whenever a mass is raised, gravity tends to act on it and make it fall, and therefore the mass has potential energy. The amount of potential energy that an elevated mass possesses is equal to the amount of kinetic energy that was required to lift the mass against gravity. When the mass has stopped elevating or is at its peak, kinetic energy is transformed into potential energy. When the mass falls, the potential energy is transformed back into kinetic energy as the mass accelerates. In human gait, kinetic energy must be expended in order to raise the body's mass, which is concentrated at the body's COG. The higher the COG is raised, the greater the amount of kinetic energy that must be expended. When the COG reaches its highest point at midstance, the body has the greatest potential energy. The downward fall of the body is brought about by potential energy, but kinetic energy is required to control the fall. Since potential energy is transformed into kinetic energy during the fall, kinetic energy is available.

Exchanges between kinetic and potential energy occur throughout the gait cycle. If gait is mechanically efficient, energy is conserved and little more energy than the energy required to initiate movement is needed. If changes in the body's COG are large and abrupt, more energy must be expended; therefore, energy expenditure in human gait is often equated with movements of the body's COG. However, since energy exchanges also take place between the segments of the body during gait, a more accurate assessment of energy exchange involves a segment-by-segment calculation to measure the mechanical efficiency of an individual's gait.[21]

Positive and Negative Work

The muscles of the body supply kinetic energy. When muscles do positive work such as in a concentric contraction, they increase the total energy of the body and transfer energy to the bony components. At a cadence of 105 to 112 steps per minute, a brief burst of positive work (energy generation) occurs as the hip extensors contract concentrically between heel strike and foot flat while the knee extensors perform negative work (energy absorption) by acting eccentrically to control knee flexion during the same period.[22] Negative work is also performed by the plantarflexors as the leg rotates over the foot during the period of stance from foot flat through midstance. However, positive work of the knee extensors occurs during this period to extend the knee following foot flat.[22] Positive work of the plantarflexors and hip flexors in late stance and in early swing increases the energy level of the body. By midswing (which is midstance on the stance side) the potential energy of the body is at a maximum. On the other hand, in late swing negative work is performed by the hip extensors as they work eccentrically to decelerate the leg in preparation for initial contact. At this point there is a decrease in the total energy level of the body.

The positive energy generated by the hip muscles during concentric muscle action for normal men walking at a cadence of 109 steps per minute is approximately double the amount of energy absorbed by the hip muscles during eccentric muscle action.[4] At the ankle the positive energy generated by concentric muscle action during a single gait cycle is almost triple that of the energy absorbed by eccentric muscle action.[4] The knee, in contrast to the hip and ankle, absorbs more energy through eccentric muscle action during a gait cycle than it generates.[4] At slow and normal speeds of walking in healthy subjects, the hip flexors and extensors contribute about 25 percent of the total concentric work. The ankle plantarflexors contribute about 66 percent and the knee extensors contribute about 8 percent.[23] The sum of all of the energy increases over a given period of time results in the net positive work that has been done. Likewise, the sum of all of the energy decreases that have occurred during the same given time period yields the net negative work that has been done. The work required to move the body during gait is the absolute sum of the positive and negative energy changes of the whole body.[24] The total body energy curve gives a measure of the mechanical energy cost per distance traveled.

KINETICS

External and Internal Forces

To gain a better understanding of energy requirements and the role of the determinants during gait, it is necessary to acquire a knowledge of the forces involved. In gait, the external forces acting on the body are inertia, gravity, and the

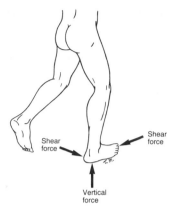

Figure 14–26. The ground reaction force (GRF) represents the force of the ground on the foot. GRFs are transmitted through the foot to the leg and rest of the body.

ground reaction force (GRF).[4] The magnitude and direction of external forces such as the GRFs may be determined through the use of instruments such as force platforms.[25] The inertial force arises from the inertial properties of the body segments. The inertial force is proportional to the acceleration of the segment but it acts in the opposite direction to the acceleration. The force of gravity acts directly downward through the center of mass of each segment. The GRF represents the force of the ground on the foot and it is equal in magnitude and opposite in direction to the force that the body applies to the floor through the foot[26] (Fig. 14–26). The GRF is probably the most important force during locomotor activities such as walking, skipping, hopping, jumping, and running. The GRFs may actually act on many points on the foot but the center of pressure (COP) is the point at which the forces are considered to act just as the COG of the body is designated as the point where the force of gravity is considered to act.

The COP moves along a path during gait and produces a characteristic pattern. The pattern for normal individuals during barefoot walking differs from the patterns using various types of footgear.[27] In barefoot walking the COP starts at the posterolateral edge of the heel at the beginning of the stance phase and moves in a nearly linear fashion through the midfoot area, remaining lateral to the midline, and then moves medially across the ball of the foot with a large concentration along the metatarsal break. The COP then moves to second and first toes during terminal stance (Fig. 14–27).

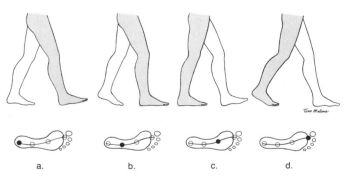

a.   b.   c.   d.

Figure 14–27. A center of pressure (COP) pathway is shown at heel strike in (a), at foot flat in (b), at the end of midstance in (c) and at toe off in (d). The COP path may vary among subjects and may be altered by different footwear.

Internal forces are created primarily by the muscles. The ligaments, tendons, joint capsules, and bony components assist the muscle by resisting, transmitting, and absorbing forces. Muscle activity can be identified by electromyography (EMG), a technique in which the electrical activity generated by an active muscle is recorded. This technique has been used extensively to determine patterns of muscle activity during gait. EMG is often used in conjunction with force plates, cinematography, and electrogoniometry to pinpoint the point in time when muscle activity occurs during the gait cycle. However, the EMG record only provides information about if and when particular muscles are acting. It does not tell why the muscles are acting or how much force the muscles are generating. EMG studies of gait are used to validate theoretical models that attempt to explain why muscles are needed to counteract the forces acting in gait, and theoretical models are developed to explain the muscle activity found by EMG.[6]

Sagittal Plane Analysis

The analyses that are presented in the following section include the location of the GRF relative to the joints of the lower extremities and kinematic data. The relationship of the ground reaction force vector (GRFV) (anterior/posterior and medial-lateral) to the joint axes of the ankle, knee, and hip is used to show the type of moment (flexion/extension, abduction/adduction) that is acting around a joint. The magnitudes of the moments that are determined by measuring the length of the moment arm (MA) (perpendicular distance of a GRFV from a joint axis) are not represented in the illustrations, that is, no attempt was made to make the distance of the GRFV from the joint axis equivalent to the actual length of the MAs. The location of the GRFV, joint positions, and muscle activity that were used to create the illustrations were derived from published studies on normal human walking.[4,5,26,28]

In the chapter on posture the concept of flexion and extension moments was reviewed. In the erect standing posture when the LOG is located at a distance from the joint axis a gravitational moment is created around the joint that threatens to disturb the equilibrium of the forces acting around that joint. To prevent motion at the joint a specific muscle or group of muscles is called into action to oppose the moment, thereby maintaining equilibrium. In a dynamic situation, such as gait, joint movement is necessary and desirable.

> *Example:* If flexion of the knee is necessary during a certain phase of gait and a flexion moment is acting at the knee, the flexion moment is desirable. Muscular activity may be required to control knee flexion. If control is necessary, than an eccentric muscle contraction of the knee extensors is required to control flexion (Fig. 14–28a). If, on the other hand, there is a flexion moment at the knee and knee extension is the desired motion, a concentric contraction of the knee extensors is necessary to oppose the flexion moment and to produce knee extension (Fig. 14–28b).

Stance Phase

The type of muscle activity (eccentric, concentric, or isometric) necessary during gait depends on the nature of the moments acting around the joints of the stance extremity and the direction of the desired motion. If a GRFV moment tends to cause movement of a bony component in a desired direction, muscle function is generally one of control or restraint (eccentric contraction).

> *Example:* In the period of gait initial contact through loading response (heel strike to foot flat), a plantarflexion moment is acting around the

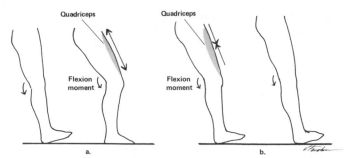

✳ *Figure 14-28.* a. There is a flexion moment at the knee produced by the GRF and knee flexion is the desired motion; therefore, an eccentric contraction of the quadriceps is necessary to control the amount of knee flexion. b. There is a flexion moment acting around the knee produced by the GRF and extension is the desired action; therefore, the quadriceps must work concentrically to counteract the flexion moment and to produce knee extension.

ankle because the resultant GRFV is posterior to the axis of the ankle joint. Plantarflexion is a desired motion in that plantarflexion is necessary to put the foot on the floor in a position to receive the body weight. However, the foot will slap the floor in an uncontrolled fashion unless muscle activity (eccentric) is used to control the plantarflexion (Fig. 14–29).

A different situation exists at the knee joint during midstance. During this period, the knee is extending as it moves from about 15° of flexion at the end of loading response to 5° of flexion at the end of midstance. A flexion moment exists at the knee, and flexion is an undesired motion. A concentric contraction of the knee extensors is necessary to oppose the flexion moment and to produce extension.

Each segment of the stance phase of gait may be examined in a similar manner by using the location of the resultant GRFV relative to a joint axis in order to determine the moments acting around a joint in the sagittal, frontal, and transverse planes. Knowledge of the desired joint motion derived from the kinematic analysis is used in conjunction with the location of the resultant GRFV to determine the type of muscle activity required to produce or control a desired motion. Sometimes the resultant GRFV changes location *within* as well as *between* periods, and as a consequence, muscle activity changes from eccentric to concentric or vice versa. These changes in the type of muscle activity account for the energy changes that occur during gait.

Example: In the period of gait from initial contact to the end of midstance, the ankle moves from the neutral position at initial contact to 15° of plan-

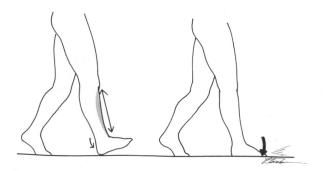

Figure 14-29. In the period of gait from heel strike to foot flat there is a plantarflexion moment acting around the ankle. Plantarflexion is the desired motion that is necessary to position the foot on the supporting surface. The dorsiflexors act eccentrically to control plantarflexion and to prevent the foot from slapping to the ground in an uncontrolled motion.

tarflexion by the end of loading response, and to 10° of dorsiflexion by the end of midstance. The GRFV changes from a location posterior to the ankle joint at initial contact to an anterior position in midstance.

Therefore, at initial contact and during loading response (heel strike to foot flat) there is a plantarflexion moment and the ankle is moving in a direction of plantarflexion. An eccentric contraction of the dorsiflexors controls the motion and negative work is done. The moment acting around the ankle changes at the end of loading response from plantar flexion to dorsiflexion. The dorsiflexion moment continues through midstance while the ankle is dorsiflexing. An eccentric contraction of the plantarflexors is necessary to control the dorsiflexion and prevent the tibia from moving forward too rapidly. The dorsiflexion moment continues until the end of the stance phase. At the end of the stance phase in preswing, plantarflexion is the desired motion. Therefore, activity of the plantarflexors changes from eccentric to concentric. The energy changes for the lower leg during this phase are the sum of the positive and negative energy changes that occurred as a result of the eccentric and concentric muscle activity.

Swing Phase

During the swing phase of gait there are no GRFs because the foot is not in contact with the ground. The swing extremity is moving in an open kinematic chain. Muscle activity is required to accelerate and decelerate the swinging extremity and to lift and hold the extremity up against the force of gravity so that the foot clears the ground and is placed in an optimum position for heel contact. Acceleration is brought about by early concentric activity of the hip flexor and knee extensor muscles. The hip flexors act concentrically to initiate the forward swing of the lower extremity during initial swing. They are inactive through midswing and terminal swing. Deceleration of the swing leg during terminal swing is accomplished primarily by eccentric muscle activity of the hip extensor and knee flexor muscles.

Frontal Plane Analysis

During the early part of the stance phase, HAT is moving forward rapidly and shifting laterally over the stance extremity. The rapid lateral shift of HAT onto the stance extremity creates demands for lateral stability at the hip, knee, and ankle (Fig. 14–30). Muscular support is essential when the body weight is being accepted by the stance extremity, partly because the hip, knee, and ankle are in loose-packed positions. Stabilization of the pelvis at the hip is provided for by activity of the gluteus medius, gluteus minimus, and the tensor fascia latae.[29] The gluteus medius on the stance side controls the lateral drop of the pelvis on the side of the swinging leg. The rapid transfer of weight and demands of single-limb balance create a valgus thrust at the knee and ankle as the body weight is accepted on the extremity.[29] The medial aspect of the knee is given dynamic support by the vastus medialis, semitendinosus, and the gracilis. These muscles counteract the valgus thrust at the knee and thereby prevent an increase in the normal physiologic valgus.[30]

At the ankle and foot, the body weight is transferred from the heel at initial contact along the lateral border of the foot during loading response. At the end of loading response all five metatarsals are weight-bearing. Subsequently, the weight is transferred across the heads of the metatarsals in terminal stance and to the great toe in preswing. The hindfoot bears weight for about 43 percent of the stance phase.[31] Pronation of the foot at the subtalar joint is initiated at heel strike mostly as a result of the heel being loaded lateral to the axis of motion.[32] Subtalar pronation continues during the first 25 percent of the stance phase in response to the acceptance of weight.[33] Pronation of the subtalar joint leaves the transverse joint mobile,

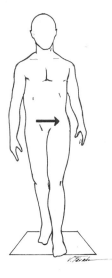

Figure 14–30. The rapid lateral weight shift in early stance creates demands for stability at the hip, knee, and ankle.

and therefore permits the foot to adapt to the supporting surface. The tibialis anterior is the only invertor active at the time of heelstrike that can restrain eversion. During loading response the valgus thrust at the ankle tends to increase the pronation of the foot, and activity of the tibialis posterior muscle is required to control the thrust toward pronation. At approximately 25 percent of the stance phase (midstance), the foot begins supinating again and returns to neutral by the end of midstance.[31]

Pronation of the foot in a weight-bearing posture (closed kinematic chain) produces a medial rotatory force on the tibia while supination produces a lateral rotatory force on the tibia.[32] Just as the position of the foot can cause tibial rotation, tibial rotation can cause a change in the position of the foot. Medial rotation of the tibia in a closed kinematic chain with the foot in a weight-bearing posture produces pronation and lateral tibial rotation causes supination.[34] The tibialis posterior, soleus, and gastrocnemius contract eccentrically to control the pronation that occurs after heelstrike and to control internal rotation of the tibia.[35] At the end of loading response and continuing through the remainder of the stance phase (midstance, terminal stance, preswing) the foot is supinating. By heel off the foot has formed a rigid lever and enhances the pulley action for extrinsic muscles.[36] In preswing the toes are weight-bearing.[34] The only muscles active during the swing phase are the tibialis anterior, extensor digitorum longus, and extensor hallicus longus.

During the middle of the stance phase the demands for medial-lateral stability are somewhat diminished as the valgus thrust decreases. The tensor fascia latae muscle that began activity at loading response continues to provide stabilization of the pelvis during midstance through terminal stance. Activity of the gluteus medius muscle diminishes during midstance and no activity of these muscles is found in preswing once the opposite limb has contacted the ground.[11] The hip adductors (magnus and longus) begin acting in terminal stance and are active eccentrically in preswing to restrain the lateral weight shift onto the opposite extremity. At the knee, the activity of the dynamic stabilizers (semitendinosus, gracilis, and vastus medialis) ceases at midstance as the valgus thrust diminishes. During the last part of stance the weight is being shifted back onto the contralateral extremity, and the hip adductors and the ankle plantarflexors help to control the weight shift. Table 14–2 provides a summary of transverse rotations in the frontal plane for the pelvis, femur, and tibia.

Table 14–2. SUMMARY OF TRANSVERSE ROTATIONS AT PELVIS, FEMUR, AND TIBIA (RIGHT LOWER EXTREMITY)*

	Initial Contact	Loading Response	Midstance	Terminal Stance	Preswing
Percent of stance gait cycle	0%	12%	31%	50%	62%
Pelvis	Left side begins to move forward	Left side moving forward	Neutral	Left side moving forward	Left side moving forward
Femur	Medial rotation	Medial rotation	Lateral rotation	Lateral rotation	Lateral rotation
Tibia	Medial rotation	Medial rotation	Lateral rotation	Lateral rotation	Lateral rotation

	Initial Swing	Midswing	Terminal Swing
Percent of swing gait cycle	75%	82%	100%
Pelvis	Right side moving forward	Neutral	Right side moving forward
Femur	Medial rotation	Medial rotation	Medial rotation
Tibia	Medial rotation	Medial rotation	Medial rotation

*The femur and tibia medially rotate to about 10–20% of stance and then begin laterally rotating to the end of preswing. At initial swing the femur and tibia begin medially rotating again. The degree of transverse rotation that occurs during gait as well as the point in the gait where the rotation occurs varies according to the gait speed. Individual variations also are common.

Table 14–3. SAGITTAL PLANE ANALYSIS (FIG. 14–31)

Heel Strike (Initial Contact) to Foot Flat (End of the Loading Response)

Joint	Motion	Ground Reaction Force	Moment	Muscle	Contraction
Hip	Flexion: 30–25°	Anterior	Flexion	Gluteus maximus Hamstrings Adductor magnus	Isometric to eccentric
Knee	Flexion: 0–15°	Anterior to posterior	Extension to flexion	Quadriceps	Concentric to eccentric
Ankle	Plantarflexion: 0–15°	Posterior	Plantarflexion	Tibialis anterior Extensor digitorum longus Extensor hallucis longus	Eccentric to lower foot to the supporting surface and to control ankle plantarflexion

Frontal Plane Analysis°

Joint	Motion	Muscle Activity
Pelvis	Forwardly rotated position on right side of pelvis at initial contact Left side of pelvis begins to move forward	
Hip	Medial rotation of the femur on the pelvis	
Knee	Valgus thrust with increasing valgus Medial rotation of the tibia	Gracilis, vastus medialis, semitendinosus Long head of biceps femoris to control medial rotation of tibia
Ankle	Valgus thrust with increasing pronation Subtalar joint pronation reaches a maximum at the end of loading response Transverse tarsal pronation	Eccentric contraction of tibialis posterior to control valgus thrust on foot.
Thorax	Right side of thorax is in posterior position at initial contact and begins moving forward	
Shoulder	Right shoulder is slightly behind right hip and moving forward	

°Reference limb is the right lower extremity.

Summary: Lower-Extremity Joint and Muscle Activity

Stance Phase

Tables 14–3 through 14–6 and Figures 14–31 to 14–34 present a summary of joint and muscle activity for one lower extremity during the stance phase of gait. The tables include the joint position in degrees, resultant GRFV, the moment, type of muscle action, and muscle activity (as determined by EMG). The reader should

Table 14–4. SAGITTAL PLANE ANALYSIS (FIG. 14–32)°

Foot Flat (End of Loading Response) to Midstance (End of Midstance)

Joint	Motion	Ground Reaction Force	Moment	Muscle	Contraction
Hip	Extension: 25–0° 20° flexion–0°	Anterior to posterior	Flexion to extension	Gluteus maximus	Concentric to no activity
Knee	Extension: 15–5° 15–5° flexion	Posterior to anterior	Flexion to extension	Quadriceps	Concentric to no activity
Ankle	15° of plantarflexion to 5–10° of dorsiflexion	Posterior to anterior	Plantarflexion to dorsiflexion	Soleus Gastrocnemius Plantarflexors	Eccentric

Frontal Plane Analysis

Joint	Motion	Muscle Activity
Pelvis	Right side rotating backward to reach neutral at midstance. Lateral tilting toward the swinging extremity	Hip abductors are active to prevent excessive lateral tilting (gluteus medius tensor fascia lata)
Hip	Medial rotation of the femur on the pelvis continues to a neutral position at midstance. Adduction moment continues throughout single support	Minimal or no activity
Knee	There is a reduction in valgus thrust and the tibia begins to rotate laterally	Minimal or no activity
Ankle-foot	The foot begins to move in the direction of supination from its pronated position at the end of loading response. The foot reaches a neutral position at midstance	The tibialis posterior helps to produce supination
Thorax	Right side moving forward to neutral	
Thorax	Translating right to neutral	
Shoulder	Moving forward	

°In the period described above, the hip is extending from about 30° of flexion to about 5° of flexion. The line of gravity shifts from anterior to posterior and the moment acting around the hip changes from a flexion to an extension moment. The hip extensors cease their activity at or around midstance because the moment is in the desired direction and momentum appears to be sufficient to carry the body forward. Muscle activity at the hip at midstance is mostly abductor activity to stabilize the pelvis. The quadriceps also ceases its activity as the femur advances over the tibia.

Table 14–5. SAGITTAL PLANE ANALYSIS (FIG. 14–33)°

Midstance (Middle of Midstance) to Heel Off (Prior to End of Terminal Stance)

Joint	Motion	Ground Reaction Force	Moment	Muscle	Contraction
Hip	Extension: 0° of flexion to 10–20° of hyperextension	Posterior	Extension	Hip flexors	Eccentric
Knee	Extension: 5° of flexion to 0°	Posterior to anterior	Flexion to extension	No activity	
Ankle	Plantarflexion: 5° dorsiflexion to 0°	Anterior	Dorsiflexion	Soleus plantarflexors	Eccentric to concentric
Toes (MTP)	Extension: 0–30° of hyperextension			Flexor hallicus longus and brevis Abductor hallicus Abductor digiti quinti Interossei Lumbricals	

Frontal Plane Analysis

Joint	Motion	Muscle Activity
Pelvis	Right side moving posteriorly from neutral position	Minimal or no muscle activity
Hip	Lateral rotation of femur and adduction	Inconsistent hip adductor activity
Knee	Lateral rotation of tibia	No activity
Ankle-foot	Supination of subtalar joint increases	Concentric plantarflexor activity
Thorax	Right side moving forward	
Shoulder	Right shoulder moving forward	

°A small extension moment exists at the hip and knee and no muscle activity is required at the knee to maintain the knee in extension. The momentum of HAT appears to assist in keeping the knee extended. The dorsiflexion moment at the ankle reaches a peak toward the end of this period and the plantar flexors are active to control the tibia and to raise the heel. Toe extension occurs as a result of a closed-chain response to heel rise.

realize that although joint positions and muscle actions are supposed to represent the average of a normal population, these averages may vary among different investigators.

An examination of the moments acting at the lower extremity during the stance phase show that for the majority of the phase the algebraic sum of all extensor (positive) moments and flexor (negative) moments acting at the hip, knee, and ankle is a positive or extensor moment. Winter[6] calls this quantification of the total limb synergy a **support moment** and he has found the extensor support moment to be consistent for all walking speeds for both normal individuals and persons with dis-

Table 14–6. SAGITTAL PLANE ANALYSIS (FIG. 14–34)*

Joint	Motion	Gravity Line	Moment	Muscle	Contraction
			Heel Off (End of Terminal Stance) to Toe Off (End of Preswing)		
Hip	Flexion: 20° of hyperextension to 0°	Posterior	Extension to neutral	Iliopsoas Adductor magnus Adductor longus	Concentric
Knee	Flexion: 0–30° of flexion	Posterior	Flexion	Quadriceps	Eccentric to no activity
Ankle	Plantarflexion: 0–20° of plantarflexion	Anterior	Dorsiflexion	Gastrocnemius Soleus Peroneus brevis Peroneus longus Flexor hallucis longus	Concentric to no activity
Toes (MTP)	Extension: 50–60° of hyperextension			Abductor hallucis Abductor digit quinti Flexor digitorum brevis Flexor hallucis brevis, Interossei Lumbricals	Closed-chain response to increasing plantarflexion at the ankle.

Frontal Plane Analysis

Joint	Motion	Muscle Activity
Pelvis	Left side moving forward until left heel contact (right toe off). Lateral tilting to the swing side ceases as the contralateral extremity enters its stance phase and the period of double support begins	The hip adductors control eccentrically
Hip	Abduction occurs as the weight is shifted onto the opposite extremity. Lateral rotation of femur	
Knee	Inconsistent Lateral rotation tibia	
Foot/Ankle	The weight is shifted to the toes and at toe off only the first toe is in contact with the supporting surface. Supination of subtalar joint	Plantarflexors
Thorax	Translation to the left	
Shoulder	Moving forward	

*Activity in the gastrocnemius and the soleus ceases after the heel leaves the ground. The other plantarflexors cease activity in the order in which they are listed in the sagittal analysis above.

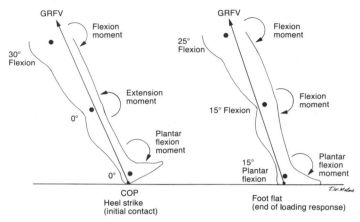

Figure 14–31. The period of gait from heel strike (initial contact) to foot flat (end of loading response). The ground reaction force vectors (GRFV) are indicated by the straight arrows. The curved clockwise arrows indicate flexion moments, and the curved counter clockwise arrows indicate extension moments. Notice that the GRFV changes from a position anterior to the axis of the knee at heel strike to posterior to the axis of the knee at foot flat.

abilities. Hip and knee moments may vary considerably among individuals but the net moment remains an extensor moment. The extensor moment keeps the leg from collapsing during the stance phase. If the knee, hip, or ankle is excessively flexed, a larger-than-normal extensor moment will be generated at another joint so that the net moment remains an extensor moment and the limb is kept from collapsing. The support moment changes from a net extensor to a net flexor moment at late stance (55 to 60 percent of the gait cycle), which continues into early swing. The flexor pattern achieves liftoff, weight shifting, and toe clearance. In late swing a net extensor moment appears again, presumably to assist in the final positioning of the limb for heel contact.[6,37]

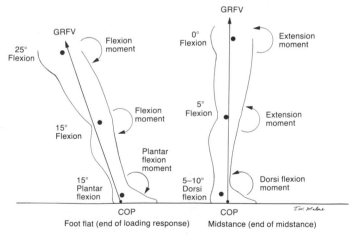

Figure 14–32. The period of gait from foot flat (end of loading response) to midstance (end of midstance).

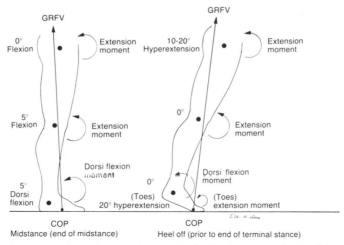

Figure 14–33. The period of gait from midstance (end of midstance) to heel off (prior to end of terminal stance).

Swing Phase

Tables 14–7 and 14–8 and Figures 14–35 and 14–36 present a summary of joint and muscle activity during the swing phase. In the swing phase of gait the primary functions of the swing extremity muscles are to maintain a certain joint position, to accelerate or decelerate the swinging extremity, to ensure toe/foot clearance, and to ensure that the foot is positioned for heel strike.

At the ankle, the tibialis anterior, extensor digitorum longus, and extensor hallicus longus contract concentrically to move the foot from the plantarflexed position at toe off to a position of neutral in midswing. These muscles then act isometrically to maintain the ankle in a neutral position throughout the swing phase.[32]

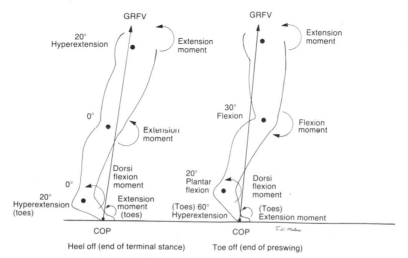

Figure 14–34. The period of gait from heel off (end of terminal stance) to toe off (end of preswing).

Table 14–7. SAGITTAL PLANE ANALYSIS (FIG. 14–35)°

Acceleration (Initial Swing through Midswing)			
Joint	Motion	Muscle	Contraction
Hip	Flexion: 0–20° of flexion to 30° of flexion	Iliopsoas Gracilis Sartorius	Concentric
Knee	Flexion: 30° of flexion to 60° of flexion Extension: 60° of flexion to 30° flexion	Biceps femoris Sartorius Gracilis	Concentric
Ankle	Dorsiflexion: 20° plantarflexion to neutral	Tibialis anterior Extensor digitorum longus Extensor hallicus longus	Concentric

Frontal Plane Analysis			
Joint	Motion	Muscle	Contraction
Pelvis	Lateral pelvic tilt to the right (right side drops). Right side moving forward	Left gluteus medius	
Hip	Rotation from lateral to medial rotation		
Knee	From lateral to medial rotation		
Foot/Ankle	Unweighted subtalar joint returns to slight supination		
Thorax	Right side moving posteriorly		
Shoulder	Right side moving posteriorly		

°Concentric contractions of the hip flexors occur to ensure adequate hip flexion to bring the extremity forward and accomplish foot clearance. The femur and tibia attain the greatest amount of lateral rotation at the beginning of initial swing and then begin to rotate medially.

The knee continues to flex after toe off and reaches a maximum flexion angle of 60° at the end of initial swing. At midswing the knee is in about 30° of flexion and by terminal swing the knee is fully extended. Quadriceps femoris activity, which began as an eccentric contraction in preswing in order to control flexion at the knee, changes after maximum flexion is attained to a brief concentric contraction that initiates forward acceleration of the tibia. Momentum then appears to carry the tibia forward. In the terminal swing period the hamstrings contract eccentrically to control the forward motion of the lower extremity. At the very end of the phase (terminal swing) the knee extensors contract to stabilize the knee in extension, in preparation for heel strike.

The hip is moving from neutral at toe off to about 20° of flexion in early swing. The hip flexes to about 30° of flexion by the end of midswing and is maintained in this position until the end of the swing phase. The hip flexors, which were active to control hip extension at toe off, contract concentrically to initiate swing. The flexors show no activity through midswing and terminal swing. However, the adductor magnus and adductor longus may act to maintain hip flexion, in addition to their

Table 14–8. SAGITTAL PLANE ANALYSIS (FIG. 14–36)*

Midswing Through Deceleration (Terminal Swing)

Joint	Motion	Muscle	Contraction
Hip	Hip remains at 30° flexion	Gluteus maximus	Eccentric
Knee	Extension: 30° flexion to 0°	Quadriceps	Concentric
		Hamstrings	Eccentric
Ankle	Ankle remains in neutral	Tibialis anterior	Isometric
		Extensor digitorum longus	Isometric
		Extensor hallicus longus	Isometric

Frontal Plane Analysis

Joint	Motion	Muscle Activity
Pelvis	Right side moving anteriorly	
Hip	Lateral tilting to the left medial rotation	Right gluteus medius
Knee	Medial rotation	
Ankle		
Thorax	Right side moving posteriorly	
Shoulder	Right shoulder moving posteriorly	

*The hip remains in a position of 30° flexion while the pelvis rotates forward to increase step length. The momentum of the swinging extremity is restrained by eccentric contraction of the hamstrings while full knee extension is ensured by a brief concentric contraction of the quadriceps. The ankle is maintained in a neutral position to ensure ground clearance and readiness for initial contact. The medial rotation of the thigh and tibia continue through terminal swing into the first part of the stance phase.

function of keeping the extremity near the midline. During terminal swing the hamstrings contract eccentrically to control the forward progression of the lower extremity.

The motor strategies that are employed by healthy individuals to oppose the moments show a considerable amount of variability among individuals and even in the same individual from one bout of walking to another. This variability occurs even though the kinematics of gait and total limb synergies (support moments)

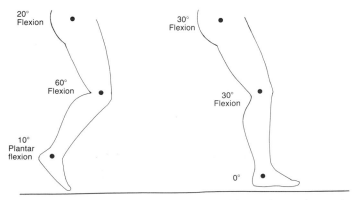

Figure 14–35. The period of gait from acceleration (initial swing) to midswing (midswing).

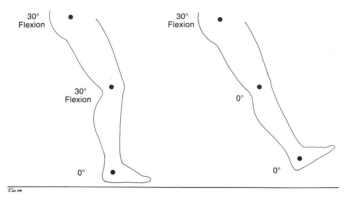

Figure 14–36. The period of gait from midswing (midswing) through terminal deceleration (terminal swing).

appear to be similar.[38] The variety of motor patterns used suggest that peripheral feedback from joint receptors has a strong influence on gait. Input from the visual system that provides information about obstacles in the walking path also appears to affect muscle activity during gait. The fact that muscle activity in the hip flexors and ankle dorsiflexors appears in advance of the appropriate movement may indicate the existence of a feedforward or preparatory system of control that is based on input from the visual system.[6]

KINEMATICS AND KINETICS OF THE TRUNK AND UPPER EXTREMITIES

Trunk

The trunk remains essentially in the erect position during normal walking. In faster walking the trunk slightly inclines forward. Rotation of the trunk is slight and occurs primarily in a direction opposite to the direction of pelvic rotation (Fig. 14–37). As the pelvis rotates forward with the swinging lower extremity, the thorax on the opposite side rotates forward as well. This trunk motion helps to prevent excess body motion and to counterbalance rotation of the pelvis. However, Stokes[39] in a study of treadmill walking found that the movements and interactions of the trunk and pelvis were extremely complex when translatory and rotatory movements of the trunk were considered along with anterior and posterior pelvic tilting, lateral pelvic tilting, and rotation.[39] Although relatively few EMG studies have been done on the trunk muscles during gait, it has been shown that the erector spinae exhibit two periods of activity. The first burst of activity occurs at heel strike, while the second occurs at toe off. Supposedly the erector spinae are active in order to prevent the trunk from falling forward owing to the flexion moment at the hip that is present at each of the bursts of activity. Other muscles that have been found to be active are the quadratus lumborum and the rectus abdominis, although there appear to be conflicting opinions among investigators regarding the activity of these muscles.

Upper Extremities

While the lower extremities are moving alternately forward and backward, the arms are swinging rhythmically. However, the arm swinging is opposite to that of the legs and pelvis but similar to that of the trunk (Fig. 14–38). The right arm swings forward with the forward swing of the left lower extremity, while the left

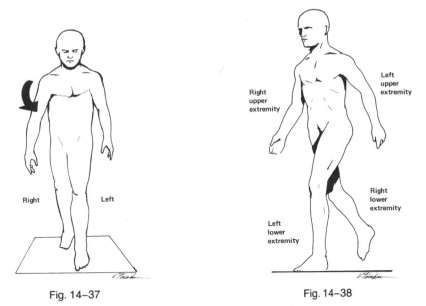

Fig. 14–37   Fig. 14–38

Figure 14–37. Trunk rotation in normal gait is slight and occurs in the opposite direction of pelvic rotation. The right side of the pelvis is rotating in a posterior direction, while the right side of the trunk is rotating in an anterior direction.

Figure 14–38. Arm swinging in gait is opposite to lower extremity motion. The right upper extremity swings forward at the same time that the right lower extremity is moving posteriorly. The right upper extremity and left lower extremity are both moving forward at the same time.

arm swings backward. This swinging of the arms provides a counterbalancing action to the forward swinging of the leg and helps to decelerate rotation of the body, which is imparted to it by the rotating pelvis. The total ROM at the shoulder is not very large. At normal free velocities the ROM is only approximately 30° (24° extension and 6° of flexion).

The normal shoulder motion is the result of the combined effects of gravity and muscle activity. During the *forward* portion of arm swinging, the following medial rotators are active: subscapularis, teres major, and latissimus dorsi.[41] In *backward* swing the middle and posterior deltoid are active throughout, while the latissimus dorsi and teres major are active only during the first portion of backward swing.[40] The supraspinatus, trapezius,[40] and posterior and middle deltoid[41] are active in both backward and forward swing. It is interesting to note that little or no activity was reported in the shoulder flexors in these studies.[40,41] It would appear that during forward swing the medial rotators are acting eccentrically to control external rotation of the arm at the shoulder, while the posterior deltoid may be acting eccentrically to restrain the forward swing. The latissimus dorsi and teres major as well as the posterior deltoid may then act concentrically to produce the backward swing. The role of the middle deltoid is unclear, although it has been suggested that it functions to keep the arm abducted so that it may clear the side of the body.[41] Activity in all muscles increases as the speed of gait increases.[41]

STAIR AND RUNNING GAITS

Stair Gait

Ascending and descending stairs are common forms of locomotion that are required for performing normal activities of daily living such as shopping, using

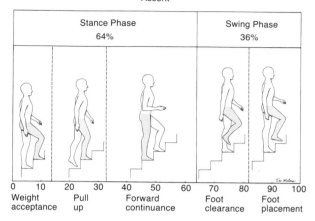

Stair Gait Cycle
Ascent

Figure 14–39. Stair gait.

public transportation, or simply getting around in a multistory home or building. Although many similarities exist between level-ground locomotion and stair locomotion, the difference between the two modes of locomotion may be significant for a patient population. The fact that a patient has adequate muscle strength and joint ROM for level walking does not ensure that the patient will be able to walk up and down stairs.

Locomotion on stairs is similar to level-ground walking in that stair gait involves both swing and stance phases in which forward progression of the body is brought about by alternating movements of the lower extremities. Also in both stair and level gait the lower extremities must balance and carry along HAT. McFayden and Winter[42] (using step dimensions of 22 cm for the stair riser and 28 cm for the tread) performed a sagittal plane analysis of stair gait. These investigators collected kinetic and kinematic data for one subject during eight trials. The stair gait cycle for stair ascent presented in Fig. 14–39 is based on data from McFayden and Winter's study.

The investigators divided the stance phase of the stair gait cycle into the three subphases and the swing phase into two subphases. The subdivisions of the stance phase are weight acceptance (WA), pull up (PU), and forward continuance (FCN). The subdivisions of the swing phase are foot clearance (FCO) and foot placement (FP). As can be seen in Figure 14–39, WA comprises approximately the first 14 percent of the gait cycle and is somewhat comparable to the heel strike through loading phase of walking gait. However, in contrast to walking gait, the point of initial contact of the foot on the stairs usually is located on the anterior portion of the foot and travels posteriorly to the middle of the foot as the weight of the body is accepted. The PU portion, which extends from approximately 14 to 32 percent of the gait cycle, is a period of single-limb support. The initial portion of PU is a time of instability, as all of the body weight is shifted onto the stance extremity when it is flexed at the hip, knee, and ankle. During this period the task is to pull the weight of the body up to the next stair level. The knee extensors are responsible for most of the energy generation required to accomplish pull up. The FCN period is from approximately 32 to 64 percent of the gait cycle and corresponds roughly to the midstance through toe-off subdivisions of walking gait. In the FCN period the ankle plantarflexors exhibit the greatest amount of energy generation.

Some of the data regarding joint ROM and muscle activity for ascending stairs that was collected by McFayden and Winter[42] is presented in Tables 14–9 to 14–11. A review of the tables demonstrates differences between level gait and stair

Table 14-9. SAGITTAL PLANE ANALYSIS OF STAIR ASCENT (FIG. 14-39)

Stance Phase—Weight Acceptance (0–14% of Stance Phase) Through Pull-Up (14–32% of Stance Phase)

Joint	Motion	Muscle	Contraction
Hip	Extension: 60–30° of flexion	Gluteus maximum Semitendinosus Gluteus medius	Concentric
Knee	Extension: 80–35° of flexion	Vastus lateralis Rectus femoris	Concentric
Ankle	Dorsiflexion: 20–25° of dorsiflexion	Tibialis anterior	Concentric
	Plantarflexion: 25–15° of dorsiflexion	Soleus Gastrocnemius	Concentric

Table 14-10. SAGITTAL PLANE ANALYSIS OF STAIR ASCENT (FIG. 14-39)

Stance Phase—Pull-Up (End of Pull-Up) Through Forward Continuance (32–64% of the Stance Phase of Gait Cycle)

Joint	Motion	Muscle	Contraction
Hip	Extension: 30–5° flexion	Gluteus maximus Gluteus medius Semitendinosus	Concentric and isometric
	Flexion: 5 to 10–20° of flexion	Gluteus maximum Gluteus medius	Eccentric
Knee	Extension: 35–10 ° of flexion	Vastus lateralis Rectus femoris	Concentric
	Flexion: 5 to 10–20° of flexion	Rectus femoris Vastus lateralis	Eccentric
Ankle	Plantarflexion: 15° of dorsiflexion to 15–10° of plantarflexion	Soleus Gastrocnemius	Concentric
		Tibialis anterior	Eccentric

Table 14-11. SAGITTAL PLANE ANALYSIS OF STAIR ASCENT (FIG. 14-39)

Swing Phase (64–100% of Gait Cycle)—Foot Clearance Through Foot Placement

Joint	Motion	Muscle	Contraction
Hip	Flexion: 10–20° to 40–60° of flexion Extension: 40–60° of flexion to 50° of flexion	Gluteus medius	Concentric
Knee	Flexion: 10° of flexion to 90–100° of flexion Extension: 90–100° of flexion to 85° of flexion	Semitendinosus Vastus lateralis Rectus femoris	Concentric Concentric
Ankle	Dorsiflexion: 10° of plantarflexion to 20° of dorsiflexion	Tibialis anterior	Concentric and isometric

gait in regard to joint ROM as well as some differences in the muscle activity required.

    Example: In Table 14–10 one can observe that considerably more hip and knee flexion are required in the initial portion of stair gait than are required in normal level-ground walking. Therefore a patient would require a greater ROM for stair climbing (the same stair dimensions and slope) than they would for normal level-ground walking. Naturally muscle activity and joint ROMs will change if stairs of other dimensions than the ones investigated by McFayden and Winter are used.

    Ascending stairs involves a large amount of positive work that is accomplished mainly through concentric action of the rectus femoris, vastus lateralis, soleus, and medial gastrocnemius. Descending stairs is achieved mostly through eccentric activity of the same muscles and involves energy absorption. The support moments during stair ascent, descent, and level walking exhibit similar patterns; however, the magnitude of the moments is greater in stair gait and consequently more muscle strength is required.

Running Gait

    Running is another locomotor activity that is similar to walking, but there are differences that need to be examined. As in the case of stair gait, a patient who is able to walk on level ground may not have the ability to run. Running requires greater balance, muscle strength, and ROM than normal walking. Greater balance is required because running is characterized not only by an absence of the double support periods observed in normal walking but also by the presence of float periods in which both feet are out of contact with the supporting surface (Fig. 14–40). The walking gait cycle presented in Figure 14–41 can be used to compare the gait cycle in walking and running gait. The percentage of the gait cycle spent in float periods will increase as the speed of running increases. Muscles must generate greater energy both to raise HAT higher than in normal walking and to balance and support HAT during the gait cycle. Muscles and joint structures also must be able to absorb more energy to accept and control the weight of HAT.

    For example, in normal level walking the magnitude of the GRFs at the COP in heel strike are approximately 70 to 80 percent of body weight and rarely exceed

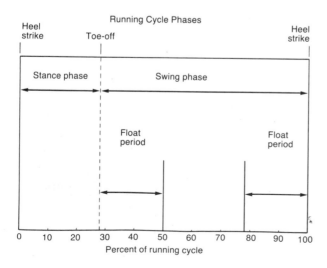

Figure 14–40. Running gait cycle.

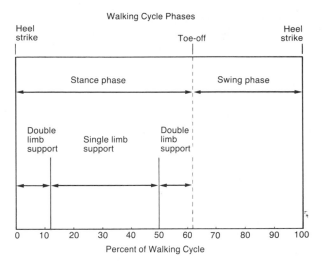

Figure 14-41. Walking gait cycle.

120 percent of body weight during the gait cycle.[32,43] However, during running, the GRFs at the COP have been shown to reach 200 percent of body weight and increase to 250 percent of body weight during the running cycle.

JOINT MOTION AND MUSCLE ACTIVITY IN RUNNING GAIT

Joint Motion

At the beginning of the stance phase of running the hip is in about 50° of flexion at heel strike and extends during the remainder of the stance phase until it reaches about 10° of hyperextension just after toe off. The hip then flexes to reach about 55° of flexion in late swing. Just prior to the end of the swing phase the hip extends slightly to 50° in preparation for heel strike.[32]

The knee is flexed to about 40° at heel strike and continues to flex to 60° during the loading response. Thereafter, the knee begins to extend, reaching 40° of flexion just prior to toe off. During the swing phase and initial float period the knee flexes to reach a maximum of approximately 125° in the middle of the swing phase. In late swing the knee extends to 40° in preparation for heel strike.[32]

The ankle is in about 10° of dorsiflexion at heel strike and rapidly dorsiflexes to reach about 25° dorsiflexion. The rapid dorsiflexion is followed immediately by plantarflexion, which continues throughout the remainder of the stance phase and into the initial part of the swing phase. Plantarflexion reaches a maximum of 25° in the first few seconds of the swing phase. Throughout the rest of the swing phase the ankle dorsiflexes to reach about 10° in late swing in preparation for heel strike.[32]

The reference extremity beings to medially rotate during the swing phase. At heel strike the extremity continues to medially rotate and the foot pronates. Lateral rotation of the stance extremity and supination of the foot begins as the swing leg passes the stance limb in midstance.

Muscle Activity

The gluteus maximus and gluteus medius are active both at the beginning of the stance phase and at the end of the swing phase. The tensor fascia lata is also active at the beginning of stance and at the end of swing but also is active between early and midswing. The adductor magnus shows activity for about 25 percent of the gait cycle from late stance through the early part of the swing phase. Activity in the iliopsoas occurs for about the same percentage of the gait cycle as the adductor longus, but iliopsoas activity also occurs during the swing phase from about 35 to 60 percent of the gait cycle.

The quadriceps muscle acts eccentrically during the first 10 percent of the stance phase to control knee flexion when the knee is flexing rapidly. The quadriceps ceases activity after the first part of stance and no activity occurs until the last 20 percent of the swing phase when concentric activity begins to extend the knee (to 40° of flexion) in preparation for heel strike.

The medial hamstrings are active at the beginning of stance and through a large part of swing. For example, the medial hamstrings are active from 18 percent to 28 percent of the stance phase, from about 40 to 58 percent of initial swing and for the last 20 percent of swing. During part of this time the knee is flexing and the hip is extending and the hamstrings may be acting to extend the hip and to control the knee. During initial swing the hamstrings are probably acting concentrically at the knee to produce knee flexion, which reaches a maximum at midswing. In late swing the hamstrings may be contracting eccentrically to control knee extension and to re-extend the hip.

A comparison of walking and running muscle activity at the ankle shows that in walking, gastrocnemius muscle activity begins just after the loading response at about 15 percent of the gait cycle and is active to about 50 percent of the gait cycle (just prior to toe off). In running, gastrocnemius muscle activity begins at heel strike and continues through the first 15 percent of the gait cycle ending at the point where activity begins in walking. The gastrocnemius becomes active again during the last 15 percent of swing.

The tibialis anterior muscle activity occurs in both stance and swing phases in walking and running. However, the total period of activity of this muscle in walking (54 percent of the gait cycle) is less than it is in running where is shows activity for about 73 percent of the gait cycle. The difference in activity of the tibialis anterior between walking and running is due partly to the differences in the length of the swing phases in the two types of gait. The swing period in walking gait is approximately 40 percent of the total gait cycle, whereas in running gait the swing phase constitutes about 62 percent of the total gait cycle. Most of the activity in the tibialis anterior during both walking and running gait is concentric or isometric action that is necessary to clear the foot in the swing phase of gait. The longer swing phase in running accounts for at least part of the difference in tibialis anterior activity between walking and running gaits. Tibialis anterior activity in the first half of the stance phase in running gait accounts for the remainder of the difference in activity in this muscle.

Summary

As a result of the efforts of many investigators, our present body of knowledge regarding human locomotion is extensive. However, gait is a very complex subject and further research is necessary to standardize methods of measuring and defining kinematic and kinetic variables, to develop inexpensive and reliable methods of analyzing gait in the clinical setting, and to augment the limited amount of knowledge available regarding kinematic and kinetic variables in children's and the elderly's gaits.

Standardization of equipment and methods used to quantify gait variables, as well as standardization of the terms used to describe these variables, would help to eliminate some of the present confusion in the literature and make it possible to compare the findings of various researchers with some degree of accuracy.

At the present time inexpensive, quantitative, and reliable clinical methods of evaluating gait are limited to time and distance variables of step length, step duration, stride length, cadence, and velocity.[44-47] These measures provide a simple means for objective assessment of a patient's status. Increases in step length and decreases in step duration may be used to document a patient's progression toward a more normal gait pattern; however, a normal gait pattern may not be appropriate for many patients. Instead a pattern that is appropriate for a patient's particular disability may be a more appropriate goal of treatment.

EFFECTS OF AGE, DISEASE, INJURY, AND MALALIGNMENT

Although information about the norms and relationships among time and distance variables are available for adult gait, little investigation of time and distance variables has been undertaken for young children. The toddler walks with a wider base of support, a decreased single-leg-support time, a shorter step length, a slower velocity, and a higher cadence in comparison to normal adult gait. A study of 3- and 5-year-old children showed that some relationships between these variables were similar to the relationships found in adult gaits.[48]

For example, as a group, the 3- and 5-year-olds showed significant increases in stride length adjusted for leg length, step length, and cadence from a slow to a free and from a free to a fast speed of gait. However, 5-year-olds differed from 3-year-olds in that they had less variability in step length adjusted for height at slow and free speeds.[48] In another study, which included children from 6 to 13 years of age, Foley and associates[49] reported that the ROM for flexion and extension of the joints of the lower extremities were almost identical to the values obtained for adults.[49] However, linear displacements, velocities, and accelerations were found to be consistently larger for these children than they were for adults.[49] Cadence, stride length, stride time, and other distance and temporal variables have been found to show variability until the child reaches 7 or 8 years of age. A gait pattern that is similar to normal adult gait is demonstrated by children from 8 to 10 years of age.[36] Many more studies need to be conducted on children's gaits to provide evaluators of human function with norms for children's gaits at different stages of development.

Effects of Age

The effects of age on gait are still being investigated.[37,50-52] The use of different age groups and levels of activity (sedentary versus active groups) among investigators has made it difficult to draw definitive conclusions about the effects of normal aging. Some investigators have found that the elderly in comparison to younger groups demonstrate a decrease in natural walking speed, shorter stride and step lengths, longer duration of double-support periods, and smaller swing to support phase ratios.[37,50,51] Hinmann and associates[50] in a study of 289 males and 149 females from 19 to 102 years of age found that between 19 and 62 years of age there was a 2.5 to 4.5 percent decline in the normal speed of walking per decade for males and females, respectively. After age 62 there was an accelerated decline in normal walking speed, that is, a 16 and 12 percent decline in walking speed for males and females, respectively.[50] Finley, Cody, and Finizie[51] in a study that compared 64- to 84-year-old women with 19- to 38-year-old women found that the older women showed increased muscle activity and a shorter step length than the younger group, but no change in relative joint motion. Winter and associates[37] compared the fit and healthy elderly with young adults and found that the natural cadence in the elderly was no different than young adults but that the stride length of the elderly was significantly shorter than young adults and that the period of double support was longer in the elderly than in young adults. Blanke and Hageman[52] on the other hand compared 12 young men between the ages of 20 and 32 years of age with 12 elderly men between the ages of 60 and 74 years of age and found no effects of aging in regard to step length, stride length, velocity, and vertical and horizontal excursions of the body's COG.[52]

Several investigators have described changes in stride length and speed of gait in the elderly. These changes may represent an attempt to make gait more stable. Falls in the elderly population are common and many of the elderly lead relatively sedentary life-styles. The inactive elderly may have some muscle atrophy due to disuse and thus be more unsure of themselves while walking. Also, the possibility exists that some of the changes in gait that have been attributed to the aging process may actually be related more to the health and physical fitness status of the individual than to his or her age.

Disease States

Although quantitative evaluation of gait using time and distance measures is being promoted for use by evaluators of human function, qualitative evaluations are useful and should be used in conjunction with the quantitative assessments. An individual's gait pattern may reflect not only his or her physical or psychologic status but also any defects or injuries in the joints or muscles of the lower extremities. Certain disease conditions such as **Parkinson's disease** produce characteristic gaits that are easily recognized by a trained observer.

The parkinsonian gait is characterized by an increased cadence, shortened stride, lack of heel strike and toe off, and diminished arm swinging. The muscle rigidity that characterizes this disease prevents normal reciprocal patterns of movement.

Another gait pattern that results from disturbed neurologic functioning is the **ataxic gait.** In this gait abnormal function of the cerebellum results in a disturbance of the normal mechanisms controlling balance, and therefore, the individual walks with an unusually large base of support. The wider-than-normal base of support creates a larger-than-normal side-to-side deviation of the COG and subsequent changes in other gait parameters.

Muscle Weakness or Paralysis

Sometimes an isolated weakness or paralysis of a single muscle will produce a characteristic gait. For example, a unilateral paralysis of the gluteus medius results in a typical gait pattern called a **gluteus medius gait.** The characteristics of this gait pattern can be deduced by reviewing the function of the gluteus medius during normal gait. Normally the gluteus medius functions to stabilize the hip and pelvis by controlling the drop of the pelvis during single-limb support, especially during the first part of the stance phase. If gluteus medius activity on the side of the stance leg is absent, the pelvis accompanied by the trunk will fall excessively on the swing side resulting in a loss of balance. To prevent the trunk and pelvis from falling to the unsupported side and to maintain HAT over the stance leg, the individual compensates by laterally bending the trunk over the stance leg. The trunk motion enables the person to maintain his or her balance by keeping the COG over the base of support and allows the swing leg to be lifted high enough to clear the ground.

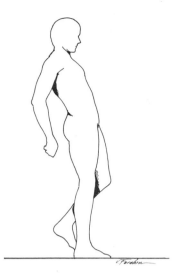

Figure 14–42. A backward lean of the trunk is used to compensate for paralysis of the gluteus maximus.

The trunk motion reduces the MA of gravity, thus reducing the need for hip abductor contraction and the concomitant compression caused by the hip abductors. The lateral trunk lean characterizes the gluteus medius gait. The use of an assistive device such as a cane on the side opposite to the paralyzed muscle reduces the need for the lateral trunk lean. The use of a cane decreases the energy required in a gluteus medius gait, but increases the energy requirements of ambulation above that of normal gait.

The gluteus maximus in normal gait provides for stability in the sagittal plane and for restraint of forward progression. This muscle helps to counteract the flexion moment at the hip in the early part of stance and restrains the forward movement of the femur in late swing in normal gait. When the gluteus maximus is paralyzed, the trunk must be thrown posteriorly at heel strike, in order to prevent the trunk from falling forward when there is a flexion moment at the hip. The backward lean is typical of a **gluteus maximus gait** (Fig. 14–42).

The quadriceps is needed during gait at initial contact and loading response when there is a flexion moment acting at the knee. Quadriceps paralysis is easily compensated for if a person has normal hip extensors and plantarflexors. The gluteus maximus and soleus muscles pull the femur and tibia, respectively, posteriorly, which results in knee extension. Additional compensation usually is accomplished by forward trunk bending and a rapid plantarflexion after initial contact. The forward shifting of the weight creates an extension moment at the knee (at initial contact and during the loading reponse period). It also places the knee in hyperextension and eliminates the need for quadriceps activity. If both the quadriceps and the gluteus maximus are paralyzed, a person may compensate for the loss by pushing the femur posteriorly with his or her hand at initial contact. The arm supports the trunk; it prevents hip flexion and also thrusts the knee into extension.

A paralysis of the plantarflexors (gastrocnemius, soleus, flexor digitorum longus, tibialis posterior, plantaris, and flexor hallucis longus) results in a **calcaneal gait** pattern.[30] This pattern is characterized by greater than normal amounts of ankle dorsiflexion and knee flexion during stance and a less-than-normal step length on the affected side. The abnormal amount of knee flexion and the fact that the soleus is not pulling the knee into extension require an abnormal amount of quadriceps activity to stabilize the knee during the stance phase. The period of single-limb support is shortened because of the difficulty of stabilizing the tibia and the knee. Step length is shorter than normal because the normal push-off segment of gait is eliminated. The normal heel off and progression to toe off are changed into a rather abrupt lift-off of the entire foot. The asymmetry of this type of gait pattern is obvious through observation and a comparison of right and left step lengths.

Asymmetries of the Lower Extremities

Asymmetries of the lower extremities may be caused by muscle paralysis, contractures of soft tissues around the joints, bony ankylosis, or developmental abnormalities. Any one or a combination of these conditions may cause either a relative or actual shortening of one extremity in comparison with the other. For example, a knee flexion contracture will cause a shortening of the affected extremity. When the affected extremity is weight-bearing, the normal extremity will be proportionately too long to swing through in a normal fashion. A method of equalizing leg lengths is necessary in order for the swing leg to swing through without hitting the floor. Apparent shortening of the too long normal extremity may be accomplished by a variety of methods. One method of shortening the normal extremity is by increasing the amount of flexion at the hip, knee, and ankle beyond what would normally be required. Other methods that produce relative shortening of the swinging leg are hip hiking (Fig. 14–43) or cicumducting the leg (Fig. 14–44). Each of these compensations makes it possible to walk, but they increase the energy requirements above normal levels.

In contrast to shortening of the normal extremity to equalize leg lengths, the person may compensate by using other parts of the body to lengthen the affected extremity. Plantarflexing the foot during stance serves to lengthen the stance extremity as does increasing the amount of pelvic rotation or pelvic tilt during

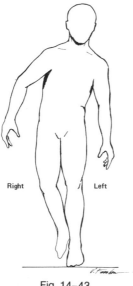

Right   Left

Fig. 14–43

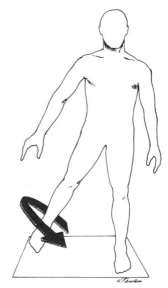

Fig. 14–44.

Figure 14–43. Hiking of the right hip during the swing phase of the right lower extremity effectively shortens the right lower extremity.

Figure 14–44. Circumduction of the right lower extremity during the right swing phase serves as a method of shortening the swing extremity, especially if knee flexion is impossible.

swing. The consequences of either muscle loss or a loss of ROM may be determined by using the model presented in Chapter 13 on posture.

Example: Paralysis of dorsiflexors. The normal function of the dorsiflexors in gait is (1) to maintain the ankle in neutral so that the heel strikes the floor at initial contact; (2) to control the plantarflexion moment at heel strike; (3) to dorsiflex the foot in initial swing; and (4) to maintain the ankle in dorsiflexion during midswing and terminal swing.

If these functions are absent one would expect that the following would occur: (1) the entire foot or the toes would strike the floor at initial contact; (2) entry into the loading response phase would be abrupt; (3) the amount of flexion at the hip and knee would have to increase in order to clear the foot in initial swing; and (4) a method of either shortening the swing leg or lengthening the stance would have to be found to clear the plantarflexed joint. (See Table 14–12.)

Table 14–12. EFFECTS OF MUSCLE PARALYSIS

Muscle	Normal Function	Effects on Gait	Possible Compensations
Dorsiflexors	1. Maintain ankle in dorsiflexion in midswing, terminal swing and at heel strike.	1. Functional lengthening of affected extremity. Toe drag during swing and lack of normal heel strike.	1. During the swing phase a functional shortening of the affected extremity can be produced by increased knee and hip flexion to prevent toe drag.
			2. During midstance a functional lengthening of the unaffected extremity can be produced by plantarflexion of the unaffected extremity.
	2. Controls plantarflexion from heel strike to foot flat.	2. Lack of control of plantarflexion from heel strike to foot flat.	2. From heel strike to foot flat a toes first position of foot at heel strike eliminates the need for dorsiflexor control.
Quadriceps	1. Helps to position leg at heel strike by maintaining knee in extension.	1. Instability at heel strike.	1. Foot flat at initial contact.
	2. Shock absorption and stability during loading response.	2. Decrease in shock absorption and stability during loading response.	2. From heel strike to foot flat an increase in trunk flexion can help to keep an extension moment at the knee.

The human body is remarkable in its ability to compensate for losses or disturbances in function. Most of the compensations that are made are performed unconsciously, and if the disturbance is slight, such as flat foot, the individual may not be aware that his or her gait pattern is in any way unusual. However, every compensation usually results in an increase in energy expenditure over the optimal and may result in excessive stress on other structures of the body.

In running, stresses are greater than in walking, so there is an accompanying increase in the likelihood of injury. In a survey of the records of 1650 running patients between the years 1978 and 1980, 1819 injuries were identified.[52] The knee was the most commonly injured site, and patellofemoral pain was the most common complaint. Increases in the Q angle, tibial torsion, and pronation of the foot are contributing causes to patellofemoral syndromes. Other injuries experienced by runners are iliotibial band syndrome and popliteal tendonitis.[54] Plantar fasciitis caused by repetitive stretching of the planter fascia between its origin at the plantar rim of the calcaneus and its insertion into the metatarsal heads is a common overuse syndrome seen in young athletes. Structural variations at any joint in the kinematic chain may alter normal gait patterns. At the hip, coxa valga, coxa vara, retroversion, or anteversion all affect gait. At the knee, genu varum, genu valgum, or genu recurvatum, and patella alta cause abnormal stresses in walking that may be magnified in running.

Coxa valga may cause alternations at the knee such as genu varum and problems at the ankle such as excessive inversion. In coxa valga abnormal weight-bearing stresses are incurred on the superior medial aspect of the femoral head. Abnormal compressive stress occurs on the medial aspect of the knee joint in genu varum. Abnormal weight-bearing stress occurs on the lateral borders of the feet in excessive inversion. These changes throughout the kinematic chain cause abnormalities in gait.

Coxa vara may lead to changes at the knee and ankle, that is, genu valgum and excessive eversion at the feet. Abnormal weight bearing in this instance would occur on the superior lateral aspect of the femoral head and excessive shearing forces would be present on the head and neck. In addition, abnormal tensile stresses would occur on the medial knee structures in genu valgum and excessive weight-bearing stresses on the medial aspects of the feet in eversion of the feet. In genu valgum the width of the base of support is considerably wider than in normal gait.

An anteverted hip may cause excessive "toeing in" during gait because of the abnormal amount of medial femoral rotation present in this condition. Conversely, a retroverted hip may cause excessive "out-toeing" during walking, because of the abnormal amount of lateral femoral rotation associated with this abnormality.

At the ankle joint, a surgical arthrodesis that fuses the trochlea of the talus in the mortise results in the imposition of greater-than-normal forces on the foot. When the plantarflexors, which are the major source of mechanical energy generation in gait, are affected, muscles at other joints must provide more energy than in normal gait.[4] For example, Winter[22] found that individuals with below-knee amputations used the gluteus maximus, semitendinosus, and knee extensors as energy generators to compensate for loss of the plantarflexors. Olney and associates[23] found that in children who had unilateral plantarflexor paralysis, the involved plantarflexors produced only 33 percent of the energy generation compared to the 66 percent produced in normal gait. Greater-than-normal hip flexor activity in these children compensated for the loss of the plantarflexors.[23]

Disturbances in the normal gait pattern cause increases in the energy cost of walking because the normal patterns of transformation from potential to kinetic energy are disturbed. Increases in muscle activity used to compensate for these disturbances lead to increases in the amount of oxygen that is consumed. In a comparison of patients who had an ankle fusion with patients who had a hip fusion, oxygen consumption for patients with the hip fusion was 32 percent greater than normal and greater than patients with the ankle fusion.[55] Pain also appears to be a factor that leads to an increase in oxygen consumption. As pain increases, oxygen consumption has been found to increase.[20] In patients with bilateral lower-extremity paralysis, walking usually involves the use of long leg braces and crutches. In this form of gait the trunk and upper extremity muscles must perform all of the work of walking and the energy cost of walking is much greater than normal. A form of electrical stimulation called functional neuromuscular stimulation (FNS) is currently being used to activate the paralyzed lower extremity muscles so that these muscles can generate energy for walking. However, the energy cost of FNS walking is still well above that of normal gait.[56]

At the foot, pes cavus and pes planus cause alterations in weight and may cause abnormal stresses at the hip or knee. In pes cavus, the weight is borne primarily on the hindfoot and metatarsal regions and the midfoot provides only minimal support.[31] In running, the metatarsals bear a disproportionate share of the weight. In pes planus, the weight is borne primarily by the midfoot rather than being distributed among the hindfoot, lateral midfoot, metatarsals, and toes, as it is in the normal walking foot.[31]

SUMMARY

The objectives of gait analysis are to identify deviations from normal and their causes. Once the cause has been determined, it is possible to take corrective action aimed at eliminating or diminishing abnormal stresses and decreasing energy expenditure. Sometimes the corrective action may be as simple as using a lift in the shoe to equalize leg lengths or developing an exercise program to increase flexibility at the hip, knee, or ankle. In other instances, corrective action may require the use of assistive devices such as braces, canes, or crutches. However, an understanding of the complexities of abnormal gait and the ability to detect abnormal gait patterns and to determine the causes of these deviations must be based on an understanding of normal structure and function. The study of human gait, like the study of human posture, illustrates the interdependence of structure and function and the infinite variety of postures and gaits available to the human species.

REFERENCES

1. Foreman, J: How to tell if you are muggable. Boston Globe, Jan 20, 1981.
2. Steindler, A: Kinesiology. Charles C Thomas, Springfield, Illinois, 1955.
3. Winter, DA: Energy assessments in pathological gait. Physiother (Canada) 30; 1978.
4. Inman, VT, Ralston, HJ, and Todd, F: Human Walking. William & Wilkins, Baltimore, 1981.
5. Professional Staff Association, Rancho Los Amigos Medical Center: Observational Gait Analysis Handbook. Downey, California, 1989.
6. Winter, DA: Biomechanics of normal and pathological gait: Implications for understanding human locomotor control. J Motor Behav 21:337–355, 1989.
7. Mann, RA: Biomechanics of running. In Mack, RP (ed): Symposium on the Foot and Leg in Running Sports. CV Mosby, St Louis, 1982.
8. Lower Limb Prosthetics, rev ed. New York University Postgraduate Medical School, New York, 1975.
9. Bampton, S: A Guide to Visual Examination of Pathological Gait. Temple University Rehabilitation Research and Training Center, Philadelphia, 1979.
10. Lamareaux, LW: Kinematic measurements in the study of human walking. Bull Prosth Res, Spring 1971.
11. Murray, MP: Gait as a total pattern of movement. Am J Phys Med 46:1, 1967.
12. Murray, MP, Drought, AB, and Kory, RC: Walking patterns of normal men. J Bone Joint Surg. 46A:335, 1964.
13. Crowinshield, RD, Brand, RA, and Johnson, RC: Effects of walking velocity and age on hip kinematics and kinetics. Bull Hosp Joint Dis, 38:1977.

14. Larsson, LE, et al: The phases of stride and their interaction in human gait. Scand J Rehab Med 12:107, 1980.
15. Lehmkuhl, D: Brunstrom's Clinical Kinesiology. FA Davis, Philadelphia, 1983.
16. Soderberg, GL and Gavel, RH: A light emitting diode system for the analysis of gait. Phys Ther 58:4, 1978.
17. Grieve, DW: Gait patterns and speed of walking. Biomedicine (Eng) 3:119, 1968.
18. Milner, M, et al: Angle diagrams in the assessment of locomotor function. South Am Med J 47:951, 1973.
19. Hersihler, C and Milner, M: Angle-angle diagrams in the assessment of locomotion. Am J Phys Med 59:3, 1980.
20. Gussoni, M, et al: Energy cost of walking with hip impairment. Phys Ther 70:295–301, 1990.
21. Winter, DA: Analysis of instantaneous energy of normal gait. J Biomech 9:253, 1976.
22. Winter, DA: Biomechanics of below knee amputee gait. J Biomech 21:361–367, 1988.
23. Olney, SJ, et al: Work and power in hemiplegic cerebral palsy gait. Phys Ther 70:431–438, 1990.
24. Wells, RP: The kinematics and energy variations of swing-through crutch gait. J Biomech 12:579, 1979.
25. Smidt, G: Methods of studying gait. Phys Ther 54:1, 1974.
26. Cerny, K: Pathomechanics of stance: Clinical concepts for analysis. Phys Ther 64:1851–1858, 1984.
27. Kotoh, Y, et al: Biomechanical analysis of foot function during gait and clinical applications. Clin Orthop Rel Res 177:23–33, 1983.
28. Skinner, SR, et al: Functional demands on the stance limb in walking. Orthopedics 8:355–361, 1985.
29. Perry, J and Hislop, H: Principles of Lower Extremity Bracing. American Physical Therapy Association, Washington, DC, 1967.
30. Perry, J: Kinesiology of lower extremity bracing. Clin Orthop 102:18, 1974.
31. Scranton, PE and McMaster, JH: Momentary distribution of forces under the foot. J Biomech 9:45, 1976.
32. Nuber, GW: Biomechanics of the foot and ankle during gait. Clin Sports Med 7:1–13, 1988.
33. Ramig, D, et al: The foot and sports medicine—Biomechanical foot faults as related to chondromalacia patellae. J Orthop Sports Phys Ther 2:2, 1980.
34. Root, ML, Arien, WP, and Weld, JH: Normal and Abnormal Function of the Foot. Clinical Biomechanics Corp., Los Angeles, 1977.
35. Sutherland, DH, Cooper, BA, and Daniel, D: The role of the plantarflexors in normal walking. J Bone Joint Surg 62:3–336, 1980.
36. Donatelli, R: Biomechanics of the Foot and Ankle. FA Davis, Philadelphia, 1990.
37. Winter, DA, et al: Biomechanical walking pattern changes in the fit and healthy elderly. Phys Ther 70:340–347, 1990.
38. Winter, DA and Yack, HJ: EMG profiles during normal walking: Stride to stride and inter-subject variability. Electroenceph Clin Neurophys (Ireland) 67:402–411, 1987.
39. Stokes, VP, Anderson, C, and Forssberg, H: Rotational and translational movement features of the pelvis and thorax during adult human locomotion. J Biomech 22:43–50, 1989.
40. Basmajian, JV: Muscles Alive, ed 4. Williams & Wilkins, Baltimore, 1979.
41. Hogue, RE: Upper extremity muscle activity at different cadences and inclines during normal gait. Phys Ther 49:9, 1969.
42. McFayden, BJ and Winter, DA: An integrated biomechanical analysis of normal stair ascent and descent. J Biomech 21:733–744, 1988.
43. Mann, RA: Biomechanics of running. In D'Ambrosia, RD and Drez, D (eds): Prevention and Treatment of Running Injuries, ed 2. Slack, New Jersey, 1989.
44. Stanic, V, et al: Standardization of kinematic gait measurements and automatic pathological gait pattern diagnostics. Scand J Rehabil Med 9:95, 1977.
45. Craik, RL and Otis, CA: Gait assessment in the clinic. In Rothstein, JM (ed): Measurement in Physical Therapy. Churchill Livingstone, London, 1985.
46. Norkin, C: Gait analysis. In O'Sullivan, S and Schmitz, TJ (eds): Physical Rehabilitation Assessment and Treatment, ed 2. FA Davis, Philadelphia, 1989.
47. Robinson, JL and Smidt, GL: Quantitative gait evaluation in the clinic. Phys Ther 61:3, 1981.
48. Rose-Jacobs, R: Development of gait at slow, free, and fast speeds in 3 and 5 year old children. Phys Ther 63:1251–1259, 1983.
49. Foley, CD, Quanbury, AO, and Steinke, T: Kinematics of normal child locomotion—a statistical study based upon TV data. J Biomech 12:1, 1979.
50. Hinmann, JE, et al: Age-related changes in speed of walking. Med Sci Sports Exercise 20:161–166, 1988.
51. Finley, FR, Cody, KA, and Finizie, RV: Locomotor patterns in elderly women. Arch Phys Med Rehabil 50:140–146, 1969.
52. Blanke, DJ and Hageman, PA: Comparison of gait of young men and elderly men. Phys Ther 69:144–148, 1989.

53. Clement, DB, et al: A survey of overuse running injuries. Phys Sports Med 9:5, 1981.
54. Taunton, JE, et al: Non-surgical management of overuse knee injuries in runners. Can J Sports Sci 12:11–18, 1987.
55. Waters, RL, et al: Comparable energy expenditure after arthrodesis of the hip and ankle. J Bone Joint Surg 70A:1032–1037, 1988.
56. Marsolais, EB and Edwards, BG: Energy costs of walking and standing with functional neuromuscular stimulation and long leg braces. Arch Phys Med Rehabil 69:243–249, 1988.

STUDY QUESTIONS

1. The stance phase constitutes what percentage of the gait cycle in normal walking? How does an increase in walking speed affect the percentage of time spent in stance?

2. What percentage of the gait cycle is spent in double support? How is double support affected by increases and decreases in the walking speed?

3. Maximum knee flexion occurs during which period of the gait cycle?

4. What is the total ROM required for normal gait at the knee, hip, and ankle?

5. How does the total ROM required for normal gait at the knee, hip, and ankle compare with the ROMs required for running and stair gait?

6. Which of the determinants of gait help to keep the vertical rise of the body's COG to a minimum?

7. Which determinant helps to minimize a drop in the body's COG?

8. Which determinant helps to keep the lateral shift of the body's COG to a minimum?

9. What is the role of the tibialis posterior during walking gait?

10. How is the swinging motion of the upper extremities related to movements of the trunk, pelvis, and lower extremities during walking gait?

11. How do the traditional terms used to describe walking gait compare with the RLA terms? Describe the similarities and differences between the terms.

12. Describe the subdivisions of the stance and swing phases of the walking gait cycle using the traditional terminology.

13. Describe the subdivisions of the stance and swing phase of the walking gait cycle using the RLA terminology.

14. What is the function of the plantarflexors during walking gait?

15. Describe the transverse rotations in the frontal plane at the pelvis, femur, and tibia during walking gait.

16. When does the foot begin supinating in normal walking gait?

17. What are the functions of the dorsiflexors in normal walking gait?

18. Compare muscle action in walking gait with muscle action in running gait.

19. Where does the GRFV fall in relation to the ankle, knee, and hip at initial contact? What type of moments are acting at the ankle, knee, and hip at initial contact? Answer the same question using different subdivisions, i.e., loading response, midstance, terminal stance, and preswing.

20. Explain valgus thrust. Identify where it occurs in the gait cycle and the muscles that help to control it.

21. Explain what would happen in walking and running if a person's plantarflexors were paralyzed. What compensations might you expect?

Index

■ ■ ■

An "f" following a page number indicates a figure, a "t" following a page number indicates a table, and an "n" following a page number indicates a footnote.